With love to
Mom and Dad
at Christmas.

Carla 1983

P9-CJH-822

10 -

IMMUNOLOGY OF REPRODUCTION

Courtesy of David M. Phillips, The Population Council, The Rockefeller University

IMMUNOLOGY OF REPRODUCTION

EDITED BY

Thomas G. Wegmann, Ph.D.
UNIVERSITY OF ALBERTA

Thomas J. Gill III, M.D.
UNIVERSITY OF PITTSBURGH

WITH

Carla D. Cumming, B.A.

Eric Nisbet-Brown, M.D., Ph.D.
UNIVERSITY OF ALBERTA

New York   Oxford

OXFORD UNIVERSITY PRESS

1983

COPYRIGHT © 1983 BY OXFORD UNIVERSITY PRESS, INC.

Library of Congress Cataloging in Publication Data

Main entry under title:
Immunology of reproduction.

Bibliography: p.
Includes index
1. Reproduction—Immunological aspects.
I. Wegmann, Thomas G. II. Gill, Thomas J.
[DNLM: 1. Fetus—Immunology. 2. Immunity—
In pregnancy. 3. Maternal-fetal exchange.
4. Placenta—Immunology. WQ 200 R424]
QLP251.I453 1982   612'.6   82-7936
ISBN 0—19—503096—6

Printing (last digit): 9 8 7 6 5 4 3 2 1

PRINTED IN THE UNITED STATES OF AMERICA

Preface

This book originated from a meeting held at the Banff Centre for the Fine Arts, May 15 to 18, 1981. Reproductive immunology is in an explosive phase of development, and recent technological advances in immunology are bringing powerful tools to bear on the important questions of the survival of the fetal allograft and the development of a vaccine to prevent or reduce fertility. We felt that it would be most timely to present an overview of where reproductive immunology is today and where it is likely to go in the near future. This book will hopefully serve as an entry into this field for graduate students, medical students, postdoctoral fellows, and physicians interested in understanding some of the fundamental issues involved, as well as a reference book for workers in the field. To this end, we have written a general introduction as a concise view of what we think are the most important and exciting issues in reproductive immunology, as well as indicating possible pitfalls in current approaches, and future directions for research. We have tried to indicate where our comments are speculative, however, so that the reader will not confuse fact with fiction. The rest of the book is divided into four sections covering the major aspects of reproductive immunology. Each section has an introduction which is meant to orient the reader to the individual chapters of that section and to integrate them. We hope that this format will provide the perspective necessary to place the individual chapters into a general framework.

As is usual in this type of endeavor, the number of people involved is too great for us to be able to acknowledge them individually. We are most grateful for their contributions, without which our task would have been much more difficult. The Reproductive Immunology Group at the University of Alberta helped to formulate the Conference and to determine the shape that this book has ultimately taken. We wish to acknowledge especially the work of Louis Hugo Francescutti, Phillip Gambel, Denise Tews, Marilyn Kobayashi, Rajgopal Raghupathy, Brenda Sandmaier, Padma Shastry, Bhagirath Singh, and Sandra Stedman. Financial support was generously provided by the Alberta Heritage Foundation for Medical Research, the Family Planning Division of Health and Welfare Canada, the National Science Foundation of the United States, and the University of Alberta Conference Fund. We also received support from Beckman Instruments Limited, Flow Laboratories Inc., Caltec Scientific Limited, Pharmacia Limited, and Reproduction Research Informations Science Limited. The Conference was held under the general aegis of The Transplantation Society, The Society for the Immunology of Reproduction, and The Canadian Association of Immunologists. It is our pleasure to acknowledge the support, encouragement, and guidance of Jeffrey House and his colleagues at Oxford University Press in New York in bringing this volume to fruition. Eric Nisbet-Brown and Carla D. Cumming jointly mastered the University of Alberta computer and were able to generate camera-ready galley proofs directly from manuscripts, as well as providing editorial advice and criticism. Kandy MacMillan and Roberta Farkas were invaluable in their intelligent, persistent, and cheerful secretarial assistance. Debbie Reinhart of the University of Alberta Textform Group, and Sylvia Chetner of the University of Alberta Health Sciences Library gave much help in the production of the book. Finally, we thank Rupert Billingham, David Billington, Michael Edidin, Susan Heyner, and Lynn Wiley for reading and criticizing parts of this book.

Contents

vii

Contributors

Nancy J. Alexander
Division of Reproductive Physiology
Oregon Regional Primate Research Center
Beaverton, Oregon

Deborah J. Anderson
Division of Immunogenetics
Sidney Farber Cancer Institute
Harvard Medical School
Boston, Massachusetts

S.C. Bell
Reproductive Immunology Group
Department of Pathology
The Medical School
University of Bristol
Bristol, England

Rupert E. Billingham
Department of Cell Biology
University of Texas Health Science Center at Dallas
Dallas, Texas

W.D. Billington
Reproductive Immunology Group
Department of Pathology
The Medical School
Univeristy of Bristol
Bristol, England

David A. Clark
Department of Medicine
McMaster University
Hamilton, Ontario

Anne Croy
Department of Biology
Brock University
St. Catharines, Ontario

C. Das
All India Institute of Medical Sciences
New Delhi, India

Bonnie S. Dunbar
Department of Cell Biology
Baylor College of Medicine
Texas Medical Center
Houston, Texas

Michael Edidin
Department of Biology
McCollum-Pratt Institute
The Johns Hopkins University
Baltimore, Maryland

Allen C. Enders
Department of Human Anatomy
School of Medicine
University of California at Davis
Davis, California

W. Page Faulk
Blond McIndoe Centre for Transplantation Biology
Queen Victoria Hospital
East Grinstead, Sussex, England

Thomas J. Gill III
Department of Pathology
University of Pittsburgh
Pittsburgh, Pennsylvania

Erwin Goldberg
Section of Biological Sciences
College of Arts and Sciences
Northwestern University
Evanston, Illinois

Victoria Gonzales-Prevatt
Section of Biological Sciences
College of Arts and Sciences
Northwestern University
Evanston, Illinois

S.K. Gupta
All India Institute of Medical Sciences
New Delhi, India

Catherine S. Hawes
Department of Obstetrics and Gynaecology
Flinders Medical Centre
Bedford Park, South Australia, Australia

Judith R. Head
Department of Cell Biology
University of Texas Health Science Center at Dallas
Dallas, Texas

Susan Heyner
Department of Biological Sciences
Philadelphia College of Pharmacy and Science
Philadelphia, Pennsylvania

W.R. Jones
Department of Obstetrics and Gynaecology
Flinders Medical Centre
Bedford Park, South Australia, Australia

Rolf Kemler
Friedrich-Miescher-Laboratorium der Max-Planck-Gellscheft
Tübingen, Federal Republic of Germany

Andrew S. Kemp
Department of Clinical Immunology
Flinders Medical Centre
Bedford Park, South Australia, Australia

Arthur C. Lee
Division of Reproductive Biology
Department of Obstetrics and Gynecology
Ohio State University
Columbus, Ohio

S.K. Manhar
All India Institute of Medical Sciences
New Delhi, India

Mark McDermott
Department of Pathology
McMaster University
Hamilton, Ontario

A. Mullick
All India Institute of Medical Sciences
New Delhi, India

R.A. Murgita
Department of Microbiology and Immunology
McGill University
Montreal, Quebec

R.K. Naz
All India Institute of Medical Sciences
New Delhi, India

A.B. Peck
Department of Pathology
University of Florida
Gainesville, Florida

John E. Powell
Division of Reproductive Biology
Department of Obstetrics and Gynecology
Ohio State University
Columbus, Ohio

Rajgopal Raghupathy
Department of Immunology
University of Alberta
Edmonton, Alberta

S. Ramakrishnan
All India Institute of Medical Sciences
New Delhi, India

Janet Rossant
Department of Biology
Brock University
St. Catharines, Ontario

David B. Searls
The Wistar Institute
Philadelphia, Pennsylvania

N. Shastri
All India Institute of Medical Sciences
New Delhi, India

Bhagirath Singh
Department of Immunology
University of Alberta
Edmonton, Alberta

O.M. Singh
All India Institute of Medical Sciences
New Delhi, India

Renata Slapsys
Department of Medicine
McMaster University
Hamilton, Ontario

G. Smith
Reproductive Immunology Group
Department of Pathology
The Medical School
University of Bristol
Bristol, England

Vernon C. Stevens
Division of Reproductive Biology
Department of Obstetrics and Gynecology
Ohio State University
Columbus, Ohio

W.H. Stimson
Department of Biochemistry
University of Strathclyde
Glasgow, Scotland

G.P. Talwar
All India Institute of Medical Sciences
New Delhi, India

A. Tandon
All India Institue of Medical Sciences
New Delhi, India

Thomas B. Tomasi
Cancer Research and Treatment Center
University of New Mexico
Albuquerque, New Mexico

Kenneth S.K. Tung
Department of Pathology
University of New Mexico
Albuquerque, New Mexico

Guy André Voisin
Centre d'Immuno-Pathologie
I.N.S.E.R.M.
Hôpital Saint-Antoine
Paris, France

Stephen S. Wachtel
Division of Cell Surface Immunogenetics
Memorial Sloan-Kettering Cancer Center
New York, New York

Thomas G. Wegmann
Department of Immunology
University of Alberta
Edmonton, Alberta

Thomas E. Wheat
Department of Biological Sciences
College of Arts and Sciences
Northwestern University
Evanston, Illinois

Hans Wigzell
Department of Immunology
Uppsala University
Uppsala, Sweden

IMMUNOLOGY OF REPRODUCTION

Chapter 1

REPRODUCTIVE IMMUNOLOGY: AN OVERVIEW

THOMAS G. WEGMANN
THOMAS J. GILL III

Reproductive immunology is a rapidly expanding field, and it is attracting an increasing number of younger scientists as well as established scientists who wish to change fields. The literature in this field may seem overwhelming on first approach because it contains a great deal of phenomenology and, unfortunately, a large amount of inconclusive data. The first chapter of this book is meant to serve as a brief primer on the current major issues in reproductive immunology and to offer some speculation on the future directions of the field. The four sections that follow provide a detailed experimental description of the various areas in reproductive immunology. The chapters have been written by experienced investigators who discuss their own research and provide a perspective on the area within their purview. Each of the sections has an introductory statement to orient the reader, to integrate the various chapters, and to pose scientific questions about the material. We hope that this introductory chapter and the introductions to the four sections will provide a more coherent view of reproductive immunology than has previously been available. Such a perspective is especially important in view of the diverse fields—immunology, reproductive biology, clinical obstetrics and gynecology, developmental biology, and genetics—in which reproductive immunology has its roots. In compiling this volume, we have tried to emphasize the need for a more systematic approach to the questions outlined in the book. Too often in the past, qualitative descriptions of various phenomena have been reported without detailed investigation into the mechanisms involved and their biological significance. The viewpoints expressed in these remarks are those of the editors and do not necessarily reflect the viewpoints of the other authors in this book or of the field in general. A list of references that provide other overviews of this field is given at the end of this chapter.

We believe that there are two main reasons for the resurgence of interest in reproductive immunology, and they appear as recurring themes throughout this volume. First, the phenomenal advances in immunology have begun to catalyze the expansion of knowledge in reproductive immunology, and this, in turn, has reopened the question of how Nature circumvents the allograft reaction to allow the survival of the fetus. This is the only known physiological situation in which there is a direct confrontation between tissues bearing a variety of foreign antigens, especially the histocompatibility antigens, and a fully competent immune system. Although this question was posed many years ago, its answer, like the Holy Grail, has remained elusive. Second, fundamental advances in reproductive immunology have profound and immediate practical consequences. The chief demographic problem of significance to the survival of the human species is overpopulation, and one of the major psychological factors affecting human welfare is the control of

3

fertility. Food shortages, pollution, and even the energy problem itself bear ultimately on the problem of population control. For instance, the latest estimate puts the population of China at between 1 and 1.2 billion people. It is extraordinary to realize that 200,000,000 people may not have been properly accounted for in a census and that possibly the country will not be self-sufficient until it has a population of half that size. Current methods of contraception fit Lewis Thomas's criteria of intermediate technologies: they are cumbersome, expensive, and not always effective. On the other hand, vaccination has a long history of safety, effectiveness, and economy in eradicating some of the major scourges of mankind, such as smallpox, polio, and diphtheria. It is a pleasant speculation that a similar approach might help to solve the population problem. These issues will be discussed in detail in the chapters that follow.

IMMUNOLOGY OF THE GAMETE

The gamete is antigenically foreign because of two classes of molecules that it carries: sperm-specific or oocyte-specific autoantigens and histocompatibility antigens. The autoantigens are presumably sequestered by the blood-testis or perhaps the blood-ovary barriers and thus are not recognized as "self" during the ontogenic development of the immune system. The prototypes of these antigens are the LDH-X antigen of sperm and the zona pellucida antigens of the ovary. The situation for oncofetal antigens is more complex because they are expressed in the testis of the adult male, on the sperm, and in the early embryo; theoretically, they are absent from the adult female. They can also be expressed on tumors that arise from germinal tissue. The prototype of this class of antigens is the F9 oncofetal antigen (or class of antigens), which is present on sperm and in the early embryo but cannot be detected shortly after implantation. It appears to be widely distributed, since it is highly crossreactive in the human and the mouse. The histocompatibility antigens—both major and minor—are expressed on sperm and, because of the extensive genetic polymorphism at these loci, may be recognized by a female which does not bear them.

Both autoantigens and histocompatibility antigens can elicit an immune response in the female and are thus potentially active in natural and artificially induced infertility and in the control of cell-cell interactions in the early stages of reproduction. The latter includes the penetration of the sperm into the ovum and cellular differentiation in the embryo. These antigens have been studied primarily by the classical serological approach using polyspecific allogeneic or xenogeneic antisera. These reagents are often ill-defined and not

4

reproducible from laboratory to laboratory. To bring these types of studies into the modern era, monoclonal antibody technology must be applied to the purification and characterization of antigens important in reproduction. Once this has been achieved, the topology of their expression and the time course of their appearance during gametogenesis and embryogenesis can be studied definitively.

These basic studies on gamete antigens also have direct relevance to the clinical problem of infertility, as illustrated by several situations. Antisperm antibodies are found in a number of cases of clinical infertility. For example, it has been known for a number of years that female prostitutes have rather high serum titers of antisperm antibodies and that they frequently have no need to resort to artificial means of contraception. Cases have been reported where women who show a local inflammatory reaction following intercourse have been found to have serum IgE antisperm antibody. Antibodies against zona pellucida autoantigens have also been associated with some cases of female infertility. A more accurate definition of these antigens by the use of monoclonal antibodies should allow a critical evaluation of the importance of antigamete antibodies in spontaneous human infertility.

The potential role of antisperm antibody in infertility has also been emphasized by the experience with the use of vasectomy as a means of fertility control. There are approximately 10 million vasectomized men in the United States, and there are about 1 million vasectomies performed each year. Approximately 50 per cent of vasectomized males have significant serum titers of antisperm antibody. The same finding has been noted in a variety of animal models, and there is evidence from some of these that vasectomy may contribute to the development of arteriosclerosis, possibly as a result of an immune complex-induced vasculitis. Whether this is true in humans remains to be seen, but the current controversy illustrates the importance of a better understanding of the response to sperm-specific antigens. One possible consequence of the formation of antisperm antibody in humans following vasectomy is that reversal is far less effective in restoring fertility than would be expected. This infertility may be due to the effects of the antisperm antibody.

One of the most interesting prospects for sperm- and ovum-specific antigens is their use in antifertility vaccines. Two factors strongly limit the usefulness of this approach: the potential danger of immunizing against autoantigens and the role of immune response genes in influencing the effectiveness of the vaccines in a particular population. First, autoimmunity must be avoided, since its long term consequences could be quite deleterious. This consideration would rule out the use of ovum-specific antigens because such an approach would involve immunization of the female against components of her own body. Likewise, vaccination of males against their own sperm might lead to undesirable autoimmune complications, such as immune complex-related pathology. Hence, if this approach is to be acceptable, it will most likely involve

5

vaccination of the female against antigens specific for the sperm or the early embryo, such as the LDH-X sperm-specific antigen or the F9 oncofetal antigen. Studies evaluating these antigens as vaccines have shown a partial reduction in fertility, but this is not sufficient because, first of all, an effective pregnancy vaccine must be close to 100 per cent effective in all immunized females. Second, the role of immune response genes in influencing the effectiveness of a given vaccine must be considered very carefully, since the vaccine will be used in a genetically heterogeneous population. For example, an antifertility vaccine may be effective only in that segment of a population which responds well to the antigen used; in time, the population would be artificially selected for low responders to the antigen. Such a problem might be circumvented by the use of a polyvalent vaccine or some of the techniques of immunological engineering, such as alteration of the physical state of the antigen to diminish the variations in the immune response to it. Nonetheless, the principal observation about pregnancy vaccines is that, rather than being selectively effective, they have simply not been effective enough. One of the reasons is probably that the routes of immunization used do not effectively immunize the reproductive tract, and thus techniques for its selective immunization must be developed before other aspects of pregnancy vaccination can be studied.

An effective antifertility vaccine would most likely be irreversible, and the consequences of this fact must be considered prior to its adoption. In practical terms, however, this consideration is unlikely to pose a serious problem, since vaccination will presumably be chosen by women wishing to terminate their reproductive capacity permanently. Tubal ligation is currently commonly used for this purpose, but has some undesirable consequences, including a certain incidence of infection. Due to lack of adequate medical support facilities for carrying out the procedure, it is also quite impractical for most of the Third World, which desperately needs effective fertility regulation. Thus, the issue of reversibility must be handled at the psychological and social levels and should not influence attempts to develop an effective antifertility vaccine.

Finally, there is the theoretical possibility that vaccination could be used to reduce the incidence of spontaneous abortion in couples wishing to have children or to prevent the development of fetuses with genetically determined developmental abnormalities. Genes affecting embryonic development and skeletal morphology have been identified in the mouse and the rat. Some of them, at least, control cell surface antigens, and mapping studies of these genes—the T/t complex in the mouse and the Grc in the rat—show that they are linked to the major histocompatibility complex (MHC). As yet, there is no clear evidence that such genes exist in humans, but some developmental anomalies—for example, craniofacial abnormalities—may be influenced by genes linked to the major histocompatibility complex (MHC). If the antigens influencing such abnormal developmental processes could be identified and shown to be expressed on the surface of sperm, a vaccine could be developed that would eliminate these abnormal gametes and thus reduce the incidence of spontaneous abortion and congenital abnormalities in offspring.

IMMUNOLOGY OF THE EMBRYO AND THE EXTRAEMBRYONIC MEMBRANES

The embryo begins life as a free-living organism in the oviduct and then implants into the endometrial surface of the uterus. Studies of the antigens involved in this process have been hampered by the exceedingly small amounts of material available for investigation, the use of polyspecific immunological reagents, and the need for highly sensitive and specific assay techniques. Nonetheless, recent studies indicate that oncofetal antigens and paternally derived major and minor histocompatibility antigens of the embryo present potential targets for the maternal immune system. One conjecture, now disproved, is that the embryo in the free-living state is protected by the zona pellucida. This appears not to be the case, since embryos can survive for prolonged periods in the free-living state without the zona and still not experience any immunological rejection. The major histocompatibility antigens make their first appearance on the inner cell mass just prior to, or at about, the time of implantation, but they appear to be missing from the trophoblast at this stage. This is apparently the first tissue differentiation in the embryo, and the expression of MHC antigens may be involved in the differentiation process at this stage. However, this still remains hypothetical, since the function of the major histocompatibility antigens has not yet been clearly defined.

The minor histocompatibility antigens can be detected on the ovum and continue to be expressed throughout embryogenesis with the curious exception of the two-cell stage, when they disappear for a brief time. This variable expression may also be involved in the process of differentiation. Indeed, the first antigen for which a biological function has been assigned with reasonable certainty is the H-Y antigen. Although the function of this antigen in the very early embryo is not understood, it is known that, later, it is primarily related to the development of the testis from the indifferent gonad in mammals. This antigen was first detected when female mice were immunized with tissue from male mice of the same inbred strain and is broadly crossreactive between species as diverse as *Homo sapiens* and *Rana pipiens*. Further genetic studies indicate, however, that the H-Y antigen is primarily associated with the heterogametic sex rather than with the male per·se; this corresponds to the XY chromosomal configuration of the male in mammals, but to the ZW configuration of the female in birds.

The antigenicity of the trophoblast is an important, and still unresolved, issue. The prevailing consensus is that the major histocompatibility antigens are absent from the trophoblast, at least up to the time of implantation and probably shortly thereafter, although minor histocompatibility antigens are definitely present at this time. Two types of experiments show that the trophoblast is special with respect to its transplantion acceptance by a foreign host. First, trophoblastic tissue from a 7½-day-old embryo survives despite an intense rejection reaction directed against the embryonic parts of the same conceptus. Second, it has recently been shown that a xenogeneic inner cell mass can be placed into an embryo containing a trophoblast syngeneic with the mother and that an interspecific chimera develops to parturtition without elicitation of an immune rejection response by the mother. If, however, the trophoblast is xenogeneic and the inner cell mass is syngeneic, the embryo is rapidly rejected. These experiments emphasize the protective role of the trophoblast in the survival of the embryo, which may be due in part to antigenic differences between the surface of the trophoblast and that of the embryo and in part to special immunosuppressive properties of the trophoblast as a tissue.

The most critical tool for further study of the topography and ontogeny of embryonic antigens and their possible role as differentiation signals is the monoclonal antibody. These not only provide large amounts of reproducible reagents but also circumvent the possibility that the antibody probe may detect antigens other than that against which it is putatively directed—a persistent problem with even the most carefully prepared alloantisera. For example, an alloantiserum could have antiviral antibodies able to crossreact with embryo stage-specific antigens and thus give a false indication of the ontogeny of these antigens. Since MHC restriction is involved in the immune response to a variety of viral antigens, these erroneous results could be further compounded by their association with MHC gene products. Monoclonal antibodies should also be useful for the study of embryological development. It has been possible to use specific antibodies to block cell surface antigens in developmental studies in such diverse species as sponges and humans. This approach has provided exceedingly important information broadening our understanding of developmental processes. The use of monoclonal antibodies will clearly advance this type of research very rapidly.

IMMUNOLOGY OF THE PLACENTA

After the blastocyst has implanted into the endometrium, the trophoblast derivatives from the embryo combine with maternal uterine decidual tissues (some of which may be bone marrow derived) to form the placenta. To determine how the fetally derived tissues of the placenta avoid maternal rejection remains one of the major objectives of this field, and there are a number of intriguing mysteries to be solved. For example, when the mouse embryo begins to implant, lymphocytes can be seen clustering about the invading trophoblastic giant cells. What antigens or other signals do they recognize, and why do they not respond with a rejection reaction? This situation is even more dramatically illustrated in the horse. Between days 36 and 38 of pregnancy, the equine conceptus gives rise to islands of trophoblast completely surrounded by maternal tissue, called endometrial cups. From approximately day 80, the cups undergo degeneration. At the first appearance of the cups, lymphocytes are clustered around them in the maternal tissue. They do not invade, however, until the time of degeneration, when massive infiltration occurs. A number of observations indicate that the lymphocytic infiltration shortens the life-span of the cups. First, cytotoxic antipaternal lymphocyte antibody almost always appears about the time that the cups first develop. This indicates that immune recognition is coextensive with the arrival of the lymphocytes. Second, if mares mate with their co-twins, to whom they exhibit skin graft tolerance (no doubt due to hemopoetic chimerism), the effective survival of the endometrial cup is greatly prolonged. Third, the rejection reaction can be accelerated by interspecific matings. The slowest reaction is seen in donkey x donkey matings. Progressively faster reactions are observed in matings of horse x horse, horse male x donkey female (hinney), and horse female x donkey male (mule). In the latter instance, the cups are destroyed within 10 to 15 days of their appearance. A curious feature of these observations is that while the cups are undergoing putative immunological destruction, morphologically identical trophoblast tissue at the maternal-fetal interface, still in contact with the rest of the fetally derived tissue, is preserved intact. How can this occur, and what distinguishes the two types of tissues other than location?

An immunoprotective role for the trophoblast is strongly suggested by a series of experiments involving interspecific mouse chimeras. *Mus caroli* fetuses survive in *Mus musculus* uteri until shortly after the placenta forms. Lymphocytic infiltration occurs along the parietal endoderm, suggesting that an immunological reaction is involved in the subsequent failure of the fetus to survive. Consistent with this view is the observation that cytotoxic lymphocytes able to kill *M. caroli* cells *in vitro* can be recovered from the moribund fetoplacental units. This presumptive immune rejection can be avoided if a *M. caroli* inner cell mass is microsurgically inserted into a *M. musculus* blastocyst and the chimera is then implanted into a *M. musculus* foster mother. This "Trojan horse" experiment results in a viable interspecific chimera. If a *M. musculus* inner cell mass is placed into a *M. caroli* blastocyst and implanted into *M. musculus*, rejection once again follows. These clever experiments suggest strongly that trophoblast somehow provides an immunoprotective barrier between the mother and the fetus. This conclusion will be strengthened if the survival of fetuses surrounded by xenogeneic trophoblast can be prolonged by maternal immunosuppression.

Two things that are clear from the preceding discussion are that the trophoblast must express some sort of histocompatibility antigen and that the reason for its persistence in an outbred population must relate to factors other than the nonexpression of such an antigen. This raises the question of the antigenic status of the trophoblast. It has been known for a long time that maternally-derived antipaternal antibodies can be eluted from the placenta, and it has been claimed that in some cases these eluted antibodies react with major histocompatibility complex antigens. One way of viewing these observations and, more important, of explaining how the fetus escapes being harmed by these antibodies, is that the placenta serves as an immunoabsorbent barrier. Maternal antibodies belonging to subclasses that are allowed acccess to the fetus and directed against fetal histocompatibility antigens react with these antigens at the placental barrier. It can be shown that, when labeled purified monoclonal antibodies directed against paternal histocompatibility antigens are injected into the mother, the antibodies are trapped in the placenta, provided the fetus bears the target antigen. If the fetus expresses a different H-2 type, the antibodies cross the placental barrier and enter the fetal circulation. There is also evidence that these antibodies are turned over in the placenta, either by digestion or through release as antigen-antibody complexes. The resolution of this problem awaits further experimentation. These findings naturally raise the question, what is the immunoabsorbent tissue within the placenta? The best way to answer this question avoiding *in vitro* artifact is to allow radiolabeled antibody to bind to the intact placenta and then to prepare it for autoradiography. Such studies indicate that there are two major areas of antibody binding. Most binding takes place in the lateral aspects of the placenta, at the yolk sac endoderm. Some of the antibody-binding cells in the region are esterase positive and, morphologically, appear to be macrophages. There is also much lesser, but still definite, binding to spongiotrophoblast cells which come into direct contact with the maternal circulation. These observations were confirmed in experiments in which tissue was explanted from various areas of the placenta and stained for H-2 antigen after culture *in vitro*. Again, the same two areas were positive.

A most intriguing aspect of these studies is that Class I MHC antigens appear to be in direct contact with the maternal circulation. Why, then, can the mother not mount an effective rejection reaction? Two observations may help explain the lack of rejection. The first is that the Class I antigens appear to be at a much lower density in the areas where they directly face the maternal circulation than at the lateral aspects of the placenta. The second observation is that, in the limited studies done so far, Class II MHC antigens appear to be missing from the placenta, including the maternal-fetal interface. Some work has used monoclonal antibodies, and so this conclusion is not based on low-titer conventional antisera. Although it is possible that some Class II specificities will eventually be identified at the interface, it is intriguing to consider the functional implications of their absence. Class II antigens are classically thought to serve as "helper" determinants for cytotoxic cells reactive against Class I MHC antigens. The absence of these antigens on the fetal tissues in direct contact with maternal lymphocytes, combined with a low density of Class I antigens on those tissues, could serve to direct both the humoral and the cellular immune responses away from an immune effector or killer reaction and toward an enhancing or suppressive reaction. Although this latter is purely speculative, there can be little doubt that the antigenic configuration at the maternal-fetal interface could decidedly influence the type of response directed against the fetus by the maternal immune system.

What is the situation in humans? As mentioned previously, anti-HLA antibodies can be eluted from term placentae. Also, anti-HLA antibodies can be found in cord blood, i.e., in the fetal circulation, only if they are *not* directed against the fetus. Otherwise, they are present in the maternal circulation and absent from cord blood. This is presumptive, but not direct, evidence for the placenta being an immunoabsorbent in humans. It is still unclear whether antigens encoded by the HLA complex are expressed at the human maternal-fetal interface. Some studies using monoclonal reagents have been negative, but more recent studies have revealed antigens which may be coded for by the HLA complex at this interface. There have also been claims that isolated human trophoblast cells are HLA positive, but this is as yet controversial. There is convincing evidence that choriocarcinoma, which is a malignant derivative of trophoblast, does express HLA antigens, but whether these stem from the origin of the tumor or are a matter of altered expression in oncogenesis remains to be determined. It will be of great interest to determine whether there is a real difference between human and mouse in the expression of MHC antigens at the maternal-fetal interface or whether this is merely due to technical variation.

Another area for future investigation is the determination of whether the placenta can serve as an immunoabsorbent for maternal antibody directed against non-MHC antigens. For example, multiparous mice serve as a source of anti-H-Y antibody. There is some evidence, albeit controversial, that anti-H-Y antibody can direct the development of an XY indifferent gonad into an ovarian structure. This raises the possibility that the placenta may remove maternal anti-H-Y antibody before it reaches the male fetus. Similarly, why is it that in only a fraction of Rh-incompatible pregnancies does the fetus suffer from erythroblastosis fetalis? Is Rh antigen on the placenta, and does it serve as an immunoabsorbent, which is occasionally overwhelmed by an excess of maternal antibody? The literature is confusing on the former point and silent on the latter. To summarize, we need to know the range of antigens for which the placenta can serve as an immunoabsorbent, both in humans and in other species.

If we are only beginning to understand the role of placenta in protecting the fetus from maternal humoral immunity, we have as yet no clear view of why the fetus is not destroyed by maternal cell-mediated immunity. There have been occasional reports of neonatal wastage accompanied by massive infiltration of what appear to be maternal lymphocytes mediating a graft-versus-host reaction, but these are, fortunately, rare. Although the subject is controversial, the evidence is that there is very little, if any, passage of maternal lymphocytes from the mother to the fetus. Thus, one of the functions of the placenta may be to serve as an impermeable barrier to the entry of maternal lymphocytes. To verify this, it would be valuable to study cell traffic in both directions, using genetic markers readily quantifiable in mixtures, such as isozymes which can be objectively assayed by electrophoresis. Such studies have not yet been performed.

If, as seems likely, the placenta does form an impermeable barrier to maternal lymphocytes, how does this barrier work, especially as fetal MHC antigens are present at the maternal-fetal interface? At this time, any discussion of mechanism must remain fairly speculative. Suffice it to say that such factors as antigenic configuration, immunosuppressive factors (see below), tight junctions, sialomucin coating, and a number of other possibilities have been suggested as contributing to this barrier. It seems likely that future experimentation will focus on whether allogeneic T cells able to specifically recognize fetal alloantigens are attracted to the interface, and if so, on their fate. The current explosive development in T cell cloning will, no doubt, help in this analysis. The development of specific means of breaching the maternal-fetal barrier may also help us to understand its nature. In the final analysis, these experiments will have to be done on intact placentae to ensure the physiological significance of the observations.

There is another aspect of placental function that falls under the heading of reproductive immunology. For a long time, it has been thought that the placenta itself serves as an immune organ, and there is no doubt that macrophage-like cells are present within the fetal component of the placenta. Classically these cells have been called Hofbauer cells, and morphologically they resemble macrophages and can ingest foreign particles. Recent studies have described two different types of cells in the placenta, which can be distinguished by their ability to bind antibody via Fc receptors. One is a trophoblast cell, which binds only nonaggregated IgG. Another type will bind only aggregated IgG or IgG incorporated into immune complexes. The latter cells have a macrophage-like morphology and are of obvious interest in the immunoprotection of the fetus. Until recently, most immunologists regarded the fetal liver as the origin of lymphoid cell precursors in the mouse. Recent work, however, indicates that B cell precursors appear first in the placenta, and some two to three days later in the fetal liver. Could some part of the placenta be the elusive mammalian equivalent of the Bursa of Fabricius? Clearly, there is already enough evidence to warrant further study of the placenta as a specialized component of the developing immune system.

MATERNAL IMMUNOREGULATION DURING PREGNANCY

For a long time, many reproductive immunologists believed that the fetus escapes destruction by adult maternal immune effector mechanisms because of some form of special immune regulation. Since the concept of active suppression was proposed in the early 1970's, mechanisms of this sort have been invoked to explain the survival of the fetoplacental unit in the presence of a competent maternal immune system. If this is the case, one would expect to see either decreased maternal immunocompetence or immunoregulatory factors unique to pregnancy and perhaps produced by tissues present only during pregnancy, especially the placenta. To summarize a large body of literature, there seems to be very little change in general maternal immunocompetence during pregnancy. This, of course, makes sense, because generalized immunosuppression would seem to be too risky a way to ensure the survival of the fetal allograft. There are apparently some changes in the proportions of cellular subsets within the maternal immune system, such as an increase in null cells and shifts in T and B cell ratios, although the significance of the latter is in dispute. In general, though, the capacity of the mother to mount an immune response against a variety of antigens, including fetal alloantigens, is relatively unimpaired. Indeed, an impressive immune effector response against fetal alloantigens can be triggered in the intact pregnant female by immunization with paternally-derived tissue. Such priming can be shown by both *in vivo* and *in vitro* assays, yet does not harm either the fetus or the placenta. This set of observations must be kept in mind in light of descriptions of various kinds of suppressor cells and substances which are either specific or nonspecific in nature and which are detected in the systemic immune system of pregnant females. Frequently, these suppressor cells or substances are more readily detectable in second and third pregnancies than in primiparous females, and so their presence cannot readily explain the successful outcome of the first pregnancy. A number of laboratories have observed that there is a correlation between the absence of serum blocking factors and a tendency toward habitual abortion in human pregnancy. It cannot be concluded from this, however, that the blocking factors are necessary to prevent abortion. One could equally postulate that whatever causes the abortion also causes a decrease in serum blocking factors. Indeed, immunoregulatory substances and cells measured systemically may be only a physiologically irrelevant downstream effect of important mechanisms acting in the vicinity of the fetoplacental unit. To date, relatively few studies have been carried out with this in mind, but there is a growing tendency to study immunoregulation at the placenta and its draining lymph nodes. It is possible to demonstrate suppressor cells in the paraaortic lymph nodes draining the fetoplacental unit in pregnant mice, capable of suppressing the development of killer cells. They are not specific, however, for the MHC haplotype of the fetus. A variety of substances possessing immmuno-suppressive ability *in vitro* can be found in the placenta and draining lymph nodes. A major difficulty with these observations is that it is impossible to be sure that these substances play a role *in vivo*. The extraction of these substances from the placenta and environs disrupts their normal topological relationship to maternal lymphocytes, which, as far as can be told, remain on the maternal side of the trophoblastic interface. Even if these substances do leak through the barrier and influence maternal lymphocytes, it is difficult to determine their effective concentrations within their area of action. Despite these caveats, some of these substances may nevertheless play important roles. A recent example is a placental component called PAPSI, which inhibits prostaglandin synthesis in a rather specific manner. Another substance of interest is a low-molecular-weight thymic factor induced by cortisone and having interesting immunoregulatory effects *in vitro* at reasonable concentrations. It seems clear that only a combined *in vivo* and *in vitro* approach will allow determination of the ultimate relevance of these substances to local maternal immunoregulation. The possibility remains that the trophoblast serves simply as a mechanical barrier that prevents effective maternal recognition of fetal antigens. Ideally, the *in vivo* significance of a putative suppressor substance can be demonstrated by its selective removal or blocking in the pregnant female with monoclonal antibody or some other specific inhibitor. It must then be shown that the ensuing damage is a consequence of immunological attack and is prevented by immunosuppression. Only by this type of approach can an immunoregulatory role during pregnancy be securely assigned to these substances, and to date none of the substances identified meets these criteria. It is rather more difficult to evaluate the influence of the hormones of pregnancy on the maternal immune response because their removal has other consequences quite separate from their immunological effects. Nonetheless, some of the hormones of pregnancy can influence immune responses and therefore must be considered as potentially protecting the fetus. For example, silastic implants containing progesterone can prevent the rejection of skin grafts placed over them. Again the question of physiological relevance must somehow be answered.

Although the effect of pregnancy on systemic cell-mediated immunity is minimal at best, the humoral immune response is dramatically affected. It is now well accepted that many, but not all, pregnant females make antibody against MHC and fetal antigens during pregnancy. Although these responses need clarification, two interesting facts have already emerged from studies in mice. First, not all pregnant females respond by producing antibody to fetal alloantigens, and this is undoubtedly related to immune response genes. If, however, a "nonresponder" female is primed to fetal MHC antigens before pregnancy, a secondary response is elicited by the fetoplacental unit. Second, the fetoplacental unit tends to stimulate a noncytotoxic humoral immune response to fetal MHC antigens, relative to other allograft situations. This response consists primarily (although not exclusively) of antibody of the IgG_1 subclass, which does not fix complement and which has been implicated in enhancement of tumors and other grafts. This immune deviation could be related to the antigenic configuration at the maternal-fetal interface or to some immunoregulatory influence on maternal B cells. We simply have no solid information at this time for anything but speculation regarding mechanisms. It would be interesting to survey the immune repertoire of mouse and human females at various stages of pregnancy by fusing their B and T cells with myeloma cells to generate hybridomas, but such studies have not yet been done. This would provide detailed information about the types and kinetics of the maternal immune responses elicited by the conceptus during pregnancy as well as clues to the types of immunoregulation involved.

To properly consider the immune response of the female to sperm, embryo, and fetus in the reproductive tract, and ultimately the neonate via breast milk, the mucosal immune system must come under scrutiny. Our current knowledge indicates that the IgA-secreting cells servicing these areas have their origin primarily in the gut-associated lymphoid tissues, such as Peyer's patches. IgA precursor cells are there exposed to environmental pathogens and gut-borne antigens and become immunocompetent. These cells then migrate to various organs of external secretion, such as the vagina, the uterus, the breast, and the salivary glands. Their migration and function are strongly influenced by hormonal stimulation. For example, the ability of the mucosal immune system to secrete IgA antibodies into the uterine lumen in the mouse depends upon the estrous cycle of the female and is affected by the estradiol concentration. The homing of IgA-secreting B cells to the breast is dramatically increased by injections of prolactin, and the receptivity of the breast tissues for these cells is enhanced by estrogen and progesterone. This observation has profound implications for protection of the neonate from environmental pathogens of enteric origin. The maternal gut is exposed to potentially dangerous enteric pathogens, a condition that leads to specific activation of those IgA-secreting precursor cells able to react with these pathogens. The IgA-secreting cells then migrate to the breast and secrete the relevant antibodies into the milk. The milk offers the infant protection from these pathogens. Indeed, it has been a long standing folk remedy to eject breast milk onto open wounds to avoid infection. Breast milk also contains lymphocytes able to recognize MHC antigens. At the moment, it is unclear if these lymphocytes migrate across the fetal gut wall, but evidence is accumulating that, at the very least, some type of

15

lymphokine can cross the gut and influence the immunity of the newborn. The concept of the neonatal gut wall as an immunological barrier to antibodies directed against the fetus and to the maternal antifetal lymphocytes, at least in rodents, deserves more study than has been carried out heretofore. This situation may be analogous to the other maternal-fetal interface of the placenta and may perhaps be studied in analogous ways.

In view of the importance of the mucosal immune system to processes associated with reproduction, it is surprising that attempts at immunization against pregnancy have failed to consider the requirements for effective antigenic stimulation of this system. This cannot be the case in the future because our knowledge of mucosal immunity is increasing rapidly. It is becoming apparent that mucosal and systemic immunity are quite separable and, in some instances, inversely related. For example, under some conditions oral immunization can lead to systemic tolerance at the same time as it produces mucosal immunity. Failure to consider the two systems as separate could lead to several errors. Mucosal vaccination, perhaps via the oral route, may not yield detectable systemic immunity and give the false impression that the immunization was ineffective. More realistically, systemic immunization against sperm or oncofetal antigens may have only a minimal effect on fertility, but that does not necessarily mean that the vaccine is ineffective. It may simply be necessary to learn more about directing the immune response onto the reproductive tract. Focusing on the mucosal immune system could have other practical consequences as well. One of the prerequisites for a safe antifertility vaccine is that it not induce an autoimmune reaction. Some of the oncofetal antigens potentially useful in this situation have recently been shown to be present in certain adult tissues, such as brain. This does not exclude them from consideration as potentially useful vaccines, provided that the response to them can be confined to the mucosal immune system through use of the proper route of immunization. A more intensive study of the relationship between mucosal immunity and reproductive immunology is warranted.

The problem of mucosal immunity could perhaps be completely circumvented by the development of vaccines against the hormones necessary for pregnancy. The leading candidate for this type of immune interception has been human chorionic gonadotropin (hCG). Its B chain contains determinants unique to that hormone, being absent from closely related hormones, such as luteinizing hormone. The difficulty to date has been to elicit a response specific for the unique determinants of hCG and yet potent enough to interrupt pregnancy. This has not yet been achieved. The elegant immunochemistry done on this molecule has led to the production of hCG-specific monoclonal antibodies in mice. If these reagents can be made by human cells, they might provide a very effective and safe way of inducing early abortion.

One other aspect of immune regulation unique to reproduction concerns the immunosuppressive properties of seminal plasma. A few reports have claimed that some components of seminal plasma can suppress immune responses both *in vivo* and *in vitro*, and this may be of potential importance in explaining why the female does not ordinarily respond to sperm-specific antigens, which could hypothetically lead to infertility. As yet, these substances and their modes of action are poorly understood.

In summary, investigation of the maternal response to sperm and the products of conception requires a clear defintion of the nature of the immune response, the route by which immunity is naturally induced (or can be artifically induced), and the way the response is genetically regulated. This information is important for a basic understanding of reproductive immunology and will have profound practical consequences for the development of antifertility vaccines.

IMMUNE RESPONSIVENESS OF THE FETUS

There are a number of examples in which immune reactivity in the fetus can be influenced by transplacental molecular traffic from the mother. For instance, if female mice of certain strains making an antiallotype immunoglobulin response are mated to males bearing that allotype, the offspring have the expression of that allotype suppressed. The suppression in some instances is permanent and can be adoptively transferred to other mice by T cells. In this experimental situation, a maternal response can permanently alter the immunoglobulin repetoire of the offspring. There is evidence in both animal and human systems for the transplacental passage of antigen. The antigen can interact with the fetal immune system to stimulate an immune response. This interaction is under genetic control, and it provides a means to study environmental modification of genetically determined immune responsiveness. For example, immunization of a low-responder female alters the immune responsiveness of her offspring such that it is higher than expected on the basis of the genetic capability of the strain. Conversely, preimmunization of a high-responder female decreases the immune response of her offspring. Thus, antigen passed across the placenta from the mother can interact with the developing immune system of the fetus to alter its phenotypic expression.

The reactivity of fetal lymphocytes differs in some respects from that of adult lymphocytes. There is also some evidence that they respond differently to mitogens. Of most interest to reproductive immunologists, however, is that fetal lymphocytes can release a soluble substance which blocks both mixed lymphocyte reactivity and the phytohemagglutinin responsiveness of adult lymphocytes. Whether this plays a role in fetal protection

17

from maternally induced graft-versus-host disease remains to be deterined. As stated above, it is not even clear whether maternal lymphocytes routinely get into the fetus. Fetal lymphocytes react to a variety of antigenic stimuli at a reduced level. Natural killer cells appear in neonatal mice about three weeks postpartum; the reason for the late appearance of these cells is not yet known. They simply may not be produced until this time, or they may be suppressed by factors present in the maternal or fetal environments, such as alpha-fetoprotein (see below) or interferon. Another quite intriguing possibility relates to the observation that these cells preferentially react with embryonic cells, such as stem cells. If they arise early, they could simply be overwhelmed by the presence of large numbers of stem cells in the fetal and neonatal environments or regulated to arise later to prevent an "autoimmune" battle between them and the stem cells.

One of the most fascinating and controversial substances to be proposed for the regulation of both the maternal and the fetal immune response is alpha-fetoprotein. It acts preferentially on the immunoregulation of T-dependent antigens and, interestingly, also on those reactions in which the H-2I region antigens (Class II) are involved as helper determinants in elicitation of an immune response. Ironically, this may rule alpha-fetoprotein out as an important immunoregulatory substance for the prevention of maternal immune recognition of fetal MHC antigens at the maternal-fetal interface, since the latest information is that Class II antigens are absent from the interface (see above). This would not rule out a role as an immunoregulatory molecule for other antigens, especially soluble ones presented by maternal macrophages. There is somewhat better evidence for involvement of alpha-fetoprotein in the induction of the large amount of nonspecific suppression seen in the fetal environment. Perhaps we will better understand its role when methods are developed to specifically eliminate it from the intact fetoplacental unit. If this were to activate the fetal immune system, then its role as a natural suppressor of the fetal immune response would be made more secure.

In summary, there are many possible substances from both the mother and the fetus which potentially alter fetal immune reactivity and influence the maternal-fetal immune relationship. This area will no doubt be expanded and clarified in the near future.

REFERENCES

Beer, A.E., and R.E. Billingham. 1976. *Immunobiology of Mammalian Reproduction.* New Jersey: Prentice-Hall.

Dhindsa, D.S., and G.F.B. Schumacher. 1980. *Immunological Aspects of Infertility and Fertility Regulation.* New York: Elsevier North Holland Biomedical Press.

Gill, T.J. III, and C.F. Repetti. 1979. Immunologic and genetic factors influencing reproduction. *Am. J. Pathol.*95:465.

Loke, Y.W. 1978. *Immunology and Immunopathology of the Human Foetal-Maternal Interaction.* New York: Elsevier North Holland Biomedical Press.

Matangkasombut, P. 1979. New approaches to immunological contraception. *Clin. Obstet. Gynecol.* 6:531.

Maternal Recognition of Pregnancy. 1979. CIBA Foundation Symposium 64. New York: Excerpta Medica.

Scott, J.S., and W.R. Jones. 1976. *Immunology of Human Reproduction.* New York: Academic Press.

Talwar, G.P. 1980. *Immunology of Contraception.* London: Edward Arnold Ltd.

Maternal Recognition of Pregnancy. 1979. CIBA Foundation Symposium 64. Amsterdam and New York: Excerpta Medica.

Part I

THE EMBRYO AS ANTIGEN

INTRODUCTION

The study of the immunological and genetic phenomena influencing reproduction is grounded in many diverse disciplines, and the body of knowledge emerging from these studies is coalescing into an identifiable field—reproductive immunology. Its major sources lie in transplantation immunology, since both fields address the question of the acceptance of foreign tissue by a host. Whereas transplantation immunology is concerned mainly with an artifactually inserted tissue, i.e., an organ transplant, and the cellular and humoral interactions that determine its fate, reproductive immunology is concerned with the only natural case in which cells face each other across a histocompatibility barrier. Different histocompatibility types come into direct contact and elicit an immune response, but it is one that is not destructive to the fetus. Thus, reproductive immunology and transplantation immunology view the same basic problem—the response to incompatible tissue—from different points of view which may eventually be complementary in advancing our understanding of the role of the immune system in the economy of the body and of how this system can be manipulated for therapeutic purposes.

The first section of this book is devoted to the theme of the embryo as antigen. Judith Head and Rupert Billingham review a number of recent findings in transplantation immunology, with particular emphasis on the role of the endothelium and passenger leukocytes in graft rejection and on various attempts to modify them in order to avoid rejection. They also point out the important possibility that the placenta provides an immunologically privileged site for the fetal allograft and, as such, is the major determinant of fetal protection from the maternal immune onslaught. The specific antigenic nature of the fetus and the immune response to the fetal allograft are then reviewed by Thomas Gill. Attention is focused on the antigens controlled by the major histocompatibility complex and on the role that these antigens might play in reproduction. There is clearly a maternal immune response—both cellular and humoral—to the paternal component of the fetal MHC antigens, and the magnitude of this response increases with subsequent pregnancies. The possibility is again raised that this immune response may be an important component in the success of a viviparous pregnancy, and a certain amount of immunological and genetic evidence is marshaled to support this contention. This assertion raises the novel concept that the immune response plays an important part in the normal physiology of the host and is not just a host defense mechanism. Although there is enough suggestive evidence in the literature to make this position a tenable working hypothesis, much more work is required to either prove or disprove it.

The rest of this section explores specific antigenic systems in the fetus. The ontogeny of antigen expression is a complex area to study because of the paucity of antigenic material available and, until the development of monoclonal antibodies, the polyspecific nature of the reagents that were used. Susan Heyner reviews major and minor alloantigen expression on mouse eggs and early embryos, and this review emphasizes the tremendous advances that have been made in our understanding of antigenic ontogeny as a result of the use of monoclonal antibodies. She raises the question of the possible function(s) of these antigens and then suggests that they may be differentiation signals. This question is addressed explicitly in the case of the H-Y antigen by Stephen Wachtel. There is an impressive body of experimental evidence to indicate that the H-Y antigen is directly involved in primary sex determination, i.e., in the determination of gonadal function in the heterogametic sex of a species. This antigen is remarkably well conserved during evolution, and its conservation coincides with the great importance of sexual dimorphism in evolution. This finding is the first clear case of an assigned role for a histocompatibility antigen in reproduction, and its function can be explored in greater detail by the use of monoclonal antibodies to alter sexual development at different stages of embryonic growth.

One of the difficulties in studying development using normal embryos is the limitation imposed by their small size and their relative scarcity. In order to circumvent these problems, teratocarcinoma models have been developed for studying early differentiation. Michael Edidin and David Searls discuss the findings with one of the classical teratocarcinoma cell lines, the TerC line, and its use to study the signals that control differentiation. One of the best-characterized teratocarcinoma cell lines is the one carrying the F9 antigen, which most likely is a series of closely related antigens. This antigen is found nowhere in the adult female body; it is present in the adult male only in the testicle and in the sperm, and it is found in the early embryo but disappears a few days after implantation. Like the H-Y antigen, it is broadly crossreactive between man and mouse. Rabbit antibody to the F9 antigen has been shown to prevent compaction, which occurs at a specific stage of early embryonic development. Regardless of the function of the teratocarcinoma antigens, there exists the possibility that antibodies to them may be the cause of infertility in some women and that a vaccine could be developed from these antigens which could render women temporarily or permanently infertile. Certainly there have been cases in which mice have been rendered partially infertile by immunization with F9 and other teratocarcinoma cell lines.

What is needed is a monoclonal antibody definition of these complex antigens, and Rolf Kemler describes initial attempts to achieve this, with the attendant difficulties that will be familiar to anyone who has used these techniques in an attempt to clarify a complex system of antigens.

Chapter 2

TRANSPLANTATION IMMUNOBIOLOGY REVISITED

JUDITH R. HEAD
RUPERT E. BILLINGHAM

Nineteen years have now elapsed since one of us (R.E.B.) had the temerity to review the fetal-maternal relationship as an allograft/host situation and in so doing became a reproductive immunologist (Billingham, 1964). During this time interval, the transplantation field passed through a relatively quiescent decade, but it is currently fomenting with activity based upon new concepts and *in vitro* techniques. The old notion that rejection of an allograft is an inescapable, all-or-nothing phenomenon has long been abandoned in the light of compelling evidence that the life history of a graft reflects the result of an interplay between forces favoring rejection and those favoring acceptance or tolerance in its broadest sense (see Hamburger, 1981). Great progress has been made in the elucidation of the nature and *modus operandi* of these forces and in the devising of means of controlling them at the experimental level, though clinical transplantation is still dominated by the need for a specific means of immunosuppression.

Pari passu with, but more or less independent of, the development of transplantation biology, the immunobiology of reproduction has emerged as a healthy, vigorous new discipline *sui generis*. Two new journals now record its progress—three if you include *Placenta*.

In transplantation immunology, the most important single objective has always been to devise means of making allografts succeed where they would normally naturally fail. By contrast, the central goals in reproductive immunology are to find out why one particular kind of allograft—the conceptus—normally succeeds when one would expect it to be rejected and how it can be caused to fail at will.

Mindful of the fact that many reproductive immunologists' working knowledge of transplantation, like our own, is often outdated, oversimplified, and prejudiced, the purpose of this chapter is to review some selected topics in the transplant field that may be pertinent to an understanding of some of the immunologic options or strategies that were open to Nature when she addressed herself to the allograft problem in the evolution of viviparity.

EXPRESSION OF ALLOANTIGENS AND GRAFT REJECTION

Transplantation biology had its beginnings near the turn of the century, when it was discovered that tumors and normal tissues exchanged between unrelated individuals almost always failed to survive. With the subsequent search

29

for the factors responsible for foreign tissue rejection, the era of immunogenetics was born. Early investigations defining these agents eventually led to the discovery of the major histocompatibility complex (MHC), which has proved to be by far the most complex and polymorphic locus on mammalian chromosomes (see Klein, 1975). The definitions of multiple alleles and combinations of alleles certainly provided reasons for the tremendous difficulties encountered by early transplanters in achieving successful engraftments of foreign tissues, whether normal or neoplastic. The elucidation of the now classical serologically defined Class I (SD) determinants and subsequently, during the last decade, the elucidation of the I-region-determined (Class II) antigens (Ia) represent major advances in transplantation biology. Two misleading, simplistic concepts of alloantigenic expression developed as a result of these early immunogenetic investigations:

1. that SD (Class I) determinants are present on virtually all nucleated cells and
2. Ia (Class II) antigens are restricted to B lymphocytes, a few T lymphocytes, spermatozoa, and the majority of macrophages.

Although these tenets of mammalian immunogenetics have persisted for many years, recently devised, highly sensitive techniques have revealed that alloantigenic expression is much more complex, both qualitatively and quantitatively, with significant variations between the different cells making up an intact tissue and even on the surfaces of individual cells.

For reasons of technical convenience, early analyses of alloantigen expression on cell surfaces were generally restricted to cells in suspension, which were usually obtained from various components of the lymphomyeloid complex. More recent studies on cells within intact tissues, as well as various cell types separated by gradient centrifugation using discriminant immune reagents, have demonstrated both an interesting disparity of antigen expression by cells within a tissue, and a discrete polarity in the distribution of determinants on individual cells. These differences are undoubtedly important with regard to allotransplantation of tissues and organs, but also may be relevant to the as yet ill-defined normal functions this antigen system fulfils in vertebrate species.

The elegant studies of Parr and his colleagues have revealed that murine epithelial cells from many sources in the body can essentially be divided into those with and those without detectable MHC-determined alloantigens (Parr and Kirby, 1979). Cells without alloantigens include parietal and chief cells in the stomach, lining cells of the vas deferens, and ciliated tracheal cells. The others have H-2 antigens and include intestinal and gall bladder epithelial cells, tracheal goblet cells, and cells lining the uterine lumen. Of greatest interest was the finding that these H-2 positive cells express the alloantigens on their lateral and basal membranes, but not on the apical membrane exposed to the lumen. Since the change from lateral (antigenic) to apical (nonantigenic) plasma membrane occurs at the tight junction that binds the cells to one another, this junctional complex probably prevents alloantigens from migrating in the fluid membrane to the apical surface. As well, it most likely prevents movement of apical membrane molecules, such as those making up the glycocalyx, down into lateral membranes.

Although the above studies demonstrate that some cells that were previously presumed to be positive may be completely negative with regard to Class I alloantigens, the opposite is true for the Class II antigens: the formerly accepted restricted distribution of these antigens must now be broadened. We now know, for example, that it is not only macrophages that have Ia antigens, but also many other cells of the same putative ancestry in a variety of locations: the reticulum cells found in lymph nodes and spleen, dendritic cells in the thymus, and Langerhans cells of the epidermis. It has recently been established that Class II antigens (DR in the human) occur on a wide range of epithelial cells regardless of embryonic origin, including those of the gut, salivary glands, bronchi and fallopian tube, and most interestingly, the alveolar cells of the lactating mammary gland (Klareskog et al., 1980; Natali et al., 1981). The expression of Class II-like antigens on mammary epithelial cells is dependent on prolactin, and this is undoubtedly the reason such antigens are also demonstrable on milk fat globule membranes. These antigens could conceivably function in the interaction of lymphoid cells with such epithelia. Certain cells are negative for these antigens, as will be discussed below.

These new findings concerning alloantigen distribution are clearly of importance when the transplantation of intact tissues and organs is considered. Most grafts are complex, "multiracial" structures containing specialized parenchymal cells, vascular endothelium, basement membranes, blood elements, and a mixture of cells and fibers that make up the interstitial connective tissue, or stroma. The expression of alloantigens on each of these components is relevant to both the immunogenicity of the whole tissue, i.e., its capacity to elicit an immune response by the host, and its subsequent susceptibility to rejection processes.

It has long been assumed that the parenchyma of an organ or tissue is a prime target for effectors of allograft rejection. However, recent studies have shown that the parenchymal cells are often the least immunogenic component of a tissue and some, in fact, do not have detectable MHC antigens. For example, murine thyroid follicle cells and exocrine pancreatic epithelial cells express modest levels of Class I antigens, but not Class II antigens, whereas beta cells of the pancreatic islets express neither when tested by sensitive immunoferritin techniques (Parr et al., 1980). Rat kidney tubule cells are weakly positive for Class I and Class II antigens, but alone are incapable of sensitizing allogeneic hosts to subsequent organ grafts from similar donor strains. Rat heart myocytes lack detectable Class I or Class II antigens and are not susceptible to direct destruction by cytotoxic effector cells. These findings are consistent with the general observation in the mouse that reduced expression of H-2 K/D antigens on target cells *in vitro* is associated with a decreased susceptibility to killing by cytotoxic T cells. This does not necessarily mean that such cells are totally exempt from destruction (Parthenais et al., 1979). These same cells can be killed in the presence of alloantiserum and allogeneic lymphoid cells, suggesting that antibody-dependent cellular cytotoxicity (ADCC) may be a relevant cytotoxic mechanism with regard to cells with diminished alloantigen expression.

Although MHC-determined antigens are properly recognized as the major ones with regard to graft rejection, it must not be overlooked that many other alloantigenic systems have been elucidated. The most noteworthy are the so-called minor histocompatibility antigens, which are certainly capable of inciting harmful responses, but whose distribution within tissues is virtually unknown. Another group of antigens of potential importance are the organ-specific alloantigens (Hart and Fabre, 1980). These polymorphic antigens have now been described for skin, kidney, heart, liver, and pancreas, as well as for various endocrine glands, and are associated with the parenchymal cells of the organ, often involving intracellular components (see Lalezari and Krakauer, 1980). Their expression is not linked to the MHC. Although specific alloantibodies can often be found in sera of recipients rejecting the respective organ grafts, in most cases these antibodies appear to play little or no role in the rejection process. Finally, autoantibodies are also often produced against parenchymal cells of organs undergoing rejection, but again, their role in the process does not appear to be substantial. One other antigen of interest is that associated with the renal proximal tubule brush border (Miettinen and Linder, 1976). It is also shared by other epithelial surfaces, primarily secretory and ductal, and, notably, by the allantochorionic epithelium of the rat placenta.

Clearly, the parenchymal component of many organs does not appear to be sufficiently immunogenic to account for graft rejection. The other two components that could be responsible are the endothelium of the vasculature and "passenger" leukocytes. There has recently been renewed interest in the vascular endothelium as an important antigen source in tissues and organs, and rightfully so. In transplanted organs, such as kidneys, hearts, and livers, the vasculature does, of course, represent the principal frontier of interaction between donor and host, being directly exposed to blood-borne effectors. Analysis of alloantigen expression on endothelial cells has confirmed that, in many locations, these cells bear alloantigens capable of both inciting host responses and rendering vessel walls susceptible to immune destruction. In laboratory rodents, the endothelial cells in a variety of organs including kidney and heart are well-endowed with Class I alloantigens, which appear to be present on both luminal and abluminal surfaces of the cells (Parr et al., 1980). With their substantial complement of alloantigens, it is not surprising that rat heart endothelial cells are readily killed by cytotoxic T cells, unlike the parenchymal cells of that organ (Parthenais et al., 1979). In these species, endothelial cells do not appear to contain significant amounts of Ia antigens. By contrast, in the human, capillary endothelial cells have been found to express both Class I and Class II-like (DR) antigens (Natali et al., 1981). In the kidney, these are restricted to glomerular and peritubular capillaries, and it is interesting to speculate that this significant difference in Class II antigen expression between rat and human endothelial cells may relate to the well-known difference in transplantability of this organ between the two host species. Rat kidneys are quite easily enhanced, and with some donor/host strain combinations may even survive in the absence of any host pretreatment, whereas the human kidney has proved very difficult to enhance. Reports that human umbilical cord endothelial cells can present antigen to T cells, although awaiting critical confirmation (Hirschberg et al., 1981), are relevant here.

Besides the MHC antigens, an additional polymorphic alloantigen system that is shared by endothelial cells and monocytes (EM antigens) has recently been defined. The system appears to be quite important, since appearance of EM alloantibodies correlates well with renal graft rejection (Cerilli and Brasile, 1980). Interestingly, the sera of some multiparous donors seem to contain antibodies to these antigens.

The role of endothelium in the rejection of allografts has been controversial for some time. Along with the clear demonstration of abundant alloantigens on many endothelial cells has come recent evidence from passive transfer studies, *in vitro* cytotoxicity tests, and morphologic analyses that indeed the vascular endothelium is an important, if not the primary, target in tissue and organ allografts even during first-set rejection. The elegant electron microscope study of Dvorak et al. (1979) on human skin allografts revealed that the microvasculature was the first site to show pathologic changes, which occurred predominantly in vessels "cuffed" by infiltrating mononuclear cells. The endothelial cells were clearly the primary targets—lesions affecting them always preceding parenchymal epithelial necrosis. The grafts essentially succumbed to ischemic infarction. These results nicely parallel those from *in vitro* analyses demonstrating that, when rat heart tissue was separated into its endothelial and myocardial components, only the endothelial cells were susceptible to specific cytotoxic effector cells. The myocytes were refractory, apparently because of their very low surface densities of relevant alloantigens (Parthenais et al., 1979).

Mention should be made here of new information concerning cellular mediators of primary allograft rejection. Although it has long been assumed, primarily on the basis of *in vitro* studies, that cytotoxic T cells (murine Lyt $2,3^+$) are responsible for graft rejection, careful adoptive transfer studies in conjunction with the use of monoclonal antibodies to delete specific mouse T cell subpopulations have revealed that, in fact, graft rejection is solely dependent on the Lyt 1^+ subpopulation, with no contribution from Lyt $2,3^+$ cells. Thus, at least in the mouse, skin graft rejection does appear to be closely related to delayed type hypersensitivity phenomena, as long suspected (Loveland et al., 1981).

Passive transfer experiments using target xenografts in ALG immunosuppressed murine hosts have demonstrated that, in the case of both kidneys and hearts, graft endothelium is also susceptible to damage by specific antibodies and complement. This applies to skin grafts, too, but their endothelium has two periods of refractoriness to antibody-mediated damage that may reveal much about the role of vessel walls in eventual graft destruction. First, specific antiserum administered within the first 10 to 12 days after grafting was harmless. Then, two to three weeks thereafter, the grafts became acutely sensitive to this serum treatment, but gradually became refractory once more. The extensive studies of Jooste et al. (1981) on this phenomenon indicate that the early period of insensitivity is nonspecific, reflecting the process of vascular restoration and regeneration, with immature endothelium being unresponsive to inflammatory stimuli released after binding of antibody and complement. An interesting alternative, that alloantigen expression may be diminished on immature endothelium and increased during the differentiation process, needs investigation. Fluorescent antibody studies revealed that the late refractoriness to antiserum is due to surreptitious replacement of donor endothelium by host cells in the graft's vasculature. Whole organ grafts apparently do not undergo significant replacement of their vascular endothelium since they do not show this refractoriness with time. Fluorescence studies on long-surviving renal grafts in rats do indeed reveal that endothelium still expresses donor, and not host, histocompatibility antigens (Hart et al., 1980).

All of these findings indicate the important role of graft endothelium in the rejection process. Whether this tissue also provides the stimulus for the induction of host responses and the production of effectors is a matter urgently in need of clarification.

The final graft component providing potentially significant antigens to the host is, of course, the leukocytic cells which make up the so-called passenger cell population. Tests on the immunogenicity of separated renal components revealed that parenchymal cells were consistently nonimmunogenic, whereas the passenger cells readily sensitized allogeneic recipients to donor tissue antigens (von Willebrand et al., 1981). For many years, attention has been focused upon the mobile intravascular passenger cells, consisting primarily of lymphocytes, which surely must represent a significant part of the antigen load of a fresh allograft. However, repeated demonstrations that perfusion or other methods of depleting organs of these cells often failed to prolong graft survival, were very discouraging. It is now clear that, besides intravascular leukocytes, many organs contain substantial numbers of interstitial cells that antigenic analysis reveals are rich in both Class I and Class II determinants. Such cells have been described in the thyroid and in the kidney, where they occur exclusively in cortical interstitial tissue and are dendritic in shape, resembling the Langerhans cells of the epidermis (Hart et al., 1980; Parr et al., 1980). Interestingly, long-surviving rat kidney allografts, while still containing donor antigen-positive endothelial and parenchymal cells, have lost these donor-specific, Ia-positive dendritic cells. The possibility that these cells eventually migrated out of the allograft concerned is intriguing and consistent with the possibility that they represent a substantial resident passenger cell population that may be important as either inducers or targets of host immune responses. The significance of passenger cells will be discussed further below.

The composite immunogenicity of a tissue or organ graft incluaes all of the above components. The endothelial and passenger cells are receiving increasing attention, both as possible stimuli to the host's immune system and as targets of immune effectors. However, it is clear that the antigenic status of each type of graft must be studied individually, as significant differences exist between various organs from a single individual as well as between animals of different species. Further studies using sensitive (monoclonal) reagents are certainly necessary to evaluate differential antigenicity, and careful analysis of separated components from single organs will be necessary to precisely define their relative immunogenicity in specific immunogenetic contexts.

THE FEASIBILITY OF ALTERING THE IMMUNOGENICITY OF GRAFTS

Ever since it was generally recognized that allograft incompatibility is an immunogenetic phenomenon, most transplantation biologists accepted an unspoken dogma that the antigenic makeup of grafted allogeneic tissues and cells is unchanging and unchangeable. Consequently, systematic attempts to prolong allograft survival usually involved treatment of the host to impair its capacity to react effectively against the alien antigens of the graft. The successes of this approach need no emphasis. However, since the beginnings of transplantation immunobiology, sporadic empirical attempts have been made to alter the ability of allografts to evoke effective host reactivity or, more optimistically, to enable them to override a state of sensitivity directed against them. Over the past 15 years, unequivocal evidence has accumulated that the effective immunogenicity of allografts is indeed, at least to some extent, susceptible to manipulation (Billingham, 1976; see Table 2-1). However, the *modus operandi* of the successful approaches is not yet fully understood, and some of them have not yielded consistently reproducible results.

TABLE 2-1
Treatments of Allografts which Appear to Impair
their Allogenicity

Some tumors, ovary, thyroid, parathyroid	Maintenance *in vitro*
Skin	Soaking *in vitro* in media containing
	Corticosteroid hormones
	Urethane
	Thalidomide
	Antilymphocyte globulin
	Specific alloantibody
	DNA or RNA from prospective hosts
	Serum from uremic donors
Kidney or heart	X-irradiation
	Perfusion with Concanavalin A
	Perfusion with ALG
	Perfusion with nucleic acid solutions
	prepared from prospective host or
	unrelated source
Kidney	Pretreatment of prospective organ donor
	with cyclophosphamide or other
	immunosuppressive drug(s)

* From Billingham, 1976

In clinical organ transplantation, systemically administered corticosteroids have long been in use as ancillary immunosuppressants, being capable at high dosage levels of arresting allograft reactions already in progress. More relevant is the discovery made about 30 years ago that the topical application every three days of cortisone acetate or of 9-α-fluorocortisol acetate approximately doubled the life expectancy of skin allografts in rabbits. This prompted many subsequent studies involving both topical application to tissues after transplantation and their *in vitro* treatment with steroids and other unrelated agents prior to transplantation.

Experimental analyses have indicated that the steroid agents, when administered topically, exert only a trivial effect through local absorption and systemic action; rather, their main impact seems to be exerted locally, by impairing the development of host sensitization. This occurs through interference with the afferent arc of the immunological response and, perhaps to a greater extent, interference with the local effectuation of the host response once it has been elicited (efferent inhibition).

Most relevant to the subject matter of this volume are the results of studies in which allografts have been maintained *in vitro* prior to transplantation. The idea that the culturing or maintenance of a tissue in a medium containing plasma proteins from the prospective recipient might result in some modification that would favor its survival after allotransplantation is an old one and has been put into clinical practice on several occasions dating back to 1934, nearly always with equivocal results (Stone et al., 1934). Cultured parathyroid allografts transplanted to post-thyroidectomy patients suffering from parathyroid tetany are claimed to have led to clinical improvement, but essential evidence concerning the fate of the grafts was not provided (Gaillard, 1954).

Regardless of the validity of these early claims, subsequent critical pioneer studies on normal ovarian tissue and certain malignant tissues established that short-term maintenance *in vitro*, under defined conditions that allow little or no proliferative activity, may extend their life expectancy after allotransplantation (see Jacobs and Uphoff, 1974). Analysis of the phenomenon as it applies to tumors has established that:

1. the alteration in transplantability to normally resistant hosts acquired *in vitro* is fully reversible—for example, by a single passage through a host syngeneic with the tumor;
2. grafts of the altered tumor tissue are fully susceptible to prior or concomitant specific sensitization of allogeneic hosts by allografts of the unmodified tumor or by skin grafts from its strain of origin; and
3. sustenance of the altered tumor by allogeneic hosts sometimes allows subsequent allografts of the untreated tumor to survive and the hosts to show hyporeactivity to donor-type skin grafts.

These and other findings indicate that the diminished ability of the modified tumor or ovarian allograft to induce transplantation immunity is associated with a mechanism that appears to alter the phenotypic expression of histocompatibility genes. This, in turn, may result in a specific hyporeactive state in the host, which may be the consequence of the appearance of enhancing or blocking antibodies. It was recognized that removal of passenger leukocytes from the tissue, as a consequence of the treatment *in vitro*, might contribute to its reduced immunogenicity.

One of the most important developments in this area was initiated by Lafferty and his associates (1976), who held MHC-incompatible thyroid allografts in culture prior to transplantation beneath the renal capsules of thyroidectomized murine hosts (see Talmage et al., 1975; Lafferty et al., 1976). Essentially, their work has established that, whereas fresh, MHC-incompatible thyroid allografts are normally rejected within 10 to 15 days, maintenance in culture in an atmosphere of 95 per cent O_2 and 5 per cent CO_2, prior to transplantation, prolongs their survival *pari passu* with the time held in culture, up to a limit. Between 80 and 90 per cent of the grafts maintained in culture for 26 days survived allotransplantation through a 60-to-70-day observation period. These grafts remained functionally active as measured by ^{125}I uptake and closely resembled syngeneic grafts upon histological examination. Hosts of long-established functional thyroid allografts rejected subsequent allografts of uncultured thyroid tissue from the original donor strain within 20 days, indicating that chronic exposure to the functioning allograft had not induced an unresponsive state. The original, first-set allografts subsequently succumbed to chronic reactions following this exposure to second-set grafts. That the cultured allografts did not lose their immunogenicity was clearly evidenced by the observation that they were rejected if transplanted to hosts that had been injected with relatively low dosages of donor-type peritoneal exudate cells. The belief of Lafferty and his associates that the apparent progressive weakening of allograft immunogenicity procured by culture *in vitro* is due to loss of passenger leukocytes was strengthened by a variety of additional observations. For example, treatment of the future graft donors with agents known to cause leukopenia, such as ^{60}Co irradiation or cyclophosphamide, reduced or in some situations eliminated the time necessary to maintain allografts in culture for them to enjoy subsequent longevity following allotransplantation.

The capacity of tissue culture to prolong allograft survival has been shown to extend to other endocrine tissues, and to rats as well as to mice. Lacy et al. (1979) obtained prolongation of survival of MHC-incompatible allografts of islets of Langerhans in rats following *in vitro* culture and a single injection of the recipients with ALS. By cyclophosphamide treatment of prospective islet donors and *in vitro* culture of aggregates of isolated islets, Bowen et al. (1980) were able to procure the long-term survival of this tissue in allogeneic mice. Naji et al. (1981) have reported that the survival of MHC-incompatible parathyroid allografts in rats was greatly improved by prior culture for 26 days in an oxygen-enriched environment. These latter investigators have made the interesting observation that this culture procedure was ineffective in prolonging the lives of MHC-compatible parathyroid allografts. Particularly striking is the degree of success obtained in transplanting cultured islet and thyroid tissue across a species barrier—from rat to mouse (Sollinger et al., 1977; Lacy et al., 1980).

Although one can scarcely doubt that the culture procedure reduces the effective immunogenicity of the grafts by selectively eliminating or inactivating passenger leukocytes, principally mononuclear cells, the precise mechanism involved has yet to be elucidated. Lafferty et al. (1976) have argued cogently that, in fresh allografts, the passenger monocytes and/or lymphocytes function as stimulator cells, presenting donor antigen to the host's immune system in an effectively immunogenic form to activate the immune response. In their opinion, the cells of the graft parenchyma constitute poor T cell immunogens even though their expression of Class I transplantation antigens, that renders them susceptible to rejection, is perfectly normal (see Lafferty, 1980).

The distribution of passenger cells is certainly not restricted to the vascular bed and tissue spaces of tissue and organ grafts. It is a long-established, though only recently recognized fact that significant numbers of mononuclear cells, principally lymphocytic and monocytic-like cells of adventitious origin, are normal components (minority cell populations) of most epithelia of the body, though their functional significance is still enigmatic (see Seelig and Billingham, 1980). In the epidermis, monocyte-like branched Langerhans cells constitute the most important passenger cell type and play an important role promoting contact hypersensitivity and eliciting allograft immunity directed against Ia antigens (Streilein et al., 1980b; Streilein and Bergstresser, 1980). These alloantigens are not expressed on keratinocyte surfaces. If means can be devised for purging epidermis of its Langerhans cells, its capacity to survive allotransplantation might well be dramatically improved.

Passenger cells have also been shown to make a significant contribution to the immunogenicity of experimental renal and cardiac allografts. Nevertheless, some ingenious experiments in which seemingly effective measures were taken to replace the donor passenger cell population with one of host type, prior to allotransplantation, have procured little prolongation of survival. However, as discussed previously, there are significant numbers of dendritic, Langerhans-like cells within the interstitium of organ grafts. Since Langerhans cells are very resistant to a variety of noxious agents, are capable of self-renewal within the epidermis, and are only slowly replaced within this tissue, we need to keep an open mind on the question of whether an organ graft totally devoid of passenger cells, if this could be achieved, would not indeed survive allotransplantation, resembling a cultured endocrine graft in this respect. Similarly, it is possible that any epithelium that normally maintains itself free of passenger cells would have unique allotransplantation properties.

IMMUNOLOGICAL ENHANCEMENT OF RENAL ALLOGRAFTS

The phenomenon of immunological enhancement, defined by Snell (1970) as the enhanced or prolonged growth of allografts due to the presence in the graft recipient of alloantibodies directed against the alloantigens of the graft donor, also has a history extending back to the beginning of this century. Early studies established that prior inoculation of rodent hosts with nonviable material from a particular allogeneic tumor days or weeks before, challenge with viable tumor tissue frequently resulted in progressive tumor growth, rather than its prompt regression. In the early 1950's, Kaliss and co-workers (see Kaliss, 1958) studied the phenomenon systematically, using inbred strains of mice, and established that not only could animals be actively "enhanced" by inoculation with alloantigenic material, but they could also be passively enhanced by pretreatment with antibody directed against relevant specificities of the prospective tumor allograft. By contrast with tumors, allografts of normal tissue, especially skin (Brent and Medawar, 1962), proved

39

only feebly amenable to enhancing procedures. An important variable here was the genetic disparity between donor and host, minor incompatibilities being easier to overcome than major ones.

With the advent of microvascular surgery came evidence, in the late 1960's, that permanent survival of vascularized, MHC-incompatible organ allografts could be procured by enhancement in rats. Stuart et al. (1968) reported that intravenous injection of Lewis strain rats with 10^8 (Lewis x BN)F_1 spleen cells 18 hours before grafting and administration of 1 ml of Lewis anti-BN serum just before and on alternate days for a short period after grafting, resulted in permanent survival of F_1 hybrid kidneys. Soon afterwards, French and Batchelor (1969) achieved the indefinite functional survival of F_1 hybrid rat kidneys in MHC-incompatible parental strain hosts by means of alloantiserum alone—i.e., by passive enhancement—administered at the time of grafting and for a few days thereafter.

This and subsequent work (see reviews by Batchelor, 1978; Morris, 1980; and Stuart et al., 1980) have shown that there is great variation in the extent of prolongation of allograft survival procurable by enhancement in different strain combinations and a very striking disparity between the enhanceability of grafts from homozygous allogeneic donors and from semiallogeneic donors, the former being much more difficult to enhance. This disparity cannot be accounted for on the basis of a simple gene dosage effect since an increase in the amount of antiserum will not overcome it. Furthermore, with the same donor/host strain combination and the same enhancing serum, there may be great variation in the prolongation of allograft survival obtainable with different tissues or organs of the same alien genetic origin—e.g., complete suppression of response against a kidney, moderate prolongation of survival of a cardiac graft, and only slight prolongation of a skin allograft. Although survival of renal allografts has been enhanced in rabbits, attempts to enhance similar allografts in dogs and monkeys have been disappointing. The remarkable amenity of the rat as a subject for procurement of enhancement of vascularized organ allografts appears to result from its relative resistance to antibody-mediated damage. This may be due to a defect in the complement pathway, inadequate fixation of complement by antibody in this species, or inadequate density of MHC antigens on the putative target cells.

With regard to passive enhancement, two stages are generally recognized:

1.   an induction phase and
2.   a maintenance phase that may be of indefinite length.

Grafting represents the host's initial exposure to the antigen. In the presence of donor-specific antibody, there is partial or complete suppression of the host's humoral immune response, including both IgM and subsequently IgG lymphocytotoxic antibody synthesis, and a delay in the generation of cytotoxic T cells, though the amplitude of this delayed cellular response is like that of controls. Damage to the graft and impaired function are frequently demonstrable during the second post-operative week, but the host's pathogenic response normally subsides spontaneously. There follows evidence of functional improvement with a decline in cytotoxic T cells, which ushers in the "maintenance" phase. This is characterized by a progressive, specific unresponsiveness on the part of the host, with marked inhibition of immunological memory, IgG antibody response, and cytotoxic T cell generation. Nevertheless, the capacity to mount GVH reactions in donor-strain animals and MLC reactivities *in vitro* against donor cells is retained.

The class of the enhancing antibody, though much studied, is still not finally resolved. Most of the activity probably resides in the IgG_2 subclass, and the Fc portion of the molecule appears to be necessary. Enhancement experiments involving the use of recombinant strains of mice to confront hosts with Class II and Class I compatible and incompatible allografts have provided evidence that enhancing capacity resides principally in antibodies to Ia (i.e., Class II) specificities (McKenzie and Henning, 1977). The results of absorption studies on enhancing alloantisera point in the same direction.

Rats bearing passively enhanced renal allografts show a progressive, specific decline in cellular immunity, as evidenced by increasingly prolonged survival of donor-strain skin grafts placed at greater intervals after renal transplantation. However, the eventual rejection of these skin allografts is not associated with, or followed by, destruction of the renal allografts. Attempts to compromise the well-being of long-surviving enhanced renal allografts by:

1.   challenging of the host with spleen cells or skin grafts of donor origin,
2.   adoptive immunization with syngeneic lymphoid cells, or
3.   parabiosis with an immunized syngeneic partner

have failed. Indeed, the immunized parabionts, after separation, manifested weakened reactivity toward donor-type kidney grafts, suggesting that an active process is responsible for frustrating renal allograft rejection in the enhanced animals. Batchelor (1981) recently suggested that passive enhancement protects the renal allograft against the host's primary response largely by inhibiting the humoral response: he believes that the cellular immune response *per se* is insufficient to destroy the graft.

If long-established renal allografts are transferred to new normal recipients of host genotype, many of them are not rejected acutely, as freshly procured grafts would be. Some of the hosts synthesize IgM antibodies and generate only weak T cell cytotoxicity against the graft antigens, which has been attributed to a helper T cell-independent response. Indeed, some of these secondary hosts develop an unresponsiveness similar to that characterizing the maintenance stage of passive enhancement. By contrast, when long-accepted renal allografts are transplanted to specifically sensitized hosts, they suffer acute rejection, probably because of the presence in the host of memory cells that do not require helper T cell activation for their further differentiation.

It is generally agreed that a major part of host reactivity against MHC antigens is helper T cell dependent, and many investigators are of the opinion that it is passenger cells within allografts that, through their expression of Ia antigens, are largely responsible for activating host helper cells and amplifying the response. As passenger cells are progressively lost from the graft during the inductive phase of the enhancement process, activation of T helper cells becomes inadequate for the host to mount a destructive reaction against it. Gradually the abundant antigenic material from the graft, which is incapable of activating T helper cells, induces the development of unresponsiveness by processes yet to be elucidated.

If a long-surviving enhanced kidney is replaced by a fresh kidney of the original donor type, this, too, will survive indefinitely, indicating—along with other observations—that the maintenance phase represents a potent and persistent state of immunosuppression that some authorities attribute to the activity of suppressor cells or to anti-idiotype antibodies.

From the evidence already presented, it is beyond question that, apart from changes in the host, some form of adaptation occurs progressively in a long-surviving renal allograft. Various studies, including infusion of ^{131}I-labeled alloantibody, have failed to produce evidence that there is any significant alteration of donor antigen or that there is replacement of graft endothelium by that of the host. However, the recent finding that intensely Ia-antigen positive dendritic cells are lost from the interstitium of long-term renal grafts suggests that resident passenger cells may contribute significantly to the determination of the fate of transplanted kidneys.

IMMUNOLOGICALLY PRIVILEGED SITES

For nearly a century, genetically alien grafts of a wide variety of tissues have been transplanted to anatomically unnatural sites in the body for various reasons. The longevity undoubtedly enjoyed by grafts in some of these sites, as compared with that of similar grafts transplanted to other sites, particularly orthotopically, gave rise to the concept of immunologically privileged, or favored, sites. Interest in such sites has recently been heightened:

1.   because of their possible employment to sustain small therapeutic allografts of endocrine tissue, notably islets of Langerhans and para-thyroid, and

2. because of evidence that exposure to allografts in some privileged sites may progressively and systemically weaken, on an immunologically specific basis, the host's capacity to harm the alien cells involved, inducing a quasi-tolerant or enhanced state.

The most familiar privileged sites are the anterior chamber of the eye, the brain, the hamster's cheek pouch, and possibly the testis (see Barker and Billingham, 1977). The fact that most known privileged sites are characterized by the absence of an afferent lymphatic drainage pathway, in conjunction with the known vulnerability of vascularized allografts in such sites to specific active or adoptive immunization of the host, led to the idea that the privileged status was due to an anatomical defect or incompleteness in the afferent limb of the immune response—a defect that essentially prevents antigenic material from gaining access to draining lymph nodes. Recent work suggests that the *modus operandi* of at least some privileged sites is much more complicated and more interesting than this and turns upon the modification of host response to the graft.

THE SYRIAN HAMSTER'S CHEEK POUCH

One of the intriguing features of the highly vascular wall of the hamster's cheek pouch as a privileged site is that it retains its protective capacity even after auto- or syngeneic transplantation to the skin of the trunk (Barker and Billingham, 1971). About 50 per cent of relatively large MHC (Hm.1)-incompatible allografts of ear skin fitted into superficial beds, prepared in established syngeneic grafts of pouch skin on hamsters' trunks, survive for upwards of 100 days. A considerable body of evidence suggests that the immunological uniqueness of both the intact cheek pouch and the heterotopic pouch skin grafts is indeed due to their alymphatic status. This may reflect properties of the acid mucopolysaccharide-rich areolar connective tissue that lies beneath the "dermis" of pouch skin (see Heyner, 1970). Restoration of lymphatic drainage to skin allografts sustained by pouch skin results in their prompt rejection (see Barker and Billingham, 1977). It is pertinent to add that, with the MHC-incompatible strain combinations used for these studies, all attempts to demonstrate any kind of humoral alloimmune response directed against Hm.1 determinants have been totally unsuccessful (Duncan and Streilein, 1977).

THE ANTERIOR CHAMBER OF THE EYE

The anterior chamber of the eye is the prototypic and most thoroughly studied of all privileged sites. With the aid of inbred strains of rats, the variables that determine allograft survival—including size, type of tissue, and immunogenetic disparity between donor and host—have recently been carefully evaluated by Kaplan and Streilein (1978) and various associates and by Raju and Grogan (1971) and Subba Rao and Grogan (1977).

Although this recent work (see Barker and Billingham, 1977, and Streilein and Kaplan, 1979) has confirmed the absence of lymphatic drainage, it has produced the surprising finding that both solid tissue and cellular allografts introduced intracamerally do exert a very important influence upon the host's systemic immune apparatus. Streilein and Kaplan (1979) have demonstrated that MHC-incompatible F_1 hybrid lymph node cells inoculated into the anterior chambers of parental strain rat hosts are very effective in evoking the prompt synthesis of hemagglutinating antibodies and an early capacity to reject orthotopic donor skin allografts in an accelerated manner. However, this heightened reactivity to skin grafts soon gives way to a transient phase of hyporeactivity as evidenced by a suppressed transplant rejection reaction. These investigators have presented cogent evidence that alloantigens introduced into the anterior chamber escape directly via the blood vascular route and that the presence of an intact spleen is essential for these alloantigens to procure prolonged survival of orthotopic skin allografts. They assert that the impingement of the slowly released antigen upon the spleen results in its being processed in a manner that leads to suppression of cell-mediated immunity, leaving a "protective" humoral response intact.

Subsequent studies in mice, involving inoculation of P815 mastocytoma cells (derived from a DBA/2 donor) into the anterior chambers of H-2 complex compatible, but minor loci incompatible, BALB/c hosts, have corroborated and extended these interesting findings and conclusions (Streilein et al., 1980a; Niederkorn et al., 1981). For example, intraocular allotransplantation of tumor cells resulted in :

1. their progressive local growth,
2. significant transient growth of tumor at subcutaneous inoculation sites, and
3. prolonged, often indefinite acceptance of orthotopic grafts of DBA/2 skin.

The spleen was found to play a crucial role in the development of this "anterior chamber-associated immune deviation." This was evidenced in part by the findings that in splenectomized animals, intracameral P815 tumors failed to thrive and subsequent DBA/2 skin grafts were rejected in an accelerated manner.

THE ADRENAL GLAND

In the Syrian hamster there is evidence that conditions within the adrenal cortex predispose it to extensive lesions and subtotal destruction following microbial colonization—for example, by mycobacteria or toxoplasma organisms or by *Besnoitia jellisoni*, an obligate intracellular protozoan. Experimental analysis revealed that infection of hypophysectomized hamsters did not lead to necrosis of the adrenals, though injection of ACTH did result in recurrence of necrosis of these organs. These and other observations led Frenkel (1956, 1961) to postulate that predisposition of the adrenal to necrosis following infection results from an endogenous supply of corticoid—predominantly cortisol in the hamster—which locally depresses an acquired cellular immunity to the organism.

In humans, too, bacterial or fungal infections that gain a foothold in the adrenals cause more extensive lesions in these organs than in other sites in the body, and, as in the hamster, the pathogenesis of adrenal infections seems to be corticoid-related. There is clinical evidence that hypophysectomy and pharmacologically induced adrenal hypofunction reverse the immunological defect of the adrenal.

Following the fortuitous observation that a fragment of a human parathyroid adenoma transplanted into the adrenal gland of a rat gave evidence of long-term functional survival, Kukreja et al. (1979) demonstrated that Ag-B locus incompatible allografts of parathyroid also enjoyed slight prolongation of survival in this site as compared with similar grafts transplanted intramuscularly. We feel that this work on the adrenal merits attention since it reveals a hitherto unrecognized principle that may confer immunologic privilege upon a graft site, i.e., local production of agents having nonspecific immunosuppressive properties. In the case of the adrenal, this appears to protect certain microbial organisms from an extant state of immunity.

THE TESTIS

The testis has long enjoyed a reputation as a privileged site based upon early tumor work, and this belief was recently confirmed by critical studies in which islets of Langerhans and parathyroid allografts were used in rats, emphasizing functional criteria of survival. The most widely held explanation

45

for the privileged status of the testis has been that it reflects not the absence, but rather peculiarities of its lymphatic drainage:

1. that there is an exceptionally long lymphatic pathway to the nearest draining lymph node or
2. that, in some cases, testicular lymphatic effluent fails to traverse a node at all before discharge into the blood circulation via the thoracic duct (see Barker and Billingham, 1977).

In collaboration with our colleague, W.B. Neaves, we are engaged in a reinvestigation of the privileged status of the rat's testis. We have demonstrated, using dyes and other tracing methods, that this organ consistently has a superb lymphatic drainage into regional nodes, usually the ipsilateral iliac and renal nodes. We have monitored the survival of intratesticular parathyroid allografts in parathyroidectomized hosts in terms of their capacity to maintain normal calcium levels. Most of our allografts remained functional for 20 to 25 days, considerably longer than did those placed in nonprivileged sites (median survival time, 13 days). A significant number of animals retained their grafts for more than 100 days. Whereas functional MHC-incompatible parathyroid allografts in the testes of parathyroidectomized hosts do not elicit the formation of hemagglutinins, impairment or cessation of function of the grafts is frequently associated with the appearance of these antibodies. However, in the light of findings by Naji et al. (1979), it appears that a cellular rather than a humoral immunity is responsible for rejection of allografts of parathyroid tissue.

Unlike the situation with the anterior chamber, prior splenectomy of the host does not prejudice the hospitality that the testis extends to parathyroid allografts, and no evidence has been forthcoming that long-term sustenance of allografts in this site causes any systemic weakening of the host's capacity to respond to the alloantigens concerned.

The studies on the adrenal gland have led us to consider the possibility that the privileged status of the testis reflects the immunosuppressive properties of testosterone or other steroid hormones produced locally in the interstitial tissue by Leydig cells. Circumstantial evidence sustaining this idea includes observations that:

1. testosterone has been shown to have some immunosuppressive properties in *in vitro* systems (Clemens et al., 1979);
2. orchidectomy significantly enhances a mouse's immunocompetence, particularly with regard to cell-mediated responses (Graff et al., 1969; Castro, 1974); and
3. endocrinologically active Leydig cell tumors can override certain histocompatibility barriers (see Ninnemann and Good, 1975).

One might conjecture that, in the intact testis, locally secreted testosterone provides an ancillary back up protection against the risk of autoimmunization to spermatozoal antigens following compromise of the blood-testis barrier.

EPILOGUE

This overview was designed to highlight some recent developments in transplantation biology which have the greatest potential relevance for those working with the enigmatic immunology of the maternal-fetal relationship. That these two fields share common ground and have something to learn from each other is illustrated in Table 2-2. As depicted, a host's response to the major alloantigens of a foreign graft has been well characterized. Hallmarks of that response include draining lymph node hypertrophy; production of cytotoxic antibodies; generation of sensitized T cells, including both delayed-type hypersensitivity and cytotoxic subpopulations; and the establishment of specific immunologic memory, manifested by heightened reactivity against a second graft of like alloantigenic constitution.

Similar analyses of the mother's immune response to antigens of her allogeneic fetus reveal significant differences in reactivity that may be of importance to allogeneic fetoplacental maintenance:

TABLE 2-2
Comparison of Host Responses to Conventional Allografts with Maternal Responses to Feto-Placental Allografts

	HOST RESPONSE AGAINST TISSUE ALLOGRAFT	MATERNAL RESPONSE AGAINST ALLOGENEIC FETUS
Draining Lymph Node Hypertrophy	+	± (strain dependent)
Cytotoxic and Agglutinating Antibodies	+	± ($< 30\%$)
Blocking Antibodies	±	+
Mixed Lymphocyte Reactions	+	+
Delayed Hypersensitivity T Cells	+	+
Cytotoxic T Cells	+	−
Generation of Cytotoxic T Cells	+	± systemically − locally
Suppressor Cells	±	+
Heightened Reactivity to Second-Set Allograft	+	− (hyporeactive)

47

1. Although lymph nodes draining the uterus do hypertrophy in many allogeneically pregnant females, this is certainly not a constant feature, and is apparently dependent on the strain combination involved.

2. The characteristics of alloantibody production are very different. Although it is well known that the sera of multiparous females are an excellent source of antibodies for tissue-typing, most reports nevertheless indicate that no more than 30 per cent of women make such complement-fixing antibodies. Studies in both humans and rodents have revealed, however, that "blocking"-type antibodies are often demonstrable in maternal sera and that they are capable of interfering with a variety of antipaternal responses *in vitro*.

3. Analysis of cellular responses has also revealed an interesting dichotomy. MLR to paternal antigens are intact, and sensitized delayed hypersensitivity T cells are demonstrable *in vitro*. Although T lymphocytes cytotoxic for paternal cells can be caused to appear systemically by subsequently challenging the female with paternal antigens, the lymph nodes draining her reproductive tract have a diminished capacity to generate antipaternal effector cells, and there is suggestive evidence that a suppressor cell is involved.

4. Finally, when parous females are challenged with paternal allografts, not only do they not display heightened reactivity, but many are actually hyporesponsive in their expression of cellular immunity (see Head and Billingham, 1981).

There is clearly much to be gained from transplantation studies, such as we have reviewed here, that may provide relevant information as to how the peculiar maternal immunoreactivities are induced and maintained. Investigations on privileged sites are relevant since the uterus may, in fact, represent a sort of privileged site. They may also shed light on the role of steroid hormones in aberrant immune responsiveness. The nature of antigenic stimuli responsible for inducing immune responses and recent information on the effectors of graft rejection place renewed focus on the trophoblast. The vascular endothelium of organ and tissue grafts appears to be a critical target of host blood-borne effectors early in acute rejection. The trophoblast lining intervillous spaces perfused with maternal blood serves an analogous role in hemochorial placentas. Similarly, the recent renewed interest in the role of passenger cells in the fate of allografts, along with increasing evidence that so-called graft adaptation may actually occur, could be applicable to the placenta as a target for immune effectors and to the nature of the antigenic stimulation to the mother engendering altered immune responses. Studies on the polarity of antigenic expression by epithelial cells have been extended to include the murine yolk sac, with similar results—i.e., alloantigens, though present on the cell surface, are not found on the side that is exposed to the mother (Parr, 1981). Finally, experimental models of immune enhancement may reveal how the host can be induced to form blocking factors, an aspect of maternal responses believed to be very important for ensuring fetoplacental survival.

Although all of the above areas of study may provide information applicable to maternal-fetal immunology, of tremendous interest would be examples from the clinic, paralleling specific maternal immune reactivity against fetuses as allografts. There is, in fact, an interesting situation in clinical transplantation that has proved to be strikingly similar to that obtained in pregnancy as described above. Thomas et al. (1977) have conducted a detailed analysis of long-term renal transplant recipients having kidneys with at least one HLA antigen mismatch, since they recognized the importance of discerning why the foreign kidneys survived so well in these particular recipients. The patients' responses had the following characteristics, summarized as follows:

1. They lacked complement-dependent cytotoxic serum antibody to donor target cells.
2. Sera from most contained factors that specifically blocked MLR when donor cells were the stimulators.
3. The T lymphocytes from all of them gave normal MLR against donor stimulators.
4. None had lymphocytes capable of direct cytotoxicity toward donor cells.
5. Tests of the *in vitro* generation of cytotoxic effector cells revealed that, of the recipients in the period two to eight years post-transplantation, only 47 per cent produced effector cells; moreover, none of those more than 10 years post-transplantation produced effectors.
6. Recipients' cells were capable of significantly suppressing unrelated cells reacting against donor lymphocytes in the CML assay.

Although the recipients were on maintenance doses of immunosuppressive agents, they had no generalized immunosuppression and the altered responses demonstrated were specific for donor antigens. The immune status of many of these recipients is remarkably similar to that of mothers carrying allogeneic fetuses. Although reproductive immunologists have long believed that elucidation of how the fetal allograft manipulates the maternal immune response and escapes rejection could have applications for organ replacement therapy, the above study indicates that the opposite is also true. Elucidation of how such organ recipients immunologically accomodate their mismatched organ allografts could very well supply information applicable to the maternal-fetal interaction. This is a clear example of the benefits that may be derived from a mutual interest of transplantation immunologists and reproductive immunologists in each other's affairs.

ACKNOWLEDGMENTS

The expenses of some of the work referred to in this chapter were defrayed in part by U.S. Public Health Services Grant AI-10678. The manuscript was prepared by Helen Patterson.

REFERENCES

Barker, C.F., and R.E. Billingham. 1971. The lymphatic status of hamster cheek pouch tissue in relation to its properties as a graft and as a graft site. *J. Exp. Med.* 133:620.

Barker, C.F., and R.E. Billingham. 1977. Immunologically privileged sites. *Adv. Immunol.* 25:1.

Batchelor, J.R. 1978. The riddle of kidney enhancement. *Transplantation* 26:139.

Batchelor, J.R. 1981. Immune mechanisms responsible for the prolonged kidney allograft survival in immunologic enhancement. *Transplant Proc.* 13:562.

Billingham, R.E. 1964. Transplantation immunity and the maternal-fetal relation. *N. Engl. J. Med.* 270:667, 720.

Billingham, R.E. 1976. The feasibility of altering the immunogenicity of grafts. *J. Invest. Dermatol.* 67:149.

Bowen, K.M., L. Andrus, and K.J. Lafferty. 1980. Successful allotransplantation of mouse pancreatic islets to non-immunosuppressed recipients. *Diabetes* Suppl. 29:98.

Brent, L., and P.B. Medawar. 1962. Quantitative studies on tissue transplantation immunity. V. The role of antiserum in enhancement and desensitization. *Proc. Roy. Soc. London (Biol.)* 155:392.

Burdick, J.F., P.S. Russell, and H.J. Winn. 1979. Sensitivity of long-standing xenografts of rat hearts to humoral antibodies. *J. Immunol.* 123:1732.

Castro, J.E. 1974. Orchidectomy and the immune response. *Proc. Roy. Soc. Lond. (Biol.)* 185:437.

Cerilli, J., and L. Brasile. 1980. Endothelial cell alloantigens. *Transplant. Proc.* 12:37.

Clemens, L.E., P. Siiteri, and D.P. Stites. 1979. Mechanism of immunosuppression of progesterone on maternal lymphocyte activation during pregnancy. *J. Immunol.* 122:1978.

Duncan, W.R., and J.W. Streilein. 1977. Major histocompatibility complex in Syrian hamsters: Are SD determinants present? *Transplant. Proc.* 9:571.

Dvorak, H.F., M.C. Mihm, Jr., A.M. Dvorak, B.A. Barnes, E.J. Manseau, and S.J. Galli. 1979. Rejection of first-set skin allografts in man. The microvasculature is the critical target of the immune response. *J. Exp. Med.* 150:322.

French, M.E., and J.R. Batchelor. 1969. Immunological enhancement of rat kidney grafts. *Lancet* 2:1103.

Frenkel, J.K. 1956. Effects of hormones on the adrenal necrosis produced by *Besnoitia jellisoni* in Golden hamsters. *J. Exp. Med.* 103:375.

Frenkel, J.K. 1961. Infections involving the adrenal cortex. In *The Adrenal Cortex.* H.D. Moon, ed. New York: Paul B. Hoeber, pp. 201-219.

Gaillard, P.J. 1954. Transplantation of cultivated parathyroid tissue in man. In *Preservation and Transplantation of Normal Tissues.* G.E.W. Wolstenholme and M.P. Cameron, eds. Ciba Symposium. London: Churchill, pp. 100-106.

Graff, R.J., M.A. Lappé, and G.D. Snell. 1969. Influence of the gonads and adrenal glands on the immune response to skin grafts. *Transplantation* 7:105.

Hamburger, J. 1981. The future of transplantation. *Transplant. Proc.* 13:10.

Hart, D.N.J., and J.W. Fabre. 1980. Kidney-specific alloantigen system in the rat. *J. Exp. Med.* 151:651.

Hart, D.N.J., C.G. Winearls, and J.W. Fabre. 1980. Graft adaptation: Studies on possible mechanisms in long term surviving rat renal allografts. *Transplantation* 30:73.

Head, J.R., and R.E. Billingham. 1981. Immunobiological aspects of the maternal-fetoplacental relationship. In *Clinical Aspects of Immunology*, 4th edn. P. Lachmann and D. Peters, eds. Oxford: Blackwell Scientific.

Heyner, S. 1970. The role of mucopolysaccharide in the anomalous survival of infant skin allografts in hamsters. *Transplantation* 10:278.

Hirschberg, H., H. Scott, and E. Thorsby. 1981. Human endothelial cells can present antigen to sensitized T lymphocytes *in vitro. Transplant. Proc.* 13:100.

Jacobs, B.B., and D.E. Uphoff. 1974. Immunologic modification: a basic survival mechanism. *Science* 185:582.

Jooste, S.V., R.B. Colvin, and H.J. Winn. 1981. The vascular bed as the primary target in the destruction of skin grafts by antiserum. II. Loss of sensitivity to antiserum in long term skin grafts. *J. Exp. Med.* 154:1319.

Kaliss, N. 1958. Immunological enhancement of tumor homografts in mice, a review. *Cancer Res.* 18:992.

Kaplan, H.J., and J.W. Streilein. 1978. Immune response to immunization via the anterior chamber of the eye. II. An analysis of F_1 lymphocyte-induced immune deviation. *J. Immunol.* 120:689.

Klareskog, L., U. Forsum, and P.A. Peterson. 1980. Hormonal regulation of the expression of Ia antigens on mammary gland epithelium. *Eur. J. Immunol.* 10:958.

Klein, J. 1975. *Biology of the Mouse Histocompatibility-2 Complex*. New York: Springer-Verlag.

Kukreja, S.C., P.A. Johnson, G. Ayala, E.N. Bowser, and G.A. Williams. 1979. Allotransplantation of rat parathyroid glands: effects of organ culture and transplantation into the adrenal gland. *Experientia* 35:559.

Lacy, P.E., J.M. Davie, and E.H. Finke. 1979. Prolongation of islet allograft survival following *in vitro* culture (24°C) and a single injection of ALS. *Science* 204:312.

Lacy, P.E., J.M. Davie, and E.H. Finke. 1980. Prolongation of islet xenograft survival without continuous immunosuppression. *Science* 209:283.

Lafferty, K.J. 1980. Immunogenicity of foreign tissues. *Transplantation* 29:179.

Lafferty, K.J., A. Bootes, G. Dart, and D.W. Talmage. 1976. Effect of organ culture on the survival of thyroid allografts in mice. *Transplantation* 22:138.

Lalezari, P., and H. Krakauer, eds. 1981. First National Conference on Organ Specific Alloantigens. *Transplant. Proc.* Suppl. 1 12:2.

Loveland, B.E., P.M. Hogarth, R. Ceredig, and I.F.C. McKenzie. 1981. Cells mediating graft rejection in the mouse. I. Lyt-1 cells mediate skin graft rejection. *J. Exp. Med.* 153:1044.

McKenzie, I.F.C., and M.M. Henning. 1977. Studies on immunogenicity and enhancement of alloantigens of the various regions of the H-2 complex. *Transplant. Proc.* 9:609.

Miettinen, A., and E. Linder. 1976. Membrane antigens shared by renal proximal tubules and other epithelia associated with absorption and excretion. *Clin. Exp. Immunol.* 23:568.

Morris, P.J. 1980. Suppression of rejection of organ allografts by alloantibody. *Immunol. Rev.* 49:93.

Naji, A., C.F. Barker, and W.K. Silvers. 1979. Relative vulnerability of isolated pancreatic islets, parathyroid and skin allografts to cellular and humoral immunity. *Transplant. Proc.* 11:560.

Naji, A., W.K. Silvers, and C.F. Barker. 1981. The influence of organ culture on the survival of MHC compatible and incompatible parathyroid allografts in rats. *Transplantation* 32:296.

Natali, P.G., C. Martino, V. Quaranta, M.R. Nicotra, F. Frezza, M.A. Pellegrino, and S. Ferrone. 1981. Expression of Ia-like antigens in normal human nonlymphoid tissues. *Transplantation* 31:75.

Niederkorn, J., J.W. Streilein, and J.A. Shadduck. 1981. Deviant immune responses to allogeneic tumors injected intracamerally and subcutaneously in mice. *Invest. Ophth.* 20:355.

Ninnemann, J.L., and R.A. Good. 1975. Altered allotransplantability of BALB/c Leydig cell tumor after organ culture or cell suspension. *Transplantation* 19:42.

Parr, E.L. 1981. The self-side expression of H-2 antigens on epithelial cells and the maternal-fetal relationship. *Transplant. Proc.* 13:973.

Parr, E.L., and W.N. Kirby. 1979. An immunoferritin labeling study of H-2 antigens on dissociated epithelial cells. *J. Histochem. Cytochem.* 27:1327.

Parr, E.L., K.J. Lafferty, K.M. Bowen, and I.F.C. McKenzie. 1980. H-2 complex and Ia antigens on cells dissociated from mouse thyroid glands and islets of Langerhans. *Transplantation* 30:142.

Parthenais, E., A. Soots, and P. Häyry. 1979. Sensitivity of rat heart endothelial and myocardial cells to alloimmune lymphocytes and to alloantibody-dependent cellular cytotoxicity. *Cell. Immunol.* 48:375.

Raju, S., and J.B. Grogan. 1971. Immunology of anterior chamber of the eye. *Transplant. Proc.* 3:605.

Seelig, L., and R.E. Billingham. 1980. Intraepithelial lymphocytes. *J. Invest. Dermatol.* 75:83.

Snell, G.D. 1970. Immunologic enhancement. *Surg. Gynecol. Obstet.* 130:1109.

Sollinger, H.W., P.M. Burkholder, W.R. Rasmus, and F.H. Bach. 1977. Prolonged survival of xenografts after organ culture. *Surgery* 81:74.

Stone, H.B., J.C. Owings, and G.O. Gey. 1934. Transplantation of living grafts of thyroid and parathyroid glands. *Ann. Surg.* 100:613.

Streilein, J.W., and P.R. Bergstresser. 1980. Overview: Ia antigens and Langerhans cells. *Transplantation* 30:319.

Streilein, J.W., and H.J. Kaplan. 1979. Immunologic privilege in the anterior chamber. In *Immunology and Immunopathology of the Eye.* A.M. Silverstein and P. O'Connor, eds. New York: Masson Publishing USA, pp. 174-179.

Streilein, J.W., J.Y. Niederkorn, and J.A. Shadduck. 1980a. Systemic immune unresponsiveness induced in adult mice by anterior chamber presentation of minor histocompatibility antigens. *J. Exp. Med.* 152:1121.

Streilein, J.W., G.B. Toews, and P.R. Bergstresser. 1980b. Langerhans cells: functional aspects revealed by *in vivo* grafting studies. *J. Invest. Dermatol.* 75:17.

Stuart, F.P., T. Saitoh, and F.W. Fitch. 1968. Rejection of renal allografts: specific immunologic suppression. *Science* 160:1463.

Stuart, F.P., T.J. McKearn, A. Weiss, and F.W. Fitch. 1980. Suppression of rat renal allograft rejection by antigen and antibody. *Immunol. Rev.* 49:127.

Subba Rao, D.S.V., and J.B. Grogan. 1977. Orthotopic skin graft survival in rats that have harbored skin implants in the anterior chamber of the eye. *Transplantation* 24:377.

Talmage, D.W., G. Dart, J. Radovich, and K. Lafferty. 1975. Activation of transplantation immunity. *Science* 191:385.

Thomas, J., F. Thomas, and H.M. Lee. 1977. A search for mechanisms facilitating human non-identical allograft survival. Some parameters of cell-mediated immunity in long-term human renal transplant recipients. In *Immune Effector Mechanisms in Disease.* M. Weksler et al., eds. New York: Grune & Stratton, pp. 77-104.

Von Willebrand, E., P. Häyry, A. Soots, A. Nemlander, K. Wiktorowicz, and E. Parthenais. 1981. Antigenic, immunogenic, and immunosensitive components in rat allografts. *Transplant. Proc.* 13:1099.

Chapter 3

IMMUNOGENETIC ASPECTS OF THE MATERNAL-FETAL INTERACTION

THOMAS J. GILL III

The existence of a maternal immune response to various paternally derived antigens on the fetus and placenta is apparently paradoxical, since the immune response is generally associated with host defense functions. Thus, it may either have a physiological function or be adventitious. In either event, it is closely associated with the existence of polymorphisms of the cell surface antigens controlled by the genes in the major histocompatibility complex (MHC), since these antigens elicit the most consistent and potent immune response in the mother.

Polymorphisms in the genes of the MHC exist in all species, but the biological reasons for them are not known. Many hypotheses have been proposed, none of which is based on firm evidence. One school of thought (Lengerova and Matousek, 1979) even proposes that MHC differences may just be "evolutionary noise" and may not have any biological functions at all. In pondering these problems, it must be remembered that there are extensive differences in the MHC polymorphisms of various species, but that this variation does not correlate with any significant differences in their biological capabilities.

In considering the nature of, and the rationale for, the maternal immune response to the feto-placental unit, three basic observations must be remembered. First, allogeneic pregnancies occur in completely immunocompetent females, and the litter sizes in allogeneic pregnancies are larger than they are in syngeneic pregnancies. Second, allogeneic pregnancies elicit both an antibody and a cellular immune response to some of the paternal MHC antigens. Finally, there is no consistent body of evidence for a biologically significant impairment of the immune response in the female during pregnancy.

THE IMMUNE RESPONSIVENESS OF THE MOTHER DURING PREGNANCY

The hypothesis that a decreased maternal immune response during pregnancy is the mechanism by which the fetal homograft survives has been proposed by a number of investigators (reviewed in J.M. Anderson, 1972). The putative immunosuppression has generally been attributed to altered responsiveness of the maternal lymphocytes, the pregnancy-associated plasma proteins, the pregnancy hormones, or an increase in suppressor cell activity. These studies (Table 3-1) do not show, on balance, any evidence for a consistent effect of pregnancy on immune responsiveness. None of the pregnancy-associated proteins and hormones that have been studied in any detail are immunosuppressive in *in vitro* systems, let alone under physiological conditions (reviewed in Gill and Repetti, 1979). Suppressor cell activity has been demonstrated in the fetus and in the neonate of animals and humans (Olding et al., 1977; Sigal, 1977), but no firm evidence has been produced for their having a role in the maintenance of pregnancy (reviewed in Gill and Repetti, 1979).

One of the difficulties in evaluating the cellular immune response during pregnancy is its variability with time in normal healthy individuals (Figure 3-1). If the level of stimulation when the cells are first tested is taken

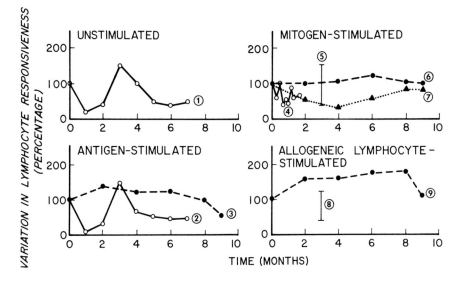

FIGURE 3-1. Longitudinal studies of lymphocyte blastogenesis as measured by the incorporation of tritiated thymidine under different conditions.

The level of incorporation at the time of the initial measurement was taken as 100%, and the other values are expressed relative to it. The standard deviations of the measurements generally fell into the range of 50 to 100%. The solid lines represent normal healthy adults, and the broken lines represent pregnant women. The data from the mothers at delivery are plotted uniformly at 9 months. Unstimulated: curve 1 (Graybill and Alford, 1976). Antigen stimulated: curve 2 is the average of stimulation by *Candida albicans*, histoplasmin, and purified protein derivative (PPD) (Graybill and Alford, 1976), and curve 3 is the average stimulation by *Candida albicans*, staphylococcus, streptokinase-streptodornase, and PPD (Birkeland and Kristoffersen, 1980b). Mitogen stimulated: curve 4 is for phytohemagglutinin (PHA) stimulation (Dionigi et al., 1973). Curve 5 is the average range over four months for PHA and pokeweed mitogen (PWM) stimulation (Osa and Weksler, 1977); curve 6 is the average for PHA and PWM stimulation (Birkeland and Kristoffersen, 1980b); and curve 7 is for Concanavalin A (Con A) stimulation (Birkeland and Kristoffersen, 1980b). Allogeneic lymphocyte stimulated: curve 8 is the range over four months (Osa and Weksler, 1977); and curve 9 is the average for stimulation by the lymphocytes of the father and the child (Birkeland and Kristoffersen, 1980a).

as 100 per cent, unstimulated and antigen-stimulated lymphocytes vary in responsiveness between 10 and 150 per cent, and mitogen-stimulated and allogeneic lymphocyte-stimulated lymphocytes vary between 40 and 150 per cent. Pregnancy appears to dampen these fluctuations, especially those at the lower limit, since the variation for lymphocytes from pregnant women in the former group is 60 to 135 per cent and that for the latter group, 100 to 180 per cent.

MATERNAL RESPONSE TO THE PATERNAL ANTIGENS OF THE FETO-PLACENTAL UNIT

Development of antigen expression in the placenta and fetus
The placental and fetal tissues express different antigens at various stages of development (Table 3-2), and blood cells pass between the mother and the fetus during pregnancy and at delivery (Table 3-3). The antigens on these tissues can elicit immune responses in the mother, and the antibodies that are

TABLE 3-1
Effect of Pregnancy on the Immunological Responsiveness of Maternal Lymphocytes

IMMUNOLOGICAL TEST	SPECIES	EFFECT OF PREGNANCY	REFERENCES
PHA stimulation	Human	Decrease	1,2,3,4
		No effect	5,6,7,8,9,10
	Mouse	Decrease	11
MLR stimulation	Human	Decrease	1,12,13
		No effect	5,9,14,15,16,17
	Mouse	Decrease	18,19
PFC†	Mouse	Decrease	20
		Increase	18,21,22
Other	Human		
Skin graft		Longer survival	23
PPD reactivity		Decrease	9,24
Cytotoxicity to HLA		Decrease	25
T cell rosetting		No effect	26

* Cells in normal adult AB plasma.
† Plaque forming cells to sheep red blood cells.
References: 1. Thong et al., 1973; 2. Blecher and Thompson, 1976; 3. Tomoda et al., 1976; 4. Strelkauskas et al., 1978; 5. Leikin, 1972; 6. Carr et al., 1973; 7. Yu et al., 1975; 8. Knoblach et al., 1976; 9. Birkeland and Kristoffersen, 1977; 10. Poskitt et al., 1977; 11. Ruppert and Richie, 1977; 12. Purtilo et al., 1972; 13. Petrocco et al., 1976; 14. Comings, 1967; 15. Cepellini et al., 1971; 16. Carr et al., 1973; 17. Lawler et al., 1975; 18. Fabris et al., 1977; 19. Hamilton and Hellström, 1977; 20. Sasaki and Ishida, 1975; 21. Humber et al., 1975; 22. Kenny and Diamond, 1977; 23. Anderson and Monroe, 1962; 24. Smith et al., 1972; 25. Taylor et al., 1974; 26. Campion and Currey, 1972.

TABLE 3-2
Expression of Antigens on Placental and Fetal Tissues*

STAGE	SPECIES	HISTOCOMPATIBILITY ANTIGENS	BLOOD GROUP ANTIGENS	OTHER
First trimester	Human	HLA on embryonic tissues	ABH on epithelia of most organs and in mucus; trophoblast negative Lewis (Le[a], Le[b]) in mucus	HCG; Fc, transferrin, insulin receptors
	Mouse	H-2 on inner cell mass† Non-H-2		T/t; teratoma antigen F9 (ca. 9 days on inner cell mass and ca. 4 days on trophoblast); Thy-1; Fc receptors; species antigen (all trophoblast); blastocyst and placenta specific antigens; Ia
Second trimester	Human	HLA on embryonic tissues	Complete regression of ABH antigens in epithelia of many organs Lewis and ABH in mucus	Fc, transferrin, insulin receptors
	Mouse	H-2 Non-H-2		§
Third trimester	Human	HLA on embryonic tissues	Lewis and ABH in mucus	Actin-like proteins on trophoblast; IgG, C3, C4, collagen and fibrinogen on basement membrane of placenta; collagen, fibrinogen and myosin on placental stroma (term); Fc, transferrin, insulin receptors; TA1, TA2
	Mouse	H-2 antigens on a variety of tissues Non-H-2		

* Adapted from Gill and Repetti (1979). See also Brent (1966), Posner (1974), Faulk et al. (1977, 1978), Galbraith et al. (1980), Szulman (1980), and Heyner (in press).

† H-2 antigens appear on the trophectoderm before implantation but are lost at implantation (Searle et al., 1976).

§ Antisera could be raised against trophoblastic antigens from the placenta at 12–14 days of gestation in the rat (Beer et al., 1972), and RT2 (Ag-C) antigens appeared at 10–12 days of gestation (Owen, 1962).

TABLE 3-3
The Passage of Blood Cells Across the Placenta During the First Pregnancy*

		DURING PREGNANCY			AT DELIVERY	
CELL	DIRECTION	FREQUENCY	POSSIBLE TRANSPORT PROCESS	FREQUENCY	PERSISTENCE OF CELLS IN MOTHER
Red cells	Fetus to mother	4–20%	Passsive	5–50%	Weeks
	Mother to fetus	Rare		12–100%†	
Granulocytes	Fetus to mother	?	?	37–80%	1 week
	Mother to fetus	?		?	
Lymphocytes	Fetus to mother	65–80%	Active	67%	> 1 year
	Mother to fetus	?		0–30%	
Platelets	Fetus to mother	?		?	
	Mother to fetus	?		45%	
Trophoblast cells and and placental anti- gens§	Fetus to mother	100%	Active	100%	Weeks

* Adapted from Gill (1977). See also Brent (1966) and Herzenberg et al. (1979).

† The higher estimates are probably more representative. The quantity of blood transferred depends, among other factors, upon the time of clamping the cord.

§ Reviewed in Brent (1966).

found most frequently and consistently are those against the antigens coded by the MHC.

Antigenicity of the placenta
The importance of the placenta as an anatomic barrier has been shown by experiments such as those in which treatment of pregnant animals with hyaluronidase alters the permeability of the placenta and allows the fetus to become tolerant to maternal antigens (Nathan et al., 1960; Najarian and Dixon, 1963). The placenta is poorly antigenic (Kirby, 1969; Bagshawe and Lawler, 1975; Searle et el., 1975), and several hypotheses for this apparent immunological inertness have been advanced: the carbohydrates of the trophoblast are poor antigens, antigens such as those coded by the MHC are lost at the time of implantation (Searle et al., 1976), or excessive amounts of antigen are produced which induce tolerance in the mother (Hulka and Brinton, 1963). None of these ideas have a firm experimental grounding. In addition, there is a substantial humoral and cellular immune response in the mother to the paternally derived antigens of the fetus (see below). The placenta reacts extensively with antibodies to Class I MHC antigens (Carlson and Wegmann, 1978; Wegmann et al., 1979, 1980), and the reactive cells are trophoblasts in the spongy zone and lymphocytes in the endodermal sinuses and surrounding tissue (D.J. Anderson et al., submitted: see Chapter 11).

The unique anatomic structure of the placenta may affect antigen presentation by the feto-placental unit and the nature of the immune response elicited by these tissues. There is also potentially a role for the placental circulation in affecting the stimulation of immunocompetent cells in the maternal blood by placental antigens. One way in which this exposure could be unusual is that the lymphocytes might receive an antigenic stimulus without the aid of macrophages, in the conventional sense in which they have been defined to function in the induction of the immune response, and this situation may alter the process of antigen recognition. The sequence of cellular events leading to the immune response and the nature of the immune response may then be different than in other tissues.

Antibody and cellular responses to fetal antigens
The mother's immune response to her conceptus involves the formation of antibodies and of cells sensitized to various tissues from the placenta and the fetus (Table 3-4). The most potent of these antigens are those coded by the MHC, and they may be presented to the mother by the passage of lymphocytes from the fetus or by their being on the placental surface; the latter point is in dispute (Faulk et al., 1977). A second major class of antibodies is that formed against red blood cell antigens, which, under certain circumstances, can lead to hemolytic disease in the newborn.

In humans, antibodies against HLA antigens are produced more frequently in successive pregnancies (Figure 3-2A, B). In all of the pregnancies, the titer of anti-HLA-A antibodies appeared to be approximately the same as that of anti-HLA-B antibodies, and ABO compatibility did not influence the prevalence or titer of the anti-HLA antibodies. The antibodies could be eluted from the placenta when the fetus had the HLA antigens, and then there was no antibody circulating in the fetus. When the fetus lacked the appropriate HLA antigens, the antibodies were present in the fetal circulation but not in the placenta (Tongio et al., 1975; Doughty and Gelsthorpe, 1976). When anti-HLA antibodies cross the placenta, they can react with fetal lymphocytes and mask 30 to 90 per cent of the paternal HLA determinants on these cells (Dumble et al., 1977). Evidence has also been presented to implicate them in the development of congenital abnormalities (Terasaki et al., 1970; Harris and Lordon, 1976) and neonatal jaundice (Carandina et al., 1977) and to implicate a deficiency of specific anti-HLA antibodies in the development of severe pre-eclampsia (Jenkins et al., 1977).

In addition to an antibody response to the fetal antigens, there is a clearly defined cell-mediated immune response. This response increases during gestation in humans (Figure 3-2C) and in rats (Figure 3-2D). No such response occurs in syngeneic pregnancies. The lymph nodes draining the uterus play a critical role in this cell-mediated immune response, since their removal prevents the production of migration inhibitory factor and leads to a significant

TABLE 3-4
Fetal and Placental Antigens Eliciting an Immune Response in the Mother

ANTIGEN TYPE	SPECIFICITY	SPECIES	TYPE OF ANTIBODY RESPONSE*	TYPE OF CELLULAR RESPONSE†	REFERENCES ANTIBODY FORMATION	REFERENCES CELLULAR IMMUNE RESPONSE
MHC	HLA-A,B,C	Human	CT, LA	CTL, MLR, LMC	1–10	11,12,13
	HLA-DR	Human	CT		10,14,15	
	H-2	Mouse	CT, LA, HA	CT, GVHR, skin rejection, rosetting, MLR	16–20	21–28
	RT1	Rat	CT, HA	MIF production	29	30,31
Placental antigens	Syncytial trophoblast	Human	Fab, blocking‡	CTL	32,33,34	35,36,37
	Mixed	Human	Blocking§	MIF production	38	38,39
Red blood cell	A, B, H	Human	CT		40,41,42	
	Rh	Human	CT	ADCC	43,44	45
Other	H-Y	Mouse	CT(S), SA		46	47,48
		Human		Enhanced placental growth		
	Non-H-2 (? type)	Mouse		GVHR		49
	Oncofetal	Mouse	ICD	CTL	50	23,24
	Immunoglobulin	Human	AAI		51,52	
		Rabbit	AAI, AId		53,54	

* CT, cytotoxic; LA, leukoagglutinating; HA, hemagglutinating; FAB, fluorescent antibody binding; CT(S), cytotoxic for sperm; SA, sperm agglutinating; ICD, immune complex deposition; AAI, anti-allotype; AId, anti-idiotype

† CTL, cytotoxic lymphocytes; MLR, mixed lymphocyte reactivity; LMT, leukocyte migration test; GVHR, graft-versus-host reaction; ADCC, antibody-dependent cell-mediated cytolysis.

‡ CTL reaction with cultured trophoblast cells.

§ MLR between maternal CTL and autologous plasma proteins.

References: 1. van Rood et al., 1958; 2. Payne, 1962; 3. Goodman and Masaitis, 1967; 4. Overweg and Engelfriet, 1969; 5. Revillard et al., 1973; 6. Rocklin et al., 1973; 7. Tongio et al., 1975; 8. Doughty and Gelsthorpe, 1976; 9. Vives et al., 1976; 10. Jeannet et al., 1977; 11. Bonnard and Lemos, 1972; 12. Finn et al., 1977; 13. Herva and Tiilikainen, 1977; 14. Winchester et al., 1975; 15. Ferrone et al., 1976; 16. Herzenberg and Gonzales, 1962; 17. Mishell et al., 1963; 18. Kaliss and Dagg, 1964; 19. Carlson and Wegmann, 1978; 20. Bell and Billington, 1980; 21. Maroni and Parrott, 1973; 22. Baines et al., 1976; 23. Hellström and Hellström, 1975a; 24. Hellström and Hellström, 1975b; 25. Hamilton et al., 1976; 26. Smith et al., 1978; 27. Chaouat and Voisin, 1979; 28. Pavia and Stites, 1979; 29. Kunz and Gill, unpublished; 30. Tofoski and Gill, 1977; 31. Gill and Kunz, 1980; 32. Hulka et al., 1963; 33. McCormick et al., 1971; 34. Taylor and Hancock, 1975; 35. Taylor et al., 1976; 36. Timonen and Saksela, 1976; 37. Toder et al., 1979; 38. Youtananukorn and Matangkasombut, 1972; 39. Youtananukorn et al., 1974; 40. Takano and Miller, 1972; 41. Szulman, 1973; 42. Fischer et al., 1977; 43. Woodrow, 1974; 44. Millison, 1979; 45. Zawodnik et al., 1976; 46. Krupen-Brown and Wachtel, 1979; 47. Toivanen and Hirvonen, 1970; 48. Gasser and Silvers, 1972; 49. Baines et al., 1976; 50. Tung, 1974; 51. Faulk et al., 1974; 52. Faulk et al., 1979; 53. Loor and Kelus, 1978; 54. Wikler et al., 1980.

decrease in litter size and uneven fetal development (Tofoski and Gill, 1977).

The immune response to pregnancy has also been defined morphologically by the enlargement of the lymph nodes draining the uterus in rats (Billingham, 1971; Mosley et al., 1975), rabbits (Scothorne and McGregor, 1955), mice (McLean et al., 1974; Chatterjee-Hasrouni et al., 1980), hamsters (Billingham, 1971; Beer and Billingham, 1974), and humans (Billingham, 1971); by proliferation of blast cells, particularly among thymus-derived populations (Parrott, 1967; Billingham, 1971; Beer and Billingham, 1974; Beer et al., 1975a, b); and by the large number of plasma cells in the medullary cords of the lymph nodes (Maroni and de Sousa, 1973). These changes were observed in allogeneic, but not syngeneic, pregnancies.

Several other types of antigens have been recognized on the human placenta, but their immunogenic potential during pregnancy is not known. Heterologous, lymphocyte-absorbed antisera have identified an oncofetal antigen designated TA-1 (Faulk et al., 1978). Unabsorbed heterologous antisera have identified trophoblastic antigens crossreactive with lymphocytes, and they have been called TA-2 antigens (Faulk et al., 1978; McIntyre and Faulk, 1979).

There are three possible explanations why the antibodies do not destroy the feto-placental unit. First, the heavy glycoprotein surface of the placenta may not be susceptible to damage by antibody and complement. Second, cytotoxic antibodies may be only a portion of the antibody response to the feto-placental unit, and a mixture of cytotoxic and noncytotoxic antibodies may prevent the attainment of the appropriate membrane surface configuration of cytotoxic antibodies needed to fix complement and to induce immunological damage. Third, the unique dynamics of blood flow in the placenta may not be conducive to the fixation of all the complement components needed to damage the placental membrane. The cellular immune response probably does not damage the fetus because relatively few maternal lymphocytes cross the placenta. It may not damage the placenta either because the cells are not cytotoxic for the placental antigens or because blocking antibodies may be produced which neutralize the cytotoxic potential of the cellular immune response.

Immunological and genetic factors influencing reproduction

There are two types of hypotheses that have been proposed to explain the acceptance of the fetal homograft. The first hypothesis, that based upon a depressed immune response in the mother during pregnancy, is not supported by firm evidence, as discussed above. The second hypothesis, that the immune response to the fetal antigens may somehow be unique and may have a beneficial effect, has a certain amount of experimental support (reviewed in Gill and Repetti, 1979). In this view, the immune response is an important factor in the normal physiology of pregnancy and may have important evolutionary survival

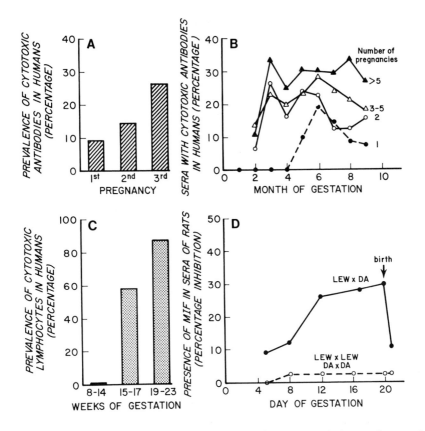

FIGURE 3-2. The humoral and cellular immune response of the mother to the paternal histocompatibility antigens of the fetus.
(a) The prevalence of cytotoxic antibodies in women with different numbers of pregnancies (Doughty and Gelsthorpe, 1976). (b) The percentage of women with cytotoxic antibodies in their sera at different times during pregnancy and after various numbers of pregnancies (Vives et al., 1976). The broken line represents the first pregnancy. (c) The prevalence of cytotoxic lymphocytes in women at different times during pregnancy (Taylor et al., 1976; Timonen and Saksela, 1976). (d) The cellular immune response in rats as measured by the production of an MIF-like substance which was detectable in the serum (Tofoski and Gill, 1977; Gill and Kunz, 1980).

value for the species. An important role for MHC polymorphisms, then, would be to improve reproductive capacity and to confer a selective advantage on the progeny (Hull, 1964a, 1964b; Clarke and Kirby, 1966; Kirby, 1968; Warburton, 1968).

Although pregnancy can occur in the absence of an immune response, as in the derivation of inbred strains, this situation is probably a special case

63

biologically and represents a maximal stress on the reproductive mechanism. There is a substantial body of circumstantial evidence to support this position. First, in the initial selection of rats, mice, or rabbits for inbreeding from wild populations, a large number of mating pairs are unable to adapt to breeding in the laboratory. Thereafter, it is common to select animals not only for the trait desired but also on the basis of their reproductive performance. Second, the development and maintenance of inbred strains is frequently difficult or impossible because of reductions in litter size and in viability of the offspring as a result of the adverse genetic or environmental effects on various components of the reproductive cycle. In some strains of inbred rats (Cramer and Gill, 1975) and mice (Lyon, 1959; Kozanowska, 1960; McCarthy, 1967), the reduction in litter size is due to post-implantation losses. Another potent factor in decreasing litter size and the source of inherited partial sterility is chromosomal aberration (Waletzky and Owen, 1942; Tyler and Chapman, 1948). Even in established inbred strains, litter sizes are smaller than those arising from allogeneic matings (Gill et al., 1979). Third, in the course of inbreeding, many of the females develop polycystic ovaries or a uterine environment inconducive to implantation either by normal fertilization or by ovum transplantation. Fourth, the mother can have difficulty with lactation and not be able to nurse beyond the first few days after delivery. Finally, there are also defects in the male, such as loss of libido, which contribute to the difficulties of inbreeding.

In exploring the nature of the maternal immunological response to the fetus, two lines of investigation have been taken. In the first approach, the nature of the uterus as an immunologically unique site has been explored in the rat by Billingham and his colleagues (Billingham, 1971; Beer and Billingham, 1974). They showed that the uterus is immunologically autonomous and that immunological memory cells develop and persist in the uterus after its initial exposure to antigen. Intrauterine sensitization with allogeneic skin grafts or cells led to a secondary response, the so-called recall flare, following local challenge with cells of the same antigenic specificity. The reaction did not occur if the cellular challenge was given in extrauterine sites. They also reported that more embryos developed in a uterine horn that had been sensitized against paternal antigens than in an unsensitized horn.

A related line of evidence argues that, since allogeneic pregnancies result in larger litters and larger mean weights of the offspring (James, 1967; Kirby, 1968; Beer and Billingham, 1974; Beer et al., 1975b; Tofoski and Gill, 1977), the activation of the immune system of the mother by the paternal antigens of the fetus brings about conditions beneficial to implantation and to intrauterine development of the fetuses. The corollary of this proposition is that abrogation of the mother's capacity to mount an immune response by the removal of the lymph nodes around the uterus should lead to the loss of the advantage that allogeneic fetuses have over syngeneic fetuses. There is considerable evidence that this is true (Beer et al., 1972; Beer and Billingham, 1974; Tofoski and Gill, 1977). There is a contrary school of thought (McLaren, 1965, 1975; Finkel and Lilly, 1971; Hetherington et al., 1976; Hetherington and Humber, 1977), however, which states that all of the advantages of the hybrid fetus are due to heterosis and not to activation of the immune system. Without a clear definition of what heterosis means, this counterargument is not decisive.

In the second approach to studying the nature of the immunological response to the fetus, immunized females have been used as a model for heterozygous pregnancies, and tolerant females as a model for homozygous pregnancies. The evidence that immunization can affect reproductive capacity in this context is conflicting (Gill and Repetti, 1979), and these studies do not provide any insights into the role of the immune response under the normal physiological conditions of pregnancy.

There are four interesting genetic situations in experimental animals and several in humans that provide evidence of the importance of the genetic relationship between the mother and her fetus for the success of the pregnancy. First, Michie and Anderson (1966) found that after 72 generations of (brother x sister) matings in the A2 strain of Wistar rats, approximately half of the skin grafts transplanted among the animals were rejected in two weeks. Attempts to breed (brother x sister) pairs that had successfully accepted skin grafts from each other were not successful. In successive generations, the frequency of two-way graft acceptance remained approximately 50 per cent. The authors explained their results by postulating a powerful genetic stabilizing mechanism capable of holding the genotypic proportions of the population constant in the face of intensive selection for homozygosity. These studies were not pursued; hence, the reproducibility of this phenomenon and its genetic basis remain problematic.

Second, Hull (1964a, 1964b) studied the effect of fetal compatibility with the mother on the genotypic proportions of fetuses born. In crosses between the C3H and Hg strains of mice, both of which had been (brother x sister) mated for at least 35 generations, the backcross mating produced offspring whose genotypic ratio deviated significantly from the expected 1:1, and the F_2 hybrid offspring differed significantly from the expected 1:2:1 ratio. These deviations were due to the lack of offspring of the same genotype as the mother and not to decreased fitness of a particular genotype. Therefore, a deleterious effect due to the interaction between the mother and her offspring was suspected. Despite variations in the ratios of homozygotes to heterozygotes, the litter sizes in all of these crosses were the same. There was no evidence of differences in fertility among the females as a possible cause for changes in genotypic frequencies. No mention was made of abnormalities in the male-to-female ratio. The effect was confirmed in a second cross (Hull, 1969).

Third, Palm (1969, 1970) studied a similar situation in rats. She found

65

an excess of heterozygous offspring in mating combinations in which the mother was derived from an inbred strain and the father was an F₁ hybrid. There were no differences in the number of heterozygous and homozygous offspring in the mating combinations where the F_1 hybrid was the mother and the inbred animal was the father. Palm postulated that the mechanism leading to the excess of animals heterozygous for the MHC was postnatal loss of homozygotes secondary to graft-versus-host disease resulting from the transfer of cells from the mother to the fetus during gestation.

Fourth, genes in the mouse and rat can influence embryonic development, body size, and tail morphology. The T/t complex in the mouse consists of a set of genes linked to the MHC which can interact to cause embryonic death at various stages .f development and can cause skeletal abnormalities, mainly in tail morphology (Bennett, 1975; Klein, 1975). The growth and reproduction complex in the rat (Grc) consists of genes (dw-3 and ft) linked to the MHC which cause embryonic death, male sterility, reduced female fertility, and decreased body size (Kunz et al., 1980). The tail anomaly lethal (Tal) gene in the rat, which is not linked to the MHC, causes embryonic death soon after implantation when homozygous, and tail anomalies when heterozygous (Hoshino et al., 1979; Schaid et al., submitted). It can also interact epistatically with the Grc to cause embryonic death (Schaid et al., submitted).

In humans, the Rh, ABO, and HLA systems can influence the outcome of gestation. The classical case of genetic differences producing disease during gestation is hemolytic disease of the newborn, which occurs because of the passage of incompatible red cells from the fetus to the mother, an immune response in the mother, and the passage of the maternal antibodies back into the fetus to destroy its red cells. In the case of Rh disease, the antigens are on the red blood cells only, and so they provide the sole target for the maternal antibodies (Woodrow, 1970). Hemolysis occurs in approximately 0.4 to 2 per cent of first Rh-incompatible pregnancies and increases with subsequent pregnancies. Incompatibility of the ABO blood groups provides considerable protection against Rh sensitization, presumably by accelerating the removal of fetal red blood cells from the maternal circulation during pregnancy and following the inevitable transplacental hemorrhage that occurs at birth. An ABO-incompatible pregnancy in which an O mother carries an A or B fetus can lead to ABO erythroblastosis or to fetal wastage (Szulman, 1973, 1980). The disease occurs almost exclusively in group O mothers because of the passage of IgG anti-A or IgG anti-B antibodies from the mother to the fetus. In contrast, group A and group B mothers have mainly IgM anti-B and IgM anti-A antibodies, respectively (Rawson and Abelson, 1960; Mollison, 1979), which do not cross the placenta in any significant amount. Mild degrees of hemolytic disease of the newborn due to ABO incompatibility are approximately twice as common as those due to Rh incompatibility. The large amount of soluble ABH glycoproteins produced by the fetus probably absorbs the antibodies transmitted across the placenta from the mother before they can damage the fetal cells, and this effect could account for the relatively low incidence of ABO erythryoblastosis.

There is reasonable evidence that the HLA and ABO systems provide the genetic basis for some cases of abortion, pregnancy abnormalities, and trophoblastic neoplasia (reviewed in Loke, 1978, Gill and Repetti, 1979, and Szulman, 1980). A correlation between HLA type and the incidence of abortions has been reported (Komlos et al., 1977). In a study of 79 couples with repeated abortions or with hydatidiform moles, a significantly higher frequency of common HLA antigens was found in the mother and the father than in controls. This finding raises the possibility that a higher incidence of abortions may occur when the fetus is homozygous for certain HLA antigens. In abortuses with normal karyotypes, there was an increased incidence of ABO incompatibility between the mother and the offspring compared with the incidence of abortuses with abnormal karyotypes (Lauritsen et al., 1975). The number of abortuses incompatible with their mothers significantly exceeded those expected from the prevailing ABO frequencies in the population. In a study of 229 cases of spontaneous abortion prior to 20 weeks of gestation (Takano and Miller, 1972), the maternal blood group was O in 52.0 per cent, whereas the prevalence of type O mothers in the populations studies was 44.5 per cent. Of the 78 maternal-fetal pairs that were examined for their blood groups in parallel, 44.7 per cent of the fetuses were incompatible with their mothers, and this frequency of incompatibility was significantly higher than expected ($P < 0.01$). In women with recurrent abnormal pregnancies of unknown etiology, there was a higher frequency of HLA-A9 and there was a higher frequency of compatibility between the mother and father for the antigens at the HLA-A locus, but not at the HLA-B locus (Gerencer et al., 1978).

There is also apparently a relationship between ABO blood group and trophoblastic neoplasia (Bagshawe et al., 1971). In 260 cases studied, the risk of developing choriocarcinoma after any form of pregnancy was related to the ABO blood groups of both the mother and the father. Women of group A mating with men of group O seemed to be at the highest risk, whereas the risk in group A women mating with group A males was the lowest. Group AB women tended to have the most rapidly progressive choriocarcinomas, and the tumors did not respond well to chemotherapy. The spontaneous regression of the trophoblast after evacuation of hydatidiform moles occurred most frequently in women mated to men of their own ABO genotype.

The evidence presented here (summarized in Tables 3-5 and 3-6) shows that a variety of immunological and genetic factors influence the health

TABLE 3-5
Summary of Immunological and Idiopathic Factors Influencing Reproduction.

FACTOR	PARAMETER	SPECIES	EFFECT
Immunological	Uterus as a privileged site	Rat	Local immune response to intrauterine sensitization
			Increased fetal development in sensitized uterus
	Pregnancy in immunized females	Mouse	Equivocal effects in females sensitized or tolerant to male MHC antigens
			Abortion by antitrophoblast or antithymocyte antisera
	Litter size and fetal weight	Mouse, rat	Allogeneic > syngeneic pregnancies
Idiopathic	Inbreeding	Mouse, rat rabbit	Early loss of many lines
			Reductions in litter size and viability
			Abnormalities of the female reproductive tract
			Difficulties in lactation
			Loss of male libido

of a pregnancy and the nature of its outcome. There is a delicately poised balance among them which is necessary for the normal reproductive process but which can be upset in a number of ways and lead to developmental abnormalities.

TABLE 3-6
Summary of Genetic Factors Influencing Reproduction

PHENOTYPIC EXPRESSION	SPECIES	CAUSE
Inability to inbreed A2 strain ("Ag-E" locus)	Rat	Selection for heterozygosity
Decreased number of offspring with maternal genotype; decreased number of males in rats only	Mouse, I B OFF, rat	Fetal histocompatibility with mother
Embryonic death, male sterility, reduced female fertility and decreased body size	Rat	Genes of the growth and reproduction complex (GRC)
Tail anomalies and embryonic death	Rat	Tail anomaly lethal (Tal) gene
	Mouse	Genes of the T/t complex
Hemolytic disease of the newborn	Human	Rh incompatibility
		ABO incompatibility
Abortion (fetal wastage)	Human	Common HLA antigens in mother and father
		ABO incompatibility between mother and father
Abnormal pregnancy of unknown etiology	Human	Excess of HLA-A9 in the mother
		Mother and father compatible at HLA-A
Trophoblastic neoplasia	Human	Blood group A mother + group O father is highest risk
		Group A mother + group A father is lowest risk
		Group AB mother has most rapidly progressive choriocarcinoma and poorest response to therapy
		Regression of trophoblast after evacuation of a mole is fastest when mother and father are ABO compatible

SUMMARY AND CONCLUSIONS

We still have relatively little insight into the nature of the maternal immune response to the paternal antigens of the feto-placental unit or into the ability of the fetus to survive in the face of brisk humoral and cellular immune responses. The reason for the survival of the fetal homograft most likely lies not in there being a general depression of maternal immune responsiveness during pregnancy, but rather in the nature of the immune response to the tissues of the placenta and the fetus. If the immune response to the fetus improves the species survival, by increasing the reproductive capacity of the species, then it is performing a normal physiological function that is not connected with its role in host defense. The presentation of the paternal components of the fetal histocompatibility antigens to the mother across the unique anatomical structure of the placenta may elicit an immune response that does not have the destructive potential of a response elicited by antigen exposure through other routes. Thus, the view of the immune response as an important part of the normal physiology of pregnancy, rather than as a host reaction which must be blocked, may lead to more productive immunological and genetic approaches to the investigation of the transplantation immunobiology of the fetal homograft.

ACKNOWLEDGMENTS

The studies in the author's laboratory were supported by grants from the National Institutes of Health (HD 08662, HD 09880, CA 18659, and GM 07349), from the Beaver County Cancer Society, and from the Tim Caracio Memorial Cancer Fund.

REFERENCES

Anderson, D.J., B.A. Sandow, R. Raghupathy, B. Singh, and T.G. Wegmann. 1981. Localization of cells constituting an immunological barrier in the mouse placenta. Submitted.

Anderson, J.M. 1972. *Nature's Transplant: The Transplantation Immunology of Viviparity.* New York: Appleton-Century-Croft.

Anderson, R.H., and C.W. Monroe. 1962. Experimental study of the behavior of adult human skin homografts during pregnancy. *Am. J. Obstet. Gynecol.* 84:1096.

Bagshawe, K., and S. Lawler. 1975. The immunogenicity of the placenta and trophoblast. In *Clinical and Experimental Immunoreproduction.* Vol. 1, *Immunology of the Tropho-*

blast. R.G. Edwards, C.W.S. Howe, and M.H. Johnson, eds. Cambridge: Cambridge Univeristy Press, pp. 171-191.

Bagshawe, K.D., G. Rawlins, M.C. Pike, and S.D. Lawler. 1971. ABO blood-groups in trophoblastic neoplasia. *Lancet* i:553.

Baines, M.G., E.A. Speers, H. Pross, and K.G. Millar. 1976. Characteristics of the maternal lymphoid response of mice to paternal strain antigens induced by homologous pregnancy. *Immunology* 31:363.

Beer, A.E., and R.E. Billingham. 1974. Host responses to intra-uterine tissue, cellular and fetal allografts. *J. Reprod. Fertil.* Suppl 21: 59.

Beer, A.E., R.E. Billingham, and S.L. Yang. 1972. Further evidence concerning the autoantigenic status of the trophoblast. *J. Exp. Med.* 135:1177.

Beer, A.E., R.E. Billingham, and J.R. Scott. 1975a. Immunogenetic aspects of implantation, placentation and feto-placental growth rates. *Biol. Reprod.* 12:176.

Beer, A.E., J.R. Scott, and R.E. Billingham. 1975b. Histoincompatibility and maternal immunological status as determinants of fetoplacental weight and litter size in rodents. *J. Exp. Med.* 142:180.

Bell, S.C., and W.D. Billington. 1980. Major anti-paternal alloantibody induced by murine pregnancy is non-complement-fixing IgG1. *Nature* 288:387.

Bennett, D. 1975. The T-locus of the mouse. *Cell* 6:441.

Billingham, R.E. 1971. The transplantation biology of mammalian gestation. *Am. J. Obstet. Gynecol.* 111:469.

Birkeland, S.A., and K. Kristoffersen. 1977. Cellular immunity in pregnancy: blast transformation and rosette formation of maternal T and B lymphocytes. *Clin. Exp. Immunol.* 30:408.

Birkeland, S.A., and K. Kristoffersen. 1980a. The fetus as an allograft: a longitudinal study of normal human pregnancies studied with mixed lymphocyte cultures between mother-father and mother-child. *Scand. J. Immunol.* 11:311.

Birkeland, S.A., and K. Kristoffersen. 1980b. Lymphocyte transformation with mitogens and antigens during normal human pregnancy: a longitudinal study. *Scand. J. Immunol.* 11:321.

Blecher, T.E., and J.J. Thompson. 1976. Comparison of uridine uptake at 24 hours with thymidine uptake at 72 hours in phyto-hemagglutinin-stimulated cultures of pregnant and other subjects. *J. Clin. Pathol.* 29:727.

Bonnard, G.D., and L. Lemos. 1972. The cellular immunity of mother versus child at delivery: sensitization in unidirectional mixed lymphocyte culture and subsequent ^{51}Cr-release cytotoxicity test. *Transplant. Proc.* 4:177.

Brent, R.L. 1966. Immunologic aspects of developmental biology. *Adv. Teratology* 1:81.

Campion, P.D., and H.L.F. Currey. 1972. Cell-mediated immunity in pregnancy. *Lancet* ii:830.

Carandina, G., L. DeRitis, P.Palazzi, and P.L. Mattiuz. 1977. Maternal lymphocytotoxic antibodies and bilirubin levels in the postpartum offspring. *Tissue Antigens* 10:348.

Carlson, G.A., and T.G. Wegmann. 1978. Paternal-strain antigen excess in semiallogeneic pregnancy. *Transplant. Proc.* 10:403.

Carr, M.C., D.P. Stites, and H.H. Fudenberg. 1973. Cellular immune aspects of the human fetal-maternal relationship. II. *In vitro* response of gravida lymphocytes to phytohemagglutinin. *Cell. Immunol.* 8:448.

Carr, M.C., D.P. Stites, and H.H. Fudenberg. 1974. Cellular immune aspects of the human fetal-maternal relationship. III. Mixed lymphocyte reactivity between related maternal and cord blood lymphocytes. *Cell. Immunol.* 11:332.

Ceppellini, R., G.D. Bonnard, F. Coppo. V.C. Miggiano, M. Pospisil, E.S. Curtoni, and M. Pellegrino. 1971. Mixed leucocyte cultures and HL-A antigens. I. Reactivity of young fetuses, newborns and mothers at delivery. *Transplant. Proc.* 3:58.

Chaouat, G., and G.A. Voisin. 1979. Regulatory T cell subpopulations in pregnancy. I. Evidence for suppressive activity of the early phase of MLR. *J. Immunol.* 122:1383.

Chatterjee-Hasrouni, S., V. Santer, and P.K. Lala. 1980. Characterization of maternal small lymphocyte subsets during allogenic pregnancy in the mouse. *Cell. Immunol.* 50:290.

Clarke, B., and D.R.S. Kirby. 1966. Maintenance of histocompatibility polymorphisms. *Nature* 211:999.

Comings, D.E. 1967. Lymphocyte transformation in response to phytohemagglutin during and following a pregnancy. *Am. J. Obstet. Gynecol.* 97:213.

Cramer, D.V., and T.J. Gill III. 1975. Maternal influence on postimplantation survival in inbred rats. *J. Reprod. Fertil.* 44:317.

Dionigi, R., A. Zonta, F. Albertario, R. Galeazzi, and G. Bellizona. 1973. *Transplantation* 16:550.

Doughty, R.W., and K. Gelsthorpe. 1976. Some parameters of lymphocyte antibody activity through pregnancy and further eluates of placental material. *Tissue Antigens* 8:43.

Dumble, L.J., B.D. Tait, S. Whittingham, and P.W. Ashton. 1977. The immunological privilege of the fetus: decreased expression of paternal HLA. *Immunogenetics* 5:345.

Fabris, N., L. Piantanelli, and G. Muzzioli. 1977. Differential effect of pregnancy or gestagens on humoral and cell-mediated immunity. *Clin. Exp. Immunol.* 28:306.

Faulk, W.P., R.M. Galbraith, G.M.P. Galbraith, R.S. Boackle, and P.M. Johnson. 1979. Evidence for immunopathology in primate pregnancies. *Protides Biol. Fluids* 26:427.

Faulk, W.P., E. van Loghem, and G.B. Stickler. 1974. Maternal antibody to fetal light chain (Inv) antigens. *Am. J. Med.* 56:393.

Faulk, W.P., R.E. Lovins, C. Yeager, and A. Temple. 1978. Antigens of human trophoblast: immunological and biochemical characterization. In *Immunological Influence on Human Fertility.* B. Boettcher, ed. New York: Academic Press, pp. 153-160.

Faulk, W.P., A.R. Sanderson, and A. Temple. 1977. Distribution of MHC antigens in human placental chorionic villi. *Transplant. Proc.* 9:1379.

Faulk, W.P., A. Temple, R.E. Lovins, and N. Smith. 1978. Antigens of human trophoblasts: a working hypothesis for their role in normal and abnormal pregnancies. *Proc. Nat. Acad. Sci. USA.* 75:1947.

Ferrone, S., M.F. Mickey, P.I. Terasaki, R.A. Reisfeld, and M.A. Pelligrino. 1976. Humoral sensitization in parous women: cytotoxic antibodies to non HL-A antigens. *Transplantation* 22:61.

Ferrone, S., M.A. Pellegrino, J.J. van Rood, and A. van Leeuwen. 1977. Functional properties of anti-B-cell antigen alloantisera. *Transplant. Proc.* 11:675.

Finkel, S.I., and F. Lilly. 1971. Influence of histoincompatibility between mother and foetus on placental size in mice. *Nature* 234:102.

Finn, R., J.C. Davis, C.A. St. Hill, and L.J. Hipkin. 1977. Feto-maternal bidirectional mixed lymphocyte reaction and survival of fetal allograft. *Lancet* ii:1200.

Fisher, K., A. Poschmann, and A. Grundmann. 1977. Hämolytische neugeborenerkrankugen infolge ABO-Unvertraglichkeit. *Z. Geburtshilfe Perinatol.* 181:227.

Galbraith, G.M.P., R.M. Galbraith, and W.P. Faulk. 1980. Immunological studies of transferrin and transferrin receptors of human placental trophoblast. *Placenta* 1:33.

Gasser, D.L., and W.K. Silvers. 1972. Genetics and immunology of sex-linked antigens. *Adv. Immunol.* 15:215.

Gerencer, M., A. Kastelan, A. Drazancic, V. Kerhin-Brkljacic, and M. Madjaric. 1978. The LHA antigens in women with recurrent abnormal pregnancies of unknown etiology. *Tissue Antigens* 12:223.

Gill, T.J. III. 1977. Chimerism in humans. *Transplant. Proc.* 9:1423.

Gill, T.J. III, and H.W. Kunz. 1980. The role of regional differences in the major histocompatibility complex in the production during pregnancy of a serum factor inhibiting macrophage migration. *J. Immunogenet.* 7:157.

Gill, T.J. III, H.W. Kunz, and C.T. Hansen. 1979. Litter sizes in inbred strains of rats (*Rattus norvegicus*). *J. Immunogenetics* 6:461. Gill, T.J. III, and C.F. Repetti. 1979. Immunologic and genetic factors influencing reproduction. *Am. J. Path.* 95:465.

Goodman, H.D., and L. Masaitis. 1967. Analysis of the isoimmune response to leucocytes. I. Maternal cytotoxic response to fetal lymphocytes. *Proc. Soc. Exp. Biol. Med.* 126:599.

Graybill, J.R., and R.H. Alford. 1976. Variability of sequential studies of lymphocyte blastogenesis in normal adults. *Clin. Exp. Immunol.* 25:28.

Hamilton, M.S., and I. Hellström. 1977. Altered immune responses in pregnant mice. *Transplantation* 23:423.

Hamilton, M.S., I. Hellström, and G. van Belle. 1976. Cell-mediated immunity to embryonic

71

antigens of syngeneically and allogeneically mated mice. *Transplantation* 21:261.

Harris, R.E., and R.E. Lordon. 1976. The association of maternal lymphocytotoxic antibodies with obstetric complications. *Obstet. Gynecol.* 43:302.

Hellström, I., and K.E. Hellstrom, K.E. 1975a. Cytotoxic effect of lymphocytes from pregnant mice on cultivated tumor cells. I. Specificity and nature of effector cells and blocking by serum. *Int. J. Cancer* 15:1.

Hellström, I., and K.E. Hellström. 1975b. Cytotoxic effect of lymphocytes from pregnant mice on cultivated tumor cells. II. Blocking and unblocking of cytotoxicity. *Int. J. Cancer.* 15:30.

Herva, E., and A. Tiilikainen. 1977. Mixed lymphocyte culture reactions at delivery and in the puerperium: effects of parity, HLA antigens and maternal serum. *Acta Pathol. Microbiol. Scand. (C)* 85:333.

Herzenberg, L.A., D.W. Bianchi, J. Schroder, H.M. Cann, and G.M. Iverson. 1979. Fetal cells in the blood of pregnant women: Detection and enrichment by fluorescence-activated cell sorting. *Proc. Nat. Acad. Sci. USA* 76:1453.

Herzenberg, L.A., and B. Gonzales. 1962. Appearance of H-2 agglutinins in outcrossed female mice. *Proc. Nat. Acad. Sci. USA* 48:570.

Hetherington, C.M., and D.P. Humber. 1977. The effect of pregnancy on lymph node weight in the mouse. *J. Immunogenet.* 4:271.

Hetherington, C.M., D.P. Humber, and A.G. Clarke. 1976. Genetic and immunologic aspects of litter size in the mouse. *J. Immunogenetics* 3:245.

Heyner, S. 1981. Immunogenetic approaches to the study of the mammalian egg and early embryo. In *Oxford Series of Reproductive Biology.* Vol. 3. C.A. Finn, ed. Oxford: Oxford University Press.

Hoshino, K., S. Oda, and Y. Kameyama. 1976. Tail anomaly lethal, Tal: a new mutant gene in the rat. *Teratology* 19: 27.

Hulka, J.F., and V. Brinton. 1963. Antibody to trophoblast during early postpartum period in toxemic pregnancies. *Am. J. Obstet. Gynecol.* 86:130.

Hulka, J.F., V. Brinton, J. Schaaf, and C. Baney. 1963. Appearance of antibodies to trophoblast during the postpartum period in normal human pregnancies. *Nature* 198:501.

Hull. P. 1964a. Equilibrium of gene frequency produced by partial incompatibility of offspring with dam. *Proc. Nat. Acad. Sci. USA* 51:461.

Hull, P. 1964b. Partial incompatibility not affecting total litter size in the mouse. *Genetics* 50:563.

Hull, P. 1969. Maternal-foetal incompatibility associated with H-3 locus in the mouse. *Heredity* 24:203.

Humber, D.P., M. Pinder, and C.M. Hetherington. 1975. The effects of pregnancy on the immunological competence of congenitally athymic (nude) mice. *Transplantation* 19:91.

James, J.A. 1967. Some effects of immunological factors on gestation in mice. *J. Reprod. Fertil.* 14:265.

Jeannet, M., C. Werner, E. Ramirez, P. Vassalli, and W.P. Faulk. 1977. Anti-HLA, anti-human "Ia-like" and MLC blocking activity of human placental IgG. *Transplant. Proc* 9:1417.

Jenkins, D.M., J. Need, and S.M. Rajah. 1977. Deficiency of specific HLA atibodies in severe pregnancy pre-eclampsia. *Clin. Exp. Immunol.* 27:485.

Kaliss, N., and M.K. Dagg. 1964. Immune response engendered in mice by multiparity. *Transplantation* 2:416.

Kenny, J.R., and M. Diamond. 1977. Immunological responsiveness to *Escherichia coli* during pregnancy. *Infect. Immunol.* 16:1974.

Kirby, D.R.S. 1968. Transplantation and pregnancy. In *Human Transplantation.* F.T. Rapaport and J. Dausset, eds. New York: Grune and Stratton, pp. 565-586.

Kirby, D.R.S. 1969. Is the trophoblast antigenic? *Transplant. Proc.* 1:53.

Klein, J. 1975. *Biology of the Mouse Histocompatibility-2 Complex.* New York: Springer-Verlag, pp. 251-274.

Knoblach, V., V. Jouja, and M. Pospisil. 1976. Feto-maternal relationships in normal pregnancy in mixed lymphocyte cultures. *Arch. Gynecol.* 220:249.

Komlos, L., R. Zamir, H. Joshua, and I. Halbrecht. 1977. Common HLA atigens in couples with repeated abortions. *Clin. Immunol. Immunopathol.* 7:330.

Kozanowska, H. 1960. Early embryonal mortality in inbred lines of mice and their crosses. *Bull. Roy. Belge. Gynecol. Obstet.* 30:719.

Krupen-Brown, K., and S.S. Wachtel. 1979. Cytotoxic and agglutinating H-Y antibodies in multiparous female mice. *Transplantation* 27:406.

Kunz, H.W., T.J. Gill III, B. Dixon, F.H. Taylor, and D.L. Greiner. 1980. Growth and reproduction complex in the rat: genes linked to the major histocompatibility complex that affect development. *J. Exp. Med.* 152:1506.

Lauritsen, J.G., N. Grunnet, and O.M. Jensen. 1975. Maternofetal ABO incompatibility as a cause of spontaneous abortion. *Clin. Genetics* 7:308.

Lawler, S.D., E.O. Ukaejiofo, and B.R. Reeves. 1975. Interaction of maternal and neonatal cells in mixed lymphocyte cultures. *Lancet* ii:1185.

Leikin, S. 1972. Depressed maternal lymphocyte response to phytohemagglutinin in pregnancy. *Lancet* ii:43.

Lengerova, A., and V. Matousek. 1979. Another look at the molecular variability of H-2 gene products. *J. Immunogenet.* 6:329.

Loke, Y.W. 1978. *Immunology and Immunopathology of the Human Foetal-Maternal Interaction.* Amsterdam: Elsevier/North-Holland, pp. 167-173.

Loor, F., and A.S. Kelus. 1978. Maternal allotype dominance and allelic exclusion in the B lineage cells of the newborn rabbit. *Eur. J. Immunol.* 8:801.

Lyon, M.F. 1959. Some evidence concerning the "mutational load" in inbred strains of mice. *Heredity* 12:341.

Maroni, E.S., and M.A.B. de Sousa. 1973. The lymphoid organs during pregnancy in the mouse: a comparison between a syngeneic and an allogenic mating. *Clin. Exp. Immunol.* 13:107.

Maroni, E.S., and D.M.V. Parrott. 1973. Progressive increase in cell-mediated immunity against paternal transplantation antigens in parous mice after multiple pregnancies. *Clin. Exp. Immunol.* 13:253.

McCarthy, J.C. 1967. The effects of inbreeding on the components of litter size in mice. *Genet. Res.* 10:73.

McCormick, J.N., W.P. Faulk, H. Fox, and H.H. Fudenberg. 1971. Immunohistological and elution studies of the human placenta. *J. Exp. Med.* 133:1.

McIntyre, J.A., and W.P. Faulk. 1979. Antigens of human trophoblast. Effects of heterologous anti-trophoblast sera on lymphoyte responses *in vitro. J. Exp. Med.* 149:824.

McLaren, A. 1965. Genetic and environmental effects on fetal and placental growth in mice. *J. Reprod. Fertil.* 9:79.

McLaren, A. 1975. Antigenic disparity: does it affect placental size, implantation or population genetics? In *Immunobiology of Trophoblast.* Clinical and Experimental Immunoreproduction Series No. 1. R.G. Edwards, C.W.S. Howe, and M.H. Johnson, eds. Cambridge: Cambridge University Press, pp. 255-273.

McLean, J.M., J.G. Mosely, and A.C.C. Gibbs. 1974. Changes in the thymus, spleen and lymph nodes during pregnancy and lactation in the rat. *J. Anat.* 118:223.

Michie, D., and J.F. Anderson. 1966. A strong selective effect associated with a histocompatibility gene in the rat. *Ann. N.Y. Acad. Sci.* 129:88.

Mishell, R.I., L.A. Herzenberg, and L.A. Herzenberg. 1963. Leukocyte agglutination in mice: detection of H-2 and non-H-2 isoantigens. *J. Immunol.* 90:628.

Mollison, P.L. 1979. *Blood Transfusion in Clinical Medicine,* 6th edn. Oxford: Blackwell, pp. 250-251.

Mosley, J.G., J.M. McLean, and A.C.C. Gibb. 1975. The response of iliac lymph nodes to the fetal allograft. *J. Anat.* 119:619.

Najarian, J.S., and F.J. Dixon. 1963. Induction of tolerance to skin homografts in rabbits by alterations of placental permeability. *Proc. Soc. Exp. Biol. Med.* 112:136.

Nathan, P., E. Gonzalez, and B.F. Miller. 1960. Tolerance to maternal skin grafts in rabbits induced by hyaluronidase. *Nature* 188:77.

Olding, L.B., R.A. Murgita, and H. Wigzell. 1977. Mitogen-stimulated lymphoid cells from human newborns suppress the proliferation of maternal lymphocytes across a cell-impermeable membrane. *J. Immunol.* 119:1109.

Osa, S.R., and M.E. Weksler. 1977. Demonstration of significant differences in the proliferative capacity of lymphocytes from normal human subjects. *Cell. Immunol.* 32:391.

Overweg, J., and C.P. Englefriet. 1969. Cytotoxic leucocyte isoantibodies formed during the first pregnancy. *Vox. Sang.* 16:97.

Owen, R.D. 1962. Earlier studies of blood groups in the rat. *Ann. N.Y. Acad. Sci.* 97:37.

Palm, J. 1969. Association of maternal genotype and excess heterozygosity for Ag-B histocompatibility antigens among male rats. *Transplant. Proc.* 1:82.

Palm, J. 1970. Maternal-fetal interactions and histocompatibility antigen polymorphisms. *Transplant. Proc.* 2:162.

Parrott, D.M.V. 1967. The response of draining lymph nodes to immunological stimulation in intact and thymectomized animals. *J. Clin. Pathol.* 20:456.

Pavia, C.S., and D.P. Stites. 1979. Humoral and cellular regulation of alloimmunity in pregnancy. *J. Immunol.* 123:2194.

Payne, R. 1962. The development and persistence of leukoagglutinins in parous women. *Blood* 19:411.

Petrucco, O.M., R.F. Seamark, K. Holmes, I.J. Forbes, and R.G. Symans. 1976. Changes in lymphocyte function during pregnancy. *Brit. J. Obstet. Gynaecol.* 83:245.

Poskitt, P.K.F., E.A. Kurt, B.B. Paul, R.J. Selvaraj, A.J. Sbarra, and G. Mitchell Jr. 1977. Response to mitogen during pregnancy and the postpartum period. *Obstet. Gynecol.* 50:319.

Posner, B.I. 1974. Insulin receptors in human and animal placental tissue. *Diabetes* 23:209.

Purtilo, D.T., H.M. Hallgren, and E.J. Yunis. 1972. Depressed maternal lymphocyte response to phytohemagglutinin in human pregnancy. *Lancet* i:769.

Rawson, A.J., and N.M. Abelson. 1960. Studies of blood group antibodies. IV. Physiochemical differences between isoanti-A,B and isoanti-A or isoanti-B. *J. Immunol.* 85:640.

Revillard, J.P., M. Robert, E. DuPont, H. Betuel, G. Rivle, and J. Traeger. 1973. Inhibition of mixed lymphocyte culture by alloantibodies in renal transplantation and in pregnancy. *Transplant. Proc.* 5:331.

Rocklin, R.E., J.E. Zuckerman, E. Alper, and J.R. David. 1973. Effect of multiparity on human maternal hypersensitivity to foetal antigen. *Nature* 241:130.

Ruppert, B.L., and E.R. Richie. 1977. Phytohemagglutinin response of murine spleen cells during pregnancy and inhibition of normal phytohemagglutinin response by pregnancy or postpartum serum. *Exp. Hematol.* 5:59.

Sasaki, K., and N. Ishida. 1975. Diminished immune response in pregnant mice. *Tohoku J. Exp. Med.* 116:391.

Schaid, D.J., H.W. Kunz, and T.J. Gill III. Submitted. Genic interaction causing embryonic mortality in the rat: epistasis between the Tal and Grc genes.

Scothorne, R.J., and I.A. McGregor. 1955. Cellular changes in lymph nodes and spleen following skin homografting in the rabbit. *J. Anat.* 89:283.

Searle, R.F., E.J. Jenkinson, and M.H. Johnson. 1975. Immunogenicity of mouse trophoblast and embryonic sac. *Nature* 255:719.

Searle, R.F., E.J. Jenkinson, and M.H. Johnson. 1976. Detection of alloantigens during preimplantation development and early trophoblast differentiation in the mouse by immunoperoxidase labeling. *J. Exp. Med.* 143:348.

Sigal, N.H. 1977. The frequency of p-azophenylarsonate and dimethyl-aminonaphthlene-sulfonate-specific B cells in neonatal and adult BALB/c mice. *J. Immunol.* 119:1129.

Smith, J.A., R.C. Burton, M. Barg, and G.F. Mitchell. 1978. Maternal alloimmunization in pregnancy. *In vitro* studies of T cell-dependent immunity to paternal alloantigens. *Transplantation* 25:216.

Smith, J.K., E.A. Caspary, and E.J. Field. 1972. Lymphocyte reactivity to antigen in pregnancy. *Am. J. Obstet. Gynecol.* 113:602.

Strelkauskas, A.J., I.J. Davies, and S. Dray. 1978. Longitudinal studies showing alteration in the levels and functional response of T and B lymphocytes in human pregnancy. *Clin. Exp. Immunol.* 32:531.

Szulman, A.E. 1973. ABO incompatibility in foetal wastage. *Res. Reprod.* 5:3.

Szulman, A.E. 1980. The ABH blood groups and development. *Curr. Top. Develop. Biol.* 14:127.

Takano, K., and J.R. Miller. 1972. ABO incompatibility as a cause of spontaneous abortion: evidence from abortuses. *J. Med. Genet.* 9:144.

Taylor, P.V., G. Gowland, K.W.Hancock, and J.S. Scott. 1976. Effect of length of gestation on maternal cellular immunity to human trophoblast antigens. *Am. J. Obstet. Gynecol.* 125:528.

Taylor, P.V., and K.W. Hancock. 1975. Antigenicity of trophoblast and possible antigen-masking effects during pregnancy. *Immunology* 28:973.

Taylor, P.A., R.A. Rachkewich, D.J. Gare, J.A. Falk, K.H. Shumak, and M.C. Crookston. 1974. Effects of pregnancy on the reactions of lymphocytes with cytotoxic antisera. *Transplantation* 17:142.

Terasaki, P.I., M.R. Mickey, J.N. Yamazaki, and D. Vredevoe. 1970. Maternal-fetal incompatibility. I. Incidence of HL-A antibodies and possible association with congenital anomalies. *Transplantation* 9:538.

Thong, Y.H., R.W. Steele, M.M. Vincent, S.A. Hensen, and J.A. Bellanti. 1973. Impaired *in vitro* cell-mediated immunity to rubella virus during pregnancy. *New Engl. J. Med.* 289:604.

Timonen, T., and E. Saksela. 1976. Cell-mediated anti-embryo cytotoxicity in human pregnancy. *Clin. Exp. Immunol.* 23:462.

Toder, V., I. Eichenbrenner, S. Amit, D. Serr, and L. Nebel. 1979. Cellular hyperreactivity to placenta in toxemia of pregnancy. *Eur. J. Obstet. Gynecol. Reprod. Biol.* 9:379.

Tofoski, J.G., and T.J. Gill III. 1977. The production of migration inhibitory factor and reproductive capacity in allogeneic pregnancies. *Am. J. Pathol.* 88:333.

Toivanen, P., and T. Hirvonen. 1970. Placental weight in human foeto-maternal incompatibility. *Clin. Exp. Immunol.* 7:533.

Tomoda, Y., M. Fuma, T. Miwa, N. Saiki, and N. Ishizuka. 1976. Cell-mediated immunity in pregnant women. *Gynecol. Invest.* 7:280.

Tongio, M.M., S. Mayer, and A. Lebec. 1975. Transfer of HL-A antibodies from the mother to the child. *Transplantation* 29:163.

Tung. K.S.K. 1974. Immune complex in the renal glomerulus during normal pregnancy: a study in the guinea pig and the mouse. *J. Immunol.* 112:186.

Tyler, W.J., and A.B. Chapman. 1948. Genetically reduced prolificacy in rats. *Genetics* 33:565.

Van Rood, J.J., J. G. Ernisse, and A. van Leeuwen. 1958. Leucocyte antibodies in sera from pregnant women. *Nature* 181:1735.

Vives, J., A. Gelabert, and R. Castillo. 1976. Decline in frequency of positive sera during last trimester. *Tissue Antigens* 7:209.

Waletzky, E., and R. Owen. 1942. A case of inherited partial sterility and embryonic mortality in the rat. *Genetics* 27:173.

Warburton, F.E. 1968. Maintenance of histocompatibility polymorphisms. *Heredity* 23:151.

Wegmann, T.G., J. Barrington-Leigh, G.A. Carlson, R. Raghypathy, and B. Singh. 1980. Quantitation of the capacity of the mouse placenta to absorb monoclonal anti-fetal H-2K antibody. *J. Reprod. Immunol.* 2:53.

Wegmann, T.G., T.R. Mosmann, G.A. Carlson, O. Olijnyk, and B. Singh. 1979. The ability of the murine placenta to absorb monoclonal anti-fetal H-2K antibody from the maternal circulation. *J. Immunol.* 123:1020.

Wikler, M., C. Demeur, G. Dewasme, and J. Urbain. 1980. Immunoregulatory role of maternal idiotypes. Ontogeny of immune networks. *J. Exp. Med.* 152:1024.

Winchester, R.J., S.M. Fu, P. Wernet, H.G. Kunkel, B. Dupont, and C. Jersild. 1975. Recognition by pregnancy serums of non-HL-A alloantigens selectively expressed on B lymphocytes. *J. Exp. Med.* 141:924.

Woodrow, J.C. 1970. Rh immunization and its prevention. *Ser. Haematol.* 3:2.

Youtananukorn, V., and P. Matangkasombut. 1972. Human maternal cell-mediated immune reaction to placental antigens. *Clin. Exp. Immunol.* 11:549.

Youtananukorn, V., P. Matangkasombut, and V. Osathanondh. 1974. Onset of human maternal cell-mediated immune reaction to placental antigens during the first pregnancy. *Clin. Exp. Immunol.* 16:593.

Yu, V.Y.H., C.A. Waller, I.C.M. MacLennan, and J.D. Baum. 1975. Lymphocyte reactivity in pregnant women and newborn infants. *Brit. Med. J.* 1:428.

Zawodnik, S.A., G.D. Bonnard, and E. Gautier. 1976. Antibody-dependent cell-mediated destruction of human erythrocytes sensitized in ABO and Rhesus fetal-maternal incompatibilities. *Pediatric Res.* 10:791.

Chapter 4

ALLOANTIGEN EXPRESSION ON MOUSE OOCYTES AND EARLY EMBRYOS

SUSAN HEYNER

When tissue grafts are exchanged between individuals in an outbreeding population, donor tissues are rejected by the recipient. This rejection is under genetic control, and the painstaking studies of Gorer in England and Snell in the United States led to the identification of the genes responsible for this phenomenon—the histocompatibility genes.

Comparative histogenetics has shown that histocompatibility genes are widespread throughout the animal kingdom. A number of invertebrate phyla show a range of histocompatibility reactions, and true allograft reactions, mediated by cells capable of specific recognition and development of immunological memory, have been described in two phyla of advanced invertebrates, the Echinodermata and the Annelida (Hildemann and Dix, 1972; Valembois, 1974). The ability to mount a true allograft reaction is found throughout the vertebrates, as shown in Figure 4-1, although the capacity to effect acute rejection of tissue grafts is found in only four groups: certain teleosts, a number of Anurans, birds, and mammals (Cohen and Collins, 1977; Klein, 1977). Therefore, the phylogenetic evidence supports the view that histocompatibility antigens, and the concomitant ability to recognize nonself existed before the evolution of viviparity.

Viviparity is the hallmark of the class Mammalia, although it occurs in a number of different vertebrate groups (Figure 4-1). Comparative zoological evidence indicates that viviparity evolved independently in different lines of descent and is found in all three classes of cold-blooded vertebrates (Amoroso, 1952). Fetal membranes almost certainly arose in the oviparous stage of vertebrate evolution, as a result of a tendency in some groups to provide their ova with large amounts of nutrient yolk material. As a consequence of possessing a large amount of yolk, cleavage is incomplete and the embryo's body begins to differentiate at the animal pole of the egg before its extraembryonic ectoderm, mesoderm, and endoderm envelop the yolky mass. When the yolk is enclosed, a primitive yolk sac is formed, which is essentially an extraembryonic organ. The development of vasculature in the mesoderm provides a means of absorbing and transporting nutrient yolk to the embryo. This primitive form of yolk sac is found in most fishes and a few amphibia. Primitive as the yolk sac appears, it has been modified in a number of fishes to provide a form of placenta that establishes an intimate connection between maternal and fetal tissues, allowing prolonged internal gestation (Amoroso, 1952; Mossman, 1974).

Viviparous amniotes have evolved a wide range of complex arrangements to provide for the development of their eggs within the body of the mother. However, all amniotes are characterized by the presence of the chorion, amnion, and allantois, in addition to the yolk sac. These fetal membranes

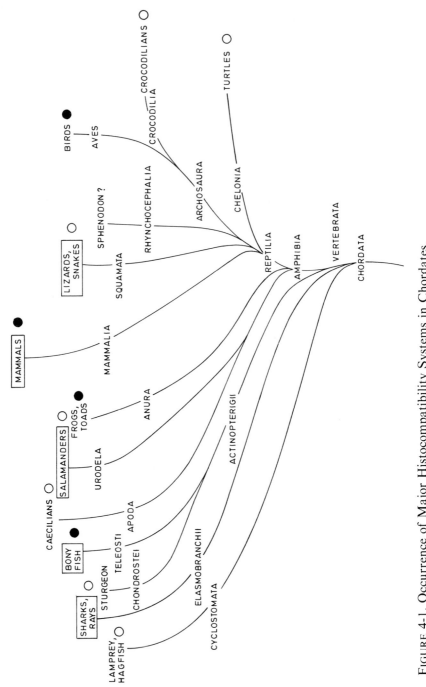

FIGURE 4-1. Occurrence of Major Histocompatibility Systems in Chordates.

have been derived from the embryo and modified to subserve the needs of the fetus during its early development, in a strikingly diverse manner. Prolonged internal gestation has been perfected in the Eutherian mammals. In this group, there is reliance on the highly developed chorioallantoic placenta, in which the fetal membranes, arranged in a multitide of adaptions to viviparity, provide the respiratory and trophic connection between fetus and mother. A consequence of the evolution of internal gestation is that the fetal and maternal tissues are in intimate juxtaposition. In all random breeding populations, pregnancy therefore results in close contact between two genetically disparate tissues since the fetus will inherit foreign, paternally derived histocompatibility genes. This situation has led a number of biologists to consider the fetus as an allograft and to propose a variety of mechanisms that would account for the paradoxical success of pregnancy (Medawar, 1953; Beer and Billingham, 1971; Barker and Billingham, 1977). Although this analogy has stimulated considerable research efforts in immunoreproduction, it has dominated the field to an undesirable extent for the past two decades. It is an oversimplification to compare the surgical exchange of tissues into prepared vascular beds with the development of a fetus separated from the maternal circulation by an extraembryonic organ.

It is instructive to look at one of the earliest attempts at viviparity, the salamander, *Salamandra salamandra*. This species has the capacity to reject skin grafts by both a cellular and a humoral reaction. When pregnant salamander females were examined for the capacity to mount an immune reaction against their embryos, it was shown that both killer cells and humoral factor(s) that inhibit the killing of isolated embryonic cells *in vitro* are produced by the females. Interestingly, the pregnancies of this species exhibit a varying degree of success, with embryos being born or "rejected" at various stages, from very immature to quite mature individuals (Chateaureynaud et al., 1979). These observations provide evidence that the immunogenicity of the embryo was a problem that had to be circumvented before viviparity could be truly successful. The universal occurrence of placentation in mammals suggests that the advantage offered to warm-blooded animals by internal gestation provided the kind of selection pressure that led to the evolution of a symbiotic *modus vivendi* between mother and fetus.

Antigen expression on the surface of embryos is of interest not only in the context of the maternal-fetal relationship. Genetically determined macromolecules on the surface of embryonic cells have been postulated as mediators of differentiation and morphogenesis (Edelman, 1976; Nicolson, 1979). It is reasonable to expect that such cell surface molecules are differentially expressed during the course of development, allowing identification of a particular temporal phase of differentiation or a particular subset of cells. The identification of these cell surface molecules, as well as the elucidation of the role they play in early embryological studies is one of the major goals of developmental biology and, further, has important implications for the study of malignant change.

In order to investigate the cell surface of the developing embryo for the presence of plasma membrane components of developmental significance, immunological methods have been widely used. Among the advantages of an immunogenetic approach to the study of cell surface components is the ability to distinguish developmentally determined and cell type-specific surface antigens. Additionally, immunological methods are highly specific, allowing studies at the single-cell level for many antigens and providing the potential of being able to quantitate very low levels of particular antigens. Once cell-specific antigens have been recognized by antibody, provided that it possesses sufficient specificity and titer, that antibody can then be used to analyze the molecular nature of the antigen, using techniques of immunoprecipitation followed by the appropriate gel analysis. Immunogenetic approaches have proved extremely useful in studies of the nervous system, where antibodies have been used to analyze interactions leading to synapse formation and myelination, as well as the earlier stages of cell migration and sorting that lead to ordered patterns of neurons (Fields, 1979). Similarly, the use of specific antibodies has allowed elegant and extensive studies on the individual programming of cell surfaces in the immune system. Studies on the thymocyte have shown that not only can the constitution of the surface phenotype be determined by the use of appropriate antibodies but further, by quantitative measurement of the interference or blocking between two antibodies applied in sequence, cell membrane surface components can be mapped (Boyse et al., 1968; Boyse and Cantor, 1978; Flaherty and Zimmerman, 1979). This approach has yielded a great deal of information on the differentiation antigens of the mouse lymphoid system in its normal and malignant states. Similarly, the use of labeled antibodies has proved valuable for mapping the cell surface distribution of antigens, particularly in sperm and nerve cells, that possess special domains (Schachner, 1979).

EARLY DEVELOPMENT OF THE MOUSE EMBRYO

The last two decades have witnessed the development of techniques that allow detailed experimental investigation of early mammalian development. The mouse has been the animal of choice for such studies because it has proved possible to fertilize mouse ova *in vitro* and to culture developmental stages up to implantation (Whittingham, 1971). Indeed, procedures have been developed that allow post-implantation development to proceed as far as the early somite stage (Hsu, 1979). The combining of cells from embryos carrying defined genetic markers has made possible critical studies on cell fate and commitment. Some of the stages of early development in the mouse are shown in Figure 4-2. Blastomeres derived from the first three cleavages are apparently totipotent, as Kelly (1975, 1977) has shown that individual cells isolated from eight-cell mouse embryos can contribute to both inner cell mass (ICM) and trophectoderm derivatives when combined with groups of genetically dissimilar blastomeres. Following the eight-cell stage, contact between adjacent blastomeres becomes extensive, with loss of resolution of cell boundaries. This phenomenon of compaction is accompanied by the formation of zonular tight junctions between cells on the outside (Ducibella and Anderson, 1975; Ducibella et al., 1975) and other changes, including those in mitochondrial conformation (reviewed by Piko, 1975) and organization of the cortical cytoskeleton (Ducibella, 1977).

Cell position appears to be the mechanism to account for the initial differentiation of the mouse embryo into trophoblast and ICM; those blastomeres retaining an inside position contribute to the ICM, whereas those on the outside are destined to form trophectoderm. This "inside-outside" hypothesis was originally advanced by Tarkowski and Wroblewska (1967) and has subsequently been supported by a number of different studies (Adamson and Gardner, 1979). The nature of the positional cues and means whereby stable differences are maintained between inside and outside cells is unknown. It is therefore potentially fruitful to follow early development in an attempt to elucidate the developmental program in terms of cell markers that might enable one to predict if there are earlier signals indicative of cueing for trophoblast or ICM commitment.

The primary trophectoderm is the first trophoblastic cell type to appear during development. Studies on the ultrastructure of the mouse embryo indicate that the trophectoderm has many of the features associated with the

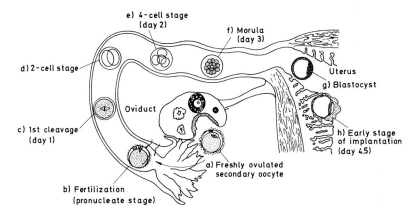

FIGURE 4-2. Ovulation and Pre-implantation Stages in the Mouse.

83

secretory epithelium, possessing microvilli on the outside, tight junctions attaching adjacent cells, and a basal lamina (Ducibella and Anderson, 1975; Ducibella et al., 1975; Hastings and Enders, 1975). The cell surfaces of the ICM and trophoblast differ, as shown by the scanning electron microscope (Calarco and Epstein, 1973). Experimental studies have shown that, by the 64-cell stage, both ICM and trophectoderm cells appear committed to separate developmental pathways. Among functional differences, trophoblast cells have been shown to possess mechanisms for sodium-dependent amino acid transport, but these properties have not been demonstrated for cells of the ICM (Borland and Tasca, 1974; DiZio and Tasca, 1977).

Mural trophectoderm subsequently differentiates to form the primary trophoblastic giant cells. Primary giant cells undergo endomitosis, resulting in very high levels of DNA (Sherman et al., 1972). Their precise function is not known, although it may include an anchoring mechanism during the early stages of implantation. Another function suggested for these cells is that of an immunological filter. This is based on the studies of Bernard et al. (1977), in which distribution of maternal immunoglobulins in the mouse uterus and embryo was studied in the days after implantation. It was shown that the embryo was surrounded by maternal immunoglobulins at the time of implantation and that the trophoblast giant cells concentrated immunoglobulins, resulting in a "barrier" of immunoglobulin-filled cells surrounding the embryo.

Trophectoderm overlying the ICM at the embryonic pole of the blastocyst is known as polar trophectoderm. This undergoes rapid proliferation to form ectoplacental cone tissue, composed of highly invasive trophoblastic cells. Secondary trophoblastic giant cells are derived from the superficial ectoplacental cone, and the central core of the placental cone differentiates to form the main fetal components of the placenta—the labyrinthine trophoblast and spongiotrophoblast.

The ICM gives rise to primitive ectoderm and primitive endoderm; based on the injection of fifth-day endoderm and ectoderm cells into fourth-day blastocysts, Gardner and Papiaoannou (1975) have shown that the most likely fate of the primitive endoderm from the fifth-day blastocyst is to form the visceral endoderm of the yolk sac. Similar transplantation experiments by the same authors have shown that ectoderm cells from fifth-day blastocysts contribute to many tissues, including the gut and its derivatives.

HISTOCOMPATIBILITY ANTIGENS

The terms "histocompatibility genes" and "histocompatibility antigens" were proposed by Snell in 1948 to describe the genes and antigens involved in acceptance and rejection of tissue grafts. In most mammalian species that have been examined critically, there is a cluster of genes designated as the major histocompatibility complex (MHC) or system. In the mouse, the MHC is composed of a gene cluster termed the H-2 complex. It is of predominant importance in determining the acceptance or rejection of tissue grafts. The H-2 region of chromosome 17 is highly complex, consisting of five tightly linked regions (Figure 4-3).

The K and D regions determine the classical Histocompatibility-2 antigens. These are also the loci whose products were first demonstrated by alloantisera and are the source of much of the serological complexity of H-2. Each individual possesses two haplotypes, a paternal and a maternal one. H-2 specificities limited to a putatively independent haplotype are termed private, and those that will crossreact with several haplotypes are termed public. There is protein structural evidence to suggest that H-2K and H-2D loci could have arisen by duplication of ancestral locus, which would explain why serological crossreactivity can be detected between H-2K and H-2D antigens (Klein, 1975; David, 1977).

The I region was first described when it was demonstrated that genes mapping in that region controlled the immune response to synthetic polypeptides (McDevitt et al., 1972). A system of antigens, the Ia antigens, is associated with this region, and since putative anti-H-2 antisera may contain anti-Ia antibodies, it is important that reactivity of H-2 sera be carefully analyzed.

A genetic map of the mouse shows that whereas the MHC is confined to a relatively short segment on chromosome 17, other histocompatibility loci, termed non-H-2 (H) loci, are found on other chromosomes as well as on chromosome 17. The precise number of these is not known, but current estimates put their number in excess of 30. Very little is known of their biological function or their chemical nature, although in accordance with Snell's definition, they are involved in tissue rejection when donor and recipient are compatible at the MHC.

It is important to realize that, as in the case of H-2 alloantisera, non-H-2 antisera raised by conventional immunization procedures in mice that

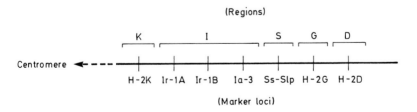

FIGURE 4-3. Map of the H-2 Region.

are compatible at the MHC may also contain antibodies to other cellular antigens. Two examples are the Ea antigens, found on erythrocyte membranes, and the Ly antigens, found on lymphocyte membranes. Although Ea and Ly antigens were originally thought to occur only on erythrocytes and lymphocytes, respectively, evidence is accumulating that their distribution may be more widespread than previously suspected (Altman and Katz, 1979).

ALLOANTIGEN EXPRESSION ON OOCYTES

There have been relatively few studies of alloantigen expression on the surface of mouse oocytes and fertilized eggs (see Table 4-1), since most investigators have studied the two-cell stage. An early study by Edidin et al. (1974) reported some positive results using complement-mediated cytotoxicity in the mouse, but Billington et al. (1977) were unable to detect H-2 antigens on the oocyte membrane using immunoperoxidase labeling. Studies on mouse oocytes at the dictyate and metaphase II stages have been carried out using multispecific alloantisera, raised in congenic strains of mice, in conjunction with indirect immunofluorescence. Serologically specific reactivity to H-2 antibody was observed on the oocyte membrane. In contrast, when zygotes were examined, they were found to exhibit variable reactivity, and H-2 expression on the surface of the two-cell embryo could not be detected (Heyner and Hunziker, 1979). These observations suggest that fertilization results in altered expression of MHC antigens, a finding in line with the observation of Johnson and Edidin (1978) that there are marked changes in the lateral mobility of the mouse oocyte membrane following fertilization.

TABLE 4-1
Serological Detection of Antigens on Oocytes

ANTIBODY DIRECTED AGAINST	METHOD OF STUDY	RESULTS	REFERENCE
H-2	Complement-mediated cytotoxicity	Some positive some negative	Edidin et al., 1974
H-2	Immunoperoxidase	Negative	Billington et al., 1977
H-2	Indirect immunofluorescence	Positive	Heyner and Hunziker, 1979
H-1[a], H-3[a], H-7[a]	Indirect immunofluorescence	Positive	Heyner et al., 1980

In a subsequent investigation, polyclonal NIH reference sera directed against K-region specificities H-2.11, 23, and 23 were compared with monoclonal H-2K^k antibody and with a broadly reactive anti-I region alloantiserum. The results are given in Table 4-2, which shows that oocytes reacted positively with NIH reference antisera directed against H-2 specificities but did not react with the broadly reactive anti-Ia^k antibodies. These results confirmed the previous observation that H-2 antigens can be detected on the surface of the mouse oocyte and excluded the possibility that the results were due to the reactivity of the products of I region genes. The absence of Class II antigens on oocytes concurs with current evidence suggesting that the tissue distribution of Ia antigens is relatively restricted. Ia antigens have been detected on B lymphocytes, some subpopulations of T lymphocytes, sperm, epidermal cells, and peritoneal macrophages (Hammerling et al., 1975). When oocytes were tested with monoclonal anti-H-2K^k antibody (11-4.1 [L. Herzenberg]), oocytes did not react in the indirect immunofluorescence test, although lymphocytes reacted positively at a dilution greater than 1 to 10000. To test the possibility that this lack of reactivity might be due to a low antigen density on the surface of the oocyte, the antibody was tested in a three-step procedure, using a goat antimouse IgG antibody as a second antibody, followed by incubation with fluorescein-conjugated goat antimouse gamma globulin. Since this amplification did not give positive results, a cocktail of four well-characterized monoclonal antibodies (11.4.1 [L. Herzenberg], 15-5-5 [D. Sachs], 100-S and 100-27 [G. Hammerling]) directed against the K region of the MHC was used in the same procedure. This resulted in a very weak positive reaction, suggesting that antigen density may be a significant problem in the detection of H-2 antigens by means of monoclonal antibodies on the surface of the oocytes.

The finding that a high-affinity, highly titered monoclonal antibody did not react with the oocyte was surprising but not entirely unexpected.

TABLE 4-2
Reactivity of Oocytes to Antibody Directed Against H-2K^k or Ia^k

STRAIN	H-2 GENOTYPE	H-2.11 and 25*	H-2.11, 23 and 25†	MONOCLONAL K^k‡	Ia^k§
AKR/J	k	+	+	0	0
C3H.HeJ	k	+	+	0	0
C58/J	k	+	+	0	0
A/J	a	+	+	0	0
PL/J	u	+	+	0	0
C57Bl/6J	b	0	0	0	0
DBA/2J	d	0	0	0	0

+, reactivity; 0, no reactivity.
* NIH D-25.
† NIH control anti-K^k.
‡ hybridoma 11.4.1.
§ A.TH anti A.TL.

Previous studies with multispecific sera had provided evidence that oocytes are rather weakly reactive (Heyner and Hunziker, 1979). Similarly, Edidin and his colleagues (1974) had reported inconsistent results with cytotoxicity on oocytes, a finding that could be explained if antigens were at a low density. Indeed, Hausman (personal communication) was unable to detect a cytotoxic effect with alloantisera directed against the MHC in the rat, although immunofluorescence and mixed hemadsorption procedures yielded positive results. If there are chemically different H-2K^k subpopulations of molecules that map in the H-2K region, as suggested by O'Neill and Parish (1980), then these results can be readily explained by low antigen density on the oocyte. In an analogous situation, Wiley (personal communication) has found that preimplantation embryos will react to multispecific antisera elicited by hCG and also to reference antisera to hCG, but not to monoclonal anti-hCG antibodies.

To examine whether synthesis of MHC antigens can be detected on dictyate-stage oocytes, sera with potential reactivities against H-2K and D antigens, as well as against Ia antigens, were raised in congenic strains of mice and tested by direct immunoprecipitation and two-dimensional (2-D) gel electrophoresis (Sawicki, Magnuson, and Epstein, personal communication). Since these methods are highly sensitive, synthesis of individual polypeptides can be detected. In addition, H-2 and Ia specificities can be distinguished from each other on the basis of differences in their isoelectric points and molecular weights. Analysis of ^{35}S-methionine-labeled immunoprecipitates on 2-D gels showed that synthesis of H-2K, D, and Ia antigens was not detectable in unfertilized eggs. One of the alloantisera tested was the same serum as that used by Heyner and Hunziker when they detected H-2 expression on the oocyte surface. Thus, while there is some evidence that MHC antigens appear to be expressed on the surface of dictyate eggs, they are apparently not synthesized at this stage of development.

Antisera directed against multiple non-H-2 specificities, and serologically characterized by the appropriate adsorptions, have been tested on dictyate and metaphase II oocytes by indirect immunofluorescence. These studies have shown that H-1^a, H-3^a, and H-7^a can be detected on the surface of the mouse oocyte. Following fertilization, these specificities can be detected on the zygote but not on the two-cell stage. When antisera are raised against minor histocompatibility antigens, it is common to also raise sera against Lyt and Ea specificities. Interestingly, both Lyt and Ea specificities can be detected on oocytes (Heyner and Hunziker, 1980; Heyner and Hunziker, unpublished data). A note of caution must be sounded with respect to the Lyt reactions, since the possibility of reactivity against an undefined histocompatibility antigen is also present (Flaherty and Bennett, 1973). Similarly, while it is known that anomalous reactions may sometimes occur because of high levels of circulating antiviral antibodies, this did not seem to be the case, since appropriate lymphocyte adsorptions and congenic strains of mice were used.

ALLOANTIGEN EXPRESSION ON EMBRYOS

The expression of histocompatibility antigens on early embryos has been rather extensively investigated (see Edidin, 1976; Solter, 1977; and Heyner, 1980). Early experiments demonstrated that transplantation of mouse tubal embryos to the kidney capsule of specifically sensitized recipients resulted in destruction of the grafts, establishing the presence of alloantigens on the embryo (Simmons and Russell, 1966). A broad range of incompatibilities was present between donor and recipient, and therefore subsequent studies utilized congenic resistant strains of mice to raise specifically-directed MHC alloantisera in order to elucidate which antigens were expressed on the surface of the embryo.

Early studies using indirect immunofluorescence (Palm et al., 1971; Muggleton-Harris and Johnson, 1976) indicated that MHC antigens could not be detected on tubal cleavage stages, or blastocysts. The earliest detection of MHC antigens by indirect immunofluorescence was around day 6 (Heyner, 1973), and, on blastocyst outgrowths, antigen expression was restricted to the ICM. These results were in agreement with transplantation studies by Patthey and Edidin (1973).

Heyner et al. (1969) showed that MHC-directed alloantisera used in complement-dependent cytotoxicity tests did not interfere with development of eight-cell embryos to the blastocyst stage. However, Krco and Goldberg (1977) used well-defined alloantisera to maternal and paternal antigens and obtained complement-mediated lysis of eight-cell embryos. The discrepancy between this result and other studies may possibly be explained as a consequence of viral antibodies present in the mouse serum, which could react with vertically transmitted viruses carried by the mice.

Webb et al. (1977) grew mouse embryos in media containing radiolabeled amino acids and used sensitive immunoprecipitation methods to show that H-2 synthesis could first be detected in the cells of the ICM in late blastocysts.

H-2 expression on trophectoderm is controversial. Although most investigators have not been able to detect MHC antigens on trophectoderm, Searle et al. (1976) used an immunoperoxidase assay to demonstrate transient expression of H-2 antigens on preimplantation blastocyst trophectoderm. Similarly, with a sensitive antiglobulin technique, trophectoderm showed variable expression of MHC antigens on blastocysts undergoing delay of implantation. After release from delay, trophectoderm was negative (Häkansson et al., 1975).

Despite discrepancies in results from different laboratories, a general consensus is that MHC antigens can first be detected on the ICM of the late blastocyst and are not present in significant amounts on trophectoderm. Most of these studies were done without taking into consideration the possibility that antisera might have been raised against Ia antigens. Indeed, analysis of many antisera shows that Ia antibodies were probably present. In studies by Delovitch et al. (1978), sensitive immunoprecipitation methods revealed that Ia antigens first appeared at day 11 of gestation and were restricted to the liver until day 16. Jenkinson and Searle (1979) confirmed these results, using immunoperoxidase labeling and examination by electron microscopy. Preimplantation blastocysts and post-implantation embryos (day 7½) were uniformly negative when tested with anti-Ia antiserum.

The ontogeny of paternal H-2 antigen expression is currently being investigated in our laboratory. Preliminary results of indirect immunofluorescence studies have shown that on day-6 and day-7 embryos, antigen expression can be detected on parietal endoderm and Reichert's membrane, but not on the tissues of the embryo proper. The possibility that this layer of antigen-reactive cells forms a protective layer around the embryo is being investigated (Heyner, Jensen, unpublished data).

The alloantigenicity of mouse tubal embryos has been ascribed in general to the presence of non-H-2 antigens on the surface of the embryo. Heyner et al. (1969) demonstrated that antibody directed against unspecified non-H-2 alloantigens reacted in a complement-dependent lysis of eight-cell mouse embryos, while in the same experimental system, specifically directed H-2 alloantibody was not cytotoxic and did not interfere with development to the blastocyst stage. Using immunofluorescence, Palm et al. (1971) demonstrated the presence of H-3 antigens, and also of Ea-6 (formerly designated H-6), on the surface of two-cell embryos. Muggleton-Harris and Johnson (1976) confirmed and extended these results; they were unable to detect H-2 antigens on tubal embryos and blastocysts, although non-H-2 antigens of maternal origin were detectable throughout preimplantation development, and alloantigens of paternal origin could first be detected around the six- to eight-cell stage.

There are two criticisms that can be leveled at early studies of non-H-2 alloantigen expression. One is that the non-H-2 alloantisera raised by conventional means in mice that were compatible at the H-2 locus had the potential to contain contaminating antibodies to other cellular antigens. A second criticism is that investigators have studied either specific alloantigens at one or two developmental stages (Palm et al., 1971) or a developmental series using nonspecified alloantisera (Muggleton-Harris and Johnson, 1976; Billington et al., 1977). In an attempt to map specific minor alloantigen expression during different stages of preimplantation development, Heyner et al. (1980) used antiserum raised in mice compatible at the MHC but differing at a number of non-H-2 loci. Detailed serological analysis revealed that one alloantibody was directed against H-3[a], H-8[a], and Ea 6.1. Adsorption with the appropriate mouse strain rendered the antibody monospecific for H-3[a], and indirect immunofluorescence tests were carried out on lymphocytes from strains that had the potential to react with H-3[a] antibody. Similar tests were carried out with an alloantiserum that was characterized as containing antibody directed against H-1[a], H-7[a], and Lyt 1.1. This alloantiserum was adsorbed to remove all reactivity except that directed against H-7[a]. Mouse developmental stages were examined for minor histocompatibility antigen expression, and the results are given in Table 4-3, which shows that for both antigens studied, H-3[a] and H-7[a], there was differential reactivity on the early preimplantation stages. While the fertilized ovum was positive in both the target strains tested, antigen expression could not be detected at the two-cell stage and was not subsequently detectable until around the eight-cell stage. Morulae were positive, as were blastocysts, with reactivity located on the surface membrane of the trophectoderm.

This pattern of differential reactivity with alloantisera directed against minor antigens is similar to the pattern of reactivity with MHC alloantisera observed by Heyner and Hunziker (1979), but is in contrast to earlier reports (Palm et al., 1971; Muggleton-Harris and Johnson, 1976) that minor alloantigens can be readily seen on the two-cell stage. An explanation of this discrepancy may be found by examining the alloantisera used in those studies. In the study of Palm et al. (1971), one of the antigens detected, H-6, has subsequently been recognized as an erythrocyte antigen, Ea-6, and, in the case of the H-3 study, the antiserum probably also contained some Ea contaminant since it had a hemagglutinating titer of 1:40. Similarly, Muggleton-Harris and Johnson (1976) used strain combinations to raise the antisera such that one cannot exclude the possibility that the reagents contained antibody directed against Ea and Lyt specificities, in addition to histocompatibility antigens.

If alloantisera used in previous studies contained antibodies directed against specificities other than minor alloantigens, this could explain why the two-cell stages appeared positive. In a preliminary study, Heyner and Hunziker (1980) showed that antibody directed apparently against an Lyt specificity reacted with two-cell embryos; similarly, it has been shown that Ea specificities can be detected on the two-cell stage (Heyner and Hunziker, unpublished data).

From this study, it appears probable that the expression of specific minor histocompatibility antigens on the surface of the oocyte represents the expression of antigens that were inserted during oogenesis. Reactivity with antibody directed against minor histocompatibility antigens at the eight-cell stage probably represents *de novo* synthesis of these antigens from the embryonic genome. The detection of minor histocompatibility antigens on trophectoderm confirmed a number of earlier studies that used immunofluorescence (Muggleton-Harris and Johnson, 1976), mixed hemadsorption (Billington et

TABLE 4-3
Serological Detection of Alloantigens on the Surface of Early Mouse Embryos

ANTIBODY DIRECTED AGAINST	TECHNIQUE	ZYGOTE	TWO CELL	4-8 CELL	MORULA	BLASTOCYST ICM	BLASTOCYST TROPHECTODERM	6-7 DAY BLASTOCYST OUTGROWTH EMBRYO	TROPHOBLAST	REFERENCE
H-2	Cytotoxicity	NT	NT	-	-	NT	-	NT	NT	1
H-2	Immunofluorescence	±	-	-	-	-	-	+	-	3,4,5,6
H-2	Mixed hemadsorption	NT	?	NT	NT	NT	+	NT	-	2
H-2	Immunoperoxidase	NT	NT			NT	+	+	-	7,8
H-2	Immunoprecipitation	NT	-		-	+	-	NT	NT	9
H-2	Isotope antiglobulin	NT	NT	NT	NT	NT	-*	NT	NT	10
Non-H-2†	Immunofluorescence	+	±	+	+	+	+	+	+	3,4,5,11
Non-H-2†	Mixed hemadsorption	NT	NT	NT	NT	NT	NT	+	+	2, 8
Non-H-2†	Immunoperoxidase	NT	NT	+	+	NT	+	+	-	7, 8
Ia	Immunoperoxidase	NT	NT	NT	NT	NT	-	-	-	12

The detection of cell surface antigens is indicated by +, inability to detect the antigen is -, and ? indicates the results are uncertain. NT means that particular stage has not been tested.

* Blastocysts undergoing experimentally-induced delay of implantation were variably positive.

† Strain combinations used for antibody production in some studies cannot exclude the possibility that antibodies to non H-2 antigens include Lyt and Ea specificities.

References: 1. Heyner et al., 1969; 2. Sellens, 1977; 3. Palm et al., 1971; 4. Heyner, 1973; 5. Muggleton-Harris and Johnson, 1976; 6. Heyner and Hunziker, 1979; 7. Searle et al., 1976; 8. Billington et al., 1977; 9. Webb et al., 1977; 10. Hakansson et al. 1975; 11. Heyner et al., 1980; 12. Jenkinson and Searle, 1979.

al., 1977; Sellens, 1978), and transplantation to specifically preimmunized hosts (Johnson, 1975) to demonstrate antigen expression on this tissue.

Trophectoderm subsequently differentiates into a variety of cell types, and therefore any study of trophoblast immunogenicity must take this into account. While evidence from experiments in which tissues have been transplanted across different histocompatibility barriers has given a rather global view of the immunogenicity of trophoblast (reviewed in Johnson, 1975), several studies have approached this problem at the cellular level.

Carter (1976, 1978) used multispecific alloantisera to demonstrate that, in the mouse, primary trophoblastic giant cells expressed maternal and paternal alloantigens, whereas secondary trophoblastic giant cells from day 8 of pregnancy did not react in the mixed hemadsorption test. After three to five days in culture, the ectoplacental cone tissue produced several layers of secondary giant cells, the outermost of which expressed antigens. This investigation is of interest since it shows that development of differential antigenic expression on trophoblast can be detected *in vitro*. The extrapolation of these results to the *in vivo* situation is complicated by the observation that the expression of cell surface antigens may be dependent on growth conditions, as suggested by the study of Cikes and Klein (1972). Nevertheless, these studies do provide evidence suggesting that antigenic expression differs between cell type and developmental stage in trophoblast. Other investigators have studied antigen expression on trophoblastic cells from later stages of pregnancy. These studies fall outside the scope of this chapter, however, and will be described in Chapters 10 and 11.

DISCUSSION

It is clear that a number of alloantigens can be detected on the mouse oocyte and early embryo. Reports that alloantigens can be identified on the unfertilized mouse ovum but not on very early cleavage stages or, in the case of MHC antigens, not until blastocyst formation suggest that antigens inserted earlier during oogenesis become diluted out as development proceeds. Immunoprecipitation studies indicate that MHC antigens are not synthesized in oocytes (Sawicki, Epstein, and Magnusen, personal communication). Thus, speculation that MHC antigens are synthesized by early cleavage embryos in a form that fails to be expressed, or in an incomplete form, is probably not well founded.

The preimplantation embryo is very vulnerable, and it is biologically advantageous that the highly immunogenic MHC antigens should not be expressed, in order to evade recognition by the maternal immune system. On the other hand, by the four- to eight-cell stage, non-H-2 histocompatibility antigens can be readily detected on mouse embryos (Heyner et al., 1980), and antigens specified by the paternal genome can also be detected (Muggleton-Harris and Johnson, 1976). Although the minor antigens have been classified as weak, they constitute a significant target of the allograft rejection reaction. The total immunogenicity of the antigens of the H-2 complex (as judged by skin allograft rejection) has been shown to be about equal to the cumulative non-H-2 immunogenicity (Graff, 1978). Therefore, the embryo can be considered a potential target for the maternal immune system from the expression of minor alloantigens alone, setting aside the question of embryo-specified antigens, such as F9 (reviewed by Jacob, 1977 and 1979) or SSEA-1 (Solter and Knowles, 1978), which represent stage-specific or tissue-specific markers of embryonic cells.

During the early cleavage stage, the embryo is surrounded by an acellullar layer largely composed of sialic acid and protein—the zona pellucida (see Chapter 23). Although the zona has been shown to be permeable to antibody (Sellens and Jenkinson, 1975), an earlier study showed that, in order to achieve complement-mediated lysis of embryos, it was necessary to remove the zona (Heyner et ai., 1969). While the protective role of the zona in terms of humoral immunity can be debated, there is little question that it constitutes an effective barrier to immunologically competent maternal cells. At implantation, however, the zona is shed and the blastocyst attaches to the uterine epithelium, which degenerates in a short time, allowing intimate contact beween trophoblast and the maternal vascular system. It has been suggested that the disappearance of MHC-specified antigens from the surface of the trophoblast at implantation, noted by Håkansson et al. (1975) and Searle et al. (1976), allows the embryo to present an outer surface lacking in antigens and therefore favors establishment of the embryo at a time when other protective mechanisms are not operative. If this is the case, and it would certainly appear a plausible argument, then the expression of minor alloantigens on the surface of the implanting blastocyst is intriguing. It is possible that the antigen dose presented to the maternal organism in this manner is insufficient to trigger a maternal immune response. It is even possible that the expression of antigens is in some way important in implantation.

It is important to remember that studies carried out in an *in vitro* system may differ significantly from the situation *in vivo*. There is evidence, for example, that the surface of implanting cells may undergo alterations in surface glycoproteins (reviewed by Schlafke and Enders, 1975), which may alter cell immunogenicity. Another suggestion, made by Jenkinson and Billington (1974), is that there may be a polarized distribution of antigens on placental epithelium such that they are expressed in the laterobasal membrane but not in the apical membrane, where they would provide a target for an alloimmune reaction. Although this hypothesis was originally advanced in an attempt to reconcile the presence of target alloantigens on yolk sac epithelial cells with the absence of a maternal immune reaction, it has been supported by recent immunoferritin-labeling studies (Parr et al., 1980). A variety of adult epithelial cells, dissociated from tissues after prefixation to preserve the native antigen distribution, have been shown to express H-2 antigens in the laterobasal membrane but not in the apical membrane. When epithelial cells were dissociated without prefixation, there was a migration of H-2 molecules into the apical membranes. Of relevance to the present discussion was the finding that the apical membrane of endodermal epithelial cells in the mouse yolk sac placenta was essentially MHC antigen-free, although variable expression could be detected in laterobasal membranes. It would be worth reexamining the distribution of non-H-2 alloantigens on trophoblast in peri- and early post-implantation embryos, to see if there is polarization at this stage.

Although the mechanisms described above may serve to avert rejection of the early embryo, there is overwhelming evidence for the presence of both humoral and cellular immune response in pregnancy (Hellstrom et al., 1969; Voisin and Chaouat, 1974; Smith, 1978; reviewed by Gill and Repetti, 1979).

The immunogenicity of alloantigens and the concomitant maternal immune response may play an important role in normal pregnancy, as evidenced by the difficulty of developing and maintaining inbred strains of laboratory animals. The expression of alloantigens may also play an important role in the maintenance of histocompatibility antigen heterozygosity, as suggested by studies in which an excess of heterozygotes resulted when an inbred female was mated to an F_1 male (Hull, 1969) and repeated reciprocal backcrosses were made between inbred females and hybrid males (Hamilton and Hellstrom, 1978). In both these studies, alloantigen heterozygote progeny resulted, in excess of those predicted from normal Mendelian segregation of these genes. The selection of such progeny may be mediated through the maternal response to fetal alloantigens. On the other hand, selection may occur at fertilization, when histocompatibility antigens on eggs and sperm could lead to selective fertilization favoring alloantigen heterozygotes, or at implantation, when immunogenetically foreign blastocysts may implant more readily than inbred blastocysts (Kirby, 1970) (see Chapter 3).

While antigens of the MHC have been subject to intense scrutiny for over three decades, virtually nothing is known of the structure or function of the minor histocompatibility antigens, with the single exception that the function of the H-Y antigen is known (see Chapter 5). In the absence of H-2 antigen expression on trophectoderm cells, these alloantigens may play an important role in the cell-cell recognition phenomena involved in implantation. The finding that these minor alloantigens can be detected on both the ICM and trophectoderm of blastocysts (Muggleton-Harris and Johnson, 1976) suggests that minor alloantigens do not cue for differentiation that distinguishes these two tissues, but could affect other aspects of differentiation. They may also

95

function in important physiological roles, such as receptors for hormones or mediators of transport of biologically important molecules. Ohno (1977) has suggested that the original role of β_2-microglobulin-MHC dimers was to act as general utility anchorage sites for regulatory proteins that direct organogenesis. He has suggested that minor alloantigens may be such proteins; however, it will be necessary to use monospecific reagents to carry out functional studies to elucidate the nature and function of these alloantigens.

REFERENCES

Adamson, E.D., and R.L. Gardner. 1979. Control of early development. *Brit. Med. Bull.* 35:113.

Altman, P.L., and D.D. Katz., eds. 1979. *Inbred and Genetically Defined Strains of Laboratory Animals.* Part 1, Mouse and Rat. Bethesda, Maryland: FASEB, p. 113.

Amoroso, E.C. 1952. In *Marshall's Physiology of Reproduction.* A.S. Parkes, ed. Vol. 2. London: Longmans, Green, p. 127.

Barker, C.F., and R.E. Billingham. 1977. Immunologically privileged sites. *Adv. Immunol.* 25:1.

Beer, A.E., and R.E. Billingham. 1971. Immunobiology of mammalian reproduction. *Adv. Immunol.* 14:1.

Bernard, O., M.A. Ripoche, and D. Bennett. 1977. Distribution of maternal immunoglobulins in the mouse uterus and embryo in the days after implantation. *J. Exp. Med.* 145:58.

Billington, W.D., E.J. Jenkinson, R.F. Searle, and M.H. Sellens. 1977. Alloantigen expression during early embryogenesis and placental ontogeny in the mouse. Immunoperoxidase and mixed hemadsorption studies. *Transplant. Proc.* 9:1371.

Borland, R.M., and R.J. Tasca. 1974. Activation of a Na$^+$ -dependent amino acid transport system in preimplantation mouse embryos. *Develop. Biol.* 30:169.

Boyse, E.A., L.J. Old, and E. Stockert. 1968. An approach to mapping of antigens of the cell surface. *Proc. Nat. Acad. Sci. USA* 60:886.

Boyse, E.A., and H. Cantor. 1978. In *The Molecular Basis of Cell-Cell Interaction.* R.A. Lerner and D. Bergsma, eds. The National Foundation March of Dimes Birth Defects: Original Article Series, Vol. 14. New York: Alan R. Liss, p. 249.

Brinster, R.L. 1971. In *Pathways to Conception.* M.I. Sherman, ed. Springfield, Illinois: Charles C. Thomas, p. 245.

Calarco, P.G., and C.J. Epstein. 1973. Cell surface changes during preimplantation development in the mouse. *Develop. Biol.* 32:208.

Carter, J. 1976. Expression of maternal and paternal antigens on trophoblast. *Nature* 262:292.

Carter, J. 1978. The expression of surface antigens on three trophoblastic tissues in the mouse. *J. Reprod. Fertil.* 54:433.

Chaouat, G., G.A. Voisin, D. Escalier, and P. Robert. 1978. Facilitation reaction (enhancing antibodies and suppresssor cells) and rejection reaction (sensitized cells) from the mother to the paternal antigens of the conceptus. *Clin. Exp. Immunol.* 35:13.

Chateaureynaud, P., M.-T. Badet, and G.A. Voisin. 1979. Antagonistic maternal immune reaction (rejection and facilitation) to the embryo in the urodele amphibian *Salamandra salamandra* Lin. *J. Reprod. Immunol.* 1:47.

Cikes, M., and G. Klein. 1972. Quantitative studies of antigen expression in cultured murine lymphoma cells. 1. Cell surface antigens in "asynchronous" cultures. *J. Nat. Cancer Inst.* 49:1599.

Cohen N., and N.H. Collins. 1977. In *The Major Histocompatibility System in Man and Animals,* D. Gotze, ed. New York: Springer-Verlag, p. 313.

David, C.S. 1977. In *The Major Histocompatibility System in Man and Animals.* D. Gotze, ed. New York: Springer-Verlag, p. 255.

Delovitch, T., J.L. Press, and H.O. McDevitt. 1978. Expression of murine Ia antigens during embryonic development. *J. Immunol.* 120:818.

Dizio, S.M., and R.J. Tasca. 1977. Sodium-dependent amino acid transport in preimplantation mouse embryos. *Develop. Biol.* 59:198.

Ducibella, T., and E. Anderson. 1975. Cell shape and membrane changes in the eight-cell mouse embryo: prerequisites for morphogenesis of the blastocyst. *Develop. Biol.* 47:45.

Ducibella, T., D.F. Albertini, E. Anderson, and J.D. Biggers. 1975. The preimplantation mammalian embryo: characterization of intercellular junctions and their appearance during development. *Develop. Biol.* 45:231.

Ducibella. T. 1977. In *Development in Mammals.* M.H. Johnson, ed. Vol. 1. Amsterdam: Elsivier/North-Holland, p. 5.

Edelman, G.M. 1976. Surface modulation in cell recognition and cell growth. *Science* 192:218.

Edidin, M. 1976a. In *The Cell Surface in Animal Embryogenesis and Development.* G. Poste and G.L. Nicolson, eds. Amsterdam: Elsevier/North-Holland, p. 127.

Edidin, M., L.R. Gooding, and M. Johnson. 1974. In *Immunological Approaches to Fertility Control.* E. Diczfausy, ed. Stockholm: Korolinska Institutet, p. 236.

Eichwald, G.M., and C. R. Silmser. 1955. Communication. *Transplant. Bull.* 2:148.

Fields, K.L. 1979. Cell type-specific antigens of cells of the central and peripheral nervous system. *Curr. Top. Develop. Biol.* 13:237.

Flaherty, L., and D. Bennett. 1973. Histocompatibilities found between congenic strains which differ at loci determining differentiation antigens. *Transplantation* 16:505.

Flaherty, L., and D. Zimmerman. 1979. Surface mapping of mouse thymocytes. *Proc. Natl. Acad. Sci. USA* 76:1990.

Gardner, R.L., and V.E. Papaioannou. 1975. In *The Early Development of Mammals.* 2nd Symposium of the British Society of Developmental Biology. M. Balls and A.E. Wild, eds. London: Cambridge University Press, p. 107.

Gill. T.J., and C.F. Repetti. 1979. Immunologic and genetics factors influencing reproduction. *Am. J. Pathol.* 95:465.

Graff, R.J. 1978. Minor histocompatibility genes and their antigens. *Transplant. Proc.* 10:701.

Häkansson, S., S. Heyner, K.G. Sundqvist, and S. Bergstrom. 1975. The presence of paternal H-2 antigens on hybrid mouse blastocysts during experimental delay of implantation and the disappearance of these antigens after onset of implantation. *Int. J. Fertil.* 20:137.

Hamilton, M.S., and I. Hellstrom. 1978. Selection for histoincompatible progeny in mice. *Biol. Reprod.* 19:267.

Hammerling, G.J., G. Mauve, E. Goldberg, and H.O. McDevitt. 1975. Tissue distribution of Ia antigens: Ia on spermatozoa, macrophages and epidermal cells. *Immunogenetics* 1:428.

Hastings, R.A., and A.C. Enders. 1975. Junctional complexes in the preimplantation rabbit embryo. *Anat. Rec.* 181:17.

Hellstrom, K.E., I. Hellstrom, and J. Brawn. 1969. Abrogation of cellular immunity to antigenically foreign mouse embryonic cells by a serum factor. *Nature* 224:914.

Heyner, S. 1973. Detection of H-2 antigens on the cells of the early mouse embryo. *Transplantation* 16:675.

Heyner, S. 1980. In *Immunological Aspects of Infertility and Fertility Regulation.* D.S. Dhindsa and G.F.B. Schumacher, eds. Amsterdam: Elsevier/North Holland, p. 183.

Heyner, S., R.L. Brinster, and J. Palm. 1969. Effect of alloantibody on pre-implantation mouse embryos. *Nature* 222:783.

Heyner, S., and R.D. Hunziker. 1979. Differential expression of alloantigens of the major histocompatibility complex on unfertilized and fertilized eggs. *Develop. Genet.* 1:69.

Heyner, S., and R.D. Hunziker. 1980. Detection of weak alloantigens on the unfertilized and preimplantation developmental stages in the mouse. *J. Supramolec. Struct.* Suppl. 4:159.

Heyner, S., R.D. Hunziker, and G.L. Zink. 1980. Differential expression of minor histocompatibility antigens on the surface of the mouse oocyte and preimplantation developmental stages. *J. Reprod. Immunol.* 2:269.

Hildemann, W.H., and T. Dix. 1972. Transplantation reactions of tropic Australian echinoderms. *Transplantation* 15:624.

Hsu, Y.-C. 1979. *In vitro* development of individually cultured whole mouse embryos from blastocyst to early somite stage. *Develop. Biol.* 68:453.

Hull, P. 1969. Maternal-foetal incompatibility associated with the H-3 locus in the mouse. *Heredity* 24:203.

Jacob, F. 1977. Mouse teratocarcinoma and embryonic antigens. *Immunol. Rev.* 33:3.

Jacob. F. 1979. Cell surface and early stages of mouse embryogenesis. *Curr. Top. Develop. Biol.* 13:117.

Jenkinson, E.J., and W.D. Billington. 1974. Differential susceptibility of mouse trophoblast and embryonic tissue to immune cell lysis. *Transplantation* 18:286.

Jenkinson, E.J., and R.F. Searle. 1979. Ia antigen expression on the developing mouse embryo and placenta. *J. Reprod. Immunol.* 1:3.

Johnson, M.H. 1975. In *Immunobiology of Trophoblast*. R.G. Edwards, C.W.S. Howe, and M.H. Johnson, eds. London: Cambridge University Press, p. 87.

Johnson, M.H., and M. Edidin. 1978. Lateral diffusion in plasma membrane of mouse egg is restricted after fertilization. *Nature* 272:448.

Kelly, S.J. 1975. In *The Early Development of Mammals*. M. Balls and A.E. Wild, eds. London: Cambridge University Press, p. 97.

Kelly, S.J. 1977. Studies of the developmental potential of 4- and 8-cell stage mouse blastomeres. *J. Exp. Zool.* 200:365.

Kirby, D.R.S. 1970. The egg and immunology. *Proc. Roy. Soc. Med.* 63:59.

Klein, J. 1975. *The Biology of the Mouse Histocompatibility-2 Complex*. New York: Springer-Verlag.

Klein, J. 1977. In *The Major Histocompatibility System in Man and Animals*. D.Gotze, ed. New York: Springer-Verlag, p. 339.

Krco, C.J., and E.H. Goldberg. 1977. Major histocompatibility antigens on preimplantation mouse embryos. *Transplant. Proc.* 9:1367.

McDevitt, H.O., B.D. Deak, D.C. Shreffler, J. Klein, J.H. Stimpfling, and G.D. Snell. 1972. Genetic control of the immune response: mapping of the Ir-1 locus. *J. Exp. Med.* 135:1259.

Medawar, P.B. 1953. Some immunological and endocrinological problems raised by the evolution of viviparity in vertebrates. *Symp. Soc. Exp. Biol.* 7:320.

Moscona, A.A. 1961. Rotation-mediated histogenetic aggregation of dissociated cells. *Exp. Cell Res.* 22:455.

Mossman, H.W. 1974. Structural changes in vertebrate foetal membranes associated with the adoption of viviparity. *Obstet. Gynecol. Ann.* 3:7.

Muggleton-Harris, A.L., and M.H. Johnson. 1976. The nature and distribution of serologically detectable alloantigens on the preimplantation mouse embryo. *J. Embryol. Exp. Morphol.* 35:59.

Nagai, Y., S. Ciccarese, and S. Ohno. 1979. The identification of human H-Y antigen and testicular transformation induced by its interaction with the receptor site of bovine foetal ovarian cells. *Differentiation* 13:155.

Nicolson, G.L. 1979. Topographic display of cell surface components and their role in transmembrane signalling. *Curr. Top. Develop. Biol.* 13:305.

Ohno, S. 1977. The original function of MHC antigens as the general plasma membrane anchorage site of organogenesis-directing proteins. *Immunol. Rev.* 35:59.

Ohno. S., Y. Nagai, and S. Ciccarese. 1978. Testicular cells lysostripped of H-Y antigen organize ovarian follicle-like aggregates. *Cytogenet. Cell Genet.* 20:351.

O'Neill, H.C., and C.R. Parish. 1980. Monoclonal antibody detection of two classes of H-2K^k molecules. *Proceedings of the 4th International Congress of Immunology*.

Palm, J., S. Heyner, and R.L. Brinster. 1971. Differential immunofluorescence of fertilized mouse eggs with H-2 and non-H-2 antibody. *J. Exp. Med.* 133:1282.

Parr, E.L., R.V. Blanden, and R.S. Tulsi. 1980. Epithelium of mouse yolk sac placenta lacks H-2 complex alloantigens. *J. Exp. Med.* 152:945.

Patthey, H.L., and M. Edidin. 1973. Evidence for the time of appearance of H-2 antigens in mouse development. *Transplantation* 15:211.

Piko, I. 1975. In *The Early Development of Mammals*. M. Balls and A.E. Wild, eds. London: Cambridge University Press, p. 167.

Schachner, M. 1979. Cell surface antigens of the nervous system. *Curr. Top. Develop. Biol.* 13:259.

Schlafke, S., and A.C. Enders. 1975. Cellular basis of interaction between trophoblast and uterus at implantation. *Biol. Reprod.* 12:41.

Searle, R.F., M.H. Sellens, J. Elson, E.J. Jenkinson, and W.D. Billington. 1976. Detection of alloantigens during preimplantation development and early trophoblast differentiation in the mouse by immunoperoxidase labelling. *J. Exp. Med.* 143:348.

Sellens, M.H. 1977. Antigen expression on early mouse trophoblast. *Nature* 269:60.

Sellens, M.H., and E.J. Jenkinson. 1975. Permeability of the mouse zona pellucida to immunoglobulin. *J. Reprod. Fertil.* 42:153.

Sherman, M.I., A. McLaren, and P.M.B. Walker. 1972. Mechanism of accumulation of DNA in giant cells of mouse trophoblast. *Nature New Biol.* 238:175.

Simmons, R.L., and P.S. Russell. 1966. Histocompatibility antigens in transplanted mouse eggs. *Ann. N.Y. Acad. Sci.* 129:35.

Smith, G. 1978. Inhibition of cell-mediated microcytotoxicity and stimulation of mixed lymphocyte reactivity by mouse pregnancy serum. *Transplantation* 26:278.

Solter, D., and B.B. Knowles. 1978. Monoclonal antibody defining a stage-specific mouse embryonic antigen (SSEA-1). *Proc. Natl. Acad. Sci. USA* 75:5565.

Tarkowski, A.K., and J. Wroblewska. 1967. Development of blastomeres of mouse eggs separated at the four- and eight-cell stage. *J. Embryol. Exp. Morphol.* 18:155.

Valembois, P. 1974. Cellular aspects of graft rejection in earthworms and other metazoa. *Contemp. Top. Immunol.* 4:121.

Voisin, G.A., and G. Chaouat. 1974. Demonstration, nature and properties of maternal antibodies fixed on placenta and directed against paternal antigens. *J. Reprod. Fertil. (Suppl.)* 21:89.

Webb, C.G., W.E. Gall, and G.M. Edelman. 1977. Synthesis and distribution of H-2 antigens in preimplantation embryos. *J. Exp. Med.* 146:923.

Whittingham, D.G. 1968. Fertilization of mouse eggs *in vitro*. *Nature* 220.592.

Zenzes, M.T., U. Wolf, E. Gunther, and W. Engel. 1978. Studies on the function of H-Y antigen: dissociation and reorganization experiments on rat gonadal tissue. *Cytogenet. Cell Genet.* 20:365.

Chapter 5

PRIMARY SEX DETERMINATION: IMMUNOGENETIC ASPECTS

STEPHEN S. WACHTEL

In 1907, Wilson reported that individual, dispersed cells of the marine sponge *Microciona prolifera*—obtained by forcing small pieces of the sponge through a fine bolting cloth—could reaggregate in the laboratory and could organize miniature conglomerates having many of the characteristics of intact functional sponges. Wilson's report generated a series of similar experiments in laboratories using dissociated tissues from "higher" organisms, such as the chicken and and the mouse. Notable among these are the experiments that Moscona conducted approximately a quarter of a century ago. Moscona (1957) obtained single-cell suspensions by mechanical dissociation of embryonic tissues after their incubation in calcium-free and magnesium-free medium containing trypsin. The cell suspensions were placed in flasks, and, in later experiments (Moscona, 1961), the flasks were rotated at low speeds to promote collision and interaction of the dispersed cells. The cells readily aggregated under those conditions, forming small clusters that could be studied histologically, and thus Moscona discovered that, in many cases, the aggregating cells had actively organized structures characteristic of the tissue or organ from which they had been derived. Dispersed cells of the embryonic chick mesonephros organized tubular and glomerular structures similar to those occurring in the mesonephros *in situ*, for example.

When cells from separate tissues were co-cultured, the cells "sorted out" according to tissue and organized distinct tissue-specific aggregates. In fact, cells from the same tissue were able to recognize one another and to associate in the reorganization of a particular structure, even when they were taken from animals belonging to different species. Thus, for example, cells of the various tissues of chick and mouse readily formed chimeric aggregates containing chick and mouse cells interspersed, provided that the cells were derived from the same tissue. It is notable that the exception involved situations in which normal cells were co-cultured with neoplastic cells.

To the extent that cells communicate with their environment and with each other via the plasma membrane, it may be inferred that the signals that govern cell-cell communication and aggregation involve the plasma membrane (Bennett et al., 1972). To the extent that these cell surface signals are tissue specific, but not species specific, it may be inferred that they are phylogenetically conservative.

H-Y ANTIGEN IS PHYLOGENETICALLY CONSERVATIVE

Histocompatibility antigens are cell surface molecules that are recognized as foreign, or "nonself", when they are present in the cells of a graft and absent in those of the host. Thus transplantation biology is concerned in large part with the histocompatibility genes (H genes), which determine the structure of histocompatibility antigens (H antigens). Generally, the H genes are characterized by multiple alleles, so that a particular H antigen, coded by a particular H locus, may have more than one structure. It is this polymorphism of structure that causes antibody formation and graft rejection, and, indeed, it is a dictum of transplantation biology that grafts may be rejected when exchanged between animals differing only with respect to a single H locus.

Of course, grafts are not expected to be sloughed when exchanged between individuals that are genetically identical—identical twins or members of the same highly inbred population of laboratory animals, for instance. Yet, that point requires a qualification: in some highly inbred strains of the mouse, rat, and rabbit, male-to-female skin grafts are rejected, whereas grafts exchanged between the other three sex combinations are accepted. Incompatibility of male-to-female grafts is due to the "male-specific" H-Y antigen, determined directly or indirectly by a gene or group of genes on the Y chromosome, which has no analogue in females.

The H-Y antigen (named by Billingham and Silvers, 1960) is remarkable for its lack of polymorphism (but see Hildemann et al., 1970; reviewed in Wachtel, 1977). Evidently H-Y, or at least part of the molecule conferring H-Y antigenicity, is the same in all male mice and in all male rats, and evidently H-Y of the mouse is the same as H-Y of the rat. Thus female mice can be sensitized to skin grafts from male mice by prior exposure to cells of the male rat (Silvers and Yang, 1973), and cells from males of the rat take up H-Y antibodies directed against H-Y antigen of the mouse (see below).

In 1971, Goldberg et al. reported that male-sensitized female mice produce sera containing H-Y antibodies that can kill mouse sperm in a complement-mediated cytotoxicity test. Specificity for the reaction was established by serologic absorption. The H-Y antisera were diluted and divided into equal portions. One portion was not absorbed, one portion was absorbed with male cells (spleen cells, for example), and one portion was absorbed with the corresponding number and type of female cells. The absorbing cells were suspended in the diluted antiserum and later discarded. If the cells contained H-Y on their membranes, they specifically absorbed the H-Y antibodies from the antiserum, which now lost its reactivity for sperm.

Goldberg's cytotoxity test laid the foundation of the serology of H-Y antigen. Several other assays for H-Y have since been developed, including the epidermal cell cytotoxicity test (Scheid et al., 1972), the mixed hemadsorption hybrid antibody test (Koo et al., 1973), the protein A mixed hemadsorption assay (Koo and Goldberg, 1978), and a new radioimmunoassay (Savikurki et al., 1981). The availability of monoclonal H-Y antisera (Koo et al., 1981) should promote the development of other new techniques. Yet all these assays rely heavily on the technique of absorption, and many of the serological observations that will be described below are based on the older assays—especially the sperm cytotoxicity test of Goldberg.

We have pointed out that H-Y antigens of mouse and rat are "homologous" according to the interspecies transplantation system of Silvers and Yang (1973). Identity or crossreactivity of male antigens of rat and mouse can also be demonstrated serologically. Thus H-Y antibodies of the mouse readily discriminate between male and female cells of the rat; after absorption with rat male cells, H-Y antisera from female mice lose their reactivity for mouse target cells in any of several assays. Yet H-Y antigen is widespread and perhaps ubiquitous in mammals, being detected directly or by absorption in cells from males of every species so far evaluated (Table 5-1). H-Y is detected, moreover, in cells from members of the heterogametic sex of every *vertebrate* species examined to date. In birds, it is the female that is the heterogametic sex (ZW), and it is the female whose cells absorb H-Y antibodies of the mouse. Among amphibians, male and female heterogametic species exist side by side; in the male heterogametic leopard frog, *Rana pipiens*, male cells (XY) absorb H-Y antibodies, and in the female heterogametic South African clawed frog,

TABLE 5-1
H-Y Antigen Among the Vertebrates

CLASS	SPECIES	SEX CHROMOSOMES FEMALE/MALE	H-Y IN	REFERENCE
Mammalia	Mouse	XX/XY	Male	Eichwald and Silmser, 1955
	Rat	XX/XY	Male	Billingham and Silvers, 1959
	Mole vole	XO/XO	Male	Nagai and Ohno, 1977
	Wood lemming	XX,XY/XY	Male	Herbst et al., 1978
	Cattle	XX/XY	Male	Ohno et al., 1976
	Human	XX/XY	Male	Koo et al., 1977a
Aves	Chicken	ZW/ZZ	Female	Müller et al., 1979
	Quail	ZW/ZZ	Female	Müller et al., 1980
Reptilia	Turtle*	ZW/ZZ†	Female	Zaborski et al., 1979
Amphibia	Leopard frog	XX/XY	Male	Wachtel et al., 1975
	Clawed frog	ZW/ZZ	Female	Wachtel et al., 1975
Osteicythyes	Medaka	XX/XY	Male	Pechan et al., 1979
	Guppy	XX/XY	Male	Müller and Wolf, 1979

* Emys orbicularis.
† Inferred on the basis of H-Y phenotype.

Xenopus laevis, female cells (ZW) absorb H-Y antibodies (Figure 5-1). From an evolutionary perspective, then, H-Y genes are very old indeed.

POSSIBLE ROLE OF H-Y IN THE DETERMINATION OF PRIMARY SEX

Mutations affecting the active site of their products are generally intolerable in genes of fundamental significance. Therefore, the phylogenetic stability of H-Y antigen, as revealed serologically, indicates a critial, sex-related function. Transplantation antigens are nonetheless, by definition, cell surface components, and we have already discussed the implicit role of phylogenetically conservative cell surface signals in the cell-cell interactions of morphogenesis and organogenesis. In collaboration with Ohno, we accordingly proposed that H-Y antigen directs the initially indifferent embryonic gonad to become a testis in male heterogametic species, such as the mouse, and an ovary in female heterogametic species, such as the chicken (see Ohno, 1976). Since Bennett and colleagues had already detected H-Y in XX males of the mouse in 1975 (published in 1977), the connection between H-Y and the testis seemed more certain at the time than the connection between the Y chromosome and the testis. (It cannot reasonably be claimed that H-Y is a secondary sex characteristic: the molecule is detected in embryos of the eight-cell stage [Krco and Goldberg, 1976; Epstein et al., 1980], in androgen-insensitive XY females exhibiting the syndrome of testicular feminization in the mouse [Bennett et al., 1975], and in humans [Koo et al., 1977b].)

Acccording to Jost (1970), sex determination and differentiation may be viewed as consisting of this sequence of events: establishment of genetic sex at fertilization, translation of genetic sex into gonadal sex during embryogenesis (primary sex determination), and translation of gonadal sex into body sex during embryogenesis and thereafter (secondary sex differentiation). We are indebted to Jost (1947) for demonstrating that the last process is mediated by testosterone secreted by the testis. In the absence of testosterone or its receptor, the embryo is feminized; castrated embryos become females regardless of genetic sex.

Hence it may be supposed that the sex-determining role of the mammalian Y chromosome is limited to the induction of the testis, further differentiation being directed by testicular secretions. It follows that the putative sex-determining role of Y-chromosome-governed H-Y antigen should also be limited to the induction of the testis. Thus the hypothesis that H-Y directs testicular organogenesis could be evaluated by testing for the expression of H-Y in subjects whose gonadal sex did not coincide with their karyotypic sex

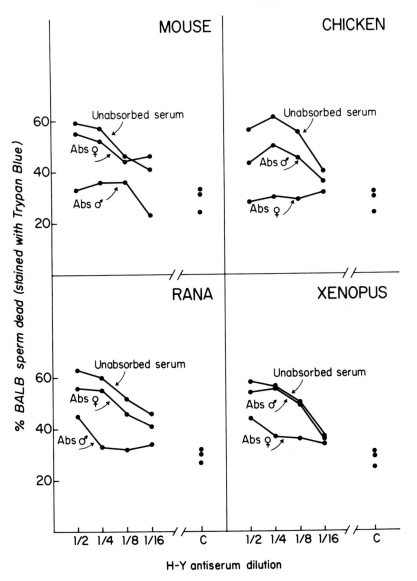

FIGURE 5-1. Phylogenetic conservation of H-Y/H-W antigen.
H-Y antibody of the mouse is absorbed by cells from heterogametic XY males of mouse and
Rana, and heterogametic ZW females of chicken and *Xenopus*. Positive absorption, manifested
as a fall in cytotoxicity (% dead sperm from males of the BALB strain of mouse), indicates the
presence of H-Y in cells used for absorption. *Abs.* signifies absorption with cells of the sex
denoted; *C* signifies control values in sperm cell suspensions containing complement but not H-Y
antiserum. After Wachtel et al., 1975.

or body sex, because the scheme requires that H-Y be present whenever testicular tissue is present, regardless of karyotype and regardless of physical appearance.

TESTING THE HYPOTHESIS *IN VIVO*

XX males and XX true hermaphrodites
Testicular development in individuals lacking a Y chromosome is known in several mammalian species. Human 46,XX males were first reported in 1964, and more than 100 such men have been described since that time (de la Chapelle, 1981). Generally, 46,XX males are characterized by small azoospermic testes. Androgen production is decreased in comparison with that in normal males; about one-third of all human XX males have female-like breasts; facial hair is often scant; and psychosexual orientation is conventional.

Human XX males are H-Y positive. An example is the pedigree containing three XX males reported by de la Chapelle et al. (1978). Evidently, maleness was inherited as an autosomal recessive trait in these subjects, raising questions about the location of the structural genes for testicular organogenesis. It is perhaps worth noting that residual expression of H-Y was indicated in the three mothers, themselves obligate heterozygotes for testis-determining genes. An explanation is that the structural genes for H-Y are normally autosomal and that residual (recessive) expression of H-Y may occur in certain leaky mutants (Müller and Wolf, 1979a). A similar condition has been reported in goats (Table 5-2). Thus XX goats that are homozygous for the *polled* gene develop testes and exhibit a range of body sex phenotypes. All XX billy goats that have been tested have been typed H-Y positive (Shalev et al., 1980). As in the corresponding human condition, residual expression of H-Y was detected in heterozygous mothers.

We have said that Sxr/-,XX male mice carrying the autosomal dominant gene, sex-reversed, are H-Y positive (Table 5-2). Sex reversal also occurs in dogs. Selden et al. (1978) described an XX male that was whelped by an XX true hermaphrodite with bilateral ovotestes. Mother and pup were both H-Y positive, suggesting that XX male sex reversal and XX true hermaphroditism are related conditions associated with abnormal inheritance of H-Y genes. As in the former examples, the condition was autosomal. Yet further genealogical analysis will be required to establish unambiguously whether the determinant genes were dominant or recessive.

True hermaphroditism has been recognized as a clinical entity since 1899; more than 400 cases are documented in the review of van Niekerk and Retief (1981). Yet certainly hermaphroditism is as old as man himself.

According to the Midrashic commentary *Breishis Rabah*, written some 1500 to 2000 years ago, primeval man was in fact an *androgynos*, that interpretation being founded on the biblical passage "male and female created He *them*, and blessed *them* and called *their* name Adam" [my italics] (Gen 5:2; and see Gen 1:27). (Perhaps the oldest reference to hermaphroditism per se occurs in the Babylonian cuneiform tablets unearthed at the site of the Royal Library at Ninevah, which date from the seventh century B.C.E. Contained in those tablets is this prediction: "When a woman gives birth to an infant whose right ear is round, there will be an *androgyne* in the house of the newborn [see review by Gordon, 1979].)

Today we define true hermaphroditism as the occurrence of ovarian and testicular tissue in the same individual. The most common karyotype in human true hermaphroditism is 46,XX; the most common gonad is the ovotestis. Of eight 46,XX true hermaphrodites studied in our laboratory, all possessed at least a single ovotestis—and all were typed H-Y positive. Yet statistical evaluation of the combined serological data revealed decreased absorption of H-Y antibodies by cells of the 46,XX true hermaphrodites, compared with absorption by cells of the 46,XY male controls. So the H-Y phenotype of the 46,XX true hermaphrodite may be represented as intermediate (H-Y ±) between the phenotypes of normal male and normal female (see Fraccaro et al., 1979). Winters et al. (1979) studied cells that had been cultured separately from the ovarian and testicular portions of a scrotal ovotestis (from a 46,XX phenotypic "male") with this result: cells derived from the ovarian portion were typed H-Y negative and cells derived from the testicular portion were typed

TABLE 5-2
H-Y in Cases of Abnormal Sexual Differentiation

SPECIES	GENETIC SEX	GONADAL SEX	BODY SEX	H-Y	REFERENCE
Mouse	Sxr/, XX	Testes	Male	+	Bennett et al., 1977
	X^{Tfm}Y	Testes	Female	+	Bennett et al., 1975
Wood lemming	XY	Ovaries	Female	–	Herbst et al., 1978
Dog	XX	Testes	Male	+	Selden et al., 1978
	XX	Ovotestes	Female	+	Selden et al., 1978
Goat	P/P,XX	Testes	Intersex*	+	Shalev et al., 1980
Cattle	XX/XY	Primitive testes	Freemartin	+	Ohno et al., 1976
Human	XX	Testes	Male	+	de la Chapelle et al., 1978
	XX	Ovotestes	Male	+	Winters et al., 1979
	XX	Ovotestis, ovary	Ambiguous	+	Saenger et al., 1976
	X^{Tfm}Y	Testes	Female	+	Koo et al., 1977b
	Xp+Y	Dysgenetic ovaries	Female	–	Bernstein et al., 1980
	XYp-	Dysgenetic ovaries	Female	–†	Rosenfeld et al., 1979

* Masculinized female.
† Possible residual expression of H-Y detected in one of three assays in this study.

H-Y positive.

So much for XX males and XX true hermaphrodites. It is sufficient to say that, in all species studied so far, H-Y is found in association with the testis, even in the absence of the Y chromosome, and that expression of H-Y occurs prior to, and independently of, development of the testis, and thus independently of any other male trait that is secondary to development of the testis.

XY females

A cousin of the species that is notorious for its occasional frenetic migrations, the Scandinavian wood lemming (*Myopus schisticolor*) is admired for its 4:1 sex ratio with a preponderance of females. Roughly half of the females have a male (XY) karyotype, yet they are fertile and anatomically indistinguishable from their "normal" XX sisters. There is this difference: XY females bear only female offspring—a phenomenon that has been attributed to a nondisjunctional event whereby the X is duplicated in the germ line and the Y eliminated. Two other points are worth mentioning. First, XY females of the wood lemming are H-Y negative, and second, the condition is inherited as an X-linked trait. Herbst et al. (1978) have now described two kinds of X chromosomes in the wood lemming, recognizable by their distinct banding patterns and denoted X and X*. Only the latter is found in XY females (X*Y) (Figure 5-2).

A comparable situation has been described in man. Bernstein et al. (1980) discovered a male karyotype in the cells of a grossly retarded phenotypic female with multiple congenital abnormalities. The Y chromosome was normal, but an additional band was located in the short arm of the X; hence the karyotype was designated 46,Xp+Y. Cells of the child were typed H-Y negative. When the child died at five years of age, autopsy disclosed hypoplastic uterus and tubes and, on sectioning of the uterine adnexae, dysgenetic ovarian stroma containing a few degenerative follicles. There was no testicular tissue.

The mother became pregnant again, and cells were drawn from the amniotic fluid for cytogenetic analysis. The cells contained a normal Y chromosome and the abnormal X; the karyotype was 46,Xp+Y. The pregnancy was terminated at 20 weeks of gestation. The fetus was a phenotypic female with multiple congenital abnormalities. Internal genitalia were those of a normal female fetus at that stage of development; the ovaries contained numerous primordial follicles; there was no trace of testicular tissue. As in the proband, cells of the fetus were typed H-Y negative in serological tests.

Computer-based videodensitometric analysis indicated the likelihood of a duplication of part of the short arm of the X in that condition. It may be inferred that mutation of the human Xp can interfere with normal testicular differentiation in man and in the wood lemming. One might argue that genes in the Y and X chromosomes are required for synthesis of H-Y antigen and testicular organogenesis in mammals generally. According to that view, mutational absence of H-Y in the XY indifferent gonad leads to development of an ovary, fertile as in the wood lemming and degenerative as in man.

Evidently, two X chromosomes are required for normal oogenesis in the human embryo. In the absence of the second X, as in 45,X Turner's syndrome, the germ cells die and the surrounding follicles become atretic. The result is failure of endocrine function in a streak gonad consisting of little more than fibrous tissue with some ovarian-like stroma. Given failure of the testis-determining mechanism, the 46,XY gonad resembles the 45,X gonad with respect to the number of X chromosomes that are present. Hence the different gonadal histologies of the proband and her fetal sibling described above

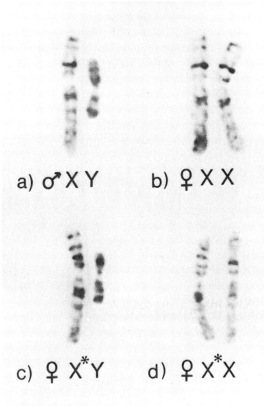

a) ♂ X Y   b) ♀ X X

c) ♀ X*Y   d) ♀ X*X

FIGURE 5-2. Sex chromosome complements of the wood lemming, *Myopus schisticolor.*
Note alternative G-banding patterns of X, found in XY males, and X*, found in XY females. From Herbst et al., 1978.

may be seen as representing different phases of the same developmental failure.

The syndrome resulting from gonadal failure in human embryos with the normal male karyotype is called 46,XY pure gonadal dysgenesis ("pure" in the absence of the stigmata of Turner's syndrome). Yet of the 30-odd cases of pure gonadal dysgenesis that have been studied serologically, about two-thirds have been typed H-Y positive and the rest H-Y negative (for examples, see Wolf, 1979). At first sight, the presence of a putative testis inducer in phenotypic females lacking the testis might seem paradoxical, but the situation is really no more paradoxical than the presence of testosterone in phenotypic females lacking the secondary characteristics of male sexuality that are normally induced by testosterone (testicular feminization syndrome). We have said that these phenotypic females are unresponsive to androgen as a result of mutational failure or absence of the nuclear cytosol androgen receptor. So testosterone is functionally absent at the target site in that condition. In the paragraphs that follow, I shall develop the notion that H-Y antigen is functionally absent in all cases of 46,XY gonadal dysgenesis, as a result of failure of synthesis of H-Y in the H-Y negative population and failure of the H-Y gonadal receptor in the H-Y positive population (see below).

XX/XY chimeric male mouse

In experimental chimeras of the mouse, produced by the fusion of two blastocysts or by the injection of cells from one blastocyst into the blastocoele of another, random combinations should give one XX/XX embryo: two XX/XY embryos : one XY/XY embryo. Now, if XX and XY gonadal cells were to develop according to inherent karyotypic dictate, one might predict a high incidence of hermaphroditism among the population of XX/XY chimeric mice but the fact is that about three-fourths of XX/XY mice become males and the remainder become females; XX/XY hermaphroditism is rare (McLaren, 1976). It seems that XY cells are able to persuade neighboring XX cells to engage in testicular organogenesis or, more specifically, that XY cells release a diffusible signal which induces testicular organogenesis in neighboring (XX) cells that take up the signal.

Could the inductive signal be H-Y antigen? To evaluate that question, Ohno et al. (1978a) studied H-Y in cells from the gonad of an XX(BALB)/XY(C3H) chimeric male mouse. Although the nongerminal tissue of the gonad consisted of a one-to-one mixture of XX(BALB) and XY(C3H) cells, as indicated by starch gel electrophoresis with staining for GPI (which differs in the two strains), at least as much H-Y was detected in cells of the chimeric gonad as was detected in cells from the gonads of normal XY males from either parental strain. The results were consistent with the view that H-Y had been disseminated in the XY gonadal cells and bound in the XX cells. There was no such evidence of transfer of H-Y in the cells of the spleen or epidermis (Table 5-3).

Two years earlier, Ohno et al. (1976) had presented data indicating a similar uptake of H-Y in XX gonadal cells of the bovine freemartin, the masculinized synchorial female twin of a bull. Freemartin gonads believed to consist mainly of XX cells were typed H-Y positive in that study. At the time, it was assumed that XY gonadal cells of the bull had entered the common circulation, migrated to the developing ovary, and released H-Y, thereby subverting ovarian development and promoting differentiation of the testis. Yet, more recent data indicate that H-Y itself can circulate in the serum of fetal bulls and fetal freemartins (Wachtel et al., 1980b, and see below).

TESTING THE HYPOTHESIS *IN VITRO*

XY sex reversal in vitro

In mammals, testicular differentiation commences in the indifferent gonad with the arrival of the primordial germ cells. Soon thereafter, the Sertoli cells appear. The Sertoli cells aggregate, thus forming the seminiferous tubules which encase the germ cells and exclude the interstitial elements (Leydig cells). These events are followed by appearance of the tunica albuginea, the thick outer covering of the testis. It should be pointed out that the germ cells are not required for testicular organogenesis, which proceeds in their absence (Coulombre and Russell, 1954). (It is remarkable that of all nucleated male cells so far studied, only the pre-meiotic germ cell is H-Y negative [Zenzes et al., 1978a; Koo et al., 1979].) Simply stated, then, early testicular organogenesis involves differentiation of the Sertoli cells and development of the seminiferous tubules.

Given the observations of Moscona (1961) as described at the start of this chapter, H-Y antigen might be expected to provide, directly or

TABLE 5-3
Ratio of XX:XY Cells and Expression of H-Y Antigen in Tissues of XX(BALB)/XY(C3H) Chimeric Male Mouse

| | RATIO OF CELLS | H-Y ANTIGEN: |
CELL TYPE	XX:XY	ESTIMATED QUANTITY IN NORMAL MALE (%)
Spleen	60:40*	50
Epidermal	55:45*	40
Germinal testicular	0:100†	100
Non-germinal testicular‡	50:50†	135

* Determined by H-2 typing.

† Determined by starch gel electrophoresis for GPI.

‡ Leydig, Sertoli, and other somatic cells (after Ohno et al., 1978a).

113

indirectly, the physical links that hold Sertoli cells together in the developing tubules (but see discussion in Urban et al., 1981). Given testicular organogenesis in XX/XY chimeric gonads, as described above, H-Y might be expected to provide a short-range inductive signal, "sex-reversing" XX Sertoli cell precursors, and to impose a program of testicular morphogenesis in elements that would normally organize a follicle. The implication is that Sertoli cells and granulosa cells of the ovarian follicle are derived from the same precursor (see Ciccarese and Ohno, 1978, and discussion below).

To evaluate *in vitro* the role of H-Y antigen, Ohno et al. (1978b) and Zenzes et al. (1978c) dispersed cells of the newborn XY testis in mice and rats and allowed the cells to reaggregate in rotary culture according to this rationale: when exposed to specific antibody in excess, plasma membrane antigens migrate to a polar cap of the cell, where they are internalized and then digested by autophagic lysosomes. Thus, in the presence of specific antibody, the cell surface is denuded, or "lysostripped", with respect to the relevant antigen. If H-Y is critical to organization of the seminiferous tubule, then tubular reaggregation should be compromised in cells denuded of H-Y.

The strategy was vindicated in practice. Untreated cells of the XY testis or cells treated with normal serum readily formed tubules after culture, but cells that had been incubated in H-Y antibody formed spherical aggregates instead. The spherical aggregates resembled early follicles. In similar experiments conducted more recently by Müller and Urban (in press), tubular organization was disrupted in H-Y antibody-treated cells but there was no evidence of follicular aggregation.

XX sex reversal in vitro
Could XX cells of the dissociated ovary be induced to organize testicular structures in reciprocal experiments? By 1979, sources of soluble H-Y antigen had been identified, examples of which are medium from cultured testicular cells and medium from the so-called Daudi cell line, which is derived from a male Burkitt lymphoma (see below). Zenzes et al. (1978b) used rat H-Y in rotary cultures with dispersed XX cells of the neonatal rat ovary and reported the appearance of tubular aggregates after 16 hours. Similar results were obtained more recently by Müller and Urban (in press). In both cases, tubule formation was impeded by addition of specific H-Y antiserum. It is worth pointing out that the H-Y antigen-mediated conversion of XX ovarian cells was correlated with the precocious appearance of the LH receptor (Müller et al., 1978b). The LH receptor is present in the neonatal rat testis at birth, but it does not normally appear in the rat ovary until six to eight days after birth. So it may be argued that one function of H-Y is to induce testis-specific differentiative programs.

In other studies, Ohno et al. (1979) cultured the intact XX primordial gonad of the fetal calf in concentrated sources of Daudi-secreted H-Y antigen.

They reported testicular transformation of the XX gonad commencing after three days with the sudden organization of seminiferous tubules and culminating after five days with the appearance of the *tunica albuginea*. There were no Leydig cells or germ cells (see Nagai et al., 1979).

H-Y AND THE CELL MEMBRANE

The nonspecific membrane anchorage site
We have pointed out that H-Y is ubiquitous, being found in all tissues and perhaps in all nucleated cells of the male, with the exception of the premeiotic germ cells. So why should a testis-inducing molecule be found in the somatic tissues? And how could a *diffusible* inducer find stable residence in the plasma membrane of the somatic cells?

As for the first question, it is conceivable that H-Y has some extra-gonadal function. It is conceivable, too, that H-Y is synthesized independently of regulation—at least in the embryo. The point is that gonadal organogenesis is influenced by physicochemical cues in the more primitive vertebrates. For example, ovarian differentiation is induced in ZZ presumptive males of the chicken by treatment with estradiol (see below) and in reptiles and amphibians, sex can be determined by the temperature of the incubating egg. Sexual differentiation is less susceptible to external influences in the higher vertebrates, and in fact, hormones and temperature exert no known influence on gonadal development in any Eutherian mammal. It could be argued that testicular differentiation of the mammalian embryo must necessarily occur independently of hormonal influence in order to escape the possible sex-reversing effects of the maternal estrogens. According to that view, ubiquitous expression of H-Y might reflect the requirement to escape hormonal (and other) regulation. Ubiquitous expression might also reflect a requirement for early synthesis of H-Y (the inducer must be present before the onset of the event that it induces). We have already seen that the molecule is present in eight-cell embryos of the mouse.

As for the question how H-Y is carried in the somatic cell membrane, there are numerous reports suggesting association of H-Y and cell surface components of the major histocompatibility complex (MHC)(H-2 of the mouse, HLA of man, etc.). H-2 or HLA cell surface components consist of a heavy chain (antigen) having a molecular weight (MW) of about 45 000 in close, noncovalent association with a smaller, lighter chain of about 12 000 MW called beta-2-microglobulin (β_2m). The heavy chain spans the plasma membrane; part is buried in the lipid portion of the membrane, and part is free at the cell surface for interaction with β_2m and with other cell surface molecules.

115

According to Ohno (1977), dimers of β_2m-MHC serve as the nonspecific anchorage sites for H-Y and for all organogenesis-directing proteins. It follows that, in the absence of its β_2m-MHC anchorage site, H-Y could not be maintained in the plasma membrane and would be released into the surrounding medium. We have discussed the release of H-Y by cultured Daudi cells. The fact is that Daudi cells have lost β_2m and HLA on their membranes. Thus, Daudi cells were typed H-Y negative in serological tests (Beutler et al., 1978), but when they were cultured together with female cells of the HeLa D98 line, which is HLA$^+$ and β_2m$^+$, the two molecules were restored in the membranes of (HeLa x Daudi) hybrids, causing restored expression of H-Y antigen. Nevertheless, the precise relationship between β_2m-HLA and H-Y remains to be ascertained, because cells of the Chevalier line, another male Burkitt lymphoma, were typed H-Y positive in the absence of HLA (Fellous et al., 1978).

The gonad-specific H-Y receptor
The gonad-specific function of an inducer molecule that is ubiquitously expressed is made feasible by postulation of the occurrence of a gonad-specific *receptor* distinct from the membrane anchorage site. The idea of a specific receptor for H-Y was developed earlier in the context of the XX/XY chimeric testis. If XX cells of the chimeric testis are H-Y positive as a consequence of having taken up H-Y disseminated by XY cells, then gonad-specific uptake of the putative inducer could be demonstrable *in vitro* according to the following consideration. Cells of the normal XX gonad in suspension, bearing vacant receptors for H-Y, should take up the molecule if it is available, and having taken up the molecule, the XX cells should acquire the H-Y positive phenotype. These H-Y-exposed cells might then be expected to absorb H-Y antibody in serological tests. Indeed, Müller et al. (1978a) reported absorption of H-Y antibody by XX ovarian cells exposed to rat epididymal fluid, a source of soluble H-Y antigen. This suggested *specific* uptake of H-Y antigen because H-Y antibody was not absorbed by cells of the female brain, epidermis, kidney, or liver when they were exposed to rat epididymal fluid.

H-Y AS A CIRCULATING MOLECULE: ANOTHER LOOK AT THE BOVINE FREEMARTIN

The gonad of the freemartin may be modified to resemble a small testis which can produce significant amounts of androgen (Short et al., 1969) and antimullerian hormone (Vigier et al., 1981). Thus masculinization of the free-martin, including development of the Wolffian ducts and suppression of the mullerian ducts, may be due to secretions of the transformed freemartin gonad. So what is it that transforms the gonad itself? As noted above, H-Y was detected in the embryonic freemartin gonad by Ohno et al. (1976), who suggested that masculinization of the gonad was due to dissemination of H-Y by migrant XY cells and consequent uptake of H-Y in the native XX population.

In the same year, however, Vigier et al. (1976) published a report showing that the freemartin stigmata could be precluded by severing of the vascular connections between heterosexual twin embryos before day 45 of gestation—but *after* the establishment of XX/XY chimerism (as evaluated in liver). Evidently, XX/XY cellular chimerism could not ensure development of the freemartin condition. Another blood-borne factor was required.

Prompted by the experience of Vigier et al., my laboratory initiated a series of studies designed to test the question of whether blood-borne H-Y might be a factor in the transformation of the freemartin gonad. We learned that H-Y was in fact prominent in the serum of fetal bulls and fetal freemartins. Thus fetal bovine ovarian cells became H-Y positive after exposure to serum of the fetal bull or fetal freemartin, but they did not become H-Y positive after exposure to serum of the normal fetal cow. Uptake of soluble H-Y was ascertained by the absorbtion of mouse H-Y antiserum with serum-treated fetal ovarian cells in the standard sperm cytotoxicity test and by a new competitive binding radioassay (CBRA).

For the CBRA, Daudi cells were cultured in medium containing tritiated leucine. The presence of H-Y in a given solution could thus be estimated by evaluation of the ability of that solution to block the uptake of Daudi-secreted tritiated H-Y in target cells from the fetal ovary. The fetal ovarian cells were first incubated with the fetal serum to be tested for soluble H-Y or with mouse testis supernatant, a demonstrated source of soluble H-Y, and then washed and incubated with radioactive Daudi culture medium. The cells were washed again and transferred to scintillation vials for enumeration of bound counts per minute (cpm).

The results, some of which are presented in Table 5-4, provide clearcut evidence that fetal calf serum from males or freemartins contains an agent that can block uptake of tritiated H-Y in the fetal ovarian target cell (the same is true of mouse testis supernatant). Since the ovarian target cells became H-Y positive following their exposure to male serum or freemartin serum (or testis supernatant or Daudi-secreted proteins), it may be argued that the agent is H-Y antigen (Wachtel et al., 1980b).

On the basis of the foregoing considerations, we concluded that H-Y can occur as a stable component of the cell membrane (in the somatic tissues), as a soluble molecule (in the serum), or as a receptor-bound substrate (in the gonad). The implication is that H-Y released in XY cells of the bull twin fetus

117

TABLE 5-4
CBRA for H-Y Antigen in Fetal Calf Serum

PRETREATMENT OF FETAL OVARIAN CELLS	cpm*	% INHIBITION†
None	16 363	0 (standard)
Female FCS	16 313	0
Male FCS	12 609	23
TS	11 938	27
Fm II‡ FCS	11 158	32
Fm III FCS	12 467	24
Fm IV FCS	12 644	23

Note: The uptake of radioactive H-Y antigen in fetal ovarian target cells is inhibited by testis supernatant, fetal bull serum or fetal freemartin serum, but not by fetal cow serum.

* Based on 10 minute counts of 2×10^6 fetal ovarian cells treated as shown and incubated with ^{3}H:H-Y (background $<$ 28 cpm).

† Given by $(x-y/x)(100)$, where x = cpm for untreated fetal ovarian cells and y=cpm for fetal ovarian cells treated as shown.

‡ Three freemartin fetuses (Fm) having estimated ages of 105 to 110 days, 155 to 170 days, and 175 to 185 days of gestation, respectively (term approximately 280 days), and designated Fm II, Fm III and Fm IV in Wachtel et al., 1980b.© MIT Press. Reprinted by permission. See Figure 5-3.

traverses the common chorionic vasculature, enters the circulation of the female twin, and engages the specific receptors of the developing XX gonad, thereby subverting ovarian development and promoting organogenesis of the testis (Figure 5-3). We have already said that Daudi-secreted H-Y induces precocious development of the testis in the indifferent gonad of the fetal calf.

BIOCHEMISTRY OF H-Y ANTIGEN

In 1979, Nagai et al. labeled Daudi cells by growing them in medium containing tritiated lysine. The Daudi-secreted proteins, which now included tritiated H-Y, were reacted with bovine fetal ovarian cells, with cells of the adult-mouse spleen or epidermis, or with cells of the newborn-mouse testis. Enumeration of tightly bound cpm in each target cell type revealed selective uptake of label in the ovarian cells. Solubilization of the ovarian cell membranes and analysis by SDS polyacrylamide gel electrophoresis (PAGE) revealed, moreover, that the bound species was a molecule of approximately 18 000 MW.

In related studies, J. Hall of our laboratory labeled Daudi cells with tritiated leucine for immunological characterization of H-Y. Daudi-secreted proteins were precipitated with H-Y antibody by the double antibody method or by the protein A method. Labeled culture medium was reacted first with specific antisera and then with goat antimouse Ig or *Staphylococcus* protein A; these reagents promote precipitation of antigen-antibody complexes.

In the experiments reported by Hall et al. (1981), precipitates were analyzed by SDS-PAGE. The gels were sliced and solubilized and the cpm per slice determined in a scintillation counter. In this system, the position of a protein on the gel is indicated by a peak of radioactivity. The molecular weight of the protein is estimated by comparing its relative mobility (position on gel) with the mobility of standard proteins of known molecular weight that have

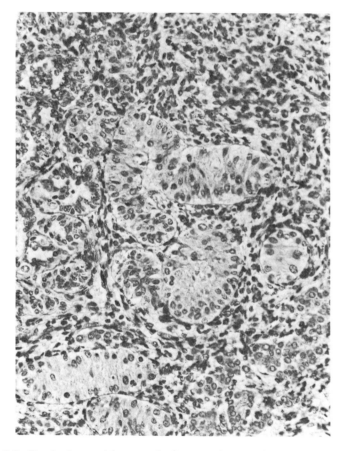

FIGURE 5-3. Testicular architecture in freemartin gonad.
Well-developed seminiferous tubules in medullary portion of gonad from freemartin fetus of approximately 105 to 110 days of gestation (Fm II in Table 5-4). Magnification: 250x. From Wachtel et al., 1980b.

been run on corresponding gels. Immunoprecipitation with H-Y antiserum yielded a molecule of 15 000 to 18 000 MW (Figure 5-4). Recovery of the molecule was obviated by absorption of the antiserum with cells from male, but not female, mice of the highly inbred C57BL/6 strain.

In another approach to the biochemical characterization of H-Y, Hall and Wachtel (1980) labeled mouse Sertoli cell plasma membrane proteins with radioactive iodine. Two-dimensional gel electrophoresis indicated specific immunoprecipitation of a pair of cell surface molecules. Their estimated molecular weights were 18 000 and 31   000, respectively. Each had an isoelectric point of from 5 to 7. It remains to be determined whether these molecules are breakdown products of a bigger molecule, whether the smaller is a subunit of the larger, whether both are able to engage the gonadal H-Y receptor, and whether both are in fact H-Y antigens.

In more recent studies, Shapiro and Erickson (1981) asked whether the H-Y antigenic determinant is a carbohydrate or a polypeptide. Mouse H-Y

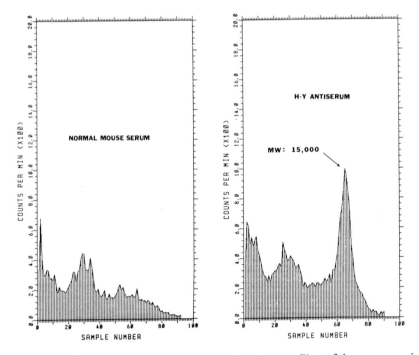

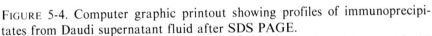

FIGURE 5-4. Computer graphic printout showing profiles of immunoprecipitates from Daudi supernatant fluid after SDS PAGE.

Anti-H-Y antibodies were fixed to Protein-A-coated Sepharose beads, and then reacted with Daudi culture medium. After treatment with SDS and β-mercaptoethanol, the PA-Sepharose was removed by centrifugation and the solubilized molecules run on polyacrylamide gels. Molecular weight standards were run on a separate gel. From Hall and Wachtel, 1980.

positive spleen cells were exposed to reagents which affect carbohydrate or polypeptide structure and then were tested for their ability to absorb H-Y antibodies in a standard epidermal cell cytotoxicity test. In fact, the spleen cells lost their ability to absorb H-Y antibodies after treatment with the glycolytic enzymes β-galactosidase and galactose oxidase, but not after treatment with the proteolytic enzymes trypsin, α-chymotrypsin, or thermolysin. Thus, unlike the situation with H-2, the antigenic integrity of H-Y seems to be at least partly dependent on a carbohydrate moiety.

GENETICS OF H-Y ANTIGEN

Koo et al. (1977a) mapped H-Y genes by studying expression of H-Y in human patients with structural modifications of the Y chromosome. Expression of H-Y was correlated with the presence of the centromeric region of the Y short arm, designated Yp. Thus, for example, H-Y was detected in a male with the karyotype 46,XYq⁻ (denoting loss of the long arm of the Y), but it was not detected in a phenotypic female with the karyotype 46,X,i(Yq) (denoting loss of the short arm and duplication, or "isochromosome", of the long arm). (In one patient, a centromeric locus in the proximal Yq could not be ruled out.) I have said that another H-Y gene was indicated in the short arm of the X chromosome in the Scandinavian wood lemming (Herbst et al., 1978) and in man (Bernstein et al., 1980). A human gene has been mapped tentatively to the distal segment, designated Xp223, by Wolf et al. (1980a).

Given the two H-Y genes, one on the Y and the other on the X, one might argue that the Y-situated gene is structural and activated by the X, or vice versa, but the situation becomes rather more complicated when one considers other data indicating the possible occurrence of structural H-Y genes on an unidentified pair of autosomes. We have already discussed the occurrence of autosomal sex-reversing genes in the mouse and goat and in man. What if these mutant genes were structural H-Y genes that had been abnormally activated in the absence of the Y?

According to the scheme favored by Wolf et al. (1980b), the structural testis-determining H-Y genes are autosomal and, under normal circumstances, are repressed in females. Repression is due to a pair of X-linked genes that are situated in the terminal segment of Xp (reputed to escape lyonization). In normal males, only one X-linked repressor gene is present; that gene is derepressed by the Y chromosomal gene; the result is synthesis of H-Y and organogenesis of the testis. Thus autosomal dominant and recessive modes of XX male sex reversal could be viewed as representing varying degrees of escape from X-linked repression in mutant H-Y structural genes.

The scheme allows for some subthreshold synthesis of H-Y in 45,X human subjects lacking a second X chromosome (and thereby lacking the repressor pair). So it is remarkable that residual expression of H-Y has now been reported in 45,X phenotypic females with Turner's syndrome (Wolf et al., 1980b) and in 39,X female mice (Engel et al., 1981). As for the latter report, readers should note that Celada and Welshons (1963) failed to detect H-Y in 39,X female mice despite use of what appears to be a particularly sensitive assay. So the precise location of th H-Y structural gene remains to be ascertained.

H-Y AND THE HETEROGAMETIC OVARY

Because mammalian H-Y antibody is absorbed by female cells in ZW female heterogametic species, such as the chicken, quail, and clawed frog (see Zaborski, 1979), it may be asked whether cross-reactive H-W antigen is the inducer of the heterogametic ovary. Given that gonadal organogenesis of the nonmammalian vertebrates is influenced by sex steroid hormones, the question arises whether hormone-mediated sex reversal of the ZZ primordium is connected with aberrant synthesis of H-W antigen. In fact, H-W antigen is detected in the estradiol-induced ZZ ovotestis of the chicken (Müller et al., 1979); and in the diethylstilbestrol-induced ZZ ovotestis of the quail (Müller et al., 1980). Yet hormone-induced expression of H-W seems to be restricted to the gonad; H-W antigen was not detected in ZZ extragonadal tissues, such as spleen.

Similar observations have been reported for the South African clawed frog (Wachtel et al., 1980a). In that species, ZZ larvae are readily sex reversed by addition of estradiol to their aquarium water. Because they are fertile, sex-reversed ZZ females can be identified by mating them with normal ZZ males. All of the progeny of such matings are (ZZ) males. But these, too, can be sex reversed by exposure to estradiol, and one can thereby generate entire populations of sex reversed ZZ females. Whereas the testes of normal ZZ males are H-W negative, the ovaries of sex-reversed ZZ females are H-W positive. However, as in quails, expression of H-W seems to be restricted to the sex-reversed gonad. The molecule was not detected in the brain or liver of sex-reversed ZZ females, nor could its presence be ascertained in the spleen.

Two other observations favor the view that H-Y (H-W) antigen is involved in organogenesis of the heterogametic ovary:

1.   Soluble H-Y antigen of the mouse testis is taken up specifically in cells of the Xenopus ZZ testis, but not in cells of the ZZ brain, liver, or spleen (Wachtel et al., 1980a).

2.   Cells of the chicken testis organize ovarian structures when cultured in H-Y-positive Daudi cell supernatant (Zenzes et al., 1980).

Why is H-W induced by sex steroid hormones in the nonmammalian vertebrates? It could be argued that the appearance of H-W is a normal prerequisite of the heterogametic ovary and that its induction by hormones is fortuitous and artifactual, or, alternatively, that synthesis of H-W is one of a *sequence* of inductive events normally triggered by hormones in the organogenesis of the heterogametic ovary.

As for the question of how H-Y can induce an XY testis in one species and a ZW ovary in another, we might consider three possibilities:

1.   H-Y and H-W antigens are only small, conservative sequences in larger molecules with alternative inductive functions.

2.   H-Y/H-W antigen induces disparate organogenetic programs in populations of cells that are committed to alternative differentiative pathways (in other words, that the response to H-Y can be different in different cells).

3.   H-Y/H-W antigens of male and female heterogametic species act in precisely the same way, linking Sertoli-cell precursors in the XY gonad and their homologous granulosa-cell precursors in the ZW gonad, and that other factors account for the ultimate divergence of the XY testis and the ZW ovary.

ACKNOWLEDGMENTS

Supported in part by grants from the National Institutes of Health (AI-11982, CA-08748, HD-00171, HD-10065) and NIH Biomedical Research Support Grant 5 S07 RR 05534.

REFERENCES

Bennett, D., E.A. Boyse, M.F. Lyon, B.J. Mathieson, M. Scheid, and K. Yanagisawa. 1975. Expression of H-Y (male) antigen in phenotypically female *Tfm*/Y mice. *Nature* 257:236.
Bennett, D., E.A. Boyse, and L.J. Old. 1972. Cell surface immunogenetics in the study of

morphogenesis. In *Cell Interactions*. L.G. Silvestri, ed. Third Lepetit Colloquium. Amsterdam: North-Holland, p. 247.

Bennett, D., B.J. Mathieson, M. Scheid, K. Yanagisawa, E.A. Boyse, S.S. Wachtel, and B.M. Cattanach. 1977. Serological evidence for H-Y antigen in *Sxr*, XX sex-reversed phenotypic males. *Nature* 265:255.

Bernstein, R., T. Jenkins, T. Dawson, J. Wagner, G. Dewald, G.C. Koo, and S.S. Wachtel. 1980. Female phenotype and multiple abnormalities in siblings with a Y-chromosome and partial X-chromosome duplication: H-Y antigen and Xg blood group findings. *J. Med. Genet.* 17:291.

Beutler, B., Y. Nagai, S. Ohno, G. Klein, and I.M. Shapiro. 1978. The HLA-dependent expression of testis-organizing H-Y antigen by human male cells. *Cell* 13:509.

Billingham, R.E., and W.K. Silvers. 1959. Inbred animals and tissue transplantation immunity. *Transplant. Bull.* 6:399.

Billingham, R.E., and W.K. Silvers. 1960. Studies on tolerance of the Y chromosome antigen in mice. *J. Immunol.* 85:14.

Celada, F., and W.J. Welshons. 1963. An immunogenetic analysis of the male antigen in mice utilizing animals with an exceptional chromosome constitution. *Genetics* 48:139.

Ciccarese, S., and S. Ohno. 1978. Two plasma membrane antigens of testicular Sertoli cells and H-2-restricted versus unrestricted lysis by female T cells. *Cell* 13:643.

Coulombre, J.C., and E.S. Russell. 1954. Analysis of the pleiotropism at the W-locus in the mouse. The effects of W and W^v substitution upon postnatal development of germ cells. *J. Exp. Zool.* 126:277.

de la Chapelle, A. 1981. The etiology of maleness in XX men. *Hum. Genet.* 58:105.

de la Chapelle, A., G.C. Koo, and S.S. Wachtel. 1978. Recessive sex-determining genes in human XX male syndrome. *Cell* 15:837.

Eichwald, E.J., and C.R. Silmser. 1955. Untitled communication. *Transplant. Bull.* 2:148.

Engel, W., B. Klemme, and A. Ebrecht. 1981. Serological evidence for H-Y antigen in XO female mice. *Hum. Genet.* 57:68.

Epstein, J., S. Smith, and B. Travis. 1980. Expression of H-Y antigen on preimplantation mouse embryos. *Tissue Antigens* 15:63.

Fellous, M., E. Gunther, R. Kemler, J. Wiels, R. Berger, J.L. Guenet, H. Jakob, and F. Jacob. 1978. Association of the H-Y male antigen with β_2-microglobulin on human lymphoid and differentiated mouse teratocarcinoma cell lines. *J. Exp. Med.* 147:58.

Fraccaro, M., L. Tiepolo, O. Zuffardi, G. Chiumello, B. Di Natale, L. Gargantini, and U. Wolf. 1979. Familial XX true hermaphroditism and the H-Y antigen. *Hum. Genet.* 48:45.

Goldberg, E.H., E.A. Boyse, D. Bennett, M. Scheid, and E.A. Carswell. 1971. Serological demonstration of H-Y (male) antigen on mouse sperm. *Nature* 232:478.

Gordon, H. 1979. Ancient ideas about sex determination. In *Genetic Mechanisms of Sexual Development*. H.L. Vallet and I.H. Porter, eds. New York: Academic Press, p. 1.

Hall, J.L., Y. Bushkin, and S.S. Wachtel. 1981. Immunoprecipitation of human H-Y antigen. *Hum. Genet.* 58:34.

Hall, J.L., and S.S. Wachtel. 1980. Primary sex determination: genetics and biochemistry. *Molec. Cell. Biochem.* 33:49.

Herbst, E.W., K. Fredga, F. Frank, H. Winking, and A. Gropp. 1978. Cytological identification of two X-chromosome types in the wood lemming (*Myopus schisticolor*). *Chromosoma* 69:185.

Hildemann, W.H., M. Morgan, and L. Frautnick. 1970. Immunogenetic components of weaker histocompatibility systems in mice. *Transplant. Proc.* 2:24.

Jost, A. 1947. Recherches sur la differenciation sexuelle de l'embryon de lapin. III. Role des gonades foetales dans la differenciation somatique. *Arch. Anat. Micr. Morphol. Exper.* 36:271.

Jost, A. 1970. Hormonal factors in the sex differentiation of the mammalian foetus. *Phil. Trans. Roy. Soc. Lond.* 259:119.

Koo, G.C., and C.L. Goldberg. 1978. A simplified technique for H-Y typing. *J. Immunol. Methods* 23:197.

Koo, G.C., L.R. Mittl, and C.L. Goldberg. 1979. Expression of H-Y antigen during spermatogenesis. *Immunogenetics* 9:293.

Koo, G.C., C.W. Stackpole, E.A. Boyse, U. Hämmerling, and M. Lardis. 1973. Topographical location of H-Y antigen on mouse spermatozoa by immunoelectronmicroscopy. *Proc. Nat. Acad. Sci. USA* 70:1502.

Koo, G.C., N. Tada, R. Chaganti, and U. Hämmerling. 1981. Application of monoclonal anti-H-Y antibody for human H-Y typing. *Hum. Genet.* 57:64.

Koo, G.C., S.S. Wachtel, K. Krupen-Brown, L.R. Mittl, W.R. Breg, M. Genel, I.M. Rosenthal, D.S. Borgaonkar, D.A. Miller, R. Tantravahi, R.R. Schreck, B.F. Erlanger, and O.J. Miller. 1977a. Mapping the locus of the H-Y gene on the human Y chromosome. *Science* 198:940.

Koo, G.C., S.S. Wachtel, P.S. Saenger, M.I. New, H. Dosik, A.P. Amarose, E. Dorus, and V. Ventruto. 1977b. H-Y antigen: expression in human subjects with the testicular feminization syndrome. *Science* 196:655.

Krco, C.J., and E.H. Goldberg. 1976. Detection of H-Y (male) antigen on 8-cell mouse embryos. *Science* 193:1134.

McLaren, A. 1976. *Mammalian Chimaeras.* Cambridge: Cambridge University Press.

Moscona, A. 1957. The development *in vitro* of chimeric aggregates of dissociated embryonic chick and mouse cells. *Proc. Nat. Acad. Sci. USA* 43:184.

Moscona, A.A. 1961. Rotation-mediated histogenic aggregation of dissociated cells: a quantifiable approach to cell interactions *in vitro. Exp. Cell. Res.* 22:455.

Müller, U., A. Guichard, M. Reyss-Brion, and D. Scheib. 1980. Induction of H-Y antigen in the gonads of male quail embryos by diethylstilbestrol. *Differentiation* 16:129.

Müller, U., and E. Urban. 1981. Reaggregation of rat gonadal cells *in vitro*: experiments on the function of H-Y antigen. *Cytogenet. Cell. Genet.* In press.

Müller, U., and U. Wolf. 1979a. Discussion note on the paper of S.S. Wachtel on H-Y antigen in the functional female. *Ann. Biol. Anim. Biochem. Biophys.* 19:1239.

Müller, U., and U. Wolf. 1979b. Cross-reactivity to mammalian anti-H-Y antiserum in teleostean fish. *Differentiation* 14:185.

Müller, U., I. Aschmoneit, M.T. Zenzes, and U. Wolf. 1978a. Binding studies of H-Y antigen in rat tissues: indications for a gonad specific receptor. *Hum. Genet.* 43:151.

Müller, U., M.T. Zenzes, T. Bauknecht, U. Wolf, J.W. Siebers, and W. Engel. 1978b. Appearance of hCG-receptor after conversion of newborn ovarian cells into testicular structures by H-Y antigen *in vitro. Hum. Genet.* 45:203.

Nagai, Y., and S. Ohno. 1977. Testis-determining H-Y antigen in XO males of the mole-vole (*Ellobius lutescens*). *Cell* 10:729.

Ohno, S. 1976. Major regulatory genes for mammalian sexual development. *Cell* 7:315.

Ohno, S. 1977. The original function of MHC antigens as the general plasma membrane anchorage site of organogenesis-directing proteins. *Immunol. Rev.* 33:59.

Ohno, S., L.C. Christian, S.S. Wachtel, and G.C. Koo. 1976. Hormone-like role of H-Y antigen in bovine freemartin gonad. *Nature* 261:597.

Ohno, S., S. Ciccarese, Y. Nagai, and S.S. Wachtel. 1978a. H-Y antigen in testes of XX(BALB)XY(C3H) chimaeric male mouse. *Arch. Andrology* 1:103.

Ohno, S., Y. Nagai, S. Ciccarese. 1978b. Testicular cells lysostripped of H-Y antigen organize ovarian follicle-like aggregates. *Cytogenet. Cell Genet.* 20:351.

Ohno, S., Y. Nagai, S. Ciccarese, and R. Smith. 1979. *In vitro* studies of gonadal organogenesis in the presence and absence of H-Y antigen. *In Vitro* 15:11.

Pechan, P., S.S. Wachtel, and R. Reinboth. 1979. H-Y antigen in the teleost. *Differentiation* 14:189.

Rosenfeld, R.G., L. Luzzatti, R.L. Hintz, O.J. Miller, G.C. Koo, and S.S. Wachtel. 1979. Sexual and somatic determinants of the human Y chromosome: studies in a 46,XYp⁻ phenotypic female. *Amer. J. Hum. Genet.* 31:458.

Saenger, P., L.S. Levine, S.S. Wachtel, S. Korth-Schutz, Y. Doberne, G.C. Koo, R.W. Lavengood, J.L. German, and M.I. New. 1976. Presence of H-Y antigen and testis in 46,XX true hermaphroditism. Evidence for Y-chromosomal function. *J. Clin. Endocrinol. Metab.* 43:1234.

Scheid, M., E.A. Boyse, E.A. Carswell, and L.J. Old. 1972. Serologically demonstrable alloantigens of mouse epidermal cells. *J. Exp. Med.* 135:938.

Selden, J.R., S.S. Wachtel, G.C. Koo, M.E. Haskins, and D.F. Patterson. 1978. Genetic basis of

XX male syndrome and XX true hermaphroditism: evidence in the dog. *Science* 201:644.

Shalev, A., R.V. Short, and J.L. Hamerton. 1980. Immunogenetics of sex determination in the polled goat. *Cytogenet. Cell Genet.* 28:195.

Shapiro, M., and R.P. Erickson. 1981. Evidence that the serologic determinant of H-Y (histocompatibility-Y) antigen is carbohydrate. *Nature* 290:503.

Short, R.V., J. Smith, T. Mann, E.P. Evans, J. Hallett, A. Fryer, and J.L. Hamerton. 1969. Cytogenetic and endocrine studies of a freemartin heifer and its bull co-twin. *Cytogenetics* 8:369.

Silvers, W.K., and S.-L. Yang. 1973. Male-specific antigen: its homology in mice and rats. *Science* 181:570.

Urban, E., M.T. Zenzes, U. Müller, and U. Wolf. 1981. Cell reorganization *in vitro* of heterosexual gonadal cocultures. *Differentiation* 18:161.

Van Niekerk, W.A., and A.E. Retief. 1981. The gonads of human true hermaphrodites. *Hum. Genet.* In press.

Vigier, B., A. Locatelli, J. Prepin, F. du Mesnil du Buisson, and A. Jost. 1976. Les premieres manifestations du "freemartinisme" chez le foetus de veau ne dependent pas du chimerisme chromosomique XX/XY. *C.R. Acad. Sci. (Paris)* 282:1355.

Vigier, B., J.-Y. Picard, J. Bezard, and N. Josso. 1981. Anti-mullerian hormone: a local or long-distance morphogenetic factor? *Hum. Genet.* 58:85.

Wachtel, S.S. 1977. H-Y antigen: genetics and serology. *Immunol. Rev.* 33:33.

Wachtel, S.S., P.A. Bresler, and S.S. Koide. 1980a. Does H-Y antigen induce the heterogametic ovary? *Cell* 20:859.

Wachtel, S.S., J.L. Hall, U. Müller, and R.S.K. Chaganti. 1980b. Serum-borne H-Y antigen in the fetal bovine freemartin. *Cell* 21:917.

Wachtel, S.S., G.C. Koo, and E.A. Boyse. 1975. Evolutionary conservation of H-Y ("male") antigen. *Nature* 254:270.

Wilson, H.V. 1907. On some phenomena of coalescence and regeneration in sponges. *J. Exp. Zool.* 5:245.

Winters, S.J., S.S. Wachtel, B.J. White, G.C. Koo, N. Javadpour, L. Loriaux, and R.J. Sherins. 1979. H-Y antigen mosaicism in the gonad of a 46,XX true hermaphrodite. *New Engl. J. Med.* 300:745.

Wolf, U. 1979. XY gonadal dysgenesis and the H-Y antigen. *Hum. Genet.* 47:269.

Wolf, U., M. Fraccaro, A. Mayerova-, T. Hecht, P. Maraschio, and H. Hameister. 1980a. A gene controlling H-Y antigen on the X chromosome. *Hum. Genet.* 54:149.

Wolf, U., M. Fraccaro, A. Mayerova-, T. Hecht, O. Zuffardi, and H. Hameister. 1980b. Turner syndrome patients are H-Y positive. *Hum. Genet.* 54:315.

Zaborski, P. 1979. Sur la constance de l'expression de l'antigene H-Y chez le sexe heterogametique de quelques Amphibiens et sur la mise en evidence d'un dimorphisme sexuel de l'expression de cet antigene chez l'Amphibien Anoure *Pelodytes punctatus* D. *C.R. Acad. Sci.* Paris 289:1153.

Zaborski, P., M. Dorizzi, and C. Pieau. 1979. Sur l'utilisation de serum anti-H-Y de Souris pour la determination du sexe genetique chez *Emys orbicularis*. L. (Testudines, Emydidae). *C.R. Acad. Sci.* Paris Ser. D. 288:351.

Zenzes, M.T., U. Müller, I. Aschmoneit, and U. Wolf. 1978a. Studies on H-Y antigen in different cell fractions of the testis during pubescence. Immature germ cells are H-Y antigen negative. *Hum. Genet.* 45:297.

Zenzes, M.T., U. Wolf, and W. Engel. 1978b. Organization *in vitro* of ovarian cells into testicular structures. *Hum. Genet.* 44:333.

Zenzes, M.T., U. Wolf, E. Gunther, and W. Engel. 1978c. Studies on the function of H-Y antigen: dissociation and reorganization experiments on rat gonadal tissue. *Cytogenet. Cell. Genet.* 20:365.

Zenzes, M.T., E. Urban, and U. Wolf. 1980. Mammalian cross-reactive H-Y antigen induces sex reversal *in vitro* in the avian testis. *Differentiation* 17:121.

Chapter 6

IMMUNOLOGICAL AND BIOCHEMICAL
CHARACTERISTICS OF THE TerC CELL LINES

MICHAEL EDIDIN
DAVID B. SEARLS

Teratocarcinomas are tumors composed of undifferentiated stem cells, termed embryonal carcinoma, intermixed with a haphazard array of differentiated tissues foreign to the site of the tumor (Stevens and Pierce, 1975). In the past decade, there has been a surge of interest in teratocarcinomas as models of early embryogenesis (Jacob, 1977). Observations on the differentiation of teratocarcinomas *in vivo* have supported a view of embryonal carcinoma cells as totipotent stem cells of embryonic character with a potential for normal differentiation. The study of embryonal carcinoma cells and their derivatives holds promise as a model for the biochemical, immunological, and genetic characterization of early embryogenesis.

Under appropriate culture conditions, embryonal carcinoma cells have been observed to differentiate to a number of tissue types, particularly nerve and muscle (Levine et al., 1974); however, the process that has been studied most intensively is the differentiation to endoderm. This occurs very early in embryogenesis and, similarly, is generally the first event in the *in vitro* differentiation of embryonal carcinoma lines (Martin, 1975). The recent discovery that low concentrations of retinoic acid can induce endodermal differentiation, even in embryonal carcinoma lines previously thought to be nullipotent (i.e., unable to differentiate *in vitro* or *in vivo*), has created a useful tool for the study of this process (Strickland and Mahdavi, 1978). While differentiated derivatives are generally not tumorigenic and cannot be maintained in culture indefinitely, teratocarcinoma-derived tumor lines of endodermal character are available as "standards" for comparison; in particular, a parietal yolk sac carcinoma, PYS-2 (Lehman et al., 1974), is often used as a companion to F9 (Bernstine et al., 1973)—a "nullipotent" embryonal carcinoma derived from the same teratocarcinoma, OTT 6050. The characterization of teratocarcinoma stem cells and their differentiated derivatives has been both biochemical and immunological. The latter approach in particular has been the basis for attempts to detect the surface features of primitive and early differentiated cells, since these features are involved in the positioning of these cells in the embryo by interaction with one another. The range of antisera which detect antigens of teratocarcinomas and normal embryos is reviewed by Solter and Knowles (1979) and Wiley (1979).

TerC, a cell line derived from teratocarcinoma 402AX, has been used extensively in immunological studies of surface antigens in our laboratory (Edidin, 1975). It was the first such cell type to be used for this purpose (Edidin et al., 1971) and is one of the few lines in current use to have been derived from a spontaneous testicular teratocarcinoma (Stevens, 1958) rather than by grafting. However, as has been noted by several reviewers (Hogan, 1977; Gachelin, 1978), TerC has not been well characterized in terms of

129

known biochemical and cell surface markers which indicate the extent and type of differentiation of teratocarcinoma lines *in vivo*. In this chapter, we describe the morphology, karyotype, and some biochemical antigenic markers of TerC and other 402-derived cell lines. The TerC phenotype is closer to that of cultured parietal endoderm and to the endoderm-like cell line PYS-2 than it is to the embryonal carcinoma cell lines F9 and nulli-SCC-1. However, the TerC phenotype differs significantly from parietal endoderm and PYS-2. It does resemble the phenotype of cells induced by retinoic acid to differentiate from embryonal carcinoma (Strickland et al., 1980). This cell differentiates further to endoderm when treated with dibutryl cyclic AMP. Strickland has speculated that the cells induced from F9 by retinoic acid treatment are a model of primitive endoderm which *in vivo* gives rise to both parietal and visceral endoderm. We believe that TerC represents a similar primitive endoderm cell.

HISTORY OF TerC

The history of the 402-teratocarcinoma-derived cell lines used in our laboratory is depicted in Figure 6-1. The original tumor arose spontaneously in a 60-day-old 129/Sv male and gave rise to two transplantable sublines, 402A and 402C (Stevens, 1956). A further subline of the latter, 402C-1684, consisted exclusively of embryonal carcinoma, which was adapted to culture as a nullipotent line, Nulli-SCC-1 (Martin and Evans, 1975). Subline 402AX

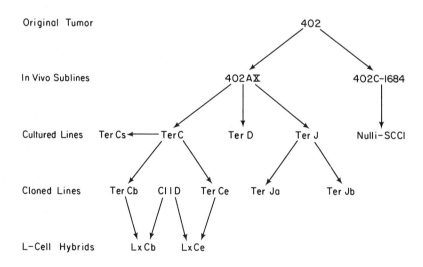

FIGURE 6-1. Origin of TerC and its clones.

gave rise to a series of cultured lines, among them TerC and TerD; the former line was immediately selected for resistance to 8-azaguanine (Gooding and Edidin, 1974). TerC was cloned on feeder layers with an efficiency exceeding 90 per cent; two of the resulting clones, TerCb and TerCe, were chosen for further study. The original TerC line was also selected for growth in suspension by repeated passage of nonadherent cells, resulting in the TerC's line (Bartlett et al., 1978). Recently, ascites-grown 402AX (from B. Pierce) was again adapted to culture as line TerJ, from which two clones were derived (TerJa and TerJb).

PHENOTYPE OF TerC AND OTHER TERATOCARCINOMA-DERIVED CELL LINES

TerC MORPHOLOGY AND KARYOLOGY

Figure 6-2 shows morphologies of selected clones of TerC and TerJ and, for comparison, PYS-2. TerJ, shown soon after cloning, forms tight epithelioid monolayers, much like PYS-2. TerJb and uncloned TerJ were very similar in appearance to TerJa. Early versions of TerC and its clones were also very flat but, over a period of time in culture, tended to become rounded and less adherent. These rounded cells come to predominate, as shown in the photograph of TerCb (about one year after cloning). TerC also demonstrated a tendency to form large cytoplasmic vesicles. Repeated subculture of nonadherent TerC gave rise to the suspension cell line, TerCs.

TABLE 6-1
Chromosome Counts for 402AX-Derived Cell Lines*

CELL TYPE	MODE	MEAN ± S.D.	METACENTRICS	PER CENT > 55	NUMBER COUNTED
TerC	40	40.2 ± 1.7	2.8 ± 1.0	9	(35)
TerCb	40	39.9 ± 1.7	3.3 ± 0.9	–	(8)
TerCe	40	39.1 ± 1.6	2.9 ± 0.3	–	(9)
TerCs	36–38	35.3 ± 3.6	5.1 ± 1.4	0	(24)
TerJ	38	39.8 ± 9.8	2.9 ± 0.9	15	(40)
TerJa	38–40	39.0 ± 5.8	2.1 ± 0.8	14	(21)
TerJb	33–37	35.2 ± 1.6	5.5 ± 1.4	8	(18)

* Metacentrics are included, but counts greater than 55 are excluded, in the calculation of means and standard deviations of chromosome numbers. Chromosome preparations and Giemsa staining were performed as described by Hsu (1973).

131

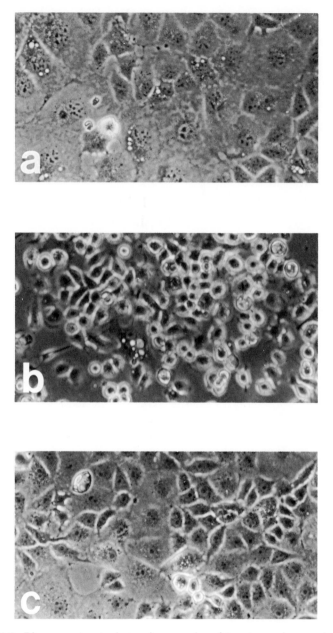

FIGURE 6-2. Phase contrast photomicrographs of teratocarcinoma-derived cell lines.
(a) PHS-2; (b) late-passsage TerCb; and (c) TerJa. Final magnifications were approximately 120x.

Chromosome counts of 402AX-derived lines are summarized in Table 6-1. TerC and its clones demonstrated a relatively consistent diploid number, whereas the suspension-selected line, TerCs, was subdiploid. TerJ and its clones were also subdiploid. Several metacentric chromosomes were observed in all lines and somewhat greater numbers were found in TerCs and TerJb. Most lines had a significant number of spreads with greater than 55 chromosomes; while it was difficult to obtain accurate counts of these populations, they were consistently subtetraploid and there was no suggestion of extensive endoreduplication.

TerC CELL SURFACE MARKERS

SSEA-1, a glycolipid-stage-specific embryonic antigen expressed on embryonal carcinoma cells but absent from their differentiated derivatives (Solter and Knowles, 1978; Nudelman et al., 1980), is defined by a monoclonal antibody which was given to us (in the form of ascites fluid from a hybridoma-bearing mouse) by D. Solter. Table 6-2 shows the results of indirect immunofluorescence using anti-SSEA-1. As expected, a large percentage of F9 and nulli-SSC-1 cells stained, and this percentage declined after retinoic acid-induced differentiation (Solter et al., 1979). The parietal yolk sac carcinoma PYS-2 was completely negative. Uncloned populations of TerC and TerJ contained small amounts of positive cells, raising the possibility of a

TABLE 6-2
Percentage of Cells Staining with Anti-SSEA-1

CELL TYPE	NUMBER COUNTED	PERCENT POSITIVE
F9	416	82.7
Nulli	137	85.4
F9 + RA*	150	17.3
Nulli + RA	102	24.5
PYS-S	356	0.0
TerC	358	2.8
TerCb	303	0.7
TerCc	385	0.3
TerJ	311	5.1
TerJa	389	0.8
TerJb	263	0.4

* Indicates treatment with retinoic acid, after Strickland and Mahdavi (1978).
Cells were stained in suspension with anti-SSEA-1 ascites fluid at 1/100 followed by a fluorescein-conjugated goat anti-mouse Ig.

small subpopulation of embyonal carcinoma in these lines. Indeed, less than 1 per cent of the cells stained after cloning; however, this proportion was never reduced to zero. These small but consistent subpopulations stained brightly, and the high cloning efficiency of the TerC line makes it extremely unlikely that these clones were each derived from the minority of positive cells in the parental lines.

Peanut agglutinin (PNA), a D-galactosyl-specific lectin (Lotan et al., 1975), has been shown to bind to embryonal carcinoma lines but not to some of their differentiated derivatives (Reisner et al., 1977). As shown in Table 6-3, fluoresceinated peanut agglutinin binds to a high percentage of both F9 and PYS-2 cells, suggesting that this marker may not be useful in distinguishing embryonal carcinoma from parietal endodermal lines *in vitro*. While F9 cells stained with somewhat greater intensity, there was no concentration of lectin at which PYS-2 was conclusively negative while F9b remained positive. Large proportions of TerC and TerJ clones also bound peanut agglutinin, as shown. In every case, peanut agglutinin binding was prevented by the addition of D-galactose. While these data were taken with harvested cells in suspension, approximately the same percentages were seen with cells stained on coverslips, indicating that harvesting procedures were not exposing new galactose residues.

H-2 ANTIGENS OF TerC

Earlier, we reported that a few per cent of TerC were H-2 positive by immunofluorescence and that large inocula of TerC slightly accelerated H-2^b skin graft rejection on congenic H-2^k mice (Edidin et al., 1974). However, the

TABLE 6-3
Percentage of Cells Staining with FITC-Peanut Agglutinin

CELL TYPE	PNA*	+ D-gal†
F-9	82	3
PYS-2	81	4
TerCb	53	3
TerCe	41	2
TerJa	73	0
TerJb	82	3

* Peanut agglutinin was the gift of Y.C. Lee. It was conjugated with fluorescein by the method of Clark and Shepard (1963). The conjugate was applied to cells at 200 μg/ml.

† D-galactose added at 100 mM.

cells are resistant to lysis by anti-H-2 antibody and complement, and TerC is able to produce tumors in a variety of allogeneic mice (Ostrand-Rosenberg et al., 1980). Also, another group did not detect H-2 antigens on endodermal derivatives of a pluripotent embryonal carcinoma (Stern et al., 1975). However, upon repeating assays for H-2 using both absorption and lytic techniques, we now found that very low levels of H-2K and H-2D molecules were indeed present on teratocarcinoma-derived primitive endodermal cell lines.

We assayed for H-2 antigens by a modification of the method of Garrido et al. (1976). Cells were exposed to antiserum plus selected rabbit complement, and cell damage was assayed in terms of uptake of ^{125}IUdR added after the incubation in antibody and complement. High-titered polyspecific anti-H-2^b serum at 1/20 dilution damaged TerC cells but did not damage F9 (Table 6-4). The sensitivity of TerC clones varied. The extremes, TerCb and TerCe, are shown in Table 6-4. PYS-2 cells were even more sensitive to anti-H-2^b sera than were TerC. Comparison of absorption of antisera by suspensions of TerC with absorption by lymphocytes, corrected for surface area of the absorbing cells, indicated that TerCb and TerCe were about 1/100 as rich in H-2^b as lymphocytes. Based on some studies of H-2 sites on lymphocytes (T. Spack, unpublished), we estimate that TerC bears about ten H-2 antigens per μM^2 of surface area. This calculation assumes, of course, that the H-2 activity of the population is not due to a small proportion of differentiated cells. We believe that our assumption is correct, since the lines tested had been clonally derived from a TerC stock. The value of about 1/100 the H-2 antigen density of a lymphocyte is close to that estimated in our earlier study (Edidin et al., 1974).

SURFACE ANTIGENS DEFINED BY TerC

We originally prepared a xenoantiserum against 402AX which contained

TABLE 6-4
Detection of H-2^b Antigens on TerC and Lines Cloned from It

DILUTION OF C3H anti-C3H.Sw ANTI-H-2^b ANTISERUM	PER CENT INHIBITION OF IUdR UPTAKE BY ANTISERUM					
	EL-4 (H-2^b)	LM (H-2^k)	TerCb	TerCe	F9	PYS-2
1/40	100 ± 1	−19 ± 15	64 ± 17	19 ± 8	9 ± 16	80 ± 10
1/200	100 ± 1	1 ± 14	67 ± 15	16 ± 10	13 ± 14	49 ± 7
1/1 000	100 ± 1	–	35 ± 8	11 ± 7	7 ± 13	22 ± 18
1/5 000	100 ± 3	–	23 ± 8	−1 ± 11	−3 ± 20	20 ± 8
1/25 000	77 ± 3	–	–	–	0 ± 17	15 ± 35

* Mean ± standard deviation. The standard for cells was cells plus culture medium plus complement.

multiple specificities defining embryonic and tumor antigens (Gooding and Edidin, 1974), and subsequent work has elucidated the chemical nature of some TerC-defined antigens (Gooding, 1976; Larraga and Edidin, 1979) and investigated cellular immune reactions against them (Bartlett et al., 1978).

The original complex xenoantisera reacted with F9 (embryonal carcinoma) cells, as well as with PYS-2 and a number of tumor cell lines (Edidin et al., 1971; Gooding and Edidin, 1974). In an attempt to further refine our analysis of the antigens of TerC, we prepared a syngeneic antiserum, 129/J anti-TerC. This serum gave a pattern of reactivity similar to that of the xenoantiserum. When tested by immunofluorescence against a panel of teratocarcinoma and other cell lines, it defined a group of four lines that reacted very well with the serum. Nearly all the cells of TerC (and of its sublines and clones), as well as F9, PYS-2, and MB2 (a blastocyst-derived line developed by M. Sherman, 1977), stained strongly with the antiserum. A second group, lines cl, ld, and Sa-I, were generally nearly 100 per cent positive with the antiserum, but stained weakly. This pattern held up when the antiserum was heat inactivated and absorbed to remove reactivity with fetal bovine serum components. Some cross-reactions of the syngeneic antiserum are indicated by the series of absorptions and reactions given in Table 6-5. The syngeneic antiserum, like the xenogenic, reacts with zygotes and cleavage-stage embryos. Reaction is abolished by absorption with TerC, but not by absorption with cl or ld.

BIOCHEMICAL MARKERS OF TerC DIFFERENTIATION

A number of biochemical markers of embryonal carcinoma cells and yolk sac cells have been described. We had previously looked at just one of these

TABLE 6-5
Cross Reactions of a Syngeneic Anti-TerC
Serum with Various Cell Lines Detected by
Immunofluorescence

	PER CENT REACTIVE TARGET CELLS			
ABSORBING CELLS	TerCs	F9	PYS-2	cl.ld
*	100	90	98	52
TerCs	0	0	0	0
F9	0	0	0	0
PYS-2	0	4	0	37
cl.ld	100	72	90	0

Anti-TerC antiserum used at 1/10 dilution.

markers, production of alpha-fetoprotein (AFP), which is a characteristic of parietal endoderm, and found TerC unreactive with antimouse AFP. Here we describe serological characterization of basement membrane production, plasminogen activator activities, alkaline phosphatase levels, and creatine phosphokinase levels in TerC clones and in authentic embryonal carcinoma and parietal endoderm lines.

An antiserum against basement membrane was obtained from B. Pierce via D. Solter. This rabbit antiserum (designated "anti-Peak I") detects Reichert's membrane (secreted by parietal endoderm *in vivo*) (Pierce et al., 1962), which is similar to a material deposited by parietal yolk sac carcinoma lines (Clark et al., 1975) or by embryonal carcinoma lines differentiating to endoderm (Solter et al., 1979) *in vitro*. The antiserum was diluted to 1/10, absorbed with an equal volume of 3T3 cells to remove any antispecies reactivity, and then tested by immunofluorescence on cells grown for three days on coverslips. The second-layer antibody was FITC-goat-antirabbit Fc.

Extracellular deposits of brightly staining material were seen with PYS-2, TerJ, and, to a lesser extent, TerC; F9 was largely negative. Despite extensive absorption with 3T3 cells, embryonal carcinomas frequently showed faint ring staining, which was, however, easily distinguishable from the extracellular material.

Plasminogen activator, which is associated with both parietal endoderm and trophectoderm during *in vitro* blastocyst outgrowth (Strickland et al., 1976) and with the appearance of parietal endoderm on embryoid bodies (Sherman et al., 1976), was assayed by the method of Searls (1980). This assay uses an artificial chromogenic substrate (S-2251, Ortho Diagostics) which is cleaved by plasmin to yield p-nitroaniline. Evolution of this product in the second step of the reaction was, in every case, dependent upon the addition of the zymogen plasminogen in the first step.

Figure 6-3 shows plasminogen activator activities associated with a variety of cell lines and midgestation tissues. As expected, activities are high in trophoblast and parietal endoderm at 11 days and low in yolk sac (which includes visceral endoderm and a layer of mesoderm). Intermediate levels are found in the embryo proper.

Plasminogen activator activity was barely detectable in embryonal carcinoma lines, but, as was first shown by Strickland and Mahdavi (1978), retinoic acid treatment resulted in a sharp increase in its production, though to levels about half those of parietal endoderm. Activities were intermediate for TerCb and somewhat lower still for TerCe. TerJb was similar to PYS-2 and appreciably more active than TerJa; the latter clone, however, demonstrated a retinoic acid inducibility reminiscent of embryonal carcinoma.

Alkaline phosphatase, which has long been associated with embryonal carcinoma in teratocarcinoma systems (Damjanov et al., 1971) (but see discussion section, below), was assayed as described by Bernstine et al. (1973), using

137

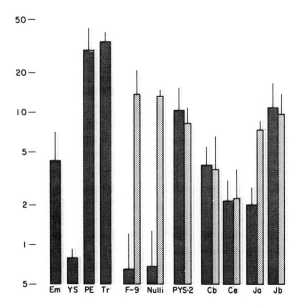

FIGURE 6-3. Plasminogen activator activities for midgestation embyonic tissues and for teratocarcinoma-derived cell lines, with or without retinoic acid treatment.

The ordinate is a logarithmic scale in units of μmoles product per hour per mg protein, as described previously (Searls, 1980). Tissues indicated are embryo proper (Em), yolk sac (YS), parietal endoderm (PE), and trophoblast (Tr). All are dissected from 11- to 12-day-old embryos, strain CD-1. Activities for embryonal carcinoma lines, PYS-2, and clones of TerC and TerJ are shown for both normal culture conditions (heavy shading) and after retinoic acid treatment (light shading).

the artificial substrate p-nitrophenyl phosphate (Sigma 104). The enzyme was demonstrated histochemically using napthol ASMX phosphate and fast blue BB salt (Sigma), as described in Barka and Anderson (1963), using cells grown on coverslips.

Alkaline phosphatase activities are shown in Figure 6-4. Considerable activities are found in parietal endoderm, and especially trophoblast; these results are in general agreement with those of Sherman and Atienza-Samols (1979). As expected, embryonal carcinoma lines showed high activity; these activities did not change appreciably upon retinoic acid-induced differentiation. Again, this result supports similar observations by Strickland and Mahdavi (1978) and by Strickland et al. (1980), and, taken together, these findings suggest that a loss of alkaline phosphatase activity may not necessarily accompany the first steps of differentiation to parietal endoderm. However, PYS-2 and clones of TerC and TerJ all show activities at least ten-fold lower than midgestation parietal endoderm. In the case of TerJ, treatment with retinoic acid increased activities markedly.

It is possible that long-term culture has resulted in a loss of alkaline phosphatase activity in these cell lines. Early uncloned TerC demonstrated considerably higher activity, and histochemical analysis revealed occasional nests of darkly staining cells. These were of approximately the same frequency as SSEA-1 reactive cells and probably could not themselves account for the overall higher activity. TerD, an early-passage line derived from 402AX at about the same time as TerC, was similar to parietal endoderm in activity yet was not perceptibly different from early TerC in morphology (data not shown).

Creatine phosphokinase, previously used to indicate the appearance of nerve and muscle in long-term *in vitro* differentiation of pluripotent embyonal carcinomas (Levine et al., 1974), was assayed by the method of Hughes (1962), according to Sigma Technical Bulletin No. 520. Values obtained were substrate dependent and were further confirmed by the retesting of sample points using an assay with an entirely independent mechanism (Sigma Technical Bulletin No. 45-UV).

Creatine phosphokinase activities are shown in Figure 6-5. Activity was relatively high in whole midgestation embryo; this is to be expected, in

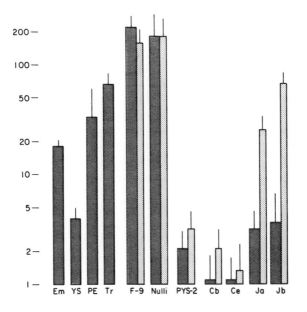

FIGURE 6-4. Alkaline phosphatase activities for embryonic tissues and teratocarcinoma-derived cell lines.
For details, see Figure 6-3. Units are μmoles per minute per mg protein.

139

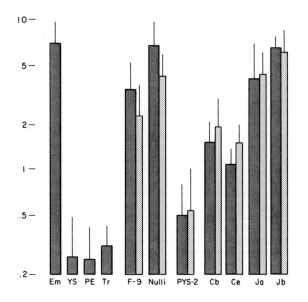

FIGURE 6-5. Creatine phosphokinase activities for embryonic tissues and teratocarcinoma-derived cell lines.
For details, see Figure 6-3. Units are μmoles per 30 minutes per mg protein.

view of the advanced development of heart and brain at this stage. Activity in all extraembryonic membranes was at the threshold of detection. The embryonal carcinoma line, however, showed appreciable activity (though these levels were still at least an order of magnitude lower than those for adult brain). Retinoic acid had slight, if any, effect. PSY-2 had 6- to 12-fold lower activity; the TerJ clones were about the same as those for embryonal carcinoma, and TerC clones were intermediate. (Again, early TerC had somewhat higher activity than long-term or cloned TerC [data not shown].)

DISCUSSION

In contrast to other types of tumors and established cell lines, embryonal carcinomas tend to remain nearly diploid even after long-term culture (see Solter and Damjanov, 1979, for review). Nevertheless, banding techniques generally reveal karyological abnormalities, and the appearance of metacentric chromosomes seems to be common (Hooper and Slack, 1977).

On the other hand, permanent differentiated lines derived from teratocarcinomas (such as PYS-2) are often aneuploid (Lehman et al., 1974). It is thus interesting that the 402AX-derived lines described here have retained a near-diploid chromosome number after several years in culture, though the appearance of metacentric chromosomes suggests that some translocation has occurred.

The secretion of basement membrane and the intermediate-to-high levels of native or retinoic acid-inducible plasminogen activator observed with 402AX-derived lines would seem to be good evidence for some parietal endodermal character to these cells. (Although plasminogen activator is also associated with trophoblast, these teratocarcinoma-derived lines do not show evidence of endoreduplication or any ability to synthesize progesterone [Searls and Edidin, unpublished]—another characteristic of trophoblast [Salomon and Sherman, 1975].)

Midgestation parietal endoderm expresses appreciably higher levels of plasminogen activator and of alkaline phosphatase activity than is found in PYS-2 or clones of TerC and TerJ. While early histochemical studies suggested that high levels of alkaline phosphatase were uniquely characteristic of the undifferentiated core of embryoid bodies (Bernstine et al., 1973; Martin and Evans, 1975), activity was subsequently detected in the outer endodermal layer, when different procedures (Wada et al., 1976) were used. As was noted before, it is possible that a loss of alkaline phosphatase activity is associated with the long-term culture.

TerC does not produce alpha-fetoprotein (L. Gooding, personal communication), a marker of visceral endoderm (Solter and Damjanov, 1979). Sherman and co-workers have found high ratios of N-acetyl-β, D-hexosaminidase activity to β-glucuronidase activity in midgestation visceral endoderm, relative to parietal endoderm (Sherman and Atienza-Samols, 1979). Early TerC and paticularly TerJ showed high ratios as well, which declined with time in culture; however, similarly high ratios were found in nullipotent embryonal carcinomas (which decreased somewhat upon retinoic acid treatment) (unpublished results). Thus, this marker is probably not useful in distinguishing visceral endoderm derived from embryonal carcinoma.

It is interesting that relatively high levels of creatine phosphokinase, another unexpected characteristic of embryonal carcinoma, were also observed in TerC and TerJ clones, but not in PYS-2. Creative phosphokinase and acetylcholinesterase were used by Levine et al. (1974) as indicators of the appearance of nerve and muscle in long-term *in vitro* differentiation of pluripotent embryonal carcinomas. When they transferred embryoid bodies from ascites to culture, acetylcholinesterase was undetectable during the first week but plasminogen activator activity increased sharply; subsequently, the former rose steadily while the latter declined. Creatine phosphokinase followed the same pattern as did acetylcholinesterase, except for a small peak of activity during the first day in culture, which then declined to baseline for the remainder of the first week. Levine et al. suggested that this transient peak was due to enzyme adsorbed to the cells from ascites fluid, since embryoid bodies cultured in suspension (to prevent differentiation) declined to, and remained at, baseline levels of activity. However, Adamson (1976), in a study of electrophoretic isoenzyme transitions, apparently found sufficient creatine phosphokinase activity in an undifferentiated teratocarcinoma line to produce a relatively intense activity stain (which corresponded to the brain [BB] form of enzyme).

A comparison of surface antigens of F9, TerC, and PYS-2 indicates similarities between TerC and both of the reference lines. SSEA-1, defining embryonal carcinoma cells, is detected on only a small subpopulation of TerC, but this population is not removed by cloning. The persistence of the subpopulation is especially interesting in light of the reported heterogeneous SSEA-1 staining of inner cell masses and of early endoderm cells in preimplantation blastocysts (Solter and Knowles, 1979). Our syngeneic anti-TerC, like previous xenogeneic antisera, reacts with F9 as well as with PYS-2, and activity is completely absorbed by any one of these cell types. Thus, these surface antigens, while they distinguish teratocarcinoma-derived cells and early embryos from normal tissues and some other cultured cell lines, do not appear to distinguish embryonal carcinoma cells from their differentiated descendents.

H-2 antigens are detectable by absorption and in a complement-mediated assay on TerC and on PYS-2. Although PYS-2 is more sensitive to anti-H-2 and complement than is TerC, it is not clear if this sensitivity reflects higher antigen densities. The membrane of PYS-2 may be more resistant to complement-mediated lysis than the membranes of TerC. However, the IUdR-uptake assay appears generally to reflect levels of surface antigens determined by absorption. Thus TerCb, which was more effectively damaged by anti-H-2^b and complement than was TerCe, also absorbed more anti-H-2 than did TerCe.

TerC shares its biochemical and serological phenotype both with embryonal carcinoma cells and with parietal endoderm and the parietal endoderm-like line, PYS-2. Recently, Strickland et al. (1980) showed that treatment of the embryonal carcinoma cell F9 with retinoic acid, followed by treatment with dibutyryl cyclic AMP, produced cells with biochemical phenotypes close to those of parietal endoderm. Treatment with retinoic acid alone gave a phenotype that was closer to parietal endoderm than to embryonal carcinoma, but was significantly different from either. Strickland (1981) has suggested that the cells differentiating from retinoic acid-treated F9 represent a primitive endoderm, capable of further differentiation to both parietal and visceral endoderm. This suggestion reinforces similar suggestions based on the analysis of the fate of normal embryos *in vivo*.

We believe that TerC cells represent a primitive endoderm, sharing some, but not all, characteristics of both parietal and visceral endoderm, and even of embryonal carcinoma. Our data and those cited at least suggest that the phenotypic changes accompanying the differentiation of embryonal carcinoma are not the result of a single coordinate "switch" event. While the failure of retinoic acid to repress creatine phosphokinase activity in embryonal carcinoma is not itself significant in the absence of turnover studies, the persistence of activity in TerC and TerJ suggests that this change may be a late event in differentiation.

It is also possible that the heterogeneity of markers is due to a mixture of cell types. Yolk sac carcinomas of human and rat origin apparently contain both parietal and visceral endoderm (reviewed in Solter and Dunjanov, 1979). In view of the clonal origin of the lines studied here, such a heterogeneity would seem to imply the existence of an endoderm stem cell which, again, is morphologically and antigenically distinct from embryonal carcinoma as well as from differentiated endoderm.

ACKNOWLEDGMENTS

Supported by Grant AI 14584 to Michael Edidin. This is part of the Ph.D. thesis work of David Searls, Contribution No. 1109 from the Department of Biology, Johns Hopkins University.

REFERENCES

Adamson, E.D. 1976. Isoenzyme transition of creatine phosphokinase, aldolase and phosphokinase, aldolase and phosphoglycerate mutase in differentiating mouse cells. *J. Embryol. Exp. Morphol.* 35:355.

Barka, T., and P.J. Anderson. 1963. *Histochemistry.* New York: Harper and Row, p. 232.

Bartlett, P.F., B.A. Fenderson, and M. Edidin. 1978. Inhibition of tumor growth mediated by lymphocytes sensitized *in vitro* to a syngeneic murine teratocarcinoma, 402AX. *J. Immunol.* 120:1211.

Bernstine, E.G., M.L. Hooper, S. Grandchamp, and B. Ephrussi. 1973. Alkaline phosphatase activity in mouse teratoma. *Proc. Nat. Acad. Sci. USA* 70:3899.

Clark, H.F., and C.C. Shepard. 1963. A dialysis technique for preparing fluorescent antibody. *Virology* 20:642.

Clark, C.C., E.A. Tomichek, T.R. Koszalka, R.R. Minor, and N.A. Kefalides. 1975. The embryonic rat parietal yolk sac. *J. Biol. Chem.* 250:5259.

Damjanov, I., D. Solter, and N. Skreb. 1971. Enzyme histochemistry of experimental embryo-derived teratocarcinoma. *Z. Krebsforsch.* 76:249.

Edidin, M. 1976. The appearance of cell-surface antigens in the development of the mouse embryo: study of cell-surface differentiation. In *Embryogenesis in Mammals*. Ciba Foundation Symposium, Vol. 40, pp. 177-197.

Edidin, M., and L.R. Gooding. 1975. Teratoma-defined and transplantation antigens in early mouse embryos. In *Teratomas and Differentiation*. M.I. Sherman and D. Solter, eds. New York: Academic Press, pp. 109-121.

Edidin, M., L.R. Gooding, and M.H. Johnson. 1974. Surface antigens of normal early embryos and a tumour model system useful for their further study. In *Immunological Approaches to Fertility*. E. Diczfalusy, ed. Stockholm: Karolinska Institutet, pp. 336-356.

Edidin, M., H.L. Patthey, E.J. McGuire, and W.D. Sheffield. 1971. An antiserum to "embryoid body" tumor cells that reacts with normal mouse embryos. In *Embryonic and Fetal Antigens in Cancer*. N.G. Anderson and J.H. Coggin Jr., eds. Oak Ridge, Tennessee: Oak Ridge National Laboratory, pp. 239-248.

Gachelin, G. 1978. The cell surface antigens of mouse embryonal carcinoma cells. *Biochem. Biophys. Acta* 516:27.

Garrido, F., V. Schirrmacher, and H. Festenstein. 1976. H-2-like specificities of foreign haplotypes on a mouse sarcoma after vaccinia virus infection. *Nature* 259:228.

Gooding, L.R. 1976. Expression of early fetal antigens on transformed mouse cells. *Cancer Res.* 36:3499.

Gooding, L.R., and M. Edidin. 1974. Cell surface antigens of a mouse testicular teratoma. Identification of an antigen physically associated with H-2 antigens on tumor cells. *J. Exp. Med.* 140:61.

Hogan, B.L.M. 1977. Teratocarcinoma cells as a model for mammalian development. In *Chemistry of Cell Differentiation* II, 15. J. Paul, ed. Baltimore: University Park Press, pp. 333-376.

Hsu, T.C. 1973. Karyology of cells in culture A. Preparation and analysis of karyotypes and idiograms. In *Tissue Culture: Methods and Applications*. P.J. Kruse and M.K. Patterson, eds., New York: Academic Press, chap. 15.

Hooper, M.L., and C. Slack. 1977. Metabolic cooperation in HGPRT$^+$ and HGPRT$^-$ embryonal carincomal cells. *Develop. Biol.* 55:271.

Hughes, B.P. 1962. A method for the estimation of serum creatine kinase and its use in comparing creatine kinase and aldolase activity in normal and pathological sera. *Clin. Chem. Acta* 7:597.

Jacob, F. 1977. Mouse teratocarcinoma and embryonic antigens . *Immunol. Rev.* 33:3.

Larraga, V., and M. Edidin. 1979. Immunochemical characterization of surface antigens of TerC, a teratocarcinoma-derived cell line. *Proc. Nat. Acad. Sci. USA* 76:2912.

Lehman, J.M., W.C. Spears, D.E. Swartzendruber, and G.B. Pierce. 1974. Neoplastic differentiation: Characteristics of cell lines derived from a murine teratocarcinoma. *J. Cell Physiol.* 84:13.

Levine, A.J., M. Torosian, A.J. Sarokhan, and A.K. Teresky. 1974. Biochemical criteria for the *in vitro* differentiation of embryoid bodies produced by a transplantable teratoma of mice. The production of acetylcholine esterase and creatine phosphokinase by teratoma cells. *J. Cell Physiol.* 84:311.

Lotan, R., E. Skutelsky, D. Danon, and N. Sharon. 1975. The purification, composition, and specificity of the anti-T lectin from peanut (*Arachis hypogea*). *J. Biol. Chem.* 250:8518.

Martin, G.R. 1975. Teratocarcinomas as a model system for the study of embryogenesis and neoplasia. *Cell* 5:229.

Martin, G.R., and M.J. Evans. 1975. Differentiation of clonal lines of teratocarcinoma cells: formation of embryoid bodies *in vitro*. *Proc. Nat. Acad. Sci. USA* 72:1441.

Nudelman, E., S.-I. Hakomori, B.B. Knowles, D. Solter, R.C. Nowinski, M.R. Tam, and W.W. Young Jr. 1980. Monoclonal antibody directed to the stage-specific embryonic antigen (SSEA-1) reacts with a branched glycosphingolipid similar in structure to the Ia antigen. *Biochem. Biophys. Res. Commun.* 97:443.

Ostrand-Rosenberg, S., T.M. Rider, and A. Twarowski. 1980. Susceptibility of allogeneic mice to teratocarcinoma 402AX. *Immunogenetics* 10:607.

Pierce, G.B., A.R. Midgely, J. Sri Ram, and J.D. Feldman. 1962. Parietal yolk sac carcinoma: clue to the histogenesis of Reichert's membrane of the mouse embryo. *Am. J. Pathol.* 41:549.

Reisner, Y., G. Gachelin, P. Dubois, J.-F. Nicolas, N. Sharon, and F. Jacob. 1977. Interaction of peanut agglutinin, a lectin specific for nonreducing terminal D-galactosyl residues, with embryonal carcinoma cells. *Develop. Biol.* 61:20.

Salomon, D.S., and M.I. Sherman. 1975. The biosynthesis of progesterone by cultured mouse midgestation trophoblast cells. *Develop. Biol.* 47:394.

Searls, D.B. 1980. An improved colorimetric assay for plasminogen activator. *Anal. Biochem.* 107:64.

Sherman, M.I. 1975. Long term culture of cells derived from mouse blastocysts. *Differentiation* 3:51.

Sherman, M.I., and S.B. Atienza-Samols. 1979. Enzyme analysis of mouse extra-embryonic tissue. *J. Embryol. Exp. Morphol.* 52:127.

Sherman, M.I., S. Strickland, and E. Reich. 1976. *Cancer Res.* 36:4208.

Solter, D., and I. Damjanov. 1979. Teratocarcinoma and the expression of oncodevelopmental genes. *Methods in Cancer Res.* 18:277.

Solter, D., and B.B. Knowles. 1978. Monoclonal antibody defining a stage-specific mouse embryonic antigen (SSEA-1). *Proc. Nat. Acad. Sci. USA* 75:5565.

Solter, D., and B.B. Knowles. 1979. Developmental stage-specific antigens during mouse embryogenesis. *Curr. Top. Develop. Biol.* 13:139.

Solter, D., L. Shevinsky, B.B. Knowles, and S. Strickland. 1979. The induction of antigenic changes in a teratocarcinoma stem cell line (F9) by retinoic acid. *Develop. Biol.* 70:515.

Stern, P., G.R. Martin, and M.J. Evans. 1975. Cell surface antigens of clonal teratocarcinoma cells at various stages of differentiation. *Cell* 6:455.

Stevens, L.C. 1958. Studies on transplantable testicular teratomas of strain 129 mice. *J. Nat. Cancer Inst.* 20:1257.

Stevens, L.C., and G.B. Pierce. 1975. In *Teratomas and Differentiation*. M.I. Sherman and D. Solter, eds. New York: Academic Press, pp. 13-14.

Strickland, S., and V. Mahdavi. 1978. The induction of differentiation in teratocarcinoma stem cells by retinoic acid. *Cell* 15:393.

Strickland, S., E. Reich, and M.I. Sherman. 1976. Plasminogen activator in early embryogenesis: enzyme production by trophoblast and parietal endoderm. *Cell* 9:231.

Strickland, S., R. Smith, and R.R. Marotti. 1980. Hormonal induction of differentiation in teratocarcinoma stem cells: generation of parietal endoderm by retinoic acid and dibutyryl cAMP. *Cell* 21:347.

Wada, H.G., S.R. Vandenberg, H.H. Sussman, W.E. Grove, and M.M. Herman. 1976. Characterization of two different alkaline phosphatases in mouse teratoma: partial purification, electrophoretic, and histochemical studies. *Cell* 9:37.

Wiley, L.M. 1979. Early embryonic cell surface antigens as developmental probes. *Curr. Top. Develop. Biol.* 13:167.

Chapter 7

CELL MARKERS IN EMBRYONIC DEVELOPMENT

ROLF KEMLER

HISTORY OF THE F9 ANTIGEN(S)

What has made the so-called F9 antigen so much more attractive to investigators than other embryonic cell surface antigens? To understand this, one has to explain a little about the ideas that guided the F9 experiments. First, it was thought that cell surface components are involved in embryonic cell differentiation processes (Bennett et al., 1971). In other words, specific cell surface structures recognize signals which, when transmitted to the genome, cause differential gene expression. Following this, one imagined a kind of cascade of membrane structures, each acting at a precise moment in embryonic development. Second, the immunological approach was chosen to characterize such cell surface components using mouse teratocarcinoma cell lines as immunogens. Syngeneic immunizations were performed on the assumption that early embryonic cell surface structures, not expressed on adult cells, are recognized as nonself by the immune system. The anti-F9 serum was obtained by immunization of 129/Sv mice with the syngeneic F9 embryonal carcinoma cell line (Artzt et al., 1973). Using this antiserum, two major results were obtained. First, the antigenic structure recognized by the anti-F9 serum was reported to be the wild-type gene product of a particular t haplotype of the T/t complex (Artzt et al., 1974). Second, biochemical characterization of the antigen detected by the anti-F9 serum showed structural similarities to H-2 antigens (Vitetta et al., 1975). Subsequent and more detailed analyses have not confirmed these results. The F9 antigen(s) seem(s) not to be the wild-type gene product of a particular t haplotype, but rather it appears that some t haplotypes change the expression of the antigenic determinants recognized by the anti-F9 serum (Kemler et al., 1976). Studies of the serological components of the anti-F9 serum demonstrate that IgM, IgG_1, and $IgG_{2a,b}$ anti-F9 antibodies react with at least three independent antigenic determinants. Moreover, the major components of the antiserum, which are the IgM anti-F9 antibodies, react with long saccharidic chains (Morello et al., 1980).

In conclusion, this type of approach, while valid, also illustrates the potential difficulties implicit in trying to specify gene products by serological methods using conventional antisera. The introduction of the myeloma fusion technique (Köhler and Milstein, 1975) has given new hope and opened new perspectives in this kind of research. This chapter summarizes our experience

149

with the myeloma fusion technique to obtain specific reagents for the study of development.

MONOCLONAL ANTIBODIES TO DEFINE CELL MARKERS

With conventional antisera against embryonal carcinoma cells, syngeneic immunization has been used in order to obtain a restricted immune response and avoid, as much as possible, polyabsorption of the sera. Monoclonal antibodies were raised against EC cells using a rather broad immune response. For this, rats were immunized with mouse EC cells and rat splenocytes fused with mouse myeloma cells. This approach is very efficient, and a high number of positive hybridomas have thus been obtained and kept as stable cell lines (Kemler et al., 1979). Several positive hybridomas have been analyzed, and it turns out that each of them produces antibodies against different antigenic structures. None of them react with an antigen present on all mouse cell types. Although each monoclonal antibody obtained in this way recognizes a cell surface component with a different distribution, it soon became obvious that these monoclonal antibodies were not good candidates for defining cell markers, for two main reasons. First, the target antigens could not be correlated with particular cell types and, second, most antigens seemed to be glycolipids and were therefore difficult to characterize biochemically. When hybridomas against EC cells were produced using an allogeneic immunization of Balb/c mice, specific monoclonal antibodies were about 100 times less frequent than the xenogeneic monoclonal antibodies mentioned above. These monoclonal antibodies seem, however, to detect surface antigens with more restricted cell distribution. One of the hybridomas produced using splenocytes of a Balb/c mouse immunized with EC cells (PCC4) secretes monoclonal antibodies which react with EC cells as well as with mouse preimplantation embryos and ectodermal cells of six-to-eight-8-day-old embryos, but not with differentiated derivatives (Kemler, in press).

To summarize this work, we conclude that the production of monoclonal antibodies using whole EC cells as an immunogen with a view to defining cell markers in the study of development has been not as fruitful as we had hoped. The myeloma fusion technique is still the method of choice in these kinds of studies, but only in combination with other analytical methods. Therefore, two-dimensional gel electrophoresis was first used to analyze protein patterns of trophectodermal and ICM cells and to compare them with those obtained from various teratocarcincoma cell lines (Brûlet et al., 1980). Trophectoderm-specific proteins, which are present in both a trophoblastoma and a parietal yolk sac cell line but absent from ICM and EC cells, were found to co-purify with preparations of intermediate filaments of trophoblast-oma cells. Hybridomas were produced against the intermediate filament preparation, and all positive monoclonal antibodies were tested against a panel of teratocarcinoma cells. Those reacting only with trophoblastoma and parietal yolk sac cells stained trophectoderm, but not ICM cells. In indirect immunofluorescence tests, these monoclonal antibodies delineate an intracellular network in attached trophoblastoma and trophectoderm cells. To identify the proteins that the monoclonal antibodies recognize, the preparation of intermediate filaments was run on a two-dimensional gel and the proteins transferred by electrophoresis to diazobenzyloxymethyl (BDM) paper (for details, see Brûlet et al., 1980). Bound monoclonal antibodies were then detected with iodine-labeled antibodies. This method may be an alternative approach, especially in cases where immunoprecipitation with monoclonal antibodies proves difficult. Three monoclonal antibodies, each delineating an intracellular network in trophoblastoma and trophectoderm cells, were thus found to react with antigenic determinants of different polypeptide chains. The reactivity pattern of the monoclonal antibodies detecting trophectoderm-specific markers was studied on cryostat sections of embryos of various stages and on adult tissue (Kemler et al., in press).

Up to day 10 of embryonic development, the three monoclonal antibodies exhibit the same reactivity: they stain exclusively trophectoderm and parietal and visceral endoderm. At day 12, the three monoclonal antibodies react with most of the epithelial cells, whatever their embryonic origin; only neuroepithelium always remains negative. Differences in the reactivity patterns of the three monoclonal antibodies were found on 14-day-old embryos and on adult tissue. The clear-cut results obtained in these experiments show an additional advantage of using monoclonal antibodies as reagents for detecting cell markers.

SUMMARY AND CONCLUSIONS

It is crucial in the analysis of embryonic cell differentiation to have specific cell markers. In general, two approaches are possible: either one can attempt to purify such markers or one can make specific probes against them. Only a few cellular markers are known for early mouse embryonic cells, and these have not yet been well purified. Rather than wasting too much effort in purifying these markers, we first tried to obtain very specific probes against them. For this purpose, the myeloma fusion technique is the method of choice. At first, we thought that this might be the best way to define unknown cell markers. However, the first experiments with monoclonal antibodies have shown this judgment to be premature.

Additional information is needed to select a cell structure against which a monoclonal antibody is desired. The combination of other analytical methods and the production of monoclonal antibodies seems to be the most promising approach.

ACKNOWLEDGMENTS

This work was supported by grants from the Centre National de la Recherche Scientifique (LA 269), the Délégation Générale à la Recherche Scientifique et Technique, the Fondation pour la Recherche Médicale Française, the Institut National de la Santé et de la Recherche Médicale, the Ligue Nationale Française contre le Cancer, and the Fondation André Meyer.

REFERENCES

Artzt, K., P. Dubois, D. Bennett, H. Condamine, C. Babinet, and F. Jacob. 1973. Surface antigens common to mouse cleavage embryos and primitive teratocarcinoma cells in culture. *Proc. Nat. Acad. Sci. USA* 70:2988.

Artzt, K., D. Bennett, and F. Jacob. 1974. Primitive teratocarcinoma cells express a differentiation antigen specified by a gene at the T-locus in the mouse. *Proc. Nat. Acad. Sci. USA* 71:811.

Bennett, D., E.A. Boyse, and L.J. Old. 1971. Cell surface immunogentics in the study of morphogenesis. In *Cell Interactions*. Third Lepetit Colloquium. Amsterdam: North Holland, pp. 247-263.

Brûlet, P., C. Babinet, R. Kemler, and F. Jacob. 1980. Monoclonal atibodies against trophectoderm specific markers during mouse blastocyst formation. *Proc. Nat. Acad. Sci. USA* 77:4113.

Kemler, R. 1981. Analysis of mouse embryonic cell differentiation. In *Fortschritte der Zoologie*. Symposium. Progress in Developmental Biology. M. Lindauer, ed. Würzburg. In press.

Kemler, R., C. Babinet, H. Condamine, G. Gachelin, J.L. Guenet, and F. Jacob. 1976. Embryonal carcinoma antigen and the T/t locus of the mouse. *Proc. Nat. Acad. Sci. USA* 73:4080.

Kemler, R., P. Brûlet, M.T. Schnebelen, J. Gaillard, and F. Jacob. 1981. Reactivity of monoclonal antibodies against intermediate filament proteins during embryonic development. *J. Embryol. Exp. Morphol.*

Kemler, R., D. Morello, and F. Jacob. 1979. Properties of some monoclonal antibodies raised against mouse embryonal carcinoma cells. In *Cell Lineage, Stem Cells and Cell Determination*. INSERM Symposium No. 10. N. Le Douarin, ed. Amsterdam: Elsevier/North-Holland Biomedical Press, pp. 101-113.

Köhler, G., and C. Milstein. 1975. Continuous cultures of fused cells secreting antibody of predefined specificity. *Nature* 256:495.

Morello, D., H. Condamine, C. Delarbre, C. Babinet, and G. Gachelin. 1980. Serological identification and cellular distribution of three F9 antigen components. *J. Exp. Med.* 152:1497.

Vitetta, E.S., K. Artzt, D. Bennett, E.A. Boyse, and F. Jacob. 1975. Structural similarities between a product of the T/t locus isolated from sperm and teratocarcinoma cells, and H-2 antigens isolated from splenoytes. *Proc. Nat. Acad. Sci. USA* 72:3215.

Part II

THE PLACENTA AS AN IMMUNOLOGICAL BARRIER

INTRODUCTION

The second section of this book deals with the privileged status of the mammalian fetal allograft from the point of view of the unique structure and function of the placenta. This organ must somehow contain an important clue to the survival of the fetal allograft, since it contains fetally derived trophoblast tissue in direct contact with the maternal circulation. The unique immunological status of the trophoblast was demonstrated in the classic experiments of Simmons and Russell, who showed that an ectopically transplanted allogeneic trophoblast of 7½ days of age is resistant to immune rejection, whereas tissue of the embryo proper is susceptible to rejection. These observations have been expanded in an elegant fashion by Rossant and her colleagues (*J. Reprod. Fertil.* 59:387, 1980 and personal communication), who demonstrated the crucial protective role of the trophoblast *in situ* using reciprocal embryo transplants of *Mus caroli* and *Mus musculus*. These two sets of experiments dramatically illustrate the unique immunological role of the placental tissue and provide the rationale for exploring this area of mammalian biology.

The chapters in this section explore in detail the anatomic and immunological possibilities for the resistance of the placenta to immunological attack. Allen Enders reviews the structure of the placenta during implantation and its subsequent growth to the mature organ. He points out that, while studies on man and mice indicate that the placenta is an extremely complex and varied organ, we miss a great deal of functionally important placental variation by confining our attention to these two species. Guy Voisin then presents a detailed overview of the various immunological phenomena that occur within the placenta. His view is that the placenta provides a means whereby the maternal immune response is set into a balance between rejection and "facilitation," perhaps by means of blocking factors or suppressor cells, and that it is this immunological legerdemain that allows the placenta to survive.

The next three chapters explore some of the antigenic properties of the placenta. W.D. Billington, S.C. Bell, and G. Smith address the critical issue of what tissues express histocompatibility antigens in the placenta and their relationship to the maternal bloodstream. The spongiotrophoblast expresses major histocompatibility antigens *in vitro*, but whether it does so *in vivo* is not known from these studies (see below). These cells are susceptible to killing *in vitro* by cytotoxic cells, but they are resistant *in vivo*, even when the pregnant mouse is primed against her fetal allograft. Thus, the anatomical relationships *in vivo* must be an important component of the resistance of the spongiotrophoblast to immunological attack. These authors also point out that the nature of the immune response to the fetal allograft varies

159

in mice of different strains and that there are probably genetic differences in responsiveness to paternal histocompatibility antigens.

Further light is shed on this situation by B. Singh, R. Raghupathy, D.J. Anderson, and T.G. Wegmann, who describe the use of monoclonal antibodies to demonstrate, on the placenta of the mouse, the existence of Class I major histocompatibility antigens (H-2K/D) but absence of Class II antigens. These experiments also indicate that the placenta can act as an immunoadsorbant and remove antibodies from the maternal circulation. This binding is not dependent on the Fc portion of the maternal immunoglobulin, since the $F(ab')_2$ fragments of the monoclonal antibodies show the same binding characteristics with the placenta. Autoradiographic studies on placentae labeled *in vivo* show that these antibodies bind to the lateral aspect of the placenta where the yolk sac inserts and to the spongiotrophoblast, which is in direct contact with the maternal circulation. These findings indicate that Class I antigens are in direct contact with the maternal circulation in the intact placenta, and they complement well the studies by Billington and his colleagues. These data indicate that the resistance of the placenta to maternal immunological attack is not simply the lack of exposure of histocompatibility antigens to the maternal circulation, at least in the case of a Class I antigen.

Finally, W. Page Faulk describes studies on a new set of trophoblastic antigens in the human and on the role of HLA antigens on the placenta. Using xenogeneic antisera, he has defined two new trophoblastic antigens, TA-1 and TA-2, which are potent inhibitors of mixed lymphocyte reactivity. Although he was unable to identify major histocompatibility antigens on the tissues at the interface between the maternal and fetal circulations in the human placenta by using monoclonal anti-HLA antibodies, he did identify a series of antigens called trophoblast-leukocyte crossreactive (TLX) antigens on the interfacing surface that might be genetically associated with the HLA complex. His studies in the human also provided further evidence for the importance of disparity in MHC antigens for fetal survival. A variety of women subject to chronic abortion have a greater sharing of HLA antigens with their mates than could be expected by chance alone. By immunizing four such women with lymphocyte-rich plasma from donors compatible for erythrocyte antigens but incompatible for HLA, Faulk was able to have them successfully bring a pregnancy to term. This striking preliminary observation provides further evidence for the importance of MHC disparity for reproductive capacity, as discussed in Chapter 3.

Chapter 8

VARYING STRUCTURAL RELATIONSHIPS DURING IMPLANTATION AND PLACENTATION OF POSSIBLE SIGNIFICANCE TO MATERNAL-FETAL IMMUNOLOGY

ALLEN C. ENDERS

As several investigators have pointed out, in considering the uniqueness of gestation as an immunological event, the principal question is why the maternal organism does not respond to the fetal membranes as a tissue transplant and consequently mobilize its cellular rejection reaction (see Billington, 1979). Moreover, since in many species transfer of IgG from the maternal organism to the fetus occurs in late pregnancy (Wild, 1979), we must also consider the possible passage of nonsurface antigens to the maternal organism, the processing of these antigens by the maternal organism, and the morphology of the developing fetal membranes and uterus, as these structures might relate to the ability of the maternal organism to mount either cellular or humoral attacks on the fetal cells that constitute at the same time a genotypic intruder and the succoring mechanism of the developing embryo. In examining the morphology of implantation and placentation, therefore, we should pay attention to those elements that intervene between fetal and maternal systems at different stages, evidence that antigens might be being processed, such as leukocytic invasion or other aspects of the inflammatory response, indications of a cellular response and localization of the mononuclear components of the immune system, and the types of relationship of the components of the fetal and maternal vascular systems with regard to both cellular and humoral passage.

From the start, it should be kept in mind that the interrelationship of fetal and maternal tissues is changing during gestation. This is not so evident in the human, where establishment of the general pattern of the placenta occurs well within the first trimester, as it is in many animals with shorter gestations, where major changes may occur even in the last third of gestation.

Second, the types of final relationship that might occur between the two vascular systems are proscribed. As has been pointed out by many authors, no placenta exists in which there is direct blood flow between fetal and maternal organisms. In addition, no endothelio-endothelial placenta exists in mammals, and no truly syndesmochorial placenta is present other than as a transitory stage. Nevertheless, the wealth of placental relationships is remarkable. Since presumably the placentae with a relationship more intimate than the epitheliochorial type are at greater risk, I shall confine my remarks largely to some species that have a hemochorial or endotheliochorial relationship in the definitive chorioallantoic placenta and a variety of yolk sac arrangements and shall consider primarily those species that have come under the surgical, glass, or diamond knives of our laboratory.

PREIMPLANTATION PERIOD

Prior to implantation, the blastocyst is situated in the uterine lumen in transient contact with the glycocalyx of the microvilli of the uterine luminal epithelium. Any leukocytic invasion that may have occurred in response to the seminal fluid or, in the case of postpartum mating, to the repair of the uterus (Padykula, 1976) has already diminished before the morulae enter the uterus and prior to loss of the zona pellucida of the blastocyst. In many species that will eventually form a hemochorial placenta, the period of time that a blastocyst without a zona pellucida is present within the uterine lumen is brief (two days or less). However, in other instances the blastocyst may be without a zona for a long time, especially in those species that exhibit a "delay" of implantation and associated blastocyst diapause (Flint et al., 1981). Some of the animals that delay retain the zona, but the armadillo, for example, may delay as long as four months without a zona pellucida (Enders, 1966). Although the rat and mouse shed their zona a day later when they delay, blastocysts may survive many days without the zona prior to implantation (Rumery and Blandau, 1971). During this time, the trophoblast at least is exposed to anything in the uterine lumen, but of course not to cellular elements in the stroma of the uterus. In the majority of species that delay, the uterine stroma appears to be a normal lamina propria devoid of inflammation but having plasma cells, mast cells, and macrophages (Given and Enders, 1978; Schlafke et al., 1981), and lymphocytes are common in the basolateral compartment between luminal epithelial cells. It is well known that serum proteins reach the uterine lumen in these stages (Bernard et al., 1977).

IMPLANTATION

It is with penetration through the uterine epithelium and subsequent association with the stroma that we would expect the ideal conditions for the sensitization of the maternal organism to any exposed antigens of the conceptus (Figure 8-1). The mechanism of penetration of the epithelium is probably not germane to a discussion of the immune response except as it relates to subsequent events. Penetration of the epithelium by intrusion or fusion is ordinarily accomplished without loss of luminal integrity. Even the facilitated sloughing of the epithelium that occurs in the rat and mouse causes little loss of luminal integrity since it occurs in an effectively isolated implantation chamber. All epithelial penetration includes adhesion to, and sharing of, junctions between trophoblast and uterine epithelial cells and is not ordinarily accompanied by all of the components of a full inflammatory response. In the rabbit, we have observed single trophoblastic knobs undergoing leukocytic infiltration, but normally this is not apparent during implantation (Enders and Schlafke, 1971). In the earliest implantation sites in the human, trophoblast has penetrated through epithelium and well into stroma (O'Rahilly, 1973). Aside from mild edema, they cannot be construed as having an inflammatory response. However, a little later, there is pronounced aggregation of polymorphonuclear leukocytes, which does not decline until decidualization is well under way and the conceptus is in the early villus stage (Enders, 1976).

In the rhesus monkey, there is a mild infiltration of eosinophils at the time of epithelial penetration, followed by pronounced edema at the periphery of the implantation site but no massive heterophil invasion. In the rat and

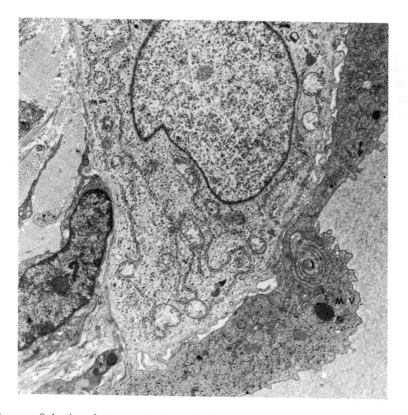

FIGURE 8-1. An electron micrograph of contact of trophoblast with stromal tissues in an early implantation site of an armadillo (*Dasypus novemcinctus*). The trophoblast (Troph) in the center is adjacent to endothelium of a maternal vascular sinus (MVS) and also to bundles of collagen and a fibroblast (left).

mouse, there is local edema but no heterophil invasion at the time of first epithelial penetration. In neither the armadillo nor the ferret could we find any sign of inflammatory response at this time.

Interestingly, in the process of implantation, the trophoblast of the majority of species appears to pause at the basal lamina of the epithelium, but no accumulation of lymphocytes occurs during this pause. Species, such as the rat, that have been shown to have lymphocytes in the basolateral compartment prior to implantation do not seem to have as many lymphocytes at this location during epithelial penetration. These limitations, however, are more ones of degree than kind, and we have seen both lymphocytes and heterophils in direct contact with trophoblast before and after epithelial penetration. It should also be noted that it is not just trophoblast that may be exposed to the maternal organism during implantation, since the neurectoderm is exposed in some blastocysts (e.g., rabbit, carnivores) and early invasion exposes the endoderm in others (guinea pig, armadillo) (Wimsatt, 1975).

STROMAL PENETRATION

It is in the association with the stroma that there is the greatest amount of modification from "normal" lamina propria. Briefly, there seem to be several different methods that strictly limit the time when either syncytial or cellular trophoblast is associated with an unmodifed lamina propria. In the rat and mouse, with a precocious decidual response, there is never any association of trophoblast with unmodified stroma. Decidualization precedes epithelial penetration and forms a cup around the implantation chamber in the first two days, restricting intercellular pathways to the implantation chamber. However, by the third day after implantation, heterophils are marginating in the vessels of the mesometrial decidua (Figure 8-2) and a few are accumulating around the residual mesometrial epithelium. From this time and throughout gestation, there is always some region of the implantation chamber where there are numerous heterophils. Most frequently, these are regions of necrosis of epithelial cells or decidual cells, but death of trophoblastic giant cells also occurs, especially during reestablishment of the uterine lumen, and, although there are no endometrial lymphatics, the myometrial lymphatics enlarge greatly (Figure 8-3; Welsh and Enders, 1981). The decidua eliminates many of the interstitial channels thought to be important in protein movement (Casley-Smith, 1980). In the rhesus monkey, the development of an epithelial plaque reaction is initiated within a day after implantation (Wislocki and Streeter, 1938). This plaque temporarily isolates all but the initial areas of trophoblast penetration from most of the lamina propria (Figure 8-4) until

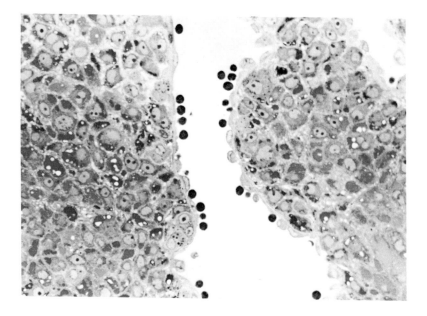

FIGURE 8-2. A vascular sinus within decidual cells in an implantation site of a rat on day 10 of pregnancy.

Note the numerous leukocytes that are apparently adherent to the walls of the vessel.

after formation of the intervillous spaces, the establishment of a typical placental circulation, and the formation of extraordinary anchoring villi (Figure 8-5). Although there is a decidual response, it is late and does not seem to pack the stroma to the extent that it does in the human.

In species that have an endotheliochorial type of placenta, such as carnivores, it might be expected that, in surrounding the maternal vessels, the trophoblast would be in prolonged contact with typical uterine stroma. However, this seems not to be the case. In carnivores, the pattern of hypertrophy of the endometrium in pregnancy results in a superficial zone rich in epithelium with little lamina propria (Buchanan, 1966). The vessels in the area forming the zonary placenta hypertrophy rapidly in apparent response to the trophoblast and become isolated within the trophoblast, separated from it by a greatly thickened basal lamina. Furthermore, the endothelial cells become nearly unrecognizable and are apparently involved in the process of extensive protein synthesis (Lawn and Chiquoine, 1965). In the cat, a few other cells remain; these are isolated decidual cells only (Amoroso, 1952; Bjorkman, 1973). In the junctional region between vessels, the trophoblast fronting on the uterus is separated from stroma by extensive uterine epithelial proliferation of the junctional zone. Only in the region of formation of hematoma is there much

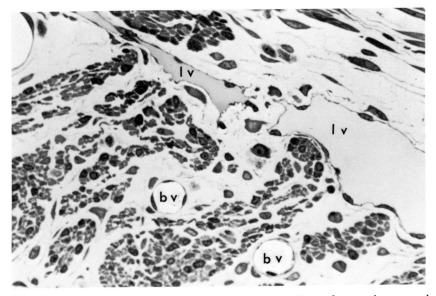

FIGURE 8-3. Lymphatic vessels (lv) in the myometrium of a rat in an early implantation stage (day 8).
Note that the lymphatics are distended, but contain lightly electron-dense plasma, whereas blood vessels (bv) are distended but clear, having been flushed during the perfusion-fixation procedures.

interaction between trophoblast and unmodified stromal elements, and this interaction is temporary (Leiser and Enders, 1980).

Many species with hemochorial chorioallantoic placentation pass through a temporary endotheliochorial condition, the maternal vessels being surrounded with trophoblast prior to loss of the endothelium. In the human this is a rapid, transient stage (Harris and Ramsey, 1966). In many bats, it is relatively slow (Enders and Wimsatt, 1968). However, two robust cellular trophoblast layers are interposed between the maternal vessels and fetal mesoderm, with the exterior layer becoming syncytial before the maternal endothelium is displaced. An interesting example to study immunologically would be the real slowpoke at this stage of implantation, *Macrotus*, which has delayed development (Bodley, 1974; Burns, 1981).

Thus, in none of the types of implantation is the placenta established as a skin graft or even as an organ transplant, with confluence of donor and recipient stroma and vascular anastomosis. In a few instances, the implantation is somewhat similar to a carcinoma or at least an adenoma. In the horse, trophoblast cells from the girdle zone invade the endometrium to form the endometrial cups, vascularized by maternal vessels (Allen et al., 1973). It is possible that trophoblast cells might secrete immunoregulatory substances that

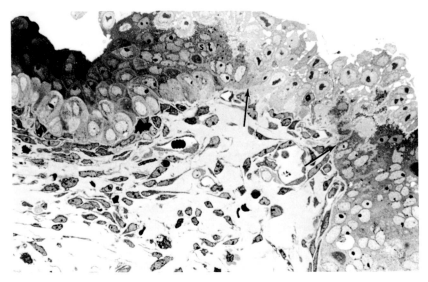

FIGURE 8-4. Implantation site of a rhesus monkey on day 10 of gestation.
There is an extensive epithelial plaque reaction (large pale cells, upper left), with unmodified stroma beneath the epithelium. In the area of penetration of trophoblast (arrows), cellular trophoblast is apparently adjacent to the basal lamina of the uterine epithelium, but without other intervening cells between trophoblast and uterine stroma. Syncytial trophoblast (st) can be seen at the lateral edges of the cellular trophoblast.

protect them in this situation or that they might, like some murine malignant cells, produce a possible anti-inflammatory or macrophage cytotoxic substance (Fauve, 1974) (see Chapters 12, 13, 14, 15 and 16). Steven and Morriss (1975) suggested that, when the endometrial cups regress, it is a rejection reaction, since there is an infiltration of mononuclear cells and eosinophil leukocytes. However, Allen (1975) has pointed out that, in a mare carrying a mule fetus, there is both low PMSG and early regression of the endometrial cups (45 to 60 days), whereas in a donkey carrying a hinny fetus, not only is there high PMSG but also, although there are many leukocytes present, the endometrial cups persist for well over 100 days. Interestingly, there is also an elevated progesterone level in the donkey carrying the hybrid conceptus (Allen, 1979), which has been suggested as another means of protection (Siiteri et al., 1977).

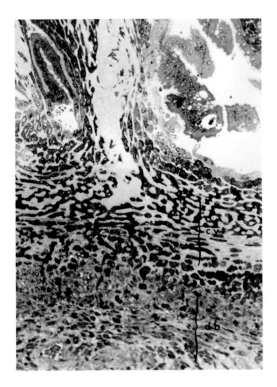

FIGURE 8-5. Anchoring villus of a placenta from a rhesus monkey on day 70 of pregnancy.
The center of the villus contains amorphous "fibrinoid" material, and this material continues onto the basal plate interspersed with cytotrophoblast cells (cyt). Decidual cells are present in the uterine stroma of the decidua basalis (db).

ESTABLISHMENT OF THE PLACENTA

In using the phrase "establishment of the placenta," we are normally considering the formation of the definitive circulatory patterns and consequently of the interhemal "membrane" or "barrier." It should be pointed out that, unlike in the human, where these changes occur largely within the first month of pregnancy, in many species the interrelationships of fetal and maternal tissues are changing throughout gestation. Thus, in the rat and mouse, inversion of the yolk sac, which takes place after reestablishment of the uterine lumen and which is probably accompanied by necrosis of some of the parietal endoderm and trophoblast, takes place in the last week of gestation (Amoroso, 1952; Welsh and Enders, 1981). In the mollosid bat *Tadari-*

da, a region of the chorioallantoic placenta changes from endothelial chorial to hemochorial with cellular trophoblast in the last days of gestation (Stephens, 1969).

This changing relationship of different aspects of the fetal membranes in relation to the uterus must be taken into consideration in analyzing either the potential antigen transfer from fetal to maternal systems or the changing varieties of isolating mechanisms that might be functional in protecting the fetal membranes from maternal response.

CHANGES DURING PREGNANCY

Trophoblast itself undergoes quite a few modifications in the course of pregnancy in most species. If the maternal organism forms an implantation chamber (rat, mouse, many bats), initial penetration of the epithelium occurs while trophoblast is cellular. It is more common to have some form of syncytial trophoblast involved in the early stages of implantation. Interestingly, this most often takes the form of syncytial masses: trophoblast knobs in the rabbit, plaques in the ferret, and implantation cones in the guinea pig (Schlafke and Enders, 1975). Even in the human, it appears that there are several masses of syncytium in the early invasion sites. In myomorph rodents, in which the initial epithelial penetration is accomplished by mural cytotrophoblast (Enders, 1975), giant cells migrate a short distance into the deciduum after it begins to decline. However, with the establishment of the hemochorial and endotheliochorial condition, trophoblast ordinarily forms a continuous syncytial layer interposed between the maternal blood and the fetal endothelium. This may take the form of a layer of syncytial trophoblast exposed directly to maternal blood over all the areas of maternal blood flow within the placenta (human, rhesus monkey, armadillo, guinea pig). In this case, there is no apparent area for intercellular passage directly from maternal blood to fetal placental stroma. In endotheliochorial placentae, the layer most closely associated with the modified maternal vessels is a continuous syncytium. In murid and cricetid rodents, the outermost layer of trophoblast is cellular but the middle layer of the hemotrichorial placenta is syncytial. In most species, the outermost layer also has cytological evidence of extensive protein synthesis (Enders, 1965).

The tremendous variety of hemochorial placentae attests to the successfulness of this vascular arrangement. Thus, there are villous and labyrinthine forms; hemomono-, di-, and trichorial forms; and syncytial and cellular forms (Table 8-1). In this relationship, trophoblast is in a unique position where, unlike normal parenchymal cells, it can both obtain materials and

171

TABLE 8-1
Variations in Definitive Interhemal Membranes in Species with Hemochorial Chorioallantoic Placentation

HEMOMONOCHORIAL

Villous (syncytial with feeder cytotrophoblast)

1. Villi in expanded placental or intervillous space (rhesus monkey, human)
2. Villi in expanded uterine blood sinuses (armadillo)

Labyrinthine

1. Syncytial trophoblast
   Continous syncytium, microvillus bays (guinea pig)
   Continous syncytium but bays with intrasyncytial lamina (chipmunk)
2. Cellular trophoblast
   Giant cells form the lining of the maternal blood space (Zapus)
   Cytotrophoblast forms wall of maternal blood space, but bays with interstitial material still present (Tadarida)

HEMODICHORIAL

Labyrinthine

1. Irregular thickness of thick superficial layer of trophoblast (rabbit)
2. Regular thick superficial syncytial layer with intrasyncytial lamina (Myotis)

HEMOTRICHORIAL

Labyrinthine

Surface cellular layer not part of macromolecular barrier. Middle syncytial layer active front on maternal blood space (rat, mouse, hamster)

secrete materials directly into the maternal blood rather than from and into the interstitial connective tissue. Even hepatocytes, with their major access to the sinuses because of the space of Disse, do not have direct access to the cells of the vascular system. Interestingly, margination of leukocytes, the first step in acute inflammation (Hoover et al., 1980), does not seem to occur in relation to trophoblast. The presence of fetal tissue in maternal blood spaces (or actually in vessels in the case of the armadillo) means that, should a fragment of tissue dislodge, it enters the maternal vascular bed even as do syncytial tabs. The reciprocal arrangement, where, for example, a tuft of vascularized endometrium is bathed in fetal blood, does not exist.

In contrast to the more common situation, in a portion of the placenta of the mollosid bat, the maternal blood is surrounded by cytotrophoblast, and trophoblast giant cells form the covering of maternal blood spaces in members of the *Dipidoidea* (King and Mossman, 1974). Cytotrophoblast is also directly exposed to maternal blood in placental hematomas or hemophagous organs, but the blood is not circulating. In the endotheliochorial placenta of the shrew, the syncytial trophoblast, although forming a continous layer in the sense of lacking cell boundaries, has numerous pore areas that would appear to be sufficient in size to allow passage of macromolecules (Wimsatt et al., 1973).

Almost all placentae have some regions of cytotrophoblast other than the feeder cytotrophoblast for syncytium. Even though these regions may be the area of slow expansion of the chorioallantoic placenta at the expense of the endometrium, these are not regions in which there is intimate vascular association. In humans and other primates, the anchoring villi are regions in which cytotrophoblast forms a portion of the basal plate (Figure 8-5). However, the presence of the decidua and the accumulation of fibrinoid and fibrin tend to isolate the stroma of the villus from "normal" maternal connective tissue. At the same time, migrating trophoblast is not isolated from the maternal system in that it directly enters maternal vessels as well as uterine stroma. It has been proposed that cytotrophoblast invades the lumen of the maternal arteries in primates and that some trophoblast (probably giant cells) also invades the surrounding smooth muscle and connective tissues (Ramsey et al., 1976). At the moment, we should reserve judgment on the intraluminal invasion, since the hypertrophy of the endothelial cells in relation to implantation renders it difficult to tell these cells from cytotrophoblast. No studies of quinacrine yellow fluorescent marking of male chromosomes have been published concerning these cells, and endothelial "candle-dripping" can occur in other species (Ford et al., 1980).

In other species as well, there is a tendency for the trophoblast unaccompanied by fetal vessels to "front" for the rest of the placental membranes in that the spongy trophoblast may form channels for maternal blood at the junctional zone, as it does in caviomorph rodents. In this group, a further isolation of the fetal membranes from the effector site of the cellular immune system is seen in the suspension of the chorioallantoic placenta from a stalk that contains blood vessels but not lymphatics (Hillemann and Gaynor, 1961). This creates, in the late pregnancy in the nutria, for example, the type of "privileged site" created artificially in alymphatic skin pedicles (see Chapter 2).

It should also be pointed out that all does not go well all of the time in all of the placentae of a multiple pregnancy, nor in a single placenta in any given pregnancy. In polytocous species (producing many young at one time), some of the sites are resorbed without deleterious effects to the remaining sites, and in one species, the plains viscacha, ordinarily all but the most distal sites are resorbed (Weir, 1971). Even in monotocous species, individual villi degenerate and become fibrotic, which again could increase the diversity of fetal antigens reaching the maternal system.

There are interesting and as yet unexplained alterations of stroma in many placentas. In the human, the early villi have an extraordinarily large number of vacuolated macrophages (Hofbauer cells) which are situated in apparent channels relatively clear of matrix in the stroma (Enders and King, 1970; Kaufmann et al., 1977; Wood, 1980). As pregnancy progresses, the channels are less obvious and the Hofbauer cells become less vacuolated and are more typical macrophages. In some insectivores and in the armadillo, the stroma in the labyrinth and villi, respectively, is packed with cells with enormous amounts of granular endoplasmic reticulum, cytologically resembling plasma cells. However, a preliminary study of such cells in the armadillo indicated that they were not producing immunoglobulin. We are thus faced with two animals, the human and the armadillo, both of which have villous hemomonochorial placentae, one of which provides evidence of a stroma especially suited to protein ingestion and the other to protein synthesis.

The passage of maternal immunoglobulins to the fetus, during late pregnancy and the early postnatal period, which imparts passive immunity to the neonate, has been discussed extensively. A few points should be kept in mind. The presence of a syncytial layer between the maternal and fetal vascular systems does not preclude the passage of protein from the maternal to the fetal organism. It does mean that their passage is subject to regulation by the trophoblast. Even exogenous protein can be transported across syncytium (e.g., Enders and Wimsatt, 1971), and the trophoblast is the known carrier of IgG in late pregnancy in the human. Although IgG binds to the trophoblast surface in the human (King, 1977), we do not yet know the mechanism of its selective survival while being transported (see, for example, the discussion in Wild, 1979; also see Chapter 11 of this volume).

The transport of IgG in the rat yolk sac points out the futility of assuming the most direct path since, in this case, the IgG, rather than being transported by trophoblast layer II of the chorioallantoic placenta, must pass into the uterine lumen to be transported by the endothelium of the inverted yolk sac placenta (Anderson, 1959). Hemoglobin derivatives are apparently transported by the yolk sac of the shrew, despite the fact that it is not inverted and a robust Reichert's membrane is interposed between the endodermal layer and the trophoblast that ingests the erythrocytes (King et al., 1978). The morphology of the mesothelial cells and their ability to transport exogenous proteins in the little brown bat suggest that, in these animals, it is the splanchnic mesothelium that is most likely to ingest luminal proteins (Enders et al., 1976).

SUMMARY AND CONCLUSIONS

There is ample opportunity for antigens of fetal origin, especially those from trophoblast, to get to the maternal system, but the period of implantation between epithelial penetration and the surrounding or penetration of maternal vessels (endotheliochorial or hemochorial, respectively) usually has the most appropriate conditions for greatest maternal sensitization. Subsequently, trophoblast directly exposed to maternal blood is generally syncytial, lacking intercellular pathways. Areas of the fetal-maternal interface where cytotrophoblast predominates are usually regions of buildup of deciduum, fibrin, or uterine epithelial proliferation, not of normal maternal connective tissue.

Lymphatics are absent from the immediate area of the placenta in those species with more intimate placentation, and intraground substance channels are often largely eliminated or restricted by decidual development. In a number of species, the area of communication between the placenta and the rest of the uterus is restricted to a stalk. Neither the surface trophoblast of hemochorial placentae nor the maternal endothelium of endotheliochorial placentae appears to become "sticky" or induce margination of lymphocytes. Consequently, it is apparent that the efferent pathways for maternal cellular response are restricted by a variety of mechanisms. Certain types of relationships appear to be proscribed:

1. There is no case of direct vascularization, and hence fetus and placenta are not a tissue transplant, unlike kidney, heart, or skin.
2. Placental villi or labyrinthine structures are not situated in unmodified lamina propria for more than a short time.

There are a number of puzzling morphological features of unknown significance to the immunological relationship:

1. What is the pathway of return of proteins from stroma to fetal vascular system in the placenta?
2. Why do all species seem to have one or more layers of trophoblast that appear extremely active in protein synthesis (even in the shrew, where the trophoblast is an apparently porous layer)?
3. What is the role of hypertrophied endothelium?
4. Why are there so many modified macrophages (Hofbauer cells) in the human placenta?
5. What is the role of the hypertrophied stromal cells present in some animals?

Perhaps, in confining most of our attention to two very different examples (muroid rodent and human), we are missing many interesting insights into the fetal-maternal immune relationship.

REFERENCES

Allen, W.R. 1975. Endocrine functions of the placenta. In *Comparative Placentation*. D.H. Steven, ed. New York: Academic Press, pp. 214-267.
Allen, W.R. 1979. Maternal recognition of pregnancy and immunological implications of trophoblast-endometrium interactions in equids. In *Maternal Recognition of Pregnancy*. Ciba Foundation Symposium 64 (new series). Amsterdam: Elsevier/North-Holland, pp. 323-346.

Allen, W.R., D.W. Hamilton, and R.M. Moor. 1973. The origin of equine endometrial cups. II. Invasion of the endometrium by trophoblast. *Anat. Rec.* 177:485.

Amoroso, E.C. 1952. Placentation. In *Marshall's Physiology of Reproduction.* A.S. Parkes, ed. Vol. 2. London: Longmans, Green, pp. 127-311.

Anderson, J.W. 1959. The placental barrier to gamma globulins in the rat. *Am. J. Anat.* 104:403.

Bernard, O., M.-A. Ripoche, and D. Bennet. 1977. Distribution of maternal immunoglobulins in the mouse uterus and embryo in the days after implantation. *J. Exp. Med.* 145:58.

Billington, W.D. 1979. The placenta and the tumor: variations on an immunological enigma. In *Placenta—A Neglected Experimental Animal.* P. Beaconsfield and C. Villee, eds. New York: Pergamon Press, pp. 267-282.

Bjorkman, N. 1973. Fine structure of the fetal-maternal area of exchange in the epitheliochorial and endotheliochorial types of placentation. *Acta Anat.* Suppl. 1, 86:1.

Bodley, H.D. 1974. The development of the chorioallantoic placental barrier in the bat *Macrotus waterhousii. Anat. Rec.* 178:313.

Buchanan, G.D. 1966. Reproduction in the ferret (*Mustela furo*). I. Uterine histology and histochemistry during pregnancy and pseudopregnancy. *Am. J. Anat.* 118:195.

Burns, J.M. 1981. Aspects of endocrine control of delay phenomena in bats with special emphasis on delayed development. In *Embryonic Diapause in Mammals.* A.P.F. Flint, M.B. Renfree, and B.J. Weir, eds. *J. Reprod. Fertil.* Suppl. 29, pp. 61-66.

Casley-Smith, J.R. 1980. The response of the microcirculation to inflammation. In *The Cell Biology of Inflammation.* G. Weissman, ed. Amsterdam: Elsevier/North-Holland, pp. 53-82.

Enders, A.C. 1965. A comparative study of the fine structure of the trophoblast in several hemochorial placentas. *Am. J. Anat.* 116:29.

Enders, A.C. 1966. The reproductive cycle of the nine-banded armadillo (*Dasypus novemcinctus*). In *Comparative Biology of Reproduction in Mammals.* I.W. Rowlands, ed. New York: Academic Press, pp. 295-310.

Enders, A.C. 1975. The implantation chamber, blastocyst and blastocyst imprint of the rat: a scanning electron microscope study. *Anat. Rec.* 182:137.

Enders, A.C. 1976. Cytology of human early implantation. *Res. Reprod.* 8:1.

Enders, A.C., and B.F. King. 1970. The cytology of Hofbauer cells. *Anat. Rec.* 167:231.

Enders, A.C., and S. Schlafke. 1971. Penetration of the uterine epithelium during implantation in the rabbit. *Am. J. Anat.* 132:219.

Enders, A.C., and W.A. Wimsatt. 1968. Formation and structure of the hemodichorial chorioallantoic placenta of the bat *(Myotis lucifugus). Anat. Rec.* 170:381.

Enders, A.C., W.A. Winsatt, and B.F. King. 1976. Cytological development of yolk sac endoderm and protein absorptive mesothelium in the little brown bat, *Myotis lucifugus. Am. J. Anat.* 146:1.

Fauve, R.M. 1974. Anti-inflammatory effects of murine malignant cells. *Proc. Nat. Acad. Sci. USA* 71:4052.

Flint, A.P.F., M.B. Renfree, and B.J. Weir, eds. 1981. Embryonic Diapause in Mammals. *J. Reprod. Fertil.* Suppl. 29.

Ford, S.P., A.S.H. Wu, and F. Stormshak. 1980. Effects of estradiol-17 and progesterone on endothelial cell morphology of ovine uterine arteries. *Biol. Reprod.* 23:135.

Given, R.L., and A.C. Enders. 1978. Mouse uterine glands during the delayed and induced implantation periods. *Anat. Rec.* 190:271.

Harris, J.W.S., and E.M. Ramsey. 1966. The morphology of human uteroplacental vasculature. *Contrib. Embryol. Carnegie Inst.* 38:43.

Hillemann, H.H., and A.I. Gaynor. 1961. The definitive architecture of placentae of nutria, *Myocastor coypus* (Molina). *Am. J. Anat.* 109:299.

Hoover, R.L., R. Folger, W.A. Haering, B.R. Ware, and M.J. Karnovsky. 1980. Adhesion of leukocytes to endothelium: roles of divalent cations, surface charge, chemotatic agents and substrates. *J. Cell Sci.* 45:73.

Kaufmann, P., J. Stark, and H.E. Stegner. 1977. The villous stroma of the human placenta. I. The ultrastructure of fixed connective tissue cells. *Cell Tissue Res.* 177:105.

King, B.F. 1977. In vitro absorption of peroxidase-conjugated IgG by human placental villi. Anat. Rec. 187:624.

King, B.F., and H.W. Mossman. 1974. The fetal membranes and unusual giant cell placenta of the jerboa (Jaculus) and jumping mouse (Zapus). Am. J. Anat. 140:405.

King, B.F., A.C. Enders, and W.A. Wimsatt. 1978. The annular hematoma of the shrew yolk-sac placenta. Am. J. Anat. 152:45.

Lawn, A.M., and A.D. Chiquoine. 1965. The ultrastructure of the placental labyrinth of the ferret (Mustela putorius furo). J. Anat. 99:47.

Leiser, R., and A.C. Enders. 1980. Light- and electron-microscopic study of the near-term paraplacenta of the domestic cat. II. Paraplacental hematoma. Acta Anat. 106:312.

O'Rahilly, R. 1973. Developmental stages in human embryos. Carnegie Institute of Washington. Publication 631.

Padykula, H.A. 1976. Cellular mechanisms involved in cyclic stromal renewal of the uterus. III. Cells of the immune response. Anat. Rec. 184:49.

Ramsey, E.M., M.L. Houston, and J.W.S. Harris. 1976. Interaction of the trophoblast and maternal tissues in three closely related primate species. Am. J. Obstet. Gynecol. 124:647.

Rumery, R.E., and R.J. Blandau. 1971. Loss of zona pellucida and prolonged gestation in delayed implantation in mice. In Biology of the Blastocyst. R.J. Blandau, ed. Chicago: University of Chicago Press, pp. 115-130.

Schlafke, S., and A.C. Enders. 1975. Cellular basis of interaction between trophoblast and uterus at implantation. Biol. Reprod. 12:41.

Schlafke, S., A.C. Enders, and R.L. Given. 1981. Cytology of the endometrium of delayed and early implantation with special reference to mice and mustelids. In Embryonic Diapause in Mammals. A.P.F. Flint, M.B. Renfree, and B.J. Weir, eds. J. Reprod. Fertil. Suppl. 29, pp. 135-141.

Siiteri, P.K., F. Febres, L.E. Clemens, R.J. Chang, B. Gondos, and D. Stites. 1977. Progesterone and maintenance of pregnancy: is progesterone nature's immunosuppressant? Ann. N.Y. Acad. Sci. 286:384.

Stephens, R.J. 1969. The development and fine structure of the allantoic placental barrier in the bat Tadarida brasiliensis cynocephala. J. Ultrastruct. Res. 28:371.

Steven, D., and G. Morriss. 1975. Development of the foetal membranes. In Comparative Placentation. D.H. Steven, ed. New York: Academic Press, pp. 214-267.

Weir, B.J. 1971. The reproductive organs of the female plains viscacha, Lagostomus maximus. J. Reprod. Fertil. 25:365.

Welsh, A.O., and A.C. Enders. 1981. Development and regression of the antimesometrial decidua of the rat uterus during normal pregnancy: a light microscopic and electron microscopic study. Anat. Rec. 199:272a.

Wild, A.E. 1979. Placental antibody transport and immunological protection—their cellular mechanisms. In Placenta—A Neglected Experimental Animal. P. Beaconsfield, and C. Villee, eds. New York: Pergamon Press, pp. 306-314.

Wimsatt, W.A. 1975. Some comparative aspects of implantation. Biol. Reprod. 12:1.

Wimsatt, W.A., A.C. Enders, and H.W. Mossman. 1973. A reexamination of the chorio-allantoic placental membrane of a shrew, Blarina brevicauda: resolution of a controversy. Am. J. Anat. 138:233.

Wislocki, G.B., and G.L. Streeter. 1938. On the placentation of the macaque (Macaca mulatta), from the time of implantation until the formation of the definitive placenta. Contrib. Embryol. Carnegie Inst. 27:1.

Wood, G.W. 1980. Mononuclear phagocytes in the human placenta. Placenta 1:113.

Chapter 9

IMMUNOLOGICAL INTERVENTIONS OF THE PLACENTA IN MATERNAL IMMUNOLOGICAL TOLERANCE TO THE FETUS

GUY ANDRÉ VOISIN

The placenta is the part of the conceptus that mammalian evolution has selected as the best means of ensuring secure developmental conditions for the fetus, allowing gestation to be beneficial to the fetus and deleterious to neither the fetus nor the mother. The placenta operates at different levels of the immune cycle (Figure 9-1) and by any of five classically described mechanisms of fetal protection against a potential maternal rejection reaction (Table 9-1).

The conceptus as an inefficient antigenic stimulus
The notion that the products of conception might be poorly antigenic was first postulated by Little in the early 1920's. The fetus is necessarily semisyngeneic to the mother—at least in the ideal case of syngeneic strains and their F_1 hybrids. There seems to be decreased expression of paternally inherited alloantigens, and expression of MHC antigens by the placenta remains controversial.

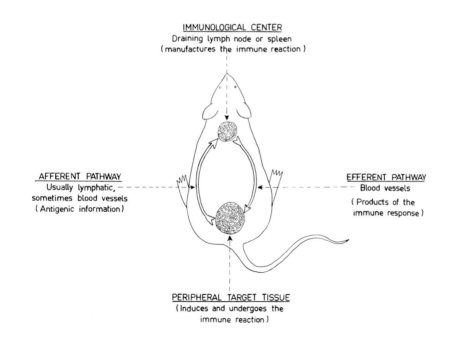

FIGURE 9-1. Cycle of the immune reaction.
The antigen-bearing target may be an artificial allograft, a fetoplacental unit, or an autoantigen-bearing organ.

181

TABLE 9-1
Classically-Cited Mechanisms of Fetal Protection Against a Potential
Maternal Rejection Reaction

	BUT
Conceptus = *non-immunogenic* (semi-syngenicity, antigenic immaturity, masking or non-expression of histocompatibility antigens)	Presence of MHC (HLA, H-2) antibodies
Uterus = *immunologically privileged site* (especially after decidualization)	Development of extrauterine pregnancies
Placenta = *anatomical barrier* (circulatory isolation, fibrin layer, sialomucin coat, trophoblast tight junctions)	Molecular and cellular traffic (both directions)
Mother = *immunologically incompetent* (non-specific immunodepression due to placental hormones and proteins)	Maternal immune reactivity practically normal
Mother = *immunologically tolerant of fetus* (cases of hyporeactivity to paternal grafts	Heterotopic fetus grafts rejected

Nonetheless, the fact that anti-MHC (HLA, H-2) antibodies specific for paternally inherited antigens are often found in the mother and the fact that heterotopic grafts of fetal tissues or organs are normally rejected indicate the presence of exposed antigens.

The uterus as as immunologically privileged site
The uterus has been compared to other immunologically privileged sites, such as the anterior chamber of the eye or the hamster cheek pouch, to which small allografts can be transplanted without rejection. Allografts do indeed show some enhancement of survival in the decidualized uterus. This effect is not seen, however, in the absence of natural or artificial decidualization (see Beer, 1979). Unlike other privileged sites, the decidualized uterus has substantial lymphatic drainage, and, furthermore, interruption of extra-uterine pregnancies is not a result of immunological factors.

The placenta as an immunological barrier
The fetus *in utero* is relatively isolated from the mother. The circulatory systems are separated by the cells of the placenta, and the trophoblastic cells themselves are firmly united by tight junctions and possess a glycoprotein covering. Nonetheless, molecules and cells can pass the placental barrier in both directions.

The pregnant mother as a nonspecifically immunosuppressed organism
The mother may become immunologically suppressed as a consequence of the physiological and endocrinological changes of pregnancy. It has been claimed that substances elaborated by the placenta play a part in this, and the cortical atrophy of the thymus in pregnancy is compatible with such a view. There is at present, however, no clear evidence for depressed immune reactivity in the pregnant female—perhaps fortunately for both mother and fetus.

The pregnant mother as specifically tolerant to her fetus
Operationally, the pregnant female behaves as if tolerant to her fetus (Breyere and Barrett, 1960; David and Volkringer, 1967). "True" tolerance, in the classical sense of an immunologically specific state of nonreactivity, does not exist in pregnancy, as the mother is able to make antipaternal antibodies and usually to reject a heterotopic tissue graft from her own fetus.

It is clear, therefore, that none of the preceding is adequate to account for the success of mammalian viviparity. These factors cannot, however, be neglected in any serious analysis and indeed merit re-examination. It is also clear that the reactivity of the maternal immune system is not invariant throughout gestation.

STAGES OF IMMUNOLOGICAL IMPORTANCE IN EMBRYONIC DEVELOPMENT AND PREGNANCY

It seems essential to consider three distinct phases in embryonic development and pregnancy (Table 9-2) at which immunological effects may be manifest.

Before implantation
After fertilization, the egg and the blastocyst, both coated by the zona pellucida, are free in the tubo-uterine lumen and are not in effective contact with

TABLE 9-2
Steps of Immunological Importance in Embryonic Development

Preimplantation
  Presence of transplantation antigens (H-2 and non-H-2) on blastocyst
  Absence of contact with mother's immune system (present in lumen; role of
  zona pellucida)

Implantation and invasion
  Intimate contact with mother
  MHC antigens not detected on external trophoblast

Post-implantation (mature placenta)
  Intimate contact with mother
  Presence of H-2 antigens (HLA ?)
  Immune response of the mother (type ?)

---

the maternal immune system. The absence or presence of MHC and non-MHC
antigens is therefore not particularly relevant at this time. The presence of
early pregnancy factor (EPF) a few hours after egg fecundation indicates that
early communication between the egg and and the mother does indeed exist.

Implantation and trophoblastic invasion
During implantation and trophoblastic invasion, there is intimate contact
between the conceptus and the mother. The cell layer of the developing
placenta, situated at the maternal-fetal interface, does not seem to express
MHC antigens in detectable form. Thus, immunological reactivity does not
pose a problem for the fetus even at this stage.

Full placentation (especially in the mature placenta)
After full placentation, the fetus can be described as a growing parabiont,
clearly expressing quantities of antigens, including those encoded by the MHC.
The placenta itself expresses both MHC and non-MHC antigens, at least in its
chorionic part. Antigen expression by trophoblast cells remains controversial.
In the human, the syncytiotrophoblast, which forms the interface between
mother and fetus, is claimed by some to be devoid of HLA antigens (see Faulk,
1980, and Chapter 12), while others have been unable to confirm this (Tongio
and Mayer 1975; Doughty and Gelsthorpe, 1976; Goodfellow et al., 1976; Loke
et al., 1980). In the mouse, mature placental spongiotrophoblast cells bear both
H-2 and non-H-2 antigens, as shown by a variety of techniques (Voisin and
Chaouat, 1974; Billington et al., 1977; Chatterjee-Hasrouni and Lala, 1979;
Wegmann et al., 1979; Chaouat et al., 1979b; see also Chapters 10 and 11).

Despite possible species differences, the pregnant female may potentially become immunized to the conceptus during this phase of full placentation. This in fact occurs, but immunization does not lead to rejection. We consider the reasons for this in the next section.

IMMUNOLOGICAL CONSIDERATIONS RELATING TO THE FATE OF THE FETUS

Rejection and facilitation reactions
It is well established that tissue allografts induce two reactions, with opposite consequences. The first, the rejection reaction, is mediated by cytotoxic T cells (T_c), delayed hypersensitivity mediating T cells (T_{DH}), and antibodies able to activate "killer" (K) cells ("null" cells, macrophages, or polymorphonuclear cells) or the complement (C') dependent cytolytic pathway. The other, which I have called the facilitation reaction (FR) (Voisin, 1962, 1971a) opposes the rejection reaction through regulatory mechanisms. It is a suppressor phenomenon and depends on cells, especially T lymphocytes (T_s); "enhancing" antibodies which do not fix complement (especially mouse IgG_1); immune complexes ("blocking factors"); and also perhaps anti-idiotypic antibodies. The interplay of factors in this double relationship may be schematized by the equation

$$IR = RR + FR.$$

It is important to understand that the FR regulates the RR at both induction and effector levels, as well as centrally and peripherally.

The fate of the antigen-bearing target cell depends on the balance between these two opposing reactions. In a typical allograft, the RR predominates and the graft is rejected. Under some circumstances, however, the FR may predominate and even lead to a state indistinguishable from immune tolerance. From the beginning, it has been our working hypothesis that the mother does react immunologically to her fetus but that the FR predominates over the RR. Although this predominance is usually too weak to allow the take of a paternal skin allograft across MHC barriers, it may permit acceptance of skin grafts across non-MHC differences (Breyere and Barrett, 1960) or of tumor allografts across the MHC (Voisin and Chaouat, 1974). One can see here the consequences of a delicate and multifactorial immunological balance.

THE IMMUNOLOGICAL ROLE OF THE PLACENTA

The trophoblast, which is situated at the interface between the maternal and fetal tissues, necessarily plays an essential role in the immunological interchange between the mother and the fetus. The placenta may be thought of as minimizing the maternal immune reaction against the conceptus, by favoring its FR component at the expense of the RR and possibly by blocking the latter at an effector level.

In my opinion, the experimental evidence supports this hypothesis. Most of what follows is devoted to demonstrating the existence of a maternal antifetal rejection reaction, the existence of a maternal facilitation reaction toward the fetus, and the crucial role of the placenta in the balance between these opposing tendencies.

THE PREGNANT MOTHER MOUNTS ELEMENTS OF A REJECTION REACTION AGAINST HER FETUS

By many criteria, the pregnant female is effectively primed against paternal-strain MHC antigens. Thus, maternal spleen cells are able to bring about the accelerated rejection of paternal-strain tumor grafts when adoptively transferred to nonpregnant maternal strain recipients in limited numbers (Figure 9-3; Chaouat et al., 1979).

Similarly, a state of delayed hypersensitivity to paternal antigens has been demonstrated in pregnancy, both in the human (Rocklin et al., 1973, 1979; Youtananukorn and Matangkasombut, 1973) and in the rat.

The adoptive transfer of H-2-compatible, non-H-2-incompatible spleen cells from Balb/c mice into newborn DBA/2 mice does not result in lethal graft-versus-host response (GVHR), except when the donors have been preimmunized. When, however, the Balb/c cells are taken from donors pregnant by DBA/2 (but not Balb/c) males, lethal GVHR ensues in more than 17 per cent of the recipients. Thus, pregnancy acts to prime the mother for a rejection reaction (Voisin et al., unpublished data).

In the present context, it is of interest to recall that preimmunization of female guinea pigs with guinea pig spermatozoa or the defined autoantigens S, P, and T (Voisin and Toullet, 1968) resulted in:

1. specific serum antibodies and delayed hypersensitivity,
2. no impairment of the fertility rate, and

3. increased abortion and fetal death with mummified fetuses and still-births (d'Almeida and Voisin, 1979).

These results demonstrate the presence of crossreactive antigens in spermatozoa and fetuses (or the placenta) capable of inducing a rejection reaction. They are perhaps analogous to some clinical situations in which antisperm antibodies are found in the sera of patients suffering habitual abortion.

Presence of embryo-specific cytotoxic cells
In mammals, the mother does not usually generate cells cytotoxic to the conceptus (Table 9-3). A major exception to this has been reported for the human, in which maternal lymphocytes cytotoxic to fetal and placental cells have been described (Youtananukorn et al., 1974; Timonen and Saksela, 1976).

By contrast, pregnant *Salamandra salamandra*, which represent an early attempt at vivparity in tetrapods, generate cells that are highly cytotoxic to the embryo. Spleen cells from pregnant females can kill, *in vitro*, up to 95 per cent of cells isolated from their own embryos. This phenomenon is specific, although crossreactivity may account for some degree of seeming nonspecificity. This difference between mammals and amphibians suggests that the placenta may prevent the induction or maturation of embryo-specific cytotoxic cells, since *Salamandra* lacks this organ. Thus, the mother in fact mounts an immune rejection reaction against her fetus, although it is completely overwhelmed by the facilitation reaction.

TABLE 9-3
Lack of Cytotoxic T Lymphocytes (CTL) in Spleens of Allopregnant Mice

	CYTOTOXIC INDEX OF SPLEEN CELLS FROM CBA MICE				
EXPERIMENT NUMBER	SPONTANEOUS RELEASE*	VIRGIN	ISOPREGNANT (OF CBA)†	ALLOPREGNANT (OF A/J)†	ALLOIMMUNIZED (WITH A/J CELLS)
1	34.9	−3.6	−3.2 (1)	0.5 (1)	36.9
2	41.6	−4.0	−2.3 (1)	−2.8 (1)	41.5
3	44.3	−11.6	−11.0 (1)	−9.7 (1)	26.3
4	42.2	−6.2	−6.7 (4)	−1.8 (4)	32.1
5	35.4	−2.7	−4.1 (4)	−5.7 (4)	23.7
6	29.3	−1.6	−0.4 (4)	4.9 (4)	38.3
7	44.5	−8.2	−4.2 (5)	−7.4 (6)	38.5
8	30.5	0.2	4.0 (5)	−0.8 (4)	26.3
9	31.2	1.8	1.1 (7)	−1.1 (4)	50.0
10	27.2	1.0	−0.7 (7)	4.1 (4)	36.3
MEAN	36.1	−3.5	−2.7	−2.0	35.0

Spleen cells from CBA mice; paternal strain A/J; ^{51}Cr labelled target = YAC 222
* YAC 222 alone; † parity of donor (parentheses)

THE MATERNAL ANTIPATERNAL FACILITATION REACTION: SPECIFIC ENHANCING ANTIBODIES

Voisin and Chaouat (1974) and Chaouat et al. (1979) described specific enhancing antibodies in the serum of pregnant female mice. These antibodies were found to be localized at the surface of the placental cells, precisely where they would be most useful. CBA (H-2^k) or C57BL/Ks (H-2^d) female mice were mated to A/J males and killed at day 15 of gestation. Immunofluorescence of frozen sections of the placentae showed a great deal of fixed immunoglobulin (Figure 9-2), which can be removed only by acid elution. This treatment dissociates immune complexes (Figure 9-3). These immunoglobulins are specific and can refix to an eluted placenta, provided it possesses paternal antigens. Quantitative immunofluorescence studies clearly showed that this refixing was not artifactual. Rhodamine-conjugated CBA anti-A/J alloimmune antibodies will bind to sections of (C57BL/Ks ♀ x A/J ♂) F$_1$ hybrid placentae in the same way as maternal immunoglobulins (Voisin and Chaouat, 1974). Immunoglobulins eluted from the placentae can also bind to paternal-strain (but not maternal) lymphocytes and thymocytes. Much less immunoglobulin is fixed to the placenta in syngeneic pregnancies. In addition, Voisin and Chaouat (1974) and Chaouat et al. (1979) showed that specific Ig fixes to spongiotrophoblast and not to the labyrinthine tissues

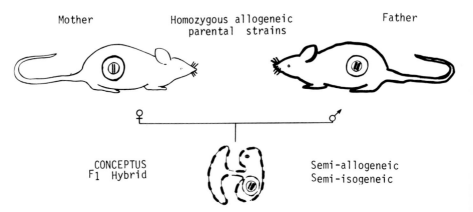

Mother　　　　Homozygous allogeneic　　　　Father
　　　　　　　　parental strains

♀ _____ ♂

CONCEPTUS　　　　　　　　Semi-allogeneic
F1 Hybrid　　　　　　　　Semi-isogeneic

FIGURE 9-2. Immunoglobulins spontaneously fixed at the periphery of trophoblast cells.
After careful perfusion of the whole organ, frozen sections of placenta are carefully washed several times and treated with fluorescein-conjugated rabbit antimouse Ig. All controls performed were found negative. Similar aspects are seen after treatment of eluted sections with placental eluates, provided the sections possess paternal antigens. From Chaouat et al., 1979b.

(Figure 9-4). The fixed immunoglobulins consist mainly of IgG_1, the antibody class most consistently associated with enhancing properties (Voisin et al., 1969; Duc et al., 1975). IgG_2 antibodies, which may also have enhancing activity (Duc et al., 1975), are found to a much lesser extent, and IgM and IgA are almost undetectable.

Antibodies eluted from the placenta are able to suppress the MLR *in vitro* (Revillard et al., 1976; Jeannet et al., 1977; Pavia and Stites, 1979). Serum from pregnant patients also shows MLR-suppressing activity (see Gusdon, 1976) and may inhibit cell-mediated immune reactions, such as macrophage migration inhibition (Youtananukorn and Matangkasombut, 1973; Pence et al., 1975). Most important, antibodies eluted from the murine placenta are endowed with specific *in vivo* enhancing properties. When injected into maternal strain recipients, they prevent rejection of paternal-strain tumor allografts (Figure 9-5), a finding confirmed in 14 different experiments.

The pregnant mother is thus able to produce enhancing antibodies, chiefly of the IgG_1 class. These antipaternal antibodies are specifically trapped in the placenta, especially on the spongiotrophoblast layer.

An indication that these antibodies might be of benefit to the fetus was obtained more than 20 years ago, when we showed that "enhancing" antibodies

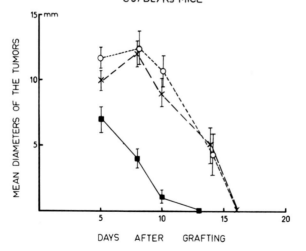

AGRESSIVE EFFECT (REJECTION REACTION) OF ALLOGENEIC PREGNANCY SPLEEN CELLS TRANSFERRED IN SMALL NUMBER ON SaI GRAFTED ON C57BL/Ks MICE

FIGURE 9-3. Aspect of placental section after acidic elution of water-non washable immunoglobulins spontaneously fixed on placenta.

Only acidic elution removes these antibodies. Similar aspects were seen after treatment of eluted sections with placental eluates when the sections did not possess paternal antigens.

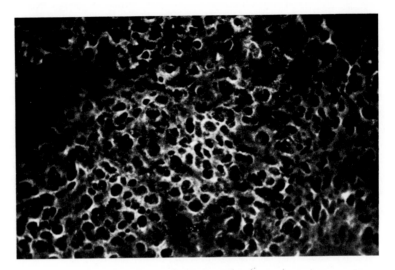

FIGURE 9-4. IgG₁ nature and spongiotrophoblast location of immunoglobulins spontaneously fixed on placenta.
Use has been made of monospecific rabbit antimouse IgG₁. S, spongiotrophoblast; L, labyrinth.
From Voisin and Chaouat, 1974.

could protect the newborn mouse from the GVH disease induced by injection of allogeneic cells (Voisin and Kinsky, 1962).

As previously described, spleen cells from Balb/c mothers pregnant by DBA/2 males (but not by Balb/c males) induced a 17 per cent incidence of lethal GVHR in Balb/c neonates. Injection of serum from allopregnant, but not isopregnant, mothers prevented the GVHR (Table 9-4). This protective effect of the serum was tentatively attributed to antibodies. A similar effect of immune serum from pregnant females has been reported in *Salamandra salamandra*, although in this instance protection is due chiefly to an anti-embryo-specific IgM—the only antibody class present in salamandra. A protein able to nonspecifically suppress spleen cell cytotoxicity *in vitro* has also been reported. We have found that anti-idiotypic antibodies directed against maternal antipaternal antibodies are present in a proportion of multiparous mice. No specific role has yet been attributed, however, to these antibodies in pregnancy (Chaouat et al., 1979a).

Finally, serum from pregnant human patients has frequently been reported to inhibit cell-mediated immunity, especially the mixed lymphocyte reaction and mitogenic proliferation, as well as *in vitro* lymphocytotoxicity. The inhibitory activity has been described as both specific and nonspecific (Pence et al., 1975; Gusdon, 1976; Kovithavongs and Dossetor, 1978; see

Rocklin et al., 1979). Nonetheless, serum factors are not the only immunoregulatory agents produced by the mother during gestation.

SUPPRESSOR CELLS

Several types of pregnancy-related suppressor cells have been described (see Chapter 16). They have been reported in the para-aortic lymph nodes (Clark and McDermott, 1978) and systemically (Chaouat and Voisin, 1979, 1980a, 1980b, 1981a, 1981b, 1981c, and to be published). The surface phenotype of the cells—T or non-T— and their specificity remain in dispute (Smith and Powell, 1977; Clark and McDermott, 1978; Chaouat and Voisin, ibid). They may act at different phases of the immune cycle, as discussed below. We have described and characterized the suppressor cells present in the spleens of allopregnant mice using in vivo and in vitro techniques.

T cells which decreased the ability of maternal-strain recipients to reject paternal-strain tumor allografts were found in the spleens of multiparous allopregnant mice (Figure 9-6; Chaouat et al., 1979b), as were T suppressor cells able to regulate the in vitro MLR. Maternal spleen cells mixed with mitomycin-treated paternal-- strain spleen cells gave an apparent secondary type of response that was soon inhibited, as demonstrated by comparison of the kinetics of the MLR obtained with responder cells from allopregnant mice with that obtained with virgin or isopregnant responding cells (Figure 9-7; Chaouat and Voisin, 1979).

Two populations of suppressor cells able to suppress MLRs between maternal- and paternal-strain cells were found in spleens from allopregnant

TABLE 9-4
Capacity of Female BALB/c Spleen Cells to Induce a Lethal GVHR in Newborn DBA/2 as a Function of In Vivo Preimmunization. Protective Capacity of a Relevant Allogestation Serum

		LETHAL GVHR IN NEWBORN DBA/2	
GROUP	ORIGIN OF BALB/c SPLEEN CELLS	NO. DEATHS/NO. INJECTED	PERCENTAGE DEATHS
1	Untreated virgin	2/55	4
2	Preimmunized virgin	12/13	92
3	Pregnant of BALB/c	20/290	5
4	Pregnant of DBA/2	42/243	17
5	Pregnant of DBA/2 and serum of same mother	17/241	7

Statistical significance:
Group 1 vs Group 5, N.S.; Group 3 vs Group 5, N.S.; Group 4 vs Group 5, $P < 0.001$; Group 1 vs Group 4, $p < 0.001$; Group 3 vs Group 4, $p < 10^{-6}$

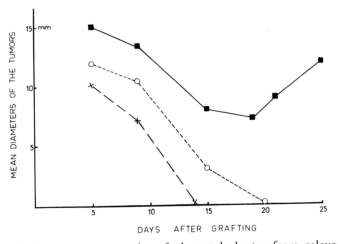

FIGURE 9-5. Enhancing properties of placental eluates from relevant alloge station.

Action of placental eluates on Sa 1 grafted onto C57BL/Ks mice; five mice per group. Origin of injected placental eluates or of transferred spleen cells: (■—■) C57BL/Ks ♀, A/J ♂; (O—O) C57BL/Ks ♀, C57BL/Ks ♂; (x—x), no treatment. From Chaouat et al., 1979b.

FIGURE 9-6. Suppressor effect on allograft rejection reaction of T cells (and not B cells) from spleens of corresponding allopregnant mice.

Allografted C57BL/Ks received 3 x 10⁷ T-enriched or 6 x 10⁷ B-enriched anti-Thy 1 + C′ treated spleen cells; four mice per group; two experiments performed. Origin of transferred C57BL/Ks spleen cells: (■—■), T-enriched spleen cells from allopregnant females (C57BL/Ks pregnant by A/J); (▼—▼), B-enriched anti-Thy 1 + C′ treated spleen cells from allopregnant females; (x—x), control (no treatment). From Chaouat et al., 1979b.

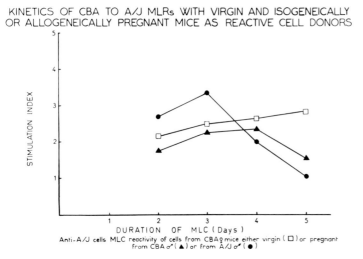

KINETICS OF CBA TO A/J MLRs WITH VIRGIN AND ISOGENEICALLY
OR ALLOGENEICALLY PREGNANT MICE AS REACTIVE CELL DONORS

DURATION OF MLC (Days)

Anti-A/J cells MLC reactivity of cells from CBA♀ mice either virgin (□) or pregnant
from CBA o⁴ (▲) or from A/J o⁴ (●)

FIGURE 9-7. Comparative kinetics of CBA versus mitomycin-treated A/J spleen cell MLR's, obtained with responder cells from virgin, isopregnant, and allopregnant (by A/J) mice.
On day 3, cells from allopregnant mice reacted significantly more strongly than each one of the two others. From day 4 on, the situation was significantly reversed. From Chaouat and Voisin, 1979; reprinted with permission.

mice. One was mitomycin-resistant and active in the inductive phase of the MLR, and the other was mitomycin-sensitive and active only when added during the proliferative phase (Figure 9-8; Chaouat and Voisin, 1979, 1980, 1981a).

Both of these suppressor cell populations are Thy 1^+, Ia^+, Ly $2,3^+$; both show genetic restriction of their inhibition of responder cells; both show specificity, although the strain distribution is complicated; and, finally, both can act through soluble factors (Table 9-5; Chaouat and Voisin, 1979, 1980, 1981a, 1981b, 1981c, and to be published).

In MLR and CML experiments, two types of suppression were detected: one weak, apparently nonspecific, and acting at the effector level; and the other, a more interesting one, acting on CTL generation. The latter cell was also Thy 1^+, Ly $2,3^+$ (Chaouat, Monnot, Hoffmann and Voisin, unpublished data). It must be emphasized that these suppressor cells were detected at the systemic (spleen) level and were found only in multiparous allopregnant mice. It therefore seems as though the pregnant mother makes an immune reaction toward the paternal antigens of her fetus. This immune reaction is much less than the usual allograft reaction, and cytotoxic T cell generation is particularly decreased. The protective immune facilitation components—enhancing antibodies and suppressor cells—are by contrast relatively increased. We shall next

193

TABLE 9-5
Identified Characteristics of the Two Populations of MLR Suppressor Cells in the Spleens of Allopregnant Mice

MLR REGULATORY SYSTEM	GENETIC REQUIREMENTS FOR SUPPRESSION		REGULATORY CELL MARKERS				SUPPRESSIVE ACTIVITY	SOLUBLE FACTORS PROPERTIES		
	IDENTITIES BETWEEN FATHER AND *IN VITRO* RETRIGGERING CELLS	COOPERATION (REG–RESP)	Thy 1	Ia	Ly	FcR		Ia SPECIFIC	I RESTRICTION	IDIOTYPE
Early inhibition (Day 0 of MLR)*	Ia (I-C, S, G) (+ non-H-2)	I-C (S,G)	+	+ (I-C, I-J)	1^-; $2,3^+$	$-?$	+	+	I-C	$+?$
Late inhibition (Day 2 of MLR)†	S,G (+I-C ?)	I-J	+	+ (I-J)	$2,3^+$	+	+			

* Mitomycin-resistant regulatory cells.
† Mitomycin-sensitive regulatory cells.

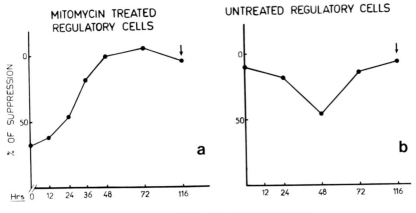

TIME OF REGULATORY CELL ADDITION TO MLC

FIGURE 9-8. Demonstration and kinetics of MLR suppression by two populations of spleen cells from allopregnant mice.
Comparison of the suppressive effect of cells from CBA mice pregnant by A/J, on CBA versus
mitomycin-treated C57BL/Ks spleen cell MLR, when they were added on different days after
having either been treated (a) or not (b) with mitomycin. Percentages of suppression as compared with cells from virgin CBA mice. MLR harvested on day 4 except for the 116 hour points
indicated by an arrow. From Chaouat and Voisin, 1981a; reprinted with permission.

consider the mechanisms of this decrease in and deviation of the immune
response and the role of the placenta in these mechanisms.

THE ROLE OF THE PLACENTA
IN IMMUNOREGULATION

The placenta intervenes at all levels of the immune reaction that the mother
mounts toward her fetus and uses all available mechanisms to prevent its
rejection. Some of these mechanisms are immunologically trivial, others are
more sophisticated. The former include a number of relatively "passive"
phenomena, such as the erection of barriers, a decrease in the expression of
histocompatibility antigens at the mother-conceptus interface, and the
suppression of local immune reactions (Table 9-6). All of these mechanisms
are immunologically nonspecific and result in a decreased immunological
reaction.

Mechanical barriers are of several types and occur at the tissue, cellular, and molecular levels. The first and most important is, of course, the
separation of maternal and fetal circulations, which prevents uncontrolled

195

exchanges between the two. The continuous trophoblastic layer, with its tight intercellular junctions, leads to an obligatory trophoblastic control over cellular and molecular feto-maternal traffic. A fibrin layer, more or less important depending on the mammalian species and the stage of gestation, forms a dense network which, even if not continuous, undoubtedly also decreases the cellular traffic.

The neuraminic acid-rich sialomucin coating of the trophoblastic microvilli at their apical (i.e., maternal) poles confers a highly negative charge to the cell surface, which could serve to repel maternal lymphocytes. These also have a negative charge, which is increased during gestation, possibly through the action of progesterone (Wrezlewicz et al., 1977).

Decreased histocompatibility antigen expression at the interface at the time of implantation has been documented. H-2 antigens have been detected on the preimplantation blastocyst and trophoblast (Håkansson et al., 1975; Searle and Jenkinson, 1978), although they may be masked by the zona pellucida (James, 1960). At the time of implantation and invasion, however, these antigens seem to disappear from the trophoblast, as judged by a number of criteria (Simmons and Russell, 1966; Jenkinson and Billington, 1974; Håkansson et al., 1975; Billington, personal communication). More recently, however, H-2 antigens have been detected at the surface of trophoblastic cells—except for external giant cells (Sellens et al., 1978; Chatterjee-Hasrouni and Lala, 1979; see Chapters 10 and 11). The presence of HLA antigens has been described on trophoblast cells (Loke et al., 1976; Lala, personal communication). Other workers have failed, however, to detect them throughout gestation, despite carefully controlled studies (Faulk and Temple, 1976; see Chapter 12).

Despite these controversies, four facts are apparent to me. First, MHC- and non-MHC-coded antigens are present on the placenta as a whole. Second, depending upon the species, and possibly due to diminished expression or to some type of masking, they are in lower or undetectable quantity at the cellular interface between the mother and the conceptus (giant cells of the spongiotrophoblast in mice and syncytiotrophoblast in humans). This is especially so at the time of implantation and invasion of the trophoblast. Third, there is an apparent difference between murine and human placentae. In the latter this apparent absence of MHC antigens may persist throughout gestation, whereas in the former, MHC antigens reappear at the surface with time. I do not consider this discrepancy hard to explain since the duration of gestation after implantation and invasion of the trophoblast in the mouse is not much greater than the allograft rejection time, whereas it is much longer in the human. This difference necessitates stricter and possibly more diversified means of preventing an immune rejection reaction in humans. Fourth, Class II MHC antigens (Ia, HLA-D) have not been detected on the murine or human trophoblast, although they have been detected on other placental cells.

In any case, the absence of Ia antigens, and the absence or decreased expression of Class I MHC antigens at the interface, are factors which may potentially decrease the immune reaction of the mother against the fetus. This type of reaction may also be minimized by some degree of local nonspecific immunosuppression.

Localized immunosuppression may be mediated by a number of agents, a few of which will be mentioned here. Substances present on the placenta may be important, viz., the two types of placental hormones previously mentioned as immunosuppressive: proteins such as human chorionic gonadotropin (hCG) and human placental lactogen (hPL), and steroids such as progesterone and estrogen; nonhormonal proteins synthesized in the placenta, for example, β_1 glycoprotein(s), SP1, PSBG, and PAPP-C; and placenta proteins PP5, and PAPP-A and -B. Additionally, proteins present in the placenta but not synthesized locally could also play a role; these include the placenta-associated α_2 glycoproteins(s) α_2 PAG, SP 3, and PPZ; placenta proteins PP-1, -2, -3, -4, -6, and -7; transcortin; and transferrin (see Part III, this volume).

More experiments need to be done before immunosuppressive properties can be assigned to the various placental substances. It must be emphasized that these substances do not act strongly enough at a systemic level to inhibit maternal immune reactivity. This does not mean that this type of substance does not play any role, but it may mean that the immunosuppression is strictly localized and confined to antigens of the conceptus. These three factors—mechanial barriers, decreasing histocompatibility antigen expression, and nonspecific local immune suppression—are by no means, however, the only ways in which the placenta regulates maternal immune reactivity.

SOPHISTICATED IMMUNOLOGICAL INTERVENTIONS OF THE PLACENTA

The placenta specifically modulates the antifetal immune response at the afferent, central, and efferent phases, chiefly by an immunoabsorption mechanism, but also by the formation of blocking immune complexes and by some kind of deviation of the immune reaction (Table 9-6).

An immunoabsorbent role for the placenta has been postulated in both the mouse and the human. We have shown above that maternal antibodies directed against paternal transplantation antigens bind to the placenta in the mouse (Voisin and Chaouat, 1974). In humans, maternal antibodies to transplantation antigens absent from the fetus were found in the fetal circulation (cord blood), and antibodies directed toward paternal transplantation antigens present in the fetus were trapped in the placenta (Tongio and Mayer, 1973, 1975; Doughty and Gelsthorpe, 1974, 1976). Placental immunoabsorp-

TABLE 9-6
Placental Interventions Contributing to the Absence of Fetal Rejection

---

Immunologically-trivial mechanisms

Building barriers
Separated circulations, fibrin layer, sialomucin coating, trophoblast tight junctions.

Non-expression of major histocompatibility complex antigens at the interface
Absent at implantation
Decreased (1/20 of lymphocyte levels) afterwards.

Non-specific local immunosuppression
Placental proteins and hormones

Sophisticated Immunological Mechanisms

Decreasing induction of immune reaction against conceptus through
1. preceding mechanisms
2. specific antibodies (masking)
3. local immune complexes (blocking)

Deviating the central immune reaction
1. Circulating immune complexes delivering signals on specific cells through bipolar bridging
2. Nonspecific substances and specific transplantation antigens
1 and 2 → specific suppressive response (facilitation reaction, i.e. suppressor cells and enhancing antibodies).

Blocking immune agents at effector level
Immunoabsorbent effect of placental antigens
Blocking effect of immune complexes on effector cells.

---

tion has been directly demonstrated in pregnant mice by passive injections of radiolabeled anti-H-2 antibodies (Wegmann et al., 1979; see Chapter 11).

The consequences of such specific absorption are important. This specific fixation leads to the masking of corresponding antigens on the placenta, making them unavailable either as sensitizers or as immunological targets. As we have seen, there is a concentration of enhancing IgG_1 antipaternal antibodies at the placental surface, i.e., at the first line of a potential immune attack, and potentially dangerous complement-fixing IgG_2 is also trapped there (Kinsky et al., 1970). Finally, there is also the formation of immune complexes (see below), which may be blocking or enhancing.

One must also consider that placental trophoblast cells have Fc receptors and may nonspecifically bind immunoglobulins, especially in the form of immune complexes.

The formation of immune complexes is by no means a simple matter because of the number of parameters that modify their composition and effects. Immune complexes fixed on the placenta were considered above.

Circulating immune complexes are a subject of more controversy. They do not seem to be more conspicuous in pathological than in normal human gestation. Their existence is expected, as a result of the association of pre-existing antibodies with surface antigens shed by the trophoblast. Furthermore, trophoblast embolism during gestation is bound to lead to immune complex formation in regions remote from the conceptus, such as the lungs.

Another important factor is the molecular composition of the complexes, especially the antigen:antibody ratio. It has long been known that immune complexes formed under conditions of antibody excess or at equivalence are usually in the form of precipitates and are removed by cells of the reticuloendothelial system. On the other hand, complexes in antigen excess become increasingly soluble. In extreme antigen excess, they seem biologically inactive, and in moderate antigen excess, they are both soluble and biologically active, i.e., able to fix and activate $C1q$, if the antibody is of a complement-fixing class. Unfortunately nothing is known, to my knowledge, of the antigen:antibody ratio of placental immune complexes.

The predominant class of antibody-forming immune complexes is of major importance. Here we have only incomplete information to guide us. For instance, IgG_1 is the main class of murine antipaternal H-2 antibody fixed on the placenta. It is probably the class of antiplacenta antibody predominantly synthesized, a fact that is important in view of its enhancing activity. Antihistocompatibility antigen IgG_2 antibodies in the form of complexes may under some circumstances lead to allograft enhancement (Duc et al., 1973, 1975; see also Voisin, 1980).

The blocking activity of immune complexes detected *in vitro* may act at both the induction and the effector phases of the immune response, and on both humoral and cellular responses. As shown with B lymphocytes (but conceivably also on other cells), an interaction of immune complexes with Fc receptors alone leads to a non-antigen-specific suppression. If, in addition, there is an interaction of the antigenic determinants with the specific receptors of the cell, antigen-specific suppression is induced. At the effector level, blocking of antibody-dependent cell-mediated cytotoxicity, of cell-mediated lymphocytotoxicity, and of delayed hypersensitivity reactions has been described (see the review by Theophilopoulos and Dixon, 1979).

Immunodeviating activity is one of the most fascinating possibilities of the immune response; for example, IgG_1 versus IgG_2 or suppressor T lymphocytes *versus* cytotoxic and/or helper T lymphocytes. It has been shown that

199

immune complexes can induce immunodeviation which affects the synthesized antibody isotypes (Vuagnat et al., 1973a, 1973b). Results have also been obtained that indicate that the placenta possesses substances that can persuade the mother's immune system to make suppressor instead of killer cells against paternal antigens (Chaouat et al., 1980).

SUMMARY AND CONCLUSIONS

There are six main interventions of the placenta with respect to the maternal antifetal reaction. Thus, the placenta acts as:

1. a relatively nonselective barrier, for both cells and molecules;
2. a relatively nonantigenic interface;
3. a generator of nonspecific immunosuppressive agents;
4. an immunoabsorbent;
5. a generator of immune cell blockade; and
6. a generator of immunodeviating substances.

These various mechanisms result in three main immunological consequences:

1. a decrease in the immune response,
2. a modification of its quality, and
3. protection of the conceptus against an immune response.

One may conclude from the foregoing that the pregnant mother does react, however moderately, to the alloantigens of her fetus(es). She does so according to the general scheme of an immune reaction against foreign cells, i.e., associating the elements of a rejection reaction (weak) with those of a regulatory facilitation reaction (stronger). The conceptus, through substances present in the placenta, "persuades" the immune system of the mother to elaborate predominantly enhancing antibodies and suppressor cells, i.e., the immune agents of the facilitation reaction. In addition, several mechanisms, either specific or nonspecific in nature, are concentrated at the placental level. Finally, and to broaden the view, the immunology of gestation gives the best support to the long-standing theory, mentioned at the beginning of this chapter, that an immune reaction against a foreign invader is regulated by a simultaneous homeostatic reaction, the facilitation reaction. The selective pressures of evolution have led to mechanisms for having one or the other reaction predominate, depending on whether the antigen-bearing target is necessary or detrimental to the survival of the species. In experimental or pathological situations, the reverse may happen. Fortunately, this is the exception rather than the rule.

ACKNOWLEDGMENTS

This chapter has been made possible by the work of my collaborators and colleagues, cited in the text, figures, and references. The assistance of M. Vioux as a faithful secretary-typist is gratefully acknowledged. Mr. Pinet, draftsman, Mr. Issoulie, photographer, and Mlle. Sorgue are also thanked for their help.

REFERENCES

d'Almeida, M., and G.A. Voisin. 1979. Resistance of female guinea pig fertility to efficient iso-immunization with spermatozoa autoantigens. *J. Reprod. Immunol.* 1:237.

Beer, A.E. 1979. The paradox of foeto-placental units as allografts. In *Recent Advances in Reproduction and Regulation of Fertility.* G.P. Talwar, ed. Amsterdam: Elsevier-North Holland, pp. 501-513.

Billington, W.D., E.J. Jenkinson, R.F. Searle, and M.H. Sellens. 1977. Alloantigen expression during early embryogenesis and placental ontogeny in the mouse: immunoperoxidase and mixed hemadsorption studies. *Transplant. Proc.* 9:1371.

Breyere, E.J., and M.K. Barrett. 1960. Prolonged survival of skin homografts in parous female mice. *J. Nat. Cancer Inst.* 25:1405.

Chaouat, G., and G.A. Voisin. 1980. Regulatory T cell subpopulations in pregnancy. I. Evidence for suppressive activity of the early phase of MLR. *J. Immunol.* 122:1383.

Chaouat, G., and G.A. Voisin. 1980. Regulatory T cells subpopulations in pregnancy. II. Evidence for suppressive activity of the late phase of MLR. *Immunology* 39:239.

Chaouat, G., and G.A. Voisin. 1981a. Regulatory T cells in pregnancy. III. Comparison of early acting and late acting suppressor T cells in MLR: evidence for involvement of differential T cells subsets. *Immunology* 44:393.

Chaouat, G., and G.A. Voisin. 1981b. Regulatory T cells in pregnancy. IV. Genetic characteristics and mode of action of early MLR suppressive T cells populations. *J. Immunol.* 127:1335.

Chaouat, G., and G.A. Voisin. 1981c. Regulatory T cells in pregnancy. V. Allopregnancy induced T cell suppressor factor is H-2 restricted and bears Ia determinants. *Cell. Immunol.* 62:186.

Chaouat, G., R.G. Kinsky, H.T. Duc, and P. Robert. 1979a. The possibility of anti-idiotypic activity in multiparous mice. *Ann. Immunol.* (Inst. Pasteur) 130:601.

Chaouat, G., G.A. Voisin, D. Escalier, and P. Robert. 1979b. Facilitation reaction (enhancing antibodies and suppressor cells) and rejection reaction (sensitized cells) from the mother to the paternal antigens of the conceptus. *Clin. Exp. Immunol.* 35:13.

Chatterjee-Hasrouni, S., and P.K. Lala. 1979. Localization of H-2 antigens on mouse trophoblast cells. *J. Exp. Med.* 149:1238.

Clark, D.A., and M.R. MacDermott. 1978. Impairment of host *versus* graft reaction in pregnant mice. I. Suppression of cytotoxic cell generation in lymph nodes draining the uterus. *J. Immunol.* 121:1389.

David, G., and P. Volkringer. 1967. Tolérance à l'homogreffe cutanée dans le post-partum chez la souris. *Ann. Inst. Pasteur* 113:483.

Doughty, R.W., and K. Gelsthorpe. 1976. Some parameters of lymphocyte antibody activity through pregnancy and further eluates of placental material. *Tissue Antigens* 8:43.

Duc, H.T., R.G. Kinsky, J. Kanellopoulos, and G.A. Voisin. 1975. Biological properties of transplantation immune sera. IV. Influence of the course of immunization, dilution and complexing to antigen on enhancing activity of Ig classes. *J. Immunol.* 115:1143.

Duc, H.T., R.G. Kinsky, and G.A Voisin. 1973. Efficacité préférentielle des immuns complexes dans la facilitation des allogreffes tumorales. *Ann. Immunol.* (Inst. Pasteur) 124:567.

Faulk, W.P., and A. Temple. 1976. Distribution of β_2 microglobulin and HLA in chorionic villi of human placentae. *Nature* 262:799.

Goodfellow, P.N., C.J. Barnstable, W.F. Bodmer, D. Snary, and M.J. Crumpton. 1976. Expression of HLA system antigens on placenta. *Transplantation* 22:595.

Gusdon, J.P. 1976. Maternal immune response in pregnancy. In *Immunology of Human Reproduction.* J.S. Scott, and W.R. Jones, eds. London: Academic Press, pp. 103-125.

Håkansson, S., S. Heyner, K.G. Sundqvist, and S. Bergström. 1975. The presence of paternal H-2 antigens on hybrid mouse blastocysts during experimental delay of implantation and the disappearance at these antigens after onset of implantation. *Int. J. Fertil.* 20:137.

James, D.A. 1960. Antigenicity of the blastocyst masked by the zona pellucida. *Transplantation* 8:846.

Jeannet, M., C. Werner, E. Ramirez, P. Vassalli, and W.P. Faulk. 1977. Anti-HLA, anti-human "Ia-like" and MLC blocking activity of human placental IgG. *Transplant. Proc.* 9:1417.

Jenkinson, E.J., and W.D. Billington. 1974. Differential susceptibility of mouse trophoblast and embryonic tissue to immune cell lysis. *Transplantation* 18:286.

Kinsky, R., I. Chouroulinkov, and G.A. Voisin. 1970. Complete *in vivo* destruction of allografted sarcoma 1 cells by circulating antibodies. *Transplantation* 10:450.

Kovithavongs, T., and J.B. Dossetor. 1978. Suppressor cells in human pregnancy. *Transplant. Proc.* 10:911.

Loke, Y.W., V.C. Joysey, and R. Borland. 1971. HL-A antigens on human trophoblast cells. *Nature* 232:403.

Loke, Y.W., A. Whyte, and S.P. Davies. 1980. Differential expression of trophoblast-specific membrane antigens by normal and abnormal human placentae and by neoplasms of trophoblastic and non-trophobastic origin. *Int. J. Cancer* 25:459.

Pavia, C.S., and D.P. Stites. 1979. Humoral and cellular regulation of alloimmunity in pregnancy. *J. Immunol.* 123:2194.

Pence, H., W.M. Petty, and R.E. Rocklin. 1975. Suppression of maternal responsiveness to paternal antigens by maternal plasma. *J. Immunol.* 114:525.

Revillard, J.P., J. Brochier, M. Robert, M. Bonneau, and J. Traeger. 1976. Immunologic properties of placental eluates. *Transplant. Proc.* 8:275.

Rocklin, R.E., J.L. Kitzmiller, and M.D. Kaye. 1979. Immunobiology of the maternal-foetal relationship. *Ann. Rev. Med.* 30:375.

Rocklin, R.E., J.E. Zuckerman, E. Alpert, and J.R. David. 1973. Effect of multiparity on human maternal hypersensitivity to foetal antigen. *Nature* 241:130.

Searle, R.F., and E.J. Jenkinson. 1978. Localization of trophoblast-defined surface antigens during early mouse embryogenesis. *J. Embryol. Exp. Morph.* 43:147.

Sellens, M.H., E.J. Jenkinson, and W.D. Billington. 1978. Major histocompatibility complex and non-major histocompatibility complex antigens on mouse ectoplacental cone and placental trophoblastic cells. *Transplantation* 25:173.

Simmons, R.L., and P.S. Russell. 1966. The histocompatibility antigens of fertilized mouse eggs and trophoblast. *Ann. N.Y. Acad. Sci.* 129:35.

Smith, G. 1978. Inhibition of cell mediated microcytotoxicity and stimulation of mixed lymphocte reactivity by mouse pregnancy serum. *Transplantation* 26:278.

Smith, R.N., and A.E. Powell. 1977. The adoptive transfer of pregnancy-induced unresponsiveness to male skin grafts with thymus-dependent cells. *J. Exp. Med.* 146:899.

Theophilopoulos, A.N., and F.J. Dixon. 1979. The biology and detection of immune complexes. *Adv. Immunol.* 28:89.

Timonen, T., and E. Saksela. 1976. Cell-mediated anti-embryo cytotoxicity in human pregnancy. *Clin. Exp. Immunol.* 23:462.

Tongio, M.-M., and S. Mayer. 1975. Transfer of HL-A antibodies from the mother to the child. *Transplantation* 20:163.

Voisin, G.A. 1962. Immunological tolerance to living cells, homologous disease and immunological facilitation (enhancement phenomenon). A working hypothesis allowing a unified concept. In *Symposium on Mechanisms of Immunological Tolerance* M. Hasek, A. Lengerova, and M. Vojtiskova, eds. Prague: Publishing House of the Czechoslovak Academy of Sciences, pp. 435-455.

Voisin, G.A. 1971a. Immunological facilitation, a broadening of the concept of enhancement phenomenon. *Progr. Allergy* 15:328.

Voisin, G.A. 1971b. Immunity and tolerance: a unified concept. *Cell. Immunol.* 2:670.

Voisin, G.A. 1979. Immune agents of the facilitation reaction (enhancing antibodies and suppressor cells) in pregnancy. Their possible role in the protection of the placental allograft. In *Placenta: A Neglected Experimental Animal.* P. Beaconsfield and C. Villee, eds. Oxford: Pergamon Press, pp. 283-294.

Voisin, G.A. 1980. Role of antibody classes in the regulatory facilitation reaction. *Immunol. Rev.* 49:3.

Voisin, G.A., and G. Chaouat. 1974. Demonstration, nature and properties of antibodies fixed on maternal placenta and directed against paternal antigens. *J. Reprod. Fertil.* Suppl. 21:89.

Voisin, G.A., and R. Kinsky. 1962. Protection against runting by specific treatment of newborn mice, followed by increased tolerance. In *Ciba Foundation Symposium on Transplantation.* G.W. Wolstenholme and M.P. Cameron, eds. London: J. & A. Churchill Publ., pp. 286-326.

Voisin, G.A., and F. Toullet. 1968. Etude sur l'orchite aspermatogénétique autoimmune et les autoantigènes de spermtazoïdes chez le cobaye. *Ann. Inst. Pasteur.* 114:727.

Voisin, G.A., R. Kinsky, F. Jansen, C. Bernard. 1969. Biological properties of antibody classes in transplantation immune sera. *Transplantation* 8:618.

Vuagnat, Ph., T. Neveu, and G.A. Voisin. 1973a. Immuno-deviation by passive antibody, an expression of selective immuno-depression. I. Action of guinea pig IgG$_1$ and IgG$_2$ anti-hapten antibodies. *Eur. J. Immunol.* 3:90.

Vuagnat, Ph., T. Neveu, and G.A. Voisin. 1973b. Immuno-deviation by passive antibody, an expression of selective immuno-depression. II. Action of guinea pig IgG$_1$ and IgG$_2$ anti-carrier antibodies. *J. Exp. Med.* 137:265.

Wegmann, T.G., B. Singh, and G.A. Carlson. 1979. Allogeneic placenta is a paternal strain antigen immunoabsorbent. *J. Immunol.* 122:270.

Wrezlewicz, W., E. Rashkoff, T. Lucas, N. Ramasamy, and P.N. Sawyer. 1977. Effects of human chorionic gonadotropin and anti-thymocyte globulin on the electrophoretic mobility of human lymphocytes. *Transplant. Proc.* 9:1441.

Youtananukorn, V., and P. Matangkasombut. 1973. Specific plasma factors blocking human maternal cell-mediated immune reaction to placental antigens. *Nature (New Biol.)* 242:110.

Youtananukorn, V., P. Matangkasombut, and V. Osathanondh. 1974. Onset of human maternal cell-mediated immune reaction to placental antigens during the first pregnancy. *Clin. Exp. Immunol.* 16:593.

Chapter 10

HISTOCOMPATIBILITY ANTIGENS OF MOUSE TROPHOBLAST OF SIGNIFICANCE IN MATERNAL-FETAL IMMUNOLOGICAL INTERACTIONS

W.D. BILLINGTON
S.C. BELL
G. SMITH

The antigenic status of the trophoblast is of paramount importance in relation to the survival of the conceptus as an intrauterine allograft. During the early stages of post-implantation development, the embryo is encapsulated within a trophoblastic barrier derived from the single-cell-layered trophectodermal wall of the blastocyst. Throughout the remainder of pregnancy in the mouse, the interface with maternal tissues is provided by specialized trophoblast populations within the placenta and additionally, in the latter half of gestation, by the extraembryonic yolk-sac membrane, which is exposed by the breakdown of the acellular Reichert's membrane and its covering layer of trophoblast giant cells in the nonplacental region (Billington, 1975). Thus the embryo itself is never in direct tissue contact with the potentially hostile maternal environment, and there is no evidence of any significant transfer of maternal lymphocytes to the fetal circulation in normal pregnancy (Adinolfi, 1975).

Evidence for histocompatibility antigen expression on the maternal-facing endodermal cells of the mouse yolk sac and the possible reasons for survival of this membrane *in vivo* have been considered in earlier reports by Jenkinson and Billington (1974a) and Jenkinson et al. (1975) and in the recent studies of Parr and his colleagues (1980). Evidence relating to alloantigens on mouse trophoblast and their possible role in maternal-fetal immunological interactions is presented herein.

ANTIGENICITY OF TROPHOBLAST

The alloantigenic status of mouse trophoblast has by now been examined on most stages of its development, from the trophectoderm of the preimplantation blastocyst to the major trophoblastic components of the definitive placenta. For this work, a wide range of assays has been used, many of them on short-term *in vitro* cultures of the tissue. A number of reviews have documented the earlier work in this field (e.g., Johnson, 1975; Billington, 1976; Wiley, 1979; Heyner, 1980), and the present summary concerns only those studies that have used animals and reagents that allow a clear distinction between major (H-2) and minor (non-H-2) histocompatibility antigen expression on well-defined trophoblast material (see Table 10-1).

Some caution is needed in the interpretation of studies using conventional alloantisera. Heyner (1980 and Chapter 4 of this volume) has recently drawn attention to the fact that H-2 and non-H-2 alloantisera raised by

TABLE 10-1

Histocompatibility Antigen Expression on Trophoblast Cells Throughout Development Assessed by Immunoperoxidase or Mixed Hemadsorption (Red Cell Marker) Assays

| ANTIGENS IDENTIFIED | TROPHECTODERM | | | ECTOPLACENTAL CONE TROPHOBLAST (7½ DAY) | | PLACENTAL TROPHOBLAST | | | |
| | | | | | | IMMATURE (9 day) | MATURE (13 day) | | |
	EXPANDED BLASTOCYST	ACTIVATED BLASTOCYST	BLASTOCYST OUTGROWTH	OUTER GIANT CELLS	INNER CORE CELLS	ENTIRE ORGAN	ENTIRE ORGAN	LABYRINTHINE TROPHOBLAST	SPONGIO TROPHOBLAST
H-2	+	–	–	–	++	+	++	(+)	++
Ia	–	(–)	(–)	–	ND	ND	–	–	–
non-H-2	+++	–	–/+++	–/+++	+++	ND	+++	ND	ND

+++ dense labeling; ++ medium labeling; + sparse labeling; (+) few cells only; – negative; (–) presumed negative but not determined; –/+++ strain dependent expression; ND not determined

standard immunization procedures may also contain antibodies to other cellular antigens, such as the Ea erythrocyte and Lyt lymphocyte membrane antigens, which may have a more widespread distribution than previously suspected. The possibility that contaminating antiviral antibody may contribute to any observed reactivity (Klein, 1975) is now well recognized and must also be taken into account. The use of monoclonal antibodies in conjunction with conventional antisera is advised wherever possible.

TROPHECTODERM

Non-H-2 antigens are readily detected on the preimplantation trophectoderm with a variety of *in vivo* and *in vitro* assays (Searle et al., 1974; Johnson, 1975; Muggleton-Harris and Johnson, 1976; Heyner et al., 1980). In contrast, H-2 antigens have been detected only by the very sensitive antiglobulin (Håkansson et al., 1975) and immunoperoxidase (Searle et al., 1976) techniques, probably reflecting the lower density of these determinants. The antigens disappear from the trophectoderm cell surface when the blastocyst is activated for implantation under the influence of estrogen and when it is allowed to undergo outgrowth *in vitro* (Searle et al., 1976). An extended analysis of the expression of antigens on the trophoblast cells in blastocyst outgrowths has confirmed that H-2 antigens are absent and shown that there are strain differences in the expression of non-H-2 antigens (Sellens, 1977).

There is a report of the presence of β_2-microglobulin on the trophectoderm (Håkansson and Peterson, 1976), which is of interest in relation to the low levels of H-2 expression and the absence of Ia antigens (Jenkinson and Searle, 1979).

EARLY PROLIFERATION OF TROPHOBLAST

Numerous workers have reported on the antigenicity and immunogenicity of the ectoplacental cone (EPC) trophoblast, which is conveniently dissected from the mouse conceptus at 7½ days postcoitum (see Billington, 1976). The EPC has been examined in transplantation assays and both as an intact tissue and as a monolayer outgrowth *in vitro* in a variety of assays for antigen detection. It is now known that the EPC consists of a proliferating core of diploid cells and an outer layer of polyploid, or more likely polytene, secondary giant cells (Rossant and Papaioannou, 1977).

Using intact EPCs from a number of different inbred strains with the immunoperoxidase assay at the electron microscope level, it has been shown that the outer trophoblastic giant cells express neither H-2 nor non-H-2 antigens (Searle et al., 1976). However, when the EPC is allowed to grow out *in vitro* for three days, populations of cells can be identified which label with both H-2 and non-H-2 antisera in a mixed hemadsorption assay (MHA) (Sellens et al., 1978). The trophoblastic giant cells were invariably unlabeled with antisera recognizing only H-2 specificities and, with the single exception of cells from the CBA strain of mouse (not examined in the immunoperoxidase study), were also nonreactive with non-H-2 antisera. Smaller, non-giant cells in the cultures expressed both H-2 and non-H-2 antigens in most of the strains examined. While it is tempting to assume that these small reactive cells represent those of the diploid trophoblast population, this has yet to be finally established. It can be concluded that H-2 antigens are absent from the outer trophoblastic giant cells of the EPC, that non-H-2 antigens may be expressed on these cells in certain strains of mice, and that there is a variability between strains in the expression of both H-2 and non-H-2 on the presumed diploid trophoblast population.

The immunogenicity of EPC trophoblast has been assessed by several investigators, with conflicting results (see Searle et al., 1975; and P.V. Taylor et al., 1979). The explanation possibly lies in minor differences in the test systems and in the variable expression of both the serologically detected H-2 K/D region products and the I region associated (Ia) antigens on the two trophoblast populations in the different inbred strains used. Ia antigens are not expressed on the trophoblastic giant cells of the EPC (Jenkinson and Searle, 1979), but the inner diploid population has yet to be examined.

PLACENTAL TROPHOBLAST

The EPC undergoes rapid proliferation and differentiation to give rise to the three main trophoblastic components of the definitive placenta: the labyrinthine trophoblast, over which the main maternal blood flow occurs; the spongiotrophoblast, in cellular contact with maternal decidual tissue; and trophoblast giant cells at the maternal boundary of the spongiotrophoblast. The labyrinthine tissue is seen at the ultrastructural level to be composed of three intimately opposed layers of trophoblast, the outermost of which is cellular and the two inner ones apparently syncytial (Enders, 1965). The ultimate aim must be to examine each of these trophoblast populations for histocompatibility antigen expression, but this has been approached by few studies to date and many reports have, in fact, failed to provide convincing evidence that the reactivities observed were directed against trophoblastic rather than to other fetal (or even maternal) cell populations in the placenta.

Using short-term (24-hour) cultures of cells obtained from mature (13-day) placentae in a mixed hemadsorption assay, Sellens et al. (1978) demonstrated the presence of both H-2 and non-H-2 antigens on the majority of cells in the population. The use of antisera directed against paternal H-2 specificities with F_1 hybrid matings and the microdissection of the placentae to remove the major fetal mesenchymal elements (confirmed by examination of smear preparations) provided strong evidence for trophoblastic cell involvement in the observed reactivities. Further support for expression of H-2 antigens on the trophoblastic cell component of the mature placenta was presented by Chatterjee-Hasrouni and Lala (1979), who used a radioimmunolabeling assay with monospecific antisera followed by quantitative autoradiography on placental cell suspensions. Grain counts on labeled cells indicated a level of antigen expression on 14-to-16-day material equivalent to that on adult thymocytes, with a 50 per cent increase at day 18.

In a study designed to examine the major subpopulations of placental trophoblast, Jenkinson and Owen (1980) used the MHA assay on overnight cultures of cell suspensions prepared from isolated fragments of the labyrinthine and spongiotrophoblast regions of the mature placenta. Almost all spongiotrophoblast cells expressed readily detectable levels of H-2 antigen, while the great majority of cells from the labyrinth were unlabeled. The absence of H-2I region products on spongiotrophoblast indicated that the reactivity on these cells was due to H-2K and/or D products. This was confirmed in additional experiments with monoclonal anti-H-2K^k antibody in an immunofluorescence assay. Only a few cells labeled in cultures of immature (day 9) whole placenta, suggesting that H-2 expression is only beginning at this stage of development. There is other evidence to support the conclusion that H-2 antigen "switch-on" on embryonic tissues occurs only at the midgestation stage (Jacob, 1977; Kirkwood and Billington, 1981).

The use of purified populations of trophoblast in conjunction with monospecific antisera has thus provided incontrovertible evidence for H-2 antigen expression on cultured cells of the spongiotrophoblast of the mature placenta. The only remaining question is whether or not these antigens are exposed *in vivo*. Support for this has come from two directions. Voisin and Chaouat (1974) observed immunofluorescent labeling of spongiotrophoblast in cryostat sections, and Wegmann and his colleagues (1979a, 1979b) have reported specific absorption by intact placentae of injected monoclonal anti-paternal H-2 antibody, with binding localized by autoradiography to spongiotrophoblast and not labyrinthine trophoblast (see Chapter 11, this volume). An alternative proposal is that antigens detected on cells *in vitro* may not be represented on the exposed apical surface of the cell *in vivo* (Jenkinson and Billington, 1974a; Munro, 1975). Immunoferritin labeling studies on the distribution of H-2 antigens on adult epithelial lining cells have demonstrated such polarization, with antigen expression on the laterobasal, but not the apical, membranes (Parr and Kirby, 1979; Parr et al., 1980). This is clearly an important point to establish, since it has numerous implications for maternal-fetal immunological interactions (see below).

SUSCEPTIBILITY OF TROPHOBLAST TO IMMUNE ATTACK *IN VITRO*

Several earlier studies have documented the insusceptibility of proliferating ectoplacental cone trophoblast to immune attack by cells or antibody both *in vivo* and *in vitro* (Kirby et al., 1966; Simmons and Russell, 1966; Vandeputte and Sobis, 1972; Jenkinson and Billington, 1974b). This is now largely explicable in terms of the absence of serologically detectable H-2K/D antigens on the outer trophoblastic giant cells at this stage of development (referred to above). Similar studies with placental material are difficult to interpret because of an inability to assess the specific involvement of trophoblastic elements in the anatomically complex organ (see Billington, 1976). In the light of the finding of both H-2 and non-H-2 antigens on placental trophoblast, we recently examined the susceptibility of placental (and embryonic) cells, at various stages of gestation, to killing by lymphocytes specifically sensitized to paternal alloantigens, using either whole placental cultures or isolated populations of spongiotrophoblast and labyrinthine tissue at day 13.

WHOLE PLACENTAL CULTURES

Susceptibility to effector cell attack is dependent upon gestational age (Figure 10-1). With the earliest material examined (day 9), the placental cells survived while a significant proportion of the embryonic cells were killed. However, by day 13, all tissues were susceptible to killing and preliminary results suggest that this later pattern of susceptibility can be found as early as day 10. These results correlate well with the data on H-2 antigen expression during embryogenesis and placental ontogeny, which is not unexpected since it is known that cytotoxic T cell killing primarily requires recognition of H-2K/D antigens rather than non-H-2 antigens. Killing of placental cells was always significantly less than that observed with embryonic cells from the same conceptus (but more than with syngeneic placenta), suggesting either fewer placental cells with H-2 antigenic determinants or reduced

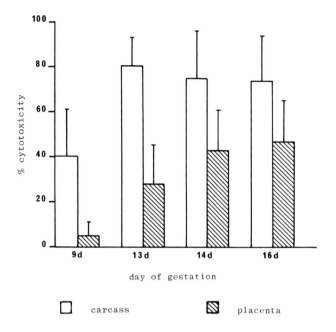

FIGURE 10-1. *In vitro* susceptibility of (C57BL/10ScSn x CBA/Ca)F_1 placental cells taken from various stages of gestation to killing by hyperimmue lymphocytes directed against paternal alloantigens.

Monolayers of placental and embryonic carcass cells were prepared from the conceptuses of C57BL/10 (H-2^b) females mated to CBA/Ca (H-2^k) males as described by Jenkinson and Owen (1980). The cultures were allowed to plate out overnight before adding effector spleen cells from either C57BL/10 virgin females or C57BL/10 hyperimmunized against CBA/Ca cells to triplicate monolayers of each cell type. After a further 48 hours, the effector cells were washed off and target cell lysis assessed using ^{3}H-uridine as a terminal label (Smith and Nicklin, 1979).

density of antigen relative to the embryonic cells. The findings of Chatterjee-Hasrouni and Lala (1979), discussed earlier, might suggest the former.

SEPARATED PLACENTAL TROPHOBLAST POPULATIONS

Although the terminal labeling assay leaves the surviving cell population for histological examination, precise identity of the cells is difficult to establish. To overcome this, samples of spongiotrophoblast and labyrinthine trophoblast were removed as described by Jenkinson and Owen (1980) and investigated

213

for their susceptibility in parallel cultures. The data shown in Table 10-2 demonstrate that the pure spongiotrophoblast population is highly susceptible to cytotoxic effector cell attack but labyrinthine material is variably affected, with four of the ten cultures showing no significant destruction. Again, this correlates well with the serological findings of Jenkinson and Owen (1980), and the killing seen in labyrinthine cultures could well be of contaminating fetal mesenchymal, rather than trophoblastic, elements.

The present observations thus demonstrate that placental spongiotrophoblast expresses histocompatibility antigens that render it susceptible to immune cell recognition and lysis *in vitro*. As indicated above, it is not yet clear whether these antigens are exposed *in vivo*, but if they are, it must be asked how this trophoblast escapes immune cell attack. The relevant question in this context would then be whether there are antipaternally directed cytotoxic T cells in the allogeneically pregnant female.

CYTOTOXIC EFFECTOR T CELLS IN PREGNANCY

Transplantation experiments using paternal skin and tumor allografts with allogeneically mated animals have failed to demonstrate the accelerated rejection that might be predicted if pregnancy resulted in maternal sensitization. Instead, there is evidence that multiparity results in prolongation of graft survival, suggesting either an absence of cytotoxic effector cells or the operation of immune enhancement (Breyere and Barrett, 1960; Beer et al., 1976; R.N. Smith and Powell, 1977). An apparent exception to this is the case where the adoptive transfer of small numbers of spleen cells from allogeneically mated multiparous females has been shown to cause tumor allograft rejection (Chaouat et al., 1979), thus implying the presence of cytotoxic cells during pregnancy. Some support for this view has come from *in vitro* studies using the cell-mediated microcytotoxicity test, where it has been demonstrated that spleen cells from multiparous, syngeneically mated mice can kill syngeneic tumor cells and embryonic fibroblast monolayers, although this does not of course involve histocompatibility antigens (Hellström and Hellström, 1975; Hamilton et al., 1976). Using a similar assay with F_1 embryonic fibroblast targets, it has been demonstrated in our laboratory that spleen cells from a small proportion of allogeneically mated primiparous mice do have cytotoxic activity, but the levels are very low and have so far been found only in C57BL/10 females (G. Smith, unpublished data). The majority of allogeneically mated mice showed no cytotoxic activity. Comparable assays

TABLE 10-2

Susceptibility of Placental and Embryonic Tissues of 13-Day Mouse Conceptus (C57BL10 x CBA)F_1 to Killing by Hyperimmune Lymphocytes Directed Against Paternal H-2 and non-H-2 Antigens (% Cytotoxicity)

EXPERIMENT NUMBER	TISSUE		
	LABYRINTH	SPONGIOTROPHOBLAST	FETAL CARCASS
1	46.5*	55.6*	42.6*
2	5.0	25.0*	39.0*
3	23.9*	37.8*	79.2*
4	50.8*	42.8*	50.0*
5	-7.0	27.0*	97.5*
6	16.8*	49.3*	57.1*
7	23.4*	21.9*	64.8*
8	14.3	17.3*	55.9*
9	0.0	28.1*	68.4*
10	75.9*	38.4*	96.8*

* Significance < 0.05

using human material have shown cytotoxic cell activity in peripheral blood leukocyte populations toward adult fibroblast, fetal lung, and placental cell targets (Douthwaite and Urbach, 1971; Taylor and Hancock, 1975; Timonen and Saksela, 1976; Gluckman and Sasportes, 1978). However, the nature of the cytotoxic cells involved in this test remains unclear, since the activity reported could be due to natural killer cells, K cells, or even armed macrophages, and not cytotoxic T cells, especially since much of the killing is not MHC restricted.

The ^{51}Cr-release assay using tumor or blast cell targets provides a better measure of cytotoxic T cell activity during pregnancy. In the majority of reports, spleen and draining lymph node cells from allogeneically mated mice investigated during first and subsequent pregnancies have been found to have little or no cytotoxic cell activity (Hamilton and Hellström, 1977; Clark and McDermott, 1978; Pavia and Stites, 1979; Gottesman and Stutman, 1980). The exception is the report of J.A. Smith et al. (1978), who found low levels of nonspecific cytotoxicity in one-third of outbred females during their first pregnancy. The lack of cytotoxic cells has been attributed to the presence of suppressor cells, both specific and nonspecific (Chaouat et al., 1980; Clark et al., 1980); however, not all workers have been able to demonstrate their universal presence (Pavia and Stites, 1979; G. Smith, 1981). Similarly conflicting results have been obtained with *in vitro* sensitization experiments using paternal alloantigen stimulators in the MLR phase, with some workers reporting elevated levels of cytotoxic activity of lymphoid cells from pregnant females relative to virgin controls (J.A. Smith et al., 1978), while others find no difference (Pavia and Stites, 1979; Wegmann et al., 1979c; Gottesman and Stutman, 1980). These conflicts may well reflect technical differences in experiments as well as strain differences in immune responsiveness.

Thus, there is no evidence for consistent maternal cytotoxic T cell activity in pregnancy. Whether this is due to afferent inhibition or efferent blockade by specific or nonspecific factors remains to be elucidated. In the meantime, it is relevant to note that the course of pregnancy is unaffected under conditions of experimentally induced antipaternal hyperimmunity, when cytotoxic T cells are undoubtedly present (Mitchison, 1953; Lanman et al., 1964; Wegmann et al., 1979c).

MATERNAL ANTIBODY RESPONSE TO PATERNALLY INHERITED HISTOCOMPATIBILITY ANTIGENS

It is relevant in the present context to consider the extent to which exposure of the mother to H-2 and non-H-2 antigens on both embryonic and spongio-trophoblast tissue results in a detectable humoral immune response. The major line of evidence that supports the contention that recognition of paternally inherited histocompatibility antigens occurs is, in fact, the detection of antipaternal alloantibodies in pregnant females of various species. Initially described in pregnant mice (Herzenberg and Gonzales, 1962), they have subsequently been identified in sera of humans (Terasaki et al., 1970), cattle (Iha et al., 1973; Newman and Hines, 1980), and horses (Allen, 1979), but so far not in pregnant rats (Beer et al., 1976; Harrison, 1976).

Although the appearance of antipaternal alloantibodies in pregnant mice is often inferred to be a ubiquitous phenomenon, examination of the literature reveals that it is apparently a property associated only with $H-2^b$ haplotype-bearing females. These "responder" strains include C57BL/6 (Mishell et al., 1963; Kaliss, 1973), C57BL/6J (Goodlin and Herzenberg, 1964; Kaliss and Dagg, 1964), and C57L/J (Kaliss and Dagg, 1964). Of 20 inbred strains examined by Kaliss (1973), only 7 produced alloantibody during pregnancy, 6 of which were $H-2^b$ strains. The remaining responder strain, $H-2^f$ haplotype, was congenic to one of the $H-2^b$ strains and thus possessed a non-H-2 genetic background identical to that of the latter. The demonstration of antipaternal alloantibody on placentae from an allogeneically mated C57BL/ks ($H-2^d$) strain (Voisin and Chaouat, 1974), which is also derived from an $H-2^b$ genetic background, may be another example of a responder strain containing an $H-2^b$ associated non-H-2 background.

Examination of $H-2^b$, $H-2^k$, and $H-2^d$ females in our laboratory has

216

also revealed only three responder strains, all of the b haplotype (Bell and Billington, 1980, 1981; S.C. Bell, unpublished observations). No antibody has been detected in pregnant CBA/J (H-2^k) (Chatterjee-Hasrouni and Lala, 1979) or Balb/c (H-2^d) females (Pavia and Stites, 1979), although very low levels of cytotoxic antibody have been reported in a portion of CBA/J (H-2^k) females (Baines et al., 1976). Thus it would appear that the capacity to respond to histocompatibility antigens with a humoral immune response is genetically determined and may possibly be associated with a responder gene which, although outside the H-2 locus, is linked to the H-2^b haplotype, as suggested by Kaliss (1973). However, only genetically homozygous females possess this ability, as indicated by the inability of F_1 hybrid females, generated by the allogeneic mating of responder strains, to produce antipaternal alloantibody (Kaliss, 1973; S.C. Bell, unpublished observations). These observations suggest that, in randomly outbred mouse populations, a humoral immune response to paternal alloantigens might be a rare phenomenon.

Can an examination of the type of antipaternal alloantibody induced by pregnancy in responder strains reveal why alloantigens stimulate a humoral immune response in such a limited number of murine strains? Any consideration of the type of alloantibody produced must take into account the type of assay used, since preferential detection of antibody populations may occur. The sensitive ^{51}Cr-release method (Snell et al., 1971) has been used to demonstrate complement-dependent cytotoxic activity, albeit of minimal titers, in multiparous C57BL/6 (H-2^b) females (Kaliss, 1973) and in a few CBA/J (H-2^k) females (Baines et al., 1976). With the dye-exclusion assay, no complement-dependent cytotoxicity has been detectable in either antibody-containing sera of multiparous C57BL (H-2^b) females (Bell and Billington, 1980, 1981) or placental eluates of allogeneically mated females (Voisin and Chaouat, 1974). In sera of multiparous C57BL females, the major portion of the alloantibody present is restricted to the IgG$_1$ subclass (Bell and Billington, 1980; Figure 10-2); and in the placental eluates, IgG$_1$ was also the predominant immunoglobulin (Voisin and Chaouat, 1974; Chaouat et al., 1979), although in both cases some IgG$_2$ was still present. It is of interest that alloantibodies induced in females by the injection of allogeneic fetal bones are essentially restricted to the IgG$_1$ isotype (Segal et al., 1980). Multiparous C57BL females appear committed to producing noncytotoxic alloantibody, since intraperitoneal injections of allogeneic cells, a route that induces complement-fixing alloantibody in virgin animals, only induced further production of pregnancy-type alloantibody (S.C. Bell, unpublished observations). It seems plausible that pregnancy-induced alloantibody appears to be primarily restricted to the non-complement-fixing, noncytotoxic IgG$_1$ subclass, but more sensitive assays might detect low cytotoxic activity attributable to low levels of the IgG$_2$ isotype.

The total spectrum of histocompatibility antigens on the conceptus, to which the humoral immune response in responder strains is directed, cannot be

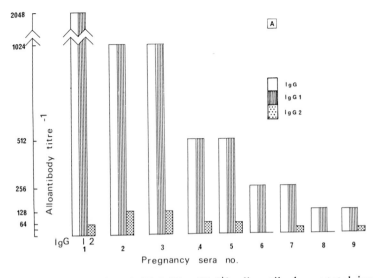

FIGURE 10-2. Levels of anti-CBA/Ca (H-2^k) alloantibody as total immuno-globulin, IgG$_1$ and IgG$_2$ subclasses in sera from nine multiparous C57BL/10 (H-2^b) females mated with CBA/Ca males.
CBA/Ca fibroblasts (1 x 10^5 cells per well of immunofluorescence tray) were cultured for 24 hours and after removal of culture medium, serial double dilutions of test sera were incubated with the monolayer for 30 minutes. Sera were diluted with culture medium. Monolayers were then incubated for 30 minutes with SRBC coated sequentially with mouse anti-SRBC sera and rabbit anti-mouse immunoglobulin, IgG$_1$ or IgG$_2$. After washing, monolayers were examined for SRBC binding. The final dilution of sera with which binding was obtained was considered to represent the alloantibody titer. Results shown are the reciprocal of the alloantibody titer obtained with: (unshaded), anti-immunoglobulin; (vertical lines), anti-IgG$_1$; (stippling), anti-IgG$_2$. From Bell and Billington, 1980; reprinted with permission.

stated with certainty. The hemagglutination assay has most commonly been used to detect antipaternal alloantibodies during pregnancy (Herzenberg and Gonzales, 1962; Goodlin and Herzenberg, 1963; Kaliss and Dagg, 1964; Kaliss, 1973) and has been reported to detect antibodies reliably only against H-2 specificities (Kaliss, 1973). These observations would suggest that, in these responder strains, anti-H-2 alloantibodies are produced, and after examination of one strain mated with an H-2 and non-H-2 incompatible male, where at least two "strong" foreign H-2 specificities were presented, the antibody detect-able by hemagglutination appeared to be directed solely against a single H-2 specificity, H-2.4 (Kaliss, 1973). However, the role of non-H-2 disparities and the recognition of non-H-2 antigen systems may be of primary importance. Responder H-2^b strains that have been examined produce antipaternal alloantibody only when mated with H-2 and non-H-2 incompatible male strains and not with congenic strains, where only H-2 differences exist (Kaliss, 1973; S.C. Bell, unpublished observations), suggesting that non-H-2 recognition must occur before anti-H-2 hemagglutinating antibodies can be synthesized.

Anti-non-H-2 alloantibodies, detectable by the leukocyte agglutination assay, are produced by C57BL/6J females when mated with an H-2 and non-H-2 incompatible strain, and it would appear that these are the predominant alloantibody populations, since absorption of anti-H-2 antibodies present did not affect the titer of antibody detectable in this assay (Mishell et al., 1963). Alloantibody in a C57BL (H-2^b) strain induced by multiparity with an H-2 and non-H-2 incompatible strain also appears to be primarily directed against non-H-2 antigens (S.C. Bell, unpublished observations). Whether this essentially anti-non-H-2 response is characteristic of alloantibody responses induced by multiparity in all responder strains or merely reflects the strong immunogenicity of foreign paternal non-H-2 antigens in the strain combinations examined hitherto remains uncertain. Pertinent to this latter consideration is the fact that the male strain (CBA) used in the mating combination we have examined is unusual in that paternal non-H-2 antigens are expressed on trophoblast populations of the implanting blastocyst and the ectoplacental cone (see Sellens et al., 1978). However, if non-H-2 antigen systems predominate, the responder gene invoked to provide a rationale for the existence of responder strains (Kaliss, 1973) could in fact represent the lack of a non-H-2 antigen system in responder strains, a system that is present in the appropriate male strain.

Further information pertaining to the basis of the strain dependence of the phenomenon of pregnancy-induced alloantibody and the nature of the alloantigenic stimuli can be obtained from examination of the kinetics of alloantibody production. Most investigators agree that in responder strains a minimum of two pregnancies is required to produce demonstrable peripheral alloantibody titers, with maximum levels being achieved after three pregnancies (Kaliss and Dagg, 1964; Kaliss, 1973; Bell and Billington, 1981; see Figure 10-3). During the second pregnancy of a C57BL (H-2^b) strain mated with a CBA/Ca (H-2^k) strain, alloantibody, assumed to represent the primary immune response, was first detectable on days 16 to 17 (Bell and Billington, 1981). Recently, however, we have examined the antipaternal alloantibody response in pregnant females of a responder strain that had been preimmunized by a single intraperitoneal injection of paternal spleen cells (R. Roe and S.C. Bell, unpublished observations), similar to a model used by G.M. Taylor (1973). Alloantibody titers in these females were boosted by allogeneic mating at an identical period during pregnancy (Figure 10-4). This strongly suggested that during normal reproductive life the alloantibody response observed in second pregnancy represents a secondary humoral immune response. Why a

219

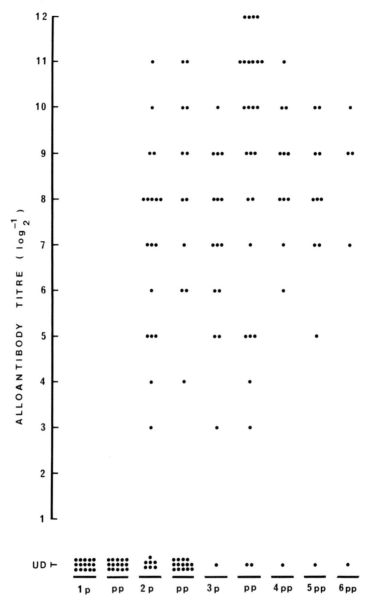

FIGURE 10-3. Antipaternal alloantibody titers with increasing multiparity. C57BL/10 females were mated with CBA/Ca males and serum samples examined during pregnancy (p) and the immediate postpartum period (pp) for up to six successful pregnancies. Anti-CBA/Ca alloantibody titers were determined using a modified absorption (Immunobead) assay employing CBA/Ca fibroblasts as target cells. UD, undetectable. From Bell and Billington, 1981; reprinted with permission.

primary response, presumably occurring in primiparous females, was undetectable is unknown, but since the alloantibody assay used does not detect IgM antibodies, a purely IgM primary response would not be observed (Bell and Billington, 1981).

Through examination of the kinetics of the secondary humoral immune response in preimmunized females subsequently challenged via the intrauterine route, it has been estimated that paternal alloantigens on the conceptus become immunogenic between days 12 and 14 of pregnancy (R. Roe and S.C. Bell, unpublished observations). The alloantigenic stimulus in preimmunized pregnant females thus almost certainly arises from the mature feto-placental unit and not from spermatozoa or the preimplantation blastocyst or early post-implantation conceptus; nor is it associated with parturition. The

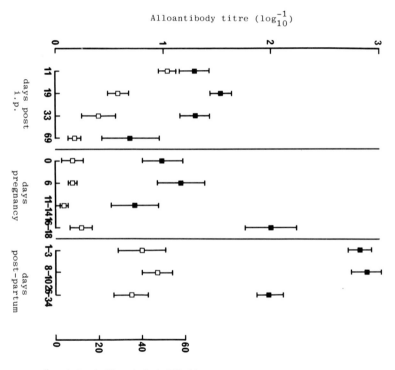

FIGURE 10-4. Humoral immune response of C57BL/10 (H-2^b) females pre-immunized to paternal alloantigens and followed by allogeneic pregnancy. C57BL/10 females were pre-immunizied with a single i.p. injection of CBA/Ca (H-2^k) spleen cells and after 69 days were mated with CBA/Ca males. Serum samples were assayed for total anti-CBA/Ca alloantibody (■) by the Immunobead method and for complement-dependent cytotoxicity (□) by the dye exclusion method. R. Roe and S.C. Bell, unpublished observations.

apparent lack of immunogenicity of fetal tissues prior to days 12 to 14 is consistent with the ontogeny of alloantigen expression (see Section I and Chapter 11). The immunogens could clearly be of either fetal or placental origin, although it should be noted that Ia antigen expression appears to be restricted to nontrophoblastic cells (Delovitch et al., 1978; Jenkinson and Searle, 1979). The recent claim for fetal nucleated cell transfer to maternal lymphoid tissue (Liegeois et al., 1981) suggests that transplacentally transmitted leukocytes may be an important, but not necessarily the sole, source.

Surprisingly, in a nonresponder strain preimmunized with paternal alloantigens, allogeneic pregnancy also induced a secondary humoral immune response during the same period observed with the responder strain (R. Roe and S.C. Bell, unpublished observations; Figure 10-5). The existence of responder and nonresponder strains does not therefore appear to be due to the differential ability of these strains to respond to alloantigens presented via the intrauterine route, at least when the animals are presensitized to alloantigens. Whether these strains differ in their ability to produce a primary immune response to alloantigens via this route remains to be determined.

Pregnancy-induced alloantibody has frequently been invoked as an important immunoregulatory agent in models to explain the survival of the allogeneic feto-placental unit. In inbred strains of mice, however, this phenomenon is clearly not ubiquitous, as many strains do not produce detectable alloantibody when mated with H-2 and non-H-2 incompatible males, and even responder strains will not produce antibody when mated with males incompatible only at the H-2 region. If pregnancy-induced antibody is considered to be directed against H-2 antigens, this would seem paradoxical since no substantial strain-dependent differences in the expression of H-2 antigen systems on the conceptus have been reported. Thus, any model implicating anti-H-2-directed alloantibody (Bernard, 1977; Segal et al., 1979; Voisin, 1980) must offer an explanation for the existence of responder and nonresponder strains and associated phenomena.

SUMMARY AND CONCLUSIONS

Mouse trophoblast is characterized throughout its development and differentiation by the absence of Ia antigens and a differential expression of H-2 and non-H-2 antigens. In the mature placenta, H-2 antigens are expressed predominantly on the spongiotrophoblast and not on the labyrinthine trophoblast population. These antigens render the trophoblast susceptible to immune cell attack *in vitro*, but it has not yet been established whether they are exposed on

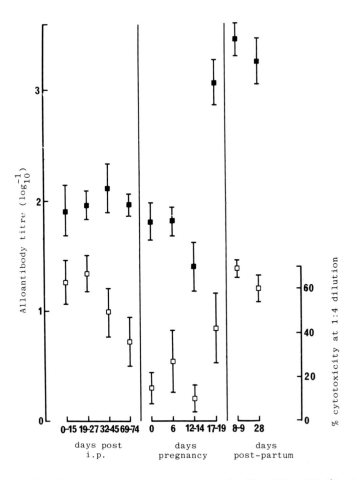

FIGURE 10-5. Humoral immune response of CBA/Ca (H-2^k) females pre-immunized to paternal alloantigens and followed by allogeneic pregnancy. CBA/Ca females were preimmunized with a single i.p. injection of C57BL/10 (H-2^b) spleen cells and after 74 days were mated with C57BL/10 males. Serum samples were assayed for total anti-C57BL/10 alloantibody (■) by the Immunobead method and for complement-dependent cytotoxicity (□) by the dye exclusion method. R. Roe and S.C. Bell, unpublished observations.

the apical membranes of the trophoblast *in situ*. Cytotoxic T cells are not consistently demonstrable in the primiparous or multiparous allogeneically pregnant female. Pregnancy-induced alloantibody is qualitatively different from that induced by experimental allografts, but few data are available on the precise specificity of this response.

It would seem crucial to determine why only a few inbred strains of mice appear to produce alloantibody in pregnancy, and what situation exists in

223

outbreeding populations, in order to establish clearly whether this humoral immune response fulfills an important immunoregulatory role in pregnancy, rather than being an irrelevant epiphenomenon consequent upon maternal recognition of the highly polymorphic histocompatibility antigen systems of the conceptus. Conflicting data frcm studies on species other than the mouse indicate the need for satisfactory reconciliation of the differences in order to uphold any unifying hypothesis for fetal allograft survival, even one involving non-antigen-specific mechanisms, or, alternatively, acceptance of the concept that evolutionary divergence can lead to independent solutions.

ACKNOWLEDGMENTS

We are indebted to our colleague Rachel Roe for allowing us to include unpublished collaborative findings, and to Fiona Dennis and Sue Watts for high-quality technical assistance. The studies in our laboratory were supported by a grant from The Rockefeller Foundation.

REFERENCES

Adinolfi, M. 1975. The human placenta as a filter for cells and plasma proteins. In *Immunobiology of Trophoblast*. R.G. Edwards, C.W.S. Howe, and M.H. Johnson, eds. Cambridge: Cambridge University Press, pp. 193-215.

Allen, W.R. 1979. Maternal recognition of pregnancy and immunological implications of trophoblast-endometrium interactions in equids. In *Maternal Recognition of Pregnancy*. Ciba Foundation Symposium 64. Amsterdam: Excerpta Medica, pp. 323-346.

Baines, M.G., E.A. Speers, H. Pross, and K.G. Millar. 1976. Characteristics of the maternal lymphoid response of mice to paternal strain antigens induced by homologous pregnancy. *Immunology* 31:363.

Beer, A.E., J.R. Head, W.G. Smith, and R.E. Billingham. 1975. Some immunoregulatory aspects of pregnancy in rats. *Transplant. Proc.* 8:267.

Bernard, O. 1977. Possible protecting role of maternal immunoglobulins on embryonic development in mammals. *Immunogenetics* 5:1.

Bell, S.C., and W.D. Billington. 1980. Major anti-paternal alloantibody induced by murine pregnancy is non-complement-fixing IgG₁. *Nature* 288:387.

Bell, S.C., and W.D. Billington. 1981. Humoral immune response in murine pregnancy. I. Anti-paternal alloantibody levels in maternal serum. *J. Reprod. Immunol.* 3:3.

Billington, W.D. 1975. Organization, ultrastructure and histochemistry of the placenta: immunological considerations. In *Immunobiology of Trophoblast*. R.G. Edwards, C.W.S. Howe, and M.H. Johnson, eds. Cambridge: Cambridge University Press, pp. 67-85.

Billington, W.D. 1976. The immunobiology of trophoblast. In *Immunology of Human Reproduction*. J.S. Scott and W.R. Jones, eds. London: Academic Press, pp. 81-102.

Breyere, E.J., and M.K. Barrett. 1960. Prolonged survival of skin homografts in parous female mice. *J. Nat. Cancer Inst.* 25:1405.

Chaouat, G., P. Monnot, M. Hoffman, and S. Chaffaux. 1980. Regulatory mechanisms of cell-mediated immunity in allogeneic pregnancy. *Am. J. Reprod. Immunol.* 1:18.

Chaouat, G., G.A. Voisin, D. Escalier, and P. Robert. 1979. Facilitation reaction (enhancing antibodies and suppressor cells) and rejection reaction (sensitized cells) from the mother to the paternal antigens of the conceptus. *Clin. Exp. Immunol.* 35:12.

Chatterjee-Hasrouni, S., and P.K. Lala. 1979. Localization of H-2 antigens on mouse trophoblast cells. *J. Exp. Med.* 149:1238.

Clark, D.A., and M.R. McDermott. 1978. Impairment of host vs graft reaction in pregnant mice. I. Suppression of cytotoxic T cell generation in lymph nodes draining the uterus. *J. Immunol.* 121:1389.

Clark, D.A., M.R. McDermott, and M.R. Szewczuk. 1980. Impairment of host-versus-graft reaction in pregnant mice. II. Selective suppression of cytotoxic T cell generation correlates with soluble suppressor activity and with successful allogeneic pregnancy. *Cell. Immunol.* 52:106.

Davies, D.A.L., and M.N. Staines. 1976. A cardinal role for I-region antigens (Ia) in immunological enhancement, and the clinical implications. *Transplant. Rev.* 30:18.

Delovitch, T.L., J.L. Press, and H.O. McDevitt. 1978. Expression of murine Ia antigens during embryonic development. *J. Immunol.* 120:818.

Douthwaite, R.M., and G.I. Urbach. 1971. *In vitro* antigenicity of trophoblast. *Am. J. Obstet. Gynecol.* 109:1023.

Enders, A.C. 1965. A comparative study of the fine structure of the trophoblast in several haemochorial placentas. *J. Anat.* 116:29.

Gluckman, J.-C., and M. Sasportes. 1978. HLA-independent cell-mediated alloimmunity detected by a microcytotoxicity assay. *Transplantation* 26:70.

Goodlin, R.C., and L.A Herzenberg. 1964. Pregnancy induced haemagglutinins to paternal H-2 antigens in muliparous mice. *Transplantation* 2:357.

Gottesman, S.R.S., and O. Stutman. 1980. Cellular immunity during pregnancy. I. Proliferative and cytotoxic reactivity of para-aortic lymph nodes. *Am. J. Reprod. Immunol.* 1:10.

Håkansson, S., S. Heyner, K.G. Sundqvist, and S. Bergstrom. 1975. The presence of paternal H-2 antigens on hybrid mouse blastocysts during experimental delay of implantation and the disappearance of these antigens after onset of implantation. *Int. J. Fertil.* 20:137.

Håkansson, S., and P.A. Peterson. 1976. Presence of β_2-microglobulin on the implanting mouse blastocyst. *Transplantation* 21:358.

Hamilton, M.S., I. Hellström, and G. van Belle. 1976. Cell-mediated immunity to embryonic antigens of syngeneically and allogeneically mated mice. *Transplantation* 21:261.

Hamilton, M.S., and I. Hellström. 1977. Altered immune responses in pregnant mice. *Transplantation* 23:423.

Harrison, M.R. 1976. Maternal immunocompetence. II. Proliferative responses of maternal lymphocytes *in vitro* and inhibition by serum from pregnant rats. *Scand. J. Immunol.* 5:881.

Hellström, I., and K.E. Hellström. 1975. Cytotoxic effect of lymphocytes from pregnant mice on cultivated tumour cells. I. Specificity, nature of effector cells and blocking by serum. *Int. J. Cancer* 15:1.

Herzenberg, L.A., and B. Gonzales. 1962. Appearance of H-2 agglutinins in outcrossed female mice. *Proc. Nat. Acad. Sci. USA* 48:570.

Heyner, S. 1980. Antigens of trophoblast and early embryo. In *Immunological Aspects of Infertility and Fertility Regulation.* D.S. Dhindsa and G.F.B. Schumacher, eds. Amsterdam: Elsevier/North-Holland, pp. 183-203.

Heyner, S., R.D. Hunziker, and G.L. Zink. 1980. Differential expression of minor histocompatibility antigens on the surface of the mouse oocyte and preimplantation developmental stages. *J. Reprod. Immunol.* 2:269.

Iha, T.H., G. Gerbrandt, W.F. Bodmer, D. McGary, and W.H. Stone. 1973. Cross-reactions of cattle lymphocytotoxic sera with HL-A and other human antigens. *Tissue Antigens* 3:291.

Jacob, F. 1977. Mouse teratocarcinoma and embryonic antigens. *Immunol. Rev.* 33:3.

Jenkinson, E.J., and W.D. Billington. 1974a. Studies on the immunobiology of mouse fetal membranes: the effect of cell-mediated immunity on yolk sac cells *in vitro. J. Reprod. Fertil.* 41:403.

Jenkinson, E.J., and W.D. Billington. 1974b. Differential susceptibility of mouse trophoblast and embryonic tissue to immune cell lysis. *Transplantation* 18:286.

Jenkinson, E.J., W.D. Billington, and J. Elson. 1975. The effect of cellular and humoral immunity on the mouse yolk sac. In *Maternofoetal Transmission of Immunoglobulins.* W.A. Hemmings, ed. Cambridge: Cambridge University Press, pp. 225-232.

Jenkinson, E.J., and V. Owen. 1980. Ontogeny and distribution of major histocompatibility complex (MHC) antigens on mouse placental trophoblast. *J. Reprod. Immunol.* 2:173.

Jenkinson, E.J., and R.F. Searle. 1979. Ia antigen expression on the developing mouse embryo and placenta. *J. Reprod. Immunol.* 1:3.

Johnson, M.H. 1975. Antigens of the peri-implantation trophoblast. In *Immunobiology of Trophoblast.* R.G. Edwards, C.W.S. Howe, and M.H. Johnson, eds. Cambridge: Cambridge Universty Press, pp. 87-112.

Kaliss, N. 1973. Immune reactions of multiparous female mice to fetal H-2 alloantigens. In *Immunology of Reproduction.* K. Bratanov, ed. Sofia: Bulgarian Academy of Sciences Press, pp. 495-511.

Kaliss, N., and M.K. Dagg. 1964. Immune response engendered in mice by multiparity. *Transplantation* 2:416.

Kirby, D.R.S., W.D. Billington, and D.A. James. 1966. Transplantation of eggs to the kidney and uterus of immunised mice. *Transplantation* 4:713.

Kirkwood, K.J., and W.D. Billington. 1981. Expression of serologically detectable H-2 antigens on mid-gestation mouse embryonic tissues. *J. Embryol. Exp. Morph.* 61:207.

Klein, P.A. 1975. Anomalous reactions of mouse alloantisera with cultured tumour cells. I. Demonstration of widespread occurrence using reference typing sera. *J. Immunol.* 115:1254.

Lanman, J.T., L. Herod, and S. Fikrig. 1964. Homograft immunity in pregnancy. Survival rates in rabbits born of ova transplanted into sensitized mothers. *J. Exp. Med.* 119:781.

Liegeois, A., M.C. Galliard, E. Ouvre, and D. Lewin. 1981. Microchimerism in pregnant mice. *Transplant. Proc.* 13:1250.

Mishell, R.I., L.A. Herzenberg, and L.A. Herzenberg. 1963. Leukocyte agglutination in mice. Detection of H-2 and non-H-2 isoantigens. *J. Immunol.* 90:628.

Mitchison, N.A. 1953. The effect on the offspring of maternal immunization in mice. *J. Genet.* 51:406.

Muggleton-Harris, A.L., and M.H. Johnson. 1976. The nature and distribution of serologically-detectable alloantigens on the pre-implantation mouse embryo. *J. Embryol. Exp. Morph.* 35:59.

Munro, A.J. 1975. Antigenic topography and the interaction of antibodies and immune cells with surface antigens. In *Immunobiology of Trophoblast.* R.G. Edwards, C.W.S. Howe, and M.H. Johnson, eds. Cambridge: Cambridge University Press, pp. 5-12.

Newman, M.J., and H.C. Hines. 1980. Stimulation of maternal anti-lymphocyte antibodies by first gestation bovine fetuses. *J. Reprod. Fertil.* 60:237.

Parr, E.L., R.V. Blanden, and R.S. Tulsi. 1980. Epithelium of mouse yolk sac placenta lacks H-2 complex alloantigens. *J. Exp. Med.* 152:945.

Parr, E.L., and W.N. Kirby. 1979. An immunoferritin labelling study of H-2 antigens on dissociated epithelial cells. *J. Histochem. Cytochem.* 27:1327.

Pavia, C.S., and D.P. Stites. 1979. Humoral and cellular regulation of alloimmunity in pregnancy. *J. Immunol.* 123:2194.

Rossant, J., and V.E. Papaioannou. 1977. The biology of embryogenesis. In *Concepts in Mammalian Embryogenesis.* M.I. Sherman, ed. Cambridge, Massachusetts: MIT Press, pp. 1-36.

Searle, R.F., E.J. Jenkinson, and M.H. Johnson. 1975. Immunogenicity of mouse trophoblast and embryonic sac. *Nature* 255:719.

Searle, R.F., M.H. Johnson, W.D. Billington, J. Elson, and S. Clutterbuck-Jackson. 1974. Investigation of H-2 and non-H-2 antigens on the mouse blastocyst. *Transplantation* 18:136.

Searle, R.F., M.H. Sellens, J. Elson, E.J. Jenkinson, and W.D. Billington. 1976. Detection of alloantigens during pre-implantation development and early trophoblast differentiation in the mouse by immunoperoxidase labelling. *J. Exp. Med.* 143:348.

Segal, S., T. Siegal, H. Altaray, A. Lev-El, Z. Nevo, L. Nebel, A. Katzenelson, and M. Feldman. 1979. Fetal bone grafts do not elicit allograft rejection because of protecting anti-Ia alloantibodies. Implication to the immune survival of fetuses in allogeneic mothers. *Transplantation* 28:88.

Segal, S., B. Tartakovsky, S. Katzav, and P. DeBaetselier. 1980. The immunologic basis of the fetal-maternal relationship. *Transplant. Proc.* 12:582.

Sellens, M.H. 1977. Antigen expression on early mouse trophoblast. *Nature* 269:60.

Sellens, M.H., E.J. Jenkinson, and W.D. Billington. 1978. Major histocompatibility complex and non-major histocompatibiliy complex antigens on mouse ectoplacental cone and placental trophoblastic cells. *Transplantation* 25:173.

Simmons, R.L., and P.S. Russell. 1966. The histocompatibility antigens of fertilized mouse eggs and trophoblast. *Ann. N.Y. Acad. Sci.* 129:35.

Smith, G. 1981. Maternal regulator cells during murine pregnancy. *Clin. Exp. Immunol.* 44:90.

Smith, J.A., R.C. Burton, M. Barg, and G.F. Mitchell. 1978. Maternal alloimmunisation in pregnancy. *In vitro* studies of T cell-dependent immunity to paternal alloantigens. *Transplantation* 25:216.

Smith, G., and S. Nicklin. 1979. ³H-uridine uptake by target monolayers as a terminal label in an *in vitro* cell-mediated cytotoxicity assay. *J. Immunol. Methods* 25:265.

Smith, R.N., and A.E. Powell. 1977. The adoptive transfer of pregnancy-induced unresponsiveness to male skin grafts with thymus-dependent cells. *J. Exp. Med.* 146:899.

Snell, G.D., P. Démant, and M. Cherry. 1971. Hemagglutination and cytotoxic studies of H-2.1 and H-2 and related specificities in the EK crossover regions. *Transplantation* 11:210.

Taylor, G.M. 1973. The level and distribution of antibody in syngeneic and allogeneic mated pregnant mice pre-immunized with H-2 alloantigens. *Immunology* 25:783.

Taylor, P.V., and K.W. Hancock. 1975. Antigenicity of trophoblast and possible antigen-masking effects during pregnancy. *Immunology* 28:973.

Taylor, P.V., K.W. Hancock, and G. Gowland. 1979. Effect of neuraminidase on immunogenicity of early mouse trophoblast. *Transplantation* 28:256.

Terasaki, P.I., M.R. Mickey, J.N. Yamazaki, and D. Vredevoe. 1970. Maternal-fetal incompatibility. I. Incidence of HL-A antibodies and possible association with congenital anomalies. *Transplantation* 9:538.

Timonen, T., and E. Saksela. 1976. Cell-mediated anti-embryo cytotoxicity in human pregnancy. *Clin. Exp. Immunol.* 23:462.

Vandeputte, M., and H. Sobis. 1972. Histocompatibility antigens on mouse blastocysts and ectoplacental cones. *Transplantation* 14:331.

Voisin, G.A. 1980. The role of antibody classes in the regulatory facilitation reaction. *Immunol. Rev.* 49:3.

Voisin, G.A., and G. Chaouat. 1974. Demonstration, nature and properties of maternal antibodies fixed on placenta and directed against paternal antigens. *J. Reprod. Fertil.* 21(suppl):89.

Wegmann, T.G., T.R. Mosmann, G.A. Carlson, O. Olijnyk, and B. Singh. 1979a. The ability of the murine placenta to absorb monoclonal antifetal H-2K antibody from the maternal circulation. *J. Immunol.* 123:1020.

Wegmann, T.G., B. Singh, and G.A. Carlson. 1979b. Allogeneic placenta is a paternal strain antigen immunoabsorbant. *J. Immunol.* 122:270.

Wegmann, T.G., C.A. Waters, D.W. Drell, and G.A. Carlson. 1979c. Pregnant mice are not primed but can be primed to fetal alloantigens. *Proc. Nat. Acad. Sci. USA.* 76:2410.

Wiley, L.M. 1979. Early embryonic cell surface antigens as developmental probes. *Curr. Top. Develop. Biol.* 13:167.

Chapter 11

THE PLACENTA AS AN IMMUNOLOGICAL BARRIER
BETWEEN MOTHER AND FETUS

BHAGIRATH SINGH
RAJGOPAL RAGHUPATHY
DEBORAH J. ANDERSON
THOMAS G. WEGMANN

One of the oldest observations in reproductive immunology is that during pregnancy, human and murine females produce antibodies, some of which are cytotoxic against the fetal major histocompatibility complex (MHC) antigens that are paternally derived. Although a good deal more needs to be known about the details of this antibody production, at least the fact that they are produced by some females is beyond dispute (Gill and Repetti, 1979). Billington et al. (Chapter 10) provide an account of some of the known factors involved in the determination of whether a female will produce antibodies against her offspring *in utero*. A number of hypotheses have been advanced over the years to account for the fate of these antibodies during pregnancy. One of the most attractive hypotheses was published by Swinburne (1970). The idea is that the antibodies, although of a subclass capable of crossing the placental barrier, nevertheless do not reach the fetus because the relevant fetal MHC antigens are present in the placenta, and can therefore serve as an immunosorbent barrier between the maternal and fetal circulations. A number of investigators have provided circumstantial evidence for this hypothesis. Voisin and Chaouat (1974) and Chaouat et al. (1979) have shown that antibodies can be eluted from the murine placenta which react specifically with paternal lymphocytes. These antibodies can also specifically enhance the growth of paternal-strain tumor cells. Similar studies by Doughty and Gelsthorpe (1974, 1976) in humans have shown that one can elute antipaternal MHC antibodies from human placentae. They showed that such antibodies were missing from cord blood when the baby had the relevant target antigens. Blocking antibodies can also be eluted from human placenta (Bonneau et al., 1973). Goodlin and Herzenberg (1964) found that antifetal H-2 agglutinins disappeared from the maternal circulation of multiparous mice during the third week of pregnancy and reappeared one week post partum. The time of disappearance coincided with the emergence of the placenta as a sizable organ. Billington et al. (1977) and Jenkinson and Owen (1980) have also demonstrated the presence of H-2 antigens on placental explants using mixed hemadsorption. More recently, Chatterjee-Hasrouni and Lala (1979) have found paternal H-2K and D antigenic specificities on dissociated placental cells identified as trophoblast cells by morphological criteria. (see Billington et al., Chapter 10, for a detailed review of this subject).

These findings suggest that the placenta removes maternal antifetal antibodies that could be potentially harmful to the fetus, but allows passage to antibodies that have protective value for the offspring. They do not, however, rule out the possibility that some of the antibodies do indeed cross the placenta but are then absorbed out by the fetus itself.

In our early studies on the fate of antipaternal MHC antibodies, we generated a defined anti-H-2 antiserum, which we showed by reactivity with congenic mice to be directed against both the left- and right-hand ends of the H-2 complex but which was not detectably reactive against non-H-2 components. When this antibody was injected into female mice bearing fetuses of the target haplotype, it rapidly disappeared from the circulation when compared with the control situation, in which the fetus did not bear the target haplotype (Wegmann and Carlson, 1977). We also observed that injection of this antibody did not cause any harm or detectable changes in the fetus or placenta, even though the antibody had *in vivo* cytotoxic effector capability (Carlson and Wegmann, 1978). In order to determine where this antibody was binding in the pregnant animals that were absorbing it, we passed the antibody over a protein A-Sepharose 4B column, labeled the eluted antibody with ^{125}I, and then partially purified it by adsorption and elution, using glutaraldehyde-fixed target leukemia cells as an immunoabsorbent (Figure 11-1). The antibody recovered generally had a specific activity of 10 to 20 per cent. Two hours after the injection of this radiolabeled antibody into pregnant mice, the only tissue in which we could detect differential binding between experimental and control animals was the placenta (Wegmann et al. 1979a). No differential binding was observed in the liver, kidney, lung, and other organs.

More recently, we have been studying the immunoabsorbent properties of the placenta by using monoclonal antibodies, which have the advantage of possessing a much higher specific activity and a well-defined and reproducible specificity. The principal monoclonal antibody we have used to date is that derived from the clone 11-4.1.72C, which originated in the laboratory of L.A. Herzenberg at Stanford University and is now available from the Salk Cell Distribution Center, San Diego. This IgG_{2a} antibody reacts with the K antigen of $H-2^k$ haplotype.

In what follows we shall describe our studies with this monoclonal antibody as well as those with monoclonal antibodies directed against the Ia region and the H-2D region. These studies give formal proof of the idea that the placenta can serve as an immunoabsorbent barrier between the maternal and fetal circulations for antifetal (paternal) antibodies. Also, by using autoradiography on pregnant mice labeled *in situ*, we have succeeded in localizing the immunoabsorbent sites in the placenta.

DIRECT EVIDENCE FOR THE PLACENTA AS AN IMMUNOABSORBENT BARRIER

Passive immunity is provided by the mother to her fetus by certain classes of antibodies which can cross the placental barrier and enter the fetal circulation (Brambell, 1970). If the placenta indeed serves as an immunoabsorbent barrier to the entry of immunoglobulin which could otherwise transit through the placenta and which is at the same time directed against fetal MHC antigens, one would then expect to see such antibodies trapped in the placenta and not getting into the fetal circulation. The uptake of radiolabeled antibodies should accordingly be high in the placenta and low in the fetus, when compared with controls in which the fetus does not bear the target antigen. In the latter case, the situation should be reversed, with the counts relatively low in the placenta and higher in the fetus. To verify this prediction, we labeled and purified the 11.4.1.72C anti-H-2K^k monoclonal antibody as described above (Figure 11-1) to 65 per cent specific activity (Wegmann et al., 1979b). It was then injected into the maternal circulation on day 13 of pregnancy. We sacrificed the animals at various times thereafter and counted the radioactivity in placentae and fetuses of both syngeneically (d x d) and allogeneically (d x k) mated animals. The results, presented in Figure 11-2, clearly conform to the predictions made from the placental immunoabsorbent hypothesis and confirm our earlier observations made using conventional anti-H-2 antibody (Wegmann et al., 1979a). Antibody binding to the target (d x k) placentae reaches a peak six to eight hours after injection of the labeled antibody and rapidly falls to background levels by three days after injection. A similar peak is not observed in the nontarget (d x d) placentae. For the fetuses, however, the situation is reversed. The non-target fetuses (d x d) contain higher levels of the labeled antibody than do the target (d x k) fetuses, as expected if the target placenta selectively absorbs the anti H-2^k antibody. There is another interesting difference between fetuses and placentae. The loss of ^{125}I is much slower in the nontarget fetus than in the target placenta. This observation led us to investigate the kinetics of antibody turnover during placental immunoabsorption.

PURIFICATION OF ANTI H-2 ANTIBODY

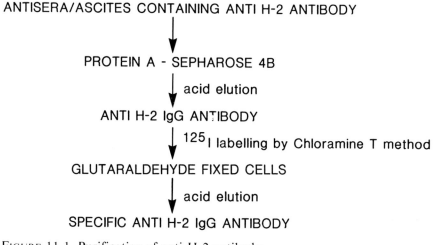

ANTISERA/ASCITES CONTAINING ANTI H-2 ANTIBODY

PROTEIN A - SEPHAROSE 4B

acid elution

ANTI H-2 IgG ANTIBODY

^{125}I labelling by Chloramine T method

GLUTARALDEHYDE FIXED CELLS

acid elution

SPECIFIC ANTI H-2 IgG ANTIBODY

FIGURE 11-1. Purification of anti-H-2 antibody.

TURNOVER KINETICS OF THE ANTIBODY

Since responder pregnant females probably produce antipaternal antibodies throughout the latter stages of pregnancy, a continuous process of turnover and re-expression of the paternal MHC antigenic determinants would bé one way of avoiding saturation of the placenta by these antibodies. We have tried to address this issue by determining whether paternal H-2K antigens in the placenta can be re-expressed after they are bound by antibody. We first determined the dosage of unlabeled anti-H-2K^k monoclonal antibody that would cause 75 per cent inhibition of the binding of radiolabeled antibody in the placenta when measured six hours after intravenous injection, by administering increasing amounts of cold monoclonal antibody along with a single dose of radiolabeled antibody (Figure 11-3). The 75 per cent inhibitory dose (50 μl of ascites) was then given either at the time of, or at various times prior to, the administration of a single dose of labeled antibody, the latter always administered on day 17 of pregnancy. The results, depicted in Figure 11-4, indicate that if one gives the cold antibody on day 15 of pregnancy (i.e., 48 hours before giving the radiolabeled antibody), there is no

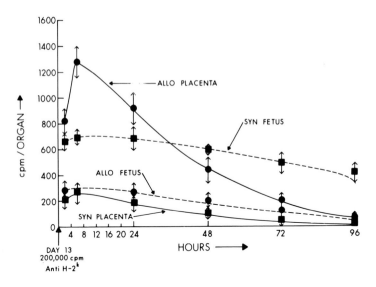

FIGURE 11-2. The relationship between the amounts of ^{125}I present at various time periods in (d x d) control and (d x k) target placentae and fetuses following a single pulse of ^{125}I-labeled antibody.
(●—●), allogeneic (d x k) placenta; (■—■), syngeneic (d x d) placenta; (●---●), allogeneic (d x k) fetus; (■---■), syngeneic (d x d) fetus.

detectable effect on the binding of radiolabeled antibody, with increasing inhibition if the cold antibody is injected later than day 15. This means that the maximal turnover time of a given antibody molecule in the placenta is 48 hours. It is not clear at this point whether this represents the appearance of new H-2K antigens or the degradation of antibody, or both. However, the fact that the antigenic capacity of the placenta is re-expressed after having been bound by antibodies is established beyond doubt. We are now investigating the fate of this antibody in the placenta. Is it digested *in situ* after placental binding (Wood et al., 1978; Wood, 1980), or is it released along with antigen as complexes into the maternal circulation (Bernard, 1977)? Preliminary experiments indicate that at least digestion *in situ* takes place (unpublished observation).

ONTOGENY OF THE IMMUNOABSORBENT CAPACITY

Since the experiments described above clearly established that the placenta is an immunoabsorbent barrier to antipaternal MHC antibodies arriving from

235

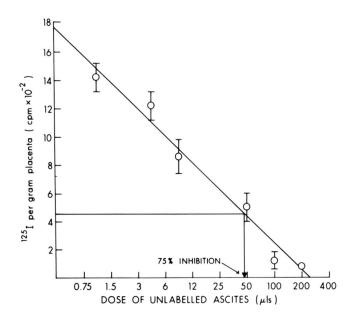

FIGURE 11-3. Determination of the dose of unlabeled antibody required to suppress the *in vivo* binding of ^{125}I-anti-H-2K[k] antibody to the (d x d) target placenta by 75 percent.
Varying amounts of the unlabeled anti-H-2K[k] antibody were mixed with 2 x 10[5] cpm of ^{125}I-labeled antibody and injected on day 17 of pregnancy into syngeneically and allogeneically pregnant mice. Six hours later, the placentae were processed and the cpm/g assessed. Control (d x d) values were subtracted from target (d x k) values prior to plotting.

the maternal side, we decided to measure the capacity of the placenta to bind these antibodies. The approach followed, first suggested by John Barrington Leigh of our department, was to inject increasing doses of the labeled antibody into pregnant females at days 10, 13, and 17 of pregnancy. The placentae were then excised at the time of peak uptake (six to eight hrs.) and the maternal blood washed out by preparation of homogenates of the placental tissue. The level of radioactivity was then determined and the dose-response curves plotted as shown in Figure 11-5. By graphing double reciprocal plots of these dose-response curves, we could calculate the immunoabsorbent capacity of the placenta in terms of the number of H-2K antigens per gram placenta (Wegmann et al., 1980a). We examined the placentae on days 10, 13, and 17 and obtained the reciprocal plots shown in Figure 11-6. It is clear that the immunoabsorbent capacity of the placenta per gram tissue increases steadily during pregnancy. The dose-response curves were replotted in double reciprocal fashion, and the number of H-2K antigenic sites were plotted as a function of time, as shown in Figure 11-7. These results indicate that the increase from day 10 to

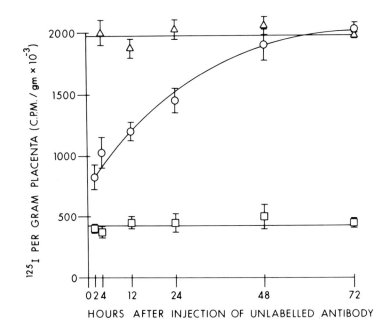

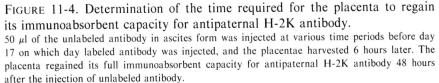

HOURS AFTER INJECTION OF UNLABELLED ANTIBODY

FIGURE 11-4. Determination of the time required for the placenta to regain its immunoabsorbent capacity for antipaternal H-2K antibody.

50 μl of the unlabeled antibody in ascites form was injected at various time periods before day 17 on which day labeled antibody was injected, and the placentae harvested 6 hours later. The placenta regained its full immunoabsorbent capacity for antipaternal H-2K antibody 48 hours after the injection of unlabeled antibody.

day 13 (p < 0.0025) and day 10 to day 17 (p < 0.001) is statistically highly significant. The increase from day 13 to day 17 is not significant (p < 0.1). Nevertheless, the observed increase was reproducible in a number of separate experiments. Clearly, the immunoabsorbent capacity of the placenta does increase with ontogeny, especially in view of the fact that we measured the increase per gram of tissue. As well, the placenta is growing rapidly from day 10 to day 13, so that the increase per feto-placental unit is even greater. We emphasize that these estimates are minimal. We assume that there is no antibody turnover, which we know to be incorrect (see above). We also assume that the mother herself does not make antibody to cover up these sites, although we have as yet no direct information on this point.

237

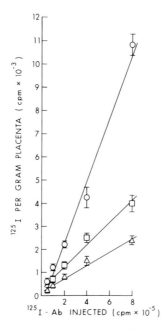

FIGURE 11-5. Comparison of the uptake of ^{125}I-labeled anti-H-2K[k] antibody by day 10 (□), day 13 (△), and day 17 (○) target (d x k) allogeneic placentae minus control (d x d) placentae.

Fc BINDING IS NOT REQUIRED FOR THE IMMUNOABSORBENT EFFECT

The next question we addressed was whether the monoclonal antipaternal H-2K antibody binds first via placental Fc receptors in order to undergo membrane transport before binding to the H-2K antigens in the placenta. This was done by removing the Fc portion of the antibody molecule by pepsin digestion, followed by protein A-Sepharose chromatography to remove residual antibody molecules bearing the Fc portion (Wegmann et al., 1979b). Polyacrylamide gel electrophoresis studies by Tim Mosmann of our department showed that this preparation contained only F(ab′)$_2$ molecules. We then radiolabeled this material with ^{125}I, purified it by the absorption-elution procedure described earlier (Figure 11-1) and used it to restudy the immunoabsorbent capacity of the placenta as well as the uptake kinetics of the placenta.

We could detect no difference in the dose-response curves for the radiolabeled F(ab′)$_2$ preparation at the time of maximal uptake (Figure 11-8)

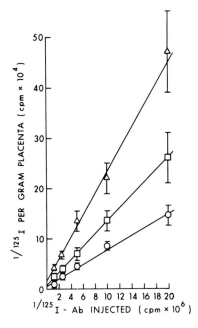

FIGURE 11-6. Determination of placental capacity for binding anti-H-2K^k antibody.

The data from Figure 11-2a were replotted in double reciprocal fashion and the straight lines obtained by a least squares fit of the data. The intercepts of the double reciprocal curves represent dose responses on day 10 (□), day 13 (△), and day 17 (O) are 1.54 ± 0.06, 0.87 ± 0.07, and 0.7 ± 0.06, respectively. The day 10 and day 13 values were significantly different and so were the values for day 10 and day 17, while the difference between day 13 and day 17 was not statistically significant.

when compared with the dose-response curves obtained with radiolabeled whole antibody. Only a very slight difference was observed in the way the placenta handles the radiolabeled F(ab')$_2$ after a single pulse (Figure 11-9). The F(ab')$_2$ anti-H-2K^k seems to be cleared slightly more rapidly than the whole antibody molecule. We therefore conclude thst Fc receptor binding is not a necessary prerequisite to H-2K antigen binding from the maternal side.

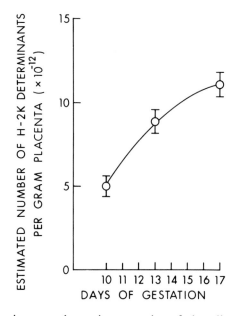

FIGURE 11-7. The immunoabsorptive capacity of the allogeneic placenta for antipaternal monoclonal H-2K antibody at different stages of ontogeny.

The numbers of H-2K^k determinants ($\pm$ S.E.) per gram were estimates for day 10, day 13, and day 17 allogeneic placentae, plotted as a function of time (see Figure 11-2b).

EFFECT OF FETECTOMY ON IMMUNOABSORPTION

Experiments by Davies and Glasser (1968) indicate that placentae derived from chronically fetectomized rats undergo striking morphological changes. There is a general loss of fetal mesenchymal elements, without much effect on the spongiotrophoblast layer, except for a large increase in the amount of maternal blood space. Another possible consequence of fetectomy could be a reduction in the capacity of the placenta to act as an immunoabsorbent. It has been suggested that fetal lymphoid cells could migrate to the placenta throughout gestation, where they might provide the placenta with its immunoabsorbent capacity (Swinburne, 1970). To determine whether the placental immunoabsorbent capacity is indeed dependent on fetal circulation, we fetectomized a series of female mice at day 11 of gestation and measured the capacity of the placenta to bind anti-H-2K^k monoclonal antibody at various intervals following fetectomy. Nonfetectomized placentae in the opposite

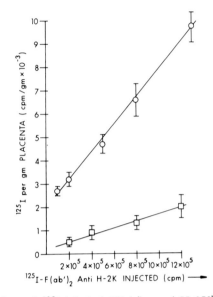

FIGURE 11-8. Binding of ^{125}I-labeled F(ab')$_2$ anti-H-2K^k to target d x k (O) and control d x d (●) placentae, on day 17 of gestation.

uterine horn were used as controls. The amount of antibody uptake per gram of placenta did not change by this procedure when examined on days 13, 15, and 17 of gestation (see Table 11-1). When examined morphologically, the placentae were found to have undergone histological changes similar to those described by Davies and Glasser (1968). These experiments suggest that elements from fetal blood and in particular the fetal mesenchyme do not participate in the placental immunoabsorbent function. They also suggest that the spongiotrophoblast may play some role in the uptake. Nevertheless, one must

TABLE 11-1
Effect of Fetectomy on the *In Vivo* Placental Immunoabsorption of Anti-paternal H-2K Antibody*

	NUMBER OF DAYS POST-FETECTOMY		
CPM/GM PLACENTA† OF FETUSES FROM	2	4	6
Fetectomized horns	1892 ± 212	2391 ± 193	2885 ± 127
Non-fetectomized horns	2020 ± 69	2423 ± 49	2624 ± 171
Control mice	1884 ± 79	2596 ± 200	2924 ± 147

* Fetectomies were performed on day 11 of gestation.

† 2 x 10^5 cpm of labeled anti H-2K^k antibody was injected per mouse, in allogeneically-mated BALB/c mice (d x k). At least five pregnant females were used to estimate each value. Each value represents cpm ^{125}I-anti-H-2^k per gram placenta ± S.D.

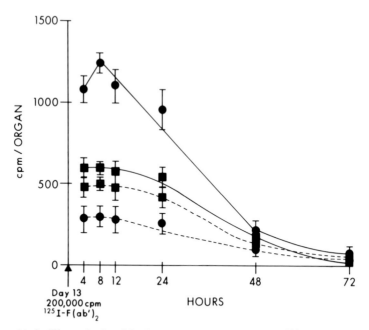

FIGURE 11-9. The relationship between the amounts of ^{125}I-F(ab')$_2$ antibody present at various time periods in allogeneic and syngenic placentae and fetuses following a single pulse of ^{125}I-labeled F(ab')$_2$ anti-H-2K^k. (●—●), allogeneic (d x k) placenta; (■—■), syngeneic (d x d) placenta; (●---●), allogeneic (d x k) fetus; (■---■), syngenic (d x d) fetus.

bear in mind that fetal cells could have migrated to the placenta before day 11 of gestation, and that these could have proliferated to form a layer of cells absorbing the antipaternal H-2K antibody from the maternal circulation.

LOCALIZATION OF IMMUNOABSORBENT SITES

In collaboration with Bruce Sandow of the Oregon Regional Primate Center, Portland, Oregon, we have carried out autoradiographic studies to localize the placental immunoabsorbent sites. We injected purified ^{125}I radiolabeled anti-H-2K^k antibody as well as its F(ab')$_2$ fragment and in some instances, antibody intrinsically labeled with ^{3}H-leucine into the circulation of female mice bearing target (d x k) or nontarget (d x d) placentae. At various intervals (one to eight hours later), the animals were anesthetized, the chest cavities opened, a cut made in the right ventricle, and the left ventricle penetrated by a 20-gauge needle attached to a 20-ml syringe containing phosphate-buffered saline. Next, 100 ml of saline was flushed through the circulatory system of the animal in order to wash away maternal blood in the placenta which might otherwise interfere with the autoradiography, as a result of contaminating nonabsorbed radiolabeled antibody. By such means, we found that we could get the amount of radioactivity in the syngeneic placentae to machine background levels. The placentae were then fixed in 10 per cent glutaraldehyde, embedded in epon-araldite, sectioned, and prepared for autoradiography. The results must be understood in the context of the controls, which showed no antibody binding whatsoever. In the target placentae, we saw binding principally in two areas. The most intense binding takes place in the lateral aspect of the placenta, where the yolk sac inserts into the lateral edges of the placenta. We observed there fairly dense labeling on both living and apparently dead cells. The dead cells were not killed by the antibody, since the same type of cells is found in control placentae not injected with the antibody. Many of the viable labeled cells have a monocyte morphology and have been shown to be esterase-positive macrophage-like cells. They are on the fetal side of Reichert's membrane and are frequently found in the crypts of Duval (see Figure 11-10).

The second major area of ^{125}I labeling which we observed is in the spongiotrophoblast zone. There, radiolabeling was seen on spongiotrophoblast cells in direct contact with maternal circulation (Figure 11-11). These cells were much less densely labeled than the ones in the lateral aspect of the placenta. Indeed, we did not detect this labeling until the emulsion had been exposed for approximately two months. Both patterns of labeling were seen at various times after injection of the antibody, ranging from one to eight hours. They were also seen on day 13, day 15, and day 17 placentae and regardless of whether one used the tritiated leucine or the ^{125}I label. Also, ^{125}I-F(ab′)$_2$ binding was identical to that of the intact antibody, confirming that Fc binding plays no role.

The most important finding in these experiments was that H-2K antigen, a strong major transplantation antigen, is present on spongiotrophoblast cells in direct contact with the maternal circulation. This contradicts long-standing speculation that MHC antigens are either missing or covered up at the fetal-maternal interface. It extends experiments of Chatterjee-Hasrouni and Lala (1979), who have claimed that cells taken from disrupted placentae and morphologically resembling trophoblast cells bear Class I MHC antigens on their surface, as detected using nonmonoclonal reagents. These investigators did not study directly the relationship between these cells and the maternal circulation. The finding that H-2K antigen is in direct contact with maternal circulation forces one to devise new concepts of how the fetus avoids rejection by the mother (see below).

243

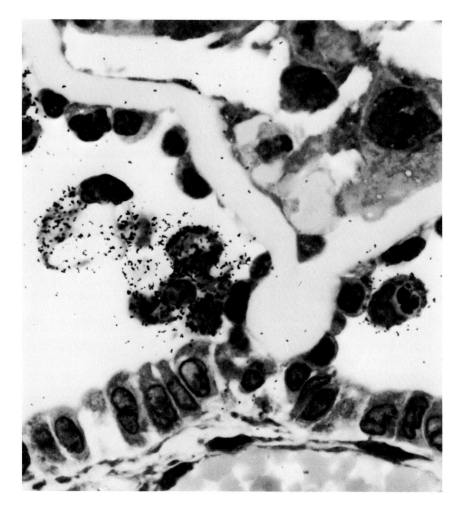

FIGURE 11-10. Autoradiogram revealing densely labeled cell of monocytic or macrophage-like appearance in an endodermal sinus of an allogeneic (Balb/c x C3H/HeJ) mouse placenta 6 hours after i.v. injection into the mother of 6.5 μg of radiolabeled anti-H-2K^k monoclonal antibody (produced by hybridoma 11-4.1 72c, from the Salk Cell Distribution Center).

Prior to injection, the antibody was separated from ascites fluid on a protein A-Sepharose column, labeled with ^{125}I by the chloramine T technique, and further purified by absorption-elution on H-2K^k positive glutaraldehyde-fixed leukemia cells. Placentae were dissected from the uterus with decidua attached, cut into wedges, fixed in 0.35 per cent glutaraldehyde in 0.1 M phosphate buffer and embedded in glycol methacrylate. Sections were cut at 2 μ m, placed on slides, and coated with Kodak NTB3 nuclear track emulsion. The autoradiograms were exposed at 4°C for 4 to 26 weeks, developed in Kodak D-19 developer, and stained with Gill's hematoxylin and Lee's methylene blue-basic fuchsin. Magnification: 980x.

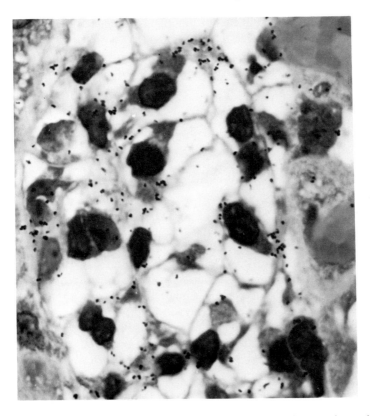

FIGURE 11-11. Autoradiogram revealing lightly labeled spongiotrophoblast cells in the spongy zone of an allogeneic (Balb/c x C3H/HeJ) mouse placenta after injection of the Balb/c mother with radiolabeled anti-H-2K^k monoclonal antibody. Magnification: 1140x.

EXPRESSION OF PATERNAL H-2D
ANTIGENS ON THE PLACENTA

We next studied the expression of paternal H-2D antigen, another Class I MHC antigen, on the placenta. The monoclonal antibody we used for these studies was anti-H-2D^k produced by clone 15-5-5S and was kindly provided by David Sachs of the National Institutes of Health, Bethesda (Epstein et al., 1980). The antibody was purified and radiolabeled as described above (Figure 11-1) to a specific activity of 59 per cent. The labeled antibody was injected in animals bearing target and nontarget placentae on days 10 and 17

245

SINGH, RAGHUPATHY, ANDERSON, WEGMANN

of gestation. The results, depicted in Figure 11-12, demonstrate that the antibody specifically binds to target placentae and that this binding increases from day 10 to day 17. The estimated number of paternal H-2D antigens obtained by the methods described above is about five times lower than the estimate for H-2K antigens at comparable states of gestation.

LACK OF CLASS II MHC ANTIGENS ON THE PLACENTA

The studies outlined above were concerned with reactivities against H-2K and H-2D antigens (Class I MHC antigens). We also looked for the presence of paternal I-region derived antigens (Class II MHC antigens) in the placenta. Monoclonal anti-I-A^k (specificity 17) antibody (belonging to the IgG$_{2a}$ subclass) was produced by clone 10-3.6, obtained from the Salk Cell Distribution Center. It was purified and ^{125}I-radiolabeled to a high specific activity (65 per cent) as described above (Figure 11-1). In this case, glutaraldehyde-fixed C3H/HeJ (H-2^k) spleen cells were used for absorption and elution. Labeled antibody was injected into pregnant animals in exactly the same manner as described for anti H-2K^k monoclonal antibody. There was no detectable uptake in the target placentae relative to control placentae (Table 11-2). Chatterjee-Hasrouni and Lala (1980) have also shown that morphologically identified trophoblast cells taken from disrupted placentae showed no binding of monoclonal anti-Ia antibodies of the same specificity. Of course, one must test other Ia specificities to see whether they aare present, but it is tempting to speculate that the placenta contains Class I but not Class II MHC antigens in direct contact with maternal circulation. This would be reminiscent of findings that pregnant rats have MIF activity in

TABLE 11-2
Lack of Expression of Paternal I-A Antigens by the Allogeneic Placenta*

DAY OF GESTATION	PREGNANCY	PLACENTAL CPM/GM ± S.D.	FETAL CPM/GM ± S.D.
13	Syngeneic (d x d)	1317 ± 247	633 ± 73
	Allogeneic (d x k)	1198 ± 231	754 ± 81
17	Syngeneic (d x d)	1414 ± 105	1394 ± 149
	Allogeneic (d x k)	1198 ± 106	1505 ± 301

* 4 x 10^5 cpm of ^{125}I-labelled anti-IA^k antibody was injected into allogeneically (d x k) and syngeneically (d x d) pregnant mice on days 13 and 17 of gestation, and the placental and fetal uptakes measured 6 hours later. Each value represents the mean of at least 5 values ± S.D.

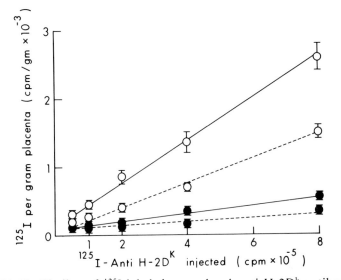

FIGURE 11-12. Binding of ^{125}I-labeled monoclonal anti-H-2D^k antibody to (d x k) and (d x d) placentae on day 10 and day 17.
(O—O), day 17 (d x k) placenta; (●—●), day 17 (d x d) placenta; (O---O), day 10 (d x k) placenta; (●---●), day 10 (d x d) placenta.;

their serum only if the mother and fetus differ by Class I antigens and not if they differ only by Class II antigens (Gill and Kunz, 1980). Similar results were obtained by Goodfellow et al. (1976), who reported the presence of HLA-A and HLA-B antigens but not D region antigens on human placenta (see also Chapter 10).

FUTURE PERSPECTIVES

From the foregoing, it seems reasonably well established that the placenta can serve as an immunoabsorbent barrier for at least Class I major histocompatibility antigens. The possibility exists that Class II antigens are not included in this type of placental function, but further study is needed on this subject. This immediately raises the question of whether the human placenta might also show similar immunoabsorbent capabilities, and indeed, not just for major histocompatibility complex antigens. One could speculate as to whether the placenta might serve as an immunoabsorbent barrier for anti-Rh antibodies and that this might explain some of the clinical observations seen with erythroblastosis fetalis.

The fact that the Class I antigens, at least in the case of H-2K and perhaps of H-2D, seem to be in direct contact with the maternal circulation leads one to wonder how the fetus avoids the cell-mediated immune response of the mother. A number of possibilities present themselves. Perhaps the putative lack of Class II antigens keeps the level of helper cell activity down to an acceptable minimum. This may be coupled with a relative resistance of tropho-blast cells to cell-mediated immunity (but see Chapter 10), which could be accomplished either by some variety of nonspecific suppressor function found in trophoblast cells, as posited by the experiments of Clark (see Chapter 16) and others. Alternatively, the trophoblast perhaps serves simply as a mechanical barrier to the entry of cellular effector cells into the fetal side of the placenta.

Perhaps relevant here are experiments on parabiosed mice. In one strain combination (DBA/2J parabiosed to the F_1 hybrid between DBA/2J and C3H/HeJ), half the pairs survive long term. Among the survivors, one sees an initial hemopoietic chimerism, followed by a complete "takeover" of both animals by the DBA/2J cell line, analogous to a bone marrow transplant (Wegmann et al., 1980b). Two features of this model are of interest here. Both parabiotic partners have cytotoxic antibody in their circulation which reacts against MHC antigens of C3H/HeJ. Also, parental T cells removed from the parabiotic F^1 hybrid spleen and exposed to C3H/HeJ alloantigen-bearing cells show almost completely normal MLR, CML, and GVH reactivities. The analogy to pregnancy is obvious. Here one also has direct exposure to fetal MHC antigens to maternal lymphocytes; anti-MHC antibody production, at least in some instances; and minimal effects on maternal T-cell-mediated immunity. In both pregnancy and parabiosis, some form of classical enhance-ment could be operative, although it is clear in the case of pregnancy that trig-gering the maternal lymphocytes against the fetus does not harm the pregnancy (Wegmann et al., 1979). All of the above experiments underscore the need to better understand the nature of the thin yet effective immunologi-cal barrier between the maternal and fetal circulations seen in the hemochorial placenta. It seems clear already, however, that the placenta has evolved mechanisms to trap and dispose of antibodies directed against Class I MHC antigens of the fetus. How the barrier functions with respect to maternal-cell-mediated immunity against the fetus stands in urgent need of further clarification.

REFERENCES

Bernard, O. 1977. Possible protective role of maternal immunoglobulins on embryonic development in mammals. *Immunogenetics* 5:1.

Billington, W.D., E.J. Jenkinson, R.F. Searle, and M.H. Sellens. 1977. Alloantigen expression during early embryogenesis and placental ontogeny in the mouse: immunoperoxidase and mixed hemadsorption studies. *Transplant. Proc.* 9:1371.

Bonneau, M., M. Latour, J.P. Revillard, M. Robert, and J. Traeger. 1973. Blocking antibodies eluted from human placenta. *Transplant. Proc.* 5:589.

Brambell, F.W. 1970. *The Transmission of Passive Immunity From Mother to Young.* Amsterdam: North Holland.

Carlson, G.A., and T.G. Wegmann. 1978. Paternal-strain antigen excess in semiallogeneic pregnancy. *Transplant. Proc.* 10:403.

Chaouat, G., G.A. Voisin, D. Excalier, and P. Robert. 1979. Facilitation reaction (enhancing antibodies and suppressor cells) and rejection reaction (sensitised cells) of mother to paternal antigens of the conceptus. *Clin. Exp. Immunol.* 35:13.

Chatterjee-Hasrouni, S., and P.K. Lala. 1979. Localization of H-2 antigens on mouse trophoblast cells. *J. Exp. Med.* 149:1238.

Chatterjee-Hasrouni, S., and P.K. Lala. 1980. MHC antigens on mouse trophoblast cells. In *Abstracts: 4th International Congress of Immunology*, Paris.

Davies, J., S.R. Glasser. 1968. Histological and fine structural observations on the placenta of the rat. *Acta Anat.* 69:542.

Doughty, R.W., and K. Gelsthorpe. 1976. Some parameters of lymphocyte antibody activity through pregnancy and further eluates of placental material. *Tissue Antigens* 8:43.

Doughty, R.W., and K. Gelsthorpe. 1974. An initial investigation of lymphocyte antibody activity through pregnancy and in eluates prepared from placental material. *Tissue Antigens* 4:291.

Epstein, S.L., K. Ozato, and D.H. Sachs. 1980. Blocking of allogeneic cell-mediated lympholysis by monoclonal antibodies to H-2 antigens. *J. Immunol.* 125:129.

Gill, T.J. III, and C.F. Repetti. 1979. Immunologic and genetic factors influencing reproduction. *Am. J. Pathol.* 95:465.

Gill, T.J. III, and H.W. Kunz. 1980. The role of regional differences in the MHC in the production during pregnancy of a serum factor inhibiting macrophage migration. *J. Immunogenet.* 7:157.

Goodfellow, P.N., C.J. Barnstable, W.F. Bodmer, D. Snary, and M.J. Crumpton. 1976. Expression of HLA system antigens on placenta. *Transplantation* 22:595.

Goodlin, R.C., and L.A. Herzenberg. 1964. Pregnancy induced hemagglutinins to paternal H-2 antigens in multiparous mice. *Transplantation* 2:357.

Jenkinson, E.J., and V. Owen. 1980. Ontogeny and distribution of major histocompatibility complex (MHC) antigens on mouse placental trophoblast.*J. Reprod. Immunol.* 2:173.

Swinburne, L.M. 1970. Leucocyte antigens and placental sponge. *Lancet* ii:562.

Voisin, G.A., and G. Chaouat. 1974. Demonstration, nature and properties of maternal antibodies fixed on placenta and directed against paternal antigens. *J. Reprod. Fertil.* 21:(Suppl.)89.

Wegmann, T.G., J. Barrington Leigh, G.A. Carlson, T.R. Mosmann, R. Raghupathy, and B. Singh. 1980. Quantitation of the capacity of the mouse placenta to absorb monoclonal anti-fetal H-2K antibody. *J. Reprod. Immunol.* 2:53.

Singh. 1980. Quantitation of the capacity of the mouse placenta to absorb monoclonal anti-fetal H-2K antibody. *J. Reprod. Immunol.* 2:53.

Wegmann, T.G., and G.A. Carlson. 1977. Alloengeic pregnancy as immunoabsorbent. *J. Immunol.* 119:1659.

Wegmann, T.G., B. Singh, and G.A. Carlson. 1979a. Allogeneic placenta is a paternal strain antigen immunoabsorbent. *J. Immunol.* 122:270.

Wegmann, T.G., T.R. Mosmann, G.A. Carlson, O. Olijnyk, and B. Singh. 1979b. The ability of the murine placenta to absorb monoclonal anti-fetal H-2K antibody from the maternal circulation. *J. Immunol.* 123:1020.

Wegmann, T.G., C.A. Waters, D.W. Drell, and G.A. Carlson. 1979. Pregnant mice are not primed but can be primed to fetal alloantigen. *Proc. Nat. Acad. Sci. USA* 76:2410.

Wegmann, T.G., J. Rosovsky, G.A. Carlson, E. Diener, and D.W. Drell. 1980. Models for the production of stable hematopoietic chimerism across major histocompatibility barriers in adults. *J. Immunol.* 125:1751.

Wood, G.W. 1980. Immunohistological identification of macrophages in murine placenta, yolk-sac membranes and pregnant uteri. *Placenta* 1:309.

Wood, G., J. Reynard, E. Kirshanan, and L. Pacela. 1978. Immunobiology of the human placenta. 1. IgG Fc receptors in trophoblastic villi. *Cell. Immunol.* 35:191.

Chapter 12

IMMUNOBIOLOGY OF HUMAN EXTRAEMBRYONIC MEMBRANES

W. PAGE FAULK

For those interested in human reproduction, it concentrates the mind to ask how the embryo is able to survive uninterrupted for nine months in functional and anatomical contact with genetically different maternal uterine cells. It is equally challenging to ask how any fertile mammalian egg is able to approximate its extraembryonic membranes to the allogeneic mother and yet manage to avoid rejection. These questions are compounded by evolutionary considerations which indicate that mammals are comparative latecomers in the history of the animal kingdom and that animals such as sponges (Hildemann et al., 1979) and other primitive metazoans (Solomon and Horton, 1977) had developed a complex system of allogeneic recognition and rejection reactions by the time mammals appeared on the evolutionary scene. If systems for the recognition of nonself antigens anteceded placentation, how then was it possible to fly in the face of this immunological progress by developing a new system of reproduction dependent on the coexistence of allogeneic tissues and the live birth of progeny (viviparity) rather than upon the laying and hatching of fertilized eggs? In other words, for viviparity to develop, major exceptions challenging the preceding thousands of millions of years of evolutionary experience had to occur, and it is the nature of these exceptions that has so successfully escaped precise definition.

The field of obstetrics is littered with question marks, for practically nothing is known of the pathogenesis of spontaneous abortion, congenital abnormalities, preeclamptic toxemia, premature rupture of membranes, etc. Just because these happen to be diseases of unknown etiology does not imply that they have anything to do with immunology, but one may wish to consider that the mother is a normal, healthy person who happens to be pregnant. The phenomenon of her pregnancy challenges the entirety of immunological experience, placing immunological processes as reasonable forces in the maintenance of her underlying pathophysiology. The relative merits of this position have been discussed in several recent reviews (Gill and Repetti, 1979; Rocklin et al., 1979; Faulk and Johnson, 1980; Faulk and Fox, 1981), and so the purpose of this chapter is to examine the possible role of extraembryonic membranes in the success and failure of the materno-fetal relationship in human pregnancy. These membranes in the early human embryo are shown in Figure 12-1.

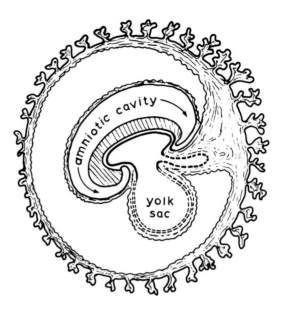

FIGURE 12-1. Schematic drawing of human extraembryonic membranes.
Out-pouching into body stalk is the allantois. Early chorionic villi on exterior are covered by trophoblast.

ONTOGENY OF EXTRAEMBRYONIC MEMBRANES

Like many other sexually reproducing species, the human embryo begins as
an egg that is fertilized in the upper part of the fallopian tube by a single
spermatozoon, following which it begins a series of mitotic cleavages. The
resulting blastomeres have been shown from experiments in other mammals
to be unspecialized. If separated and allowed to develop independently, each
blastomere can produce a whole individual with a complete set of tissues and
organs. This cluster of protodermal cells, the morula, is held together and
protected by the zona pellucida (Aitken and Richardson, 1980) and swept
toward the uterus by cilia lining the oviduct. After about four days, it
reaches the uterus, where the zona pellucida disappears within 12 hours. The
morula remains in uterine fluid for two to three days and then is converted
into a unilaminar blastocyst by a process of cavitation. This is characterized
by the development of a split between the outer and inner layers of cells, a
split that spreads in all directions until it has detached the surface cells from
the inner cells everywhere but at the boundaries of the inner cell mass
(ICM). Accumulation of fluid in the blastocoele stretches and flattens the

cells except where they remain attached to the ICM, creating a distinction between the extraembryonic trophoblast and the still undifferentiated embryonic protodermal cells of the ICM.

Some embryologists have been committed to an idea that embryonic and extraembryonic materials are separated by the first cleavage (Hertig, 1968), but this belief is discredited by the observation of Nicholas and Hall (1934), who showed that each of the first two blastomeres of the rat is totipotent, and by the subsequent demonstration by Kelly (1977) that the four- and eight-cell stages of the mouse are also totipotent. Spindle (1978) has separated the ICM from the outer layer and shown that the inner cells in mice remain totipotent until 69 to 93 hours in tissue culture after the two-cell stage. During the totipotent period, the ICM is capable of regenerating the lost outer layer as well as of producing embryonic and extraembryonic tissues. Culture of the outer cells gives rise to nothing but ectodermal vesicles (Snow, 1973; Rossant, 1975). The results of these experiments seem to deny any separation of embryonic and extraembryonic material during early cleavage and indicate instead that extraembryonic membranes arise from the ICM and that this differentiation pathway is not reversible (Gardner and Rossant, 1976).

The zygote establishes contact with uterine mucosa on the seventh day and responds by inserting pseudopodal anchors from chorionic ectoderm between the uterine epithelial cells. It next produces, from cytotrophoblast, a new tissue called syncytiotrophoblast, which is formed by the fusion of cell membranes to build a syncytium, in which the nuclei are scattered at random through a continuous cytoplasm. A single layer of endodermal cells then begins to appear below the amniotic ectoderm to complete the ICM by the eighth day. During the ninth day, endoderm proliferates until it has completely lined the vesicle, converting most of the blastocoele into a primitive yolk sac. During the tenth and eleventh days, the contiguity between amnion and cytotrophoblast remains undisturbed but the proliferating endoderm, after completing the primary yolk sac, continues shedding cells to form a reticulum between its well-formed lining and the cytotrophoblast. During the thirteenth day, most of the primary yolk sac disappears except that portion most closely associated with the medullary plate (Luckett, 1978). This is retained to give rise to the definitive yolk sac, gut, and allantois. Finally, as the dwindling primary yolk sac and its reticular extensions shrink away from the cytotrophoblast, they are replaced by mesoderm, which by the fifteenth day has formed a continuous membrane around the outside of the amniotic cavity and definitive yolk sac, separating for the first time the amniotic ectoderm from the cytotrophoblastic layer.

AMNION

Amniotic ectoderm eventually lines the entire interior of the amniotic cavity, the structure which later in pregnancy contributes several protective and supportive services thought to be important in fetal survival as well as in maintenance of the extraembryonic allograft. The amniotic ectoderm, which covers the embryo's third branchial arch and the groove that adjoins the arch on its caudal side, makes a contribution to the thymic cortex (Crisan, 1935), and this bit of ectoderm is essential to the development of Hassall's corpuscles within the thymus (Norris, 1938). The concept of a functional relationship between amniotic ectoderm and Hassall's bodies is supported by studies in nude mice, for abortive development of the ectodermal component of the thymus in the embryo of these animals has been shown to result in abnormal Hassall's bodies and profound immunological deficiencies (Cordier and Heremans, 1975). The thymus is thus not extraembryonic in origin, but it cannot play its central immunological role without amniotic ectoderm as well as a supply of stem cells from the fetal yolk sac, the next extraembryonic membrane to be discussed.

YOLK SAC

The fetal yolk sac in some fishes and in all reptiles, birds, and mammals is an extraembryonic extension of the midgut. In birds and mammals, it is known to give rise to stem cells which migrate first to the fetal liver and then to the central lymphoid organs, the thymus, and the bursa of Fabricius (or its mammalian equivalent), where they finally mature and diversify. These cells then seed to the peripheral lymphoid structures to form the cells of the immunological system. Moore and Owen (1967) first described the details of this cell traffic in chickens, using chromosomal markers and parabiosis. It has been possible to confirm some aspects of their work, but the results of recent research on frogs (Volpe et al., 1981) and birds (le Douarin and Jotereau, 1980) indicate that some stem cells may be embryonic rather than extraembryonic in origin. Although the yolk sac in humans is vestigial, there is a stage in human development when it is robust enough to carry on its historic role of harboring all of the primordial germ cells until it releases them through the umbilical cord to their final home in the mesodermal genital ridges. It similarly has perhaps maintained its role as a source of blood cells, including those stem cells which eventually differentiate and diversify to form the cells of the immunological system.

ALLANTOIS

The allantois was originally a storage depot for nitrogenous waste in eggs enclosed within a shell or membrane, but it progressively became overloaded and distended as the period of gestation was extended. It was a type of urinary bladder, but this expanded outside the body wall and eventualy established contact with ectodermal chorion, the trophoblastic surface of which in humans is closely applied to maternal blood. Here the bladder emptied itself into the mother's blood, from which the maternal kidneys removed and disposed of the fetal effluent. The allantoic placenta thus arose for waste disposal, but the connections that served this function also enabled it to utilize maternal supplies of nutrients, oxygen, and immunoglobulins for the embryo. Although embryologists have not pointed out the possibility of a relationship between these two areas, the bursa of Fabricius and the allantois in birds arise from very similar locations in the hindgut, raising the possibility that cells of the allantoic mesoderm may be associated with a bursal equivalent function for B lymphocyte development in mammals (Faulk, 1981).

That ontogeny may recapitulate phylogeny in B cell production has been supported by the results of a recent study by Melchers and Abramzynk (1980) of precursor B lymphocytes in the blood of mouse embryos. These researchers defined pre-B cells as those which, on adoptive transfer with antigen into irradiated syngeneic recipients, will not yield an immediate B cell response of antigen-specific, immunoglobulin-secreting, plaque-forming cells, but will do so after a period of several days in the host. They also state that precursor B cells in bone marrow and fetal liver show delayed mitogenic responses to lipopolysaccharide. Using this approach, they were able to identify pre-B cells in the blood of embryonic mice at day 10, which then become most abundant at day 12. The earliest pre-B cells came from the placenta, and yet another wave appeared between days 13 and 19 in fetal liver: neither yolk sac nor liver contained pre-B cells at the time they first appeared in the placenta. Later still, cells appeared in the bone marrow and continued to come from bone marrow and liver throughout life. The placenta of mice and nearly all mammals is an allantoic derivative from the hindgut near the site of origin of the bursa of Fabricius. Thus, the observations of Melchers and Abramzynk (1980) support the speculation of Faulk and McCrady (1981) that the mammalian equivalent of the bursa of Fabricius is the allantois and that the splanchnopleuric mesoderm adjoining its endodermal lining is the earliest site of maturation of stem cells into pre-B and B cells.

THE MATERNO-FETAL INTERFACE

PLACENTA

There are many differences and some similarities in placental structure and function in most animals (Amoroso, 1952; Enders, 1965; see Chapter 8), although it is not clear why these variations should exist. This may be related to the evolutionary need for chorionic ectoderm to develop means to evade maternal recognition and rejection, and different mammals may have done this in different ways (Ramsey et al., 1976). Nonetheless, there are several basic solutions to the problems of allogeneic coexistence, perhaps typified by the report that spleen cells from mother salamanders are cytotoxic for cells from their embryos and that this reaction is inhibited by maternal serum (Chateaureynaud et al., 1979). Unlike transplantation research on other organs, it is necessary to work within one species when studying the immunology of extraembryonic membranes, though it imposes an awkward impediment for those interested in human pregnancy. It is of central importance to establish precisely the boundaries of the materno-fetal interface, as allogeneic tissues coexist only at this junction. Placentae are in fact classified according to the type of tissues that compose the interface. In humans, this is predominantly formed by a specialized derivative of chorionic ectoderm, called the syncytiotrophoblast (Brambell, 1970). The syncytial surface contains species-specific antigens as well as an array of receptors for biologically active molecules, such as IgG, insulin, and transferrin (Faulk and Galbraith, 1979a). The presence of transferrin receptors, in particular, at the materno-fetal interface has attracted the attention of many investigators from several different fields of interest.

The significance of transferrin receptors is not known, but their presence at the operational interface of the mother and fetus suggests that they are of importance in the biology of human pregnancy (Figure 12-2). It is possible that they participate in biological functions besides iron transport (Faulk and Galbraith, 1979b). This could occur in at least three ways. First, maternal transferrin bound to fetal receptors may restrict the amount of available iron within the placental intervillous spaces. This may nonspecifically increase resistance to certain iron-dependent infectious processes. Second, trophoblastic transferrin may exert a local effect on maternal lymphocyte proliferation. Antigen- or mitogen-stimulated lymphocytes accumulate at the G_1 phase of the cell cycle in the absence of transferrin and proceed normally through S phase when human transferrin is added (Tormey and Mueller, 1972; Tormey et al., 1972). This may locally impede maternal cell-mediated immune reactions within the placental bed. A third possible role of trophoblast transferrin receptors is analogous to a mechanism utilized by certain parasites, such as schistosomes, which bind host proteins, presumably to cover and disguise the parasite's own antigens (Smithers et al., 1969; Goldring et al., 1977). The transferrin receptors of trophoblast (the "parasite") may exert a similar effect by binding transferrin from the mother (the "host"). This receptor, like the trophoblast antigens identified on certain cell lines in culture (Faulk et al., 1979a), appears on many transformed cells (Galbraith et al., 1980; Yeh and Faulk, 1981). Interestingly, it is also identified on leukemia and lymphoma cells (Yeh et al., 1981) and may serve a metabolic function for transformed and malignant cells, such as adenocarcinoma of the human breast (Faulk et al., 1980b).

Another characteristic of the materno-fetal interface in human placentae is the fate of syncytiotrophoblastic microvilli that extend from the syncytium into maternal blood. These microvilli are of fetal origin, and with age they break away and enter the maternal circulation (Ikle, 1964; Attwood and Park, 1961). Membrane fragments or microvilli continuously enter the mother, and entire clusters or sprouts of trophoblast pass from the placenta into her lungs, where they stimulate neither inflammatory reactions nor cellular infiltrates indicative of allogeneic recognition (Park, 1965). Thus, the mother is being continuously tranfused with allogeneic tissue, but is the trophoblast really allogeneic to the mother? It differentiates from the blastocyst and as such should

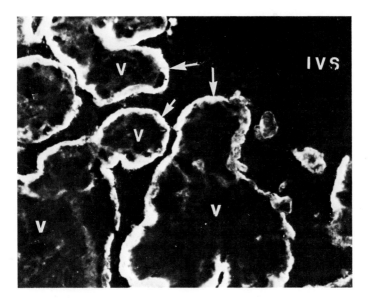

FIGURE 12-2. Transferrin reactivity (arrows) with maternal transferrin from blood in the intervillous spaces (IVS) bound to fetal transferrin receptors in syncytiotrophoblastic plasma membranes of chorionic villi (V).

259

be treated as a nonself antigen by the mother. Since histocompatibility antigens form the basis of the present concept of allogenicity, it should be stressed that the human syncytiotrophoblast does not contain these antigens (Faulk and Temple, 1976; Goodfellow et al., 1976; Faulk et al., 1977a). This is rather more than an academic point, for the absence of MHC antigens at the materno-fetal interface is so unlike the situation at the donor-recipient inter-face in all other grafts that forced analogies between the two are liable to be spurious.

In addition to the hemochorial interface within placentae, materno-fetal cellular contacts also occur between endothelium and endovascular tro-phoblast in uterine spiral arteries (Robertson, 1976), at the basal plate, and, most extensively, between uterine tissues and cytotrophoblast of the amnio-chorion (Bourne, 1962). The fetal cell which forms contact with maternal tis-sues at all of these sites is the trophoblast. In contrast, there is no single maternal cell type at these several interfaces. Indeed, blastocysts implanted ectopically have developed normally, suggesting that the trophoblast is respons-ible for its own success, rather than its being related to the uterine environ-ment. Several recent observations have extended the idea that the immunologi-cal success of the materno-fetal interface depends upon the trophoblast, and these findings primarily relate to the ability of normal trophoblast to regulate the presentation of histocompatibility antigens. One of the more recent discov-eries is the exciting observation that abnormal trophoblastic tissues contain MHC antigens (Shaw et al., 1979; Yamashita et al., 1979; Trowsdale et al., 1980). It is not known how trophoblast suppresses its ability to synthesize transplantation antigens, but if this is related to its ability to survive in an allo-geneic host, it would seem possible to learn how normal trophoblast suppresses the genes controlling the synthesis and membrane presesntation of MHC anti-gens, by, for instance, hybridizing trophoblast with other cells.

AMNIOCHORION

Although some of the materno-fetal interface is formed where the syncytio-trophoblastic plasma membrane makes contact with maternal blood in the placental bed, most of the interface is formed by the contact of amniochorion with maternal uterine tissues. The anatomy of this relationship has been extensively reviewed by Hoyes (1975). The actual contact is between chorionic cytotrophoblast and decidual cells, making it important to know whether transplantation antigens are present on the fetal cytotrophoblast, particularly in light of the absence of these antigens from syncytiotrophoblast in the placenta. Experiments done here by Hsi and colleagues (1981) with mouse monoclonal antibodies to HLA and β_2- microglobulin have shown that neither antigen can be identified on cytotrophoblastic tissues of the amnio-chorion. The chorionic component of this tissue is contiguous to endometrial cells which express both HLA and β_2-microglobulin surface antigens without evidence of cellular infiltrates or inflammation. It is not known how these allogeneic tissues coexist, but their ability to do so may relate to recent reports of trophoblast-specific glycoproteins which inhibit mixed leukocyte cultures (MLC) reactions without affecting lymphocyte responses to poke-weed mitogen or phytohemagglutinin (McIntyre and Faulk, 1979a).

A great deal of basic research has been done to define the develop-mental pathway of syncytiotrophoblast. Much of this work has led to the now generally accepted idea that placental syncytium differentiates from cytotro-phoblast (Boyd and Hamilton, 1970). Inasmuch as cytotrophoblast of human amniochorion does not develop a syncytium, it is important to know if syncytio-trophoblastic antigens can be identified in this tissue. It is even more important to learn if some of the placental trophoblast glycoproteins that selectively block MLC reactions can be identified in the amniochorionic cytotrophoblast. Their presence at this interface could perhaps explain the peaceful coexistence of amniochorion and maternal cells. We have used immunohistology to study human amniochorion, using heterologous antibodies to antigens of the syncytio-trophoblast which have been previously shown to block allogeneic recognition and rejection reactions *in vitro*. These investigations show that cytotrophoblast contains these biologically active antigens (Figure 12-3a), but that other cells of the amniochorion, including the amniotic epithelium, do not (Hsi et al.,

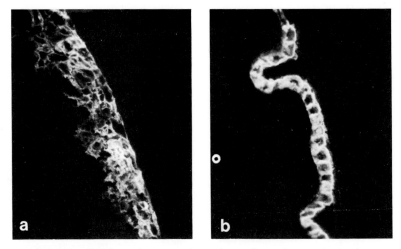

FIGURE 12-3. Distribution of extraembryonic antigens in amniochorion.
(a) Reactivity of TA1 antiserum with chorionic cytotrophoblast, and (b) pattern of AA1 anti-serum reaction with amniotic epithelium. Note no immunofluorescence of connective tissues, even though these photomicrographs were made from whole tissue sections of amniochorion.

1981). Thus, both the placental and amniochorionic interfaces lack MHC antigens and contain trophoblast membrane markers shown by *in vitro* studies to inhibit allogeneic rejection reactions. One group of these markers has been tentatively assigned the designation of Trophoblast Antigens 1 or TA1 (Faulk et al., 1978).

Given the presence of TA1 on cytotrophoblast and its absence from amnion epithelium, a query arises as to whether this epithelium manifests a characteristic antigen not found on chorionic cytotrophoblast and, if so, whether such antigens have intrinsic biological activities analogous to those of TA1 and anti-TA1 on lymphocyte responses *in vitro* (McIntyre and Faulk, 1979b). These questions pose many technical problems, for pure amniotic epithelium is difficult to prepare and the cells do not readily proliferate in primary cultures. Small amounts of amniotic epithelial plasma membranes have, however, been isolated from fresh extraembryonic membranes obtained at Caesarean section, and heterologous antisera to these cells have revealed the presence of an amnion antigen not found on either cyto- or syncytiotrophoblast (Figure 12-3b). Following the nomenclature for some of the serologically defined trophoblast antigens (Faulk et al., 1978), this marker will be tentatively referred to as Amnion Antigens 1 or AA1 in the following comments. The AA1 are identified on both reflected and placental amnion, and double-antibody studies using fluoresceinated anti-TA1 and rhodamine-labeled anti-AA1 provide tools to immunologically map the distribution of both antigens within normal and abnormal tissues. Results obtained by using fluorochrome labelled antisera to antigens of extraembryonic antigens have provided data to support the idea that the fetus develops within a cocoon of TA1, at both placental and amniochorionic interfaces. Inside this sanctuary of trophoblast antigens, the baby is shielded by AA1 on amniotic epithelium.

As described above in the section entitled "Ontogeny of Extraembryonic Membranes," ectoderm originates from the lining epithelium of the floor of the amniotic ectoderm. This can be further divided into so-called surface ectoderm (giving rise to the skin, mammary glands, etc.) and neuroectoderm, which is subdivided into neural crest (pigment cells, adrenal medulla, etc.) and neural tube derivatives (retina, central nervous system, etc.) (Moore, 1977). It is not known if developmental defects in embryonic ectoderm are mirrored by concomitant abnormalities in amniotic epithelium, but a recent study of amnion from a newborn with defective skin (epidermolysis bullosa letalis) found that some congenital problems of ectoderm are reflected in amniotic epithelium (Faulk et al., 1981). Cells of the placental amnion were morphologically abnormal and the epithelium of the reflected amnion lacked AA1, whereas a large number of normal, control membranes all proved to be positive for AA1.

Another possible use of anti-AA1 sera is as a probe to follow the differentiation of ectodermally derived tissues, such as ductal epithelium in the breast. Many studies of human tumors have established that certain cancer cells in adults may express genes responsible for the synthesis of fetal gene products, such as carcino-embryonic antigen and alpha-fetoprotein, during intrauterine development. Extraembryonic antigens have also been identified on transformed human cell lines (Faulk et al., 1979b), as well as in samples of tissue from breast carcinoma (Shah et al., 1980), but these studies used antitrophoblast rather than antiamnion sera. There are, as yet, no reports of amniotic ectodermal antigens in human cancer, but it is possible that these markers are normally present on cells of ectodermal origin and are lost following malignant transformation. Indeed, support for such an interpretation can be drawn from preliminary results from our laboratory showing AA1 on breast biopsies from benign (Hsi et al., 1981), but not malignant, lesions. Interestingly, the AV$_3$ transformed cell line of human amniotic epithelium also does not react with anti-AA1 sera, even though the original epithelial cells are positive. Finally, both malignant breast duct epithelium (Faulk et al., 1980b) and AV$_3$ cells (Faulk and Galbraith, 1979a) manifest transferrin receptors, and neither normal ducts nor fresh amniotic epithelium bears detectable amounts of these markers, suggesting an inverse relationship between the presence of this receptor and the normally present markers of extraembryonic ectoderm of the amnion.

PLACENTAL PROTEINS IN MATERNO-FETAL IMMUNITY

Many structures that may possibly be immunogenic to the mother are found on extraembryonic membranes, but not all of these are specific to the trophoblast. For example, IgG receptors are located at the materno-fetal interface (Brambell, 1970), but are also found on many other tissues. Thus, they are unlikely to be of central importance to trophoblast in allogeneic coexistence. Similar objections can be raised to certain viruses that have been ultrastructurally identified in trophoblastic tissues (Sawyer et al., 1978), but these may be important in some diseases of pregnancy, for example, in preeclamptic toxemia (Thiry et al., 1981). Other potential antigens, such as trophoblastic alkaline phosphatase, are normally limited to the trophoblast (Beckman and Beckman, 1970). This particular enzyme is delivered to maternal blood in such large amounts, however, that it is difficult to imagine it serving as an antigen, particularly if humoral immunity is involved as a component of the maternal response. Human chorionic gonadotrophin (hCG) is another potential interface antigen that has been extensively studied (Stevens, 1980). Its role in materno-fetal biology is somewhat of an enigma, for antibodies to hCG normally exist (Wass et al., 1978) and the hCG molecule is not confined to the trophoblast (Hearn, 1980). Several "new" placental proteins, such as SP1, PP5 and PAPP-A have received considerable attention as possible placental antigens (Klopper, 1980). The SP1 molecule is a soluble product of the syncytiotrophoblast, which is discharged in large amounts into maternal blood. World Health Organization collaborative studies have cast doubt on the role of this molelcule in primate reproduction, for heterologous antisera to SP1 given to pregnant baboons did not affect their pregnancies, and similar results were obtained following the administration of antisera to PP5 (Faulk, 1980). There have still been no careful studies of the biology of either protein in mammalian pregnancy. An immunologically more interesting protein is PAPP-A, thought to be involved in the blood clotting and complement systems (Bischof, 1979). Thus far, amnion and chorion are the only human extraembryonic membranes studied for unique antigens, but future research will no doubt focus attention on the yolk sac and allantois.

Some incompatible allotypic antigens stimulate maternal immune responses potentially damaging to the fetus, the most well-studied example being antibodies to the D, or rhesus, antigen (Davey, 1979). Human placentae lack D antigen (Szulman, 1973), and maternal IgG antibodies thus pass unimpeded through chorionic villi to collect in the umbilical and fetal circulations, where they destroy D-positive erythrocytes, causing the clinical syndrome of erythroblastosis fetalis. Plasma protein antigens from the fetus also enter and sometimes immunize the mother—the Gm and Km antigens of immunoglobulins, for example (Faulk et al., 1974b). This is of potential significance, for certain anti-immunoglobulin antibodies in experimental animals cause agammaglobulinaemia in the offspring of immunized mothers (Dray, 1972). Inasmuch as anti-immunoglobulin antibodies are not uncommon in pregnancy (Nathenson et al., 1971), it is unclear why agammaglobulinaemia is not more common, although it is well known that immunoglobulins are physiologically depressed during the first six months of life.

One reason that maternal antibodies to antigenic groupings on fetal immunoglobulins do not cause serious clinical disease is that they do not in fact reach the fetus. They bind and precipitate their antigens within the mesenchymal stroma of chorionic villi (Johnson et al., 1977) and thus are not free to enter the fetal circulation. This is supported by reports of an activated fragment of the first component of human complement (C1q) deposited around fetal vessels in placentae (Faulk et al., 1980c; Lindner, 1981). In addition, those complexes not coprecipitable by complement will in all probability be bound by Fc receptors, placental endothelium and Hofbauer cells being rich sources of these receptors (Johnson et al., 1976; Wood, 1980). A concept thus emerges of maternal IgG antibodies that enter but never leave the placenta, suggesting that there is a sink for immune complexes within the connective tissues of human placentae (Faulk and Johnson, 1980). Consideration of the placental sink also reveals why maternal antibodies to D antigen cause disease in the fetus, while antibodies to HLA do not, for D antigens are not found in placenta, but HLA antigens are present on all cells of chorionic villi except trophoblast (Faulk and Temple, 1976). Thus anti-HLA antibodies are presumably immunoabsorbed while anti-D antibodies undoubtedly pass unimpeded into the umbilical circulation. A similar type of placental sink seems to be present in the immunobiology of pregnancy in the mouse (see Chapter 11). These matters await direct experimental attack.

THE TROPHOBLAST

Since trophoblast forms most of the materno-fetal interface in human pregnancy, many investigators have begun to focus on structural components of the trophoblastic plasma membrane as being of possible importance in reproductive immunobiology. A useful ultracentrifugation technique for the preparation of syncytiotrophoblastic microvilli is widely used (Smith et al., 1974), although other methods are available (Carlson et al., 1976; Smith et al., 1977). Smith and Brush (1978) and Ogbimi et al. (1979) characterized several proteins and glycoprotein subunits at the interface. Faulk et al. (1978) described serological criteria for the definition of the trophoblast antigens TA1 and TA2, and these findings have been extended by Whyte and Loke (1979) and O'Sullivan et al. (1980). O'Sullivan and Faulk (1981) recently reviewed the chemistry of trophoblast surface antigens.

Finally, McIntyre and Faulk (1978, 1979a, 1979b) have shown that antisera to TA1 antigens as well as the antigens themselves are potent inhibitors of the mixed lymphocyte culture reactions *in vitro*, thought to be equivalent to the phenomenon resulting from the passage of fetal lymphocytes into the maternal circulation during normal pregnancy. Maternal antibody responses during pregnancy do not seem to be generally compromised (but see Chapter 10), and antibodies reacting with placentae, placental polysaccharides, and trophoblastic microsomes have been described (Faulk and Fox, 1981).

TROPHOBLAST-LYMPHOCYTE CROSSREACTIVE ANTIGENS

Some trophoblast antigens must be shared by lymphocytes, as all *in vitro* blocking assays thus far used in pregnancy research depend for their targets upon lymphocyte function tests, such as lymphokine release or measurement of responses to allogeneic cells in MLC reactions (Youtananukorn and Matangkasombut, 1972; McIntyre and Faulk, 1979c). The identification of trophoblast-lymphocyte crossreactive (TLX) antigens and the characterization of polymorphic patterns of reactivity for heterologous antisera to TLX have now been reported. Earlier experimental studies indicated the presence of TLX antigens (Beer et al., 1972), and it has now been shown that rabbit antisera to human syncytiotrophoblastic microvilli are cytotoxic for peripheral blood lymphocytes (PBL) (Faulk et al., 1978). In addition, common bands have been identified upon electrophoresis of solubilized preparations of PBL and trophoblast membranes (Hamilton et al., 1980), providing serological and biochemical support for TLX. Polymorphism of these antigens has been shown with the use of antisera prepared in individual rabbits to microvillous pellets of trophoblast membranes from different placentae and with the use of these antisera as typing reagents for PBL from many different donors of known HLA types (Faulk et al, 1980a). Results from this type of analysis have shown that some antisera are cytotoxic for PBL from some people but not from others and that these reactions cannot be correlated with the HLA types of the donors (McIntyre and Faulk, 1981). Since extraembryonic membranes are devoid of MHC antigens, these findings suggest that trophoblast cells have their own histocompatibility complex and that these extraembryonic antigens are manifest on PBL as TLX antigens.

Pregnancy is similar to transplantation in that both groups of patients produce anti-HLA antibodies, although the presence of these antibodies has rather different meanings. It is not clear what stimulates mothers to mount immune recognition to the presently known MHC antigens, but it is fortunate for the production of tissue-typing reagents that they do! Our knowledge about the MHC might be expanded by using trophoblast as well as lymphocytes as targets in defining the specificity of antilymphocyte reactivity in blood from pregnant mothers. This could be particularly useful in sorting out the crossreactive groups, as well as in more precisely defining sera with as yet unclassifiable patterns of reactivity. Unlike the situation in recipients of allografts, there are no diseases of pregnancy associated with maternal anti-HLA, and, other than the unexplained observation that maternal antibodies to transplantation antigens are higher during the first pregnancy when the fetus is a male (Johansen and Festenstein, 1974), they presently seem to have no clinical meaning. Whether abnormalities result from a lack of antibodies to MHC antigens is not known and would be difficult to study because of the placental sink.

Blocking factors are known to exist in both maternal blood and placental eluates (Faulk et al., 1974a; Jeannet et al., 1977), but the antigens that give rise to such blocking activity are not known. At this time, it is not even clear that maternal blocking factors are antibodies, for a paternal-specific inhibitor of the mixed lymphocyte culture reaction is found in mother's plasma but not in serum (McIntyre and Faulk, 1979d), suggesting that the factor or factors are removed or destroyed at some phase of the clotting cascade. It is believed, however, that such activity is associated with the maintenance of normal pregnancy, for it seems to be absent from the blood of chronic aborters (Rocklin et al., 1976; Stimson et al., 1979). Spontaneous abortion thus appears to result from inadequate recognition of the implanted blastocyst, a conclusion consistent with the results of immunogenetic studies showing unexpected compatibility of HLA-A and HLA-B between the mating partners in aborting couples (Komlos et al., 1977). There is a growing belief that immune recognition is vital to maternal acceptance of the implanting blastocyst and that such recognition depends upon polymorphic trophoblast antigens. If normal pregnancy requires immunological recognition for its success (McCormick et al., 1971; Faulk and Johnson, 1977), then it is reasonable to suggest that some failed pregnancies may be due to a compatibility of TLX antigens between father and mother. In this case, TLX would serve the function of a transplantation antigen that triggers the process of maternal recognition necessary to support the extraembryonic graft (Faulk and McIntyre, 1981).

The biology of these transplantation antigens is not known (see Chapter 5 for a transplantation antigen of known function), but their anatomical location at the materno-fetal interface suggests that they might be important in maternal acceptance or rejection of the trophoblastic allograft. If, however, the TLX system is involved in maintenance of human pregnancy, its polymorphism would indicate that spontaneous abortion (i.e., rejection) should be relatively common. Current studies on the frequency of blastocyst implantation, as determined by measurements of gonadotrophin concentrations in urine, have now provided information suggesting that maternal rejection within the first several days of nidation is not uncommon (Miller et al., 1980). The only estimate to date is that 43 per cent of the blastocysts are lost. It is unlikely that such early embryonic losses have much to do with B-cell function, for a certain number of days are required to mount an antibody response, but T cells capable of recognizing nonself or allogeneic antigens are normally present. It was mentioned previously that allogeneic recognition and rejection reactions can be specifically inhibited by the TA1 group of trophoblast membrane glycoproteins (Faulk et al., 1977b), and this may be a biological mechanism for blastocyst protection from maternal immunosurveillance at implantation. An antigenically identical material has been identified on the plasma membrane of certain malignant and transformed lines of human cells maintained *in vitro* (Faulk et al., 1979b; McIntyre and Faulk, 1980), but nothing is known about the intra- or extracellular events that must occur to initiate the synthesis of this group of antigens.

FAULK

POSSIBLE ROLE OF TLX ANTIGENS IN SPONTANEOUS ABORTION

Sixteen test couples with three or more clinically evident abortions, as well as control couples, had their lymphocytes HLA-typed for A, B, C, and Dr antigens. The results showed that two or fewer shared antigens were characteristic of the controls whereas three or more antigens were commonly shared between chronically aborting couples (McIntyre and Faulk, unpublished observations). Other causes of spontaneous abortion were excluded from the test group, but some of the aborting couples shared only two or less antigens whereas none of the controls shared three or more, and there were no particular HLA types associated with spontaneous abortions. Most experience in transplantation immunology indicates that HLA-matched transplants do better than unmatched ones (International Forum, 1978), but this is apparently not the case in immunogenetical considerations of spontaneous abortions, bringing into question the validity of generalizations drawn from analogies between organ transplantation and pregnancy.

By analogy with contemporary thinking on the usefulness of blood transfusions prior to renal allografting (Williams et al., 1980), consideration has been given to sensitizing chronic aborters to TLX antigens in order to mobilize their blastocyst recognition mechanism before impregnation. Interestingly, there is also a greater than expected amount of antigen sharing for HLA between father and mother in pre-eclampsia (Redman, 1980), and the disease is less common in women who have had a blood transfusion before becoming pregnant (Feeney et al., 1977). The condition can also be associated with a change of mate, similar to that observed in chronic aborters when they become pregnant by another man (Feeney, 1980). The issue of an immune component in "preeclamptic" toxemia will not be discussed here, but when these observations are considered in light of a role for trophoblast antigens in necessary maternal recognition of the blastocyst and embryo, it becomes apparent that maternal sensitization to TLX antigens through repeated blood transfusions might be a useful approach to the prevention of spontaneous abortions. This hypothesis was tested by studying four women, each of whom had had at least three spontaneous abortions of unknown cause and who shared at least three HLA antigens with their mates. Each patient received multiple infusions of leukocyte-enriched plasma from many different donors who were carefully typed and matched to be compatible for erythrocyte antigens and incompatible for HLA, the HLA incompatibility being theoretically linked with TLX antigen incompatibility. These transfusions were given before and throughout pregnancy with the result that all four mothers delivered normal babies at term, whereas none had progressed past the twelfth week in any of their previous pregnancies (Taylor and Faulk, 1981).

268

The point to be made from these preliminary results is to emphasize again the growing concept of the crucial role of trophoblast and trophoblast crossreactive antigens in the immunobiology of pregnancy. Of more direct relevance to basic concepts in human reproduction is how the chronic aborters were assisted through their pregnancies by multiple infusions of allogeneic leukocytes. A plausible explanation of this phenomenon is the need for blastocyst recognition in trophoblast survival and the role of polymorphic, TLX antigens which could be linked to HLA within the MHC. If this were the case, mating partners compatible, through consanguinity or chance, at three or more antigens of HLA would be expected to produce a blastocyst with TLX antigens that would result in lack of adequate maternal recognition and subsequent failure of the pregnancy. In other words, reproduction is favored in couples genetically different at HLA—a mechanism whereby HLA is in part responsible for the generation of diversity. Thus, not only did the specialization of chorionic ectoderm resulting in trophoblast allow the appearance of placentation and viviparity in evolution, but the same process may continue today to perpetuate genetical heterogeneity in human populations.

REFERENCES

Aitken, R.J., and D.W. Richardson. 1980. Immunization against zona pellucida antigens. In *Immunological Aspects of Reproduction and Fertility Control.* J.P. Hearn, ed. Lancaster: MTP Press, England, pp. 173-201.

Amoroso, E.D. 1952. Placentation. In *Marshall's Physiology of Reproduction.* Vol. 2. A.S. Parkes, ed. London: Longmans Green, pp. 127-311.

Attwood, H.D., and W.W. Park. 1961. Embolism to the lungs by trophoblast. *J. Obstet. Gynaecol. Br. Cwlth.* 68:611.

Beckman, G., and L. Beckman. 1970. Relation between ABO blood groups and the level of placental alkaline phosphatase in sera of pregnant women. *Hum. Heredity* 20:187.

Beer, A.E., R.E. Billingham, and S.L. Yange. 1972. Further evidence concerning the autoantigenic status of the trophoblast. *J. Exp. Med.* 135:1177.

Bischof, P. 1979. Observations on the isolation of pregnancy-associated plasma protein-A. In *Placental Proteins.* A. Klopper and T. Chard, eds. Heidelberg: Springer-Verlag, pp. 105-118.

Bourne, G. 1962. *The Human Amnion and Chorion.* London: Lloyd-Luke (Medical Books).

Boyd, J.D., and W.J. Hamilton. 1970. *The Human Placenta.* London: Macmillan Press.

Brambell, F.W.R. 1970. *The Transmission of Passive Immunity from Mother to Young.* Amsterdam: North-Holland.

Carlson, R.W., H.G. Wada, and H.H. Sussman. 1976. The plasma membrane of human placentae: isolation of microvillous membrane and characterization of protein and glycoprotein subunits. *J. Biol. Chem.* 251:4139.

Chateaureynaud, P., M.T. Badet, and G.A. Voisin. 1979. Antagonistic maternal immune reactions (rejection and facilitation) to the embryo in the urodele amphibian *Salamandra salamandra* Lin. *J. Reprod. Immunol.* 1:47.

Cordier, A.C., and J.F. Heremans. 1975. Nude mouse embryo: ectodermal nature of the primordial thymic defect. *Scand. J. Immunol.* 4:193.

Crisan, C. 1935. Die entwicklung des thyreo-parathyreo-thymischen Systems der wiessen Maus. *Z. Anat. Entwickl.-Gench.* 104:327.

Davey, M. 1979. The prevention of rhesus-isoimmunization. *Clin. Obstet. Gynecol.* 6:509.

Dray, S. 1972. Allotype suppression. In *Ontogeny of Acquired Immunity.* R. Porter and J. Knight, eds. Amsterdam: Elsevier/North-Holland, pp. 87-112.

Enders, A.C. 1965. A comparative study of the fine structure of the trophoblast in several hemochorial placentas. *Am. J. Anat.* 116:29.

Faulk, W.P., M. Jeannet, W.D. Creighton, and A. Carbonara. 1974. Immunological studies of human placentae: characterization of immunoglobulins on trophoblastic basement membranes. *J. Clin. Invest.* 54:1011.

Faulk, W.P., E. van Loghem. and G.M. Stockler. 1974a. Maternal antibody to fetal light chain (Inv) antigens. *Am. J. Med.* 56:393.

Faulk, W.P., and A. Temple. 1976. Distribution of B_2-microglobulin and HLA in chorionic villi of human placentae. *Nature* 262:799.

Faulk, W.P., A.R. Sanderson, and A. Temple. 1977. Distribution of MHC antigens on human placentae. *Transplant. Proc.* 9:1379.

Faulk, W.P., R. Lovins, C. Yeager, and A. Temple. 1977a. Antigens of human trophoblast: Immunological and biochemical characterization. In *Immunological Influence on Human Fertility.* B. Boettcher, ed. New York: Academic Press, pp. 152-160.

Faulk, W.P., and P.M. Johnson. 1977. Immunological studies of human placentae: identification and distribution of proteins in mature chorionic villi. *Clin. Exp. Immunol.* 27:365.

Faulk, W.P., A. Temple, R. Lovins, and N.C. Smith. 1978. Antigens of human trophoblast: A working hypothesis for their role in normal and abnormal pregnancies. *Proc. Nat. Acad. Sci. USA* 75:1947.

Faulk, W.P., and G.M.P. Galbraith. 1979a. Trophoblast transferrin and transferrin receptors in the host-parasite relationship of human pregnancy. *Proc. R. Soc. Lond. (Biol.)* 204:83.

Faulk, W.P., and G.M.P. Galbraith. 1979b. Transferrin and transferrin receptors of human trophoblast. In *Transport of Protein Across Biological Membranes.* W.A. Hemmings, ed. Amsterdam: Elsevier/North-Holland Biomedical Press, pp. 55-61.

Faulk, W.P., J.A. McIntyre, and C.J.G. Yeh. 1979a. Trophoblast antigens on membranes of human tumours and transformed cells. In *Protides of Biological Fluids.* H. Peeters, ed. 27:21-26.

Faulk, W.P., C. Yeager, J.A. McIntyre, and M. Ueda. 1979b. Oncofetal antigens of human trophoblast. *Proc. R. Soc. London (Biol.)*63.

Faulk, W.P. 1980. Immunological approaches to fertility control. In *Autoimmune Aspects of Endocrine Disorders.* A. Pinchara, D. Doniach, G.F. Fenzi, and L. Baschieri, eds. London: Academic Press, pp. 401-408.

Faulk, W.P., and P.M. Johnson. 1980. Immunological studies of human placentae: basic and practical implications. *Recent Adv. Clin. Immunol.* 2:1.

Faulk, W.P., J.A. McIntyre, and B. Hsi. 1980a. Transplantation analogies of the materno-fetal relationship in human pregnancy. In *Transplantation and Clinical Immunology.* J.P. Revillard, ed. Amsterdam: Excerpta Medica pp. 143-150.

Faulk, W.P., B.-L. Hsi, and P. Stevens. 1980b. Transferrin and transferrin receptors in carcinoma of the breast. *Lancet* ii:390.

Faulk, W.P., R. Jarret, M. Keane, P.M. Johnson, and R. Boackle. 1980c. Immunological studies of human placentae: complement components in immature and mature chorionic villi. *Clin. Exp. Immunol.* 40:299.

Faulk, W.P. 1981. Trophoblast and extra-embryonic membranes in the immunobiology of human pregnancy. In *Placenta: Receptors, Pathology and Toxicology. Placenta* (Supplement 3) In press.

Faulk, W.P., and E. McCrady. 1981. Role of the extra-embryonic membranes in transplantation analogies of human pregnancy. In *Encyclopaedia of Ignorance.* Vol. 3. R. Duncan and M. Weston-Smith, eds. Oxford: Pergamon Press, in press.

Faulk, W.P., and J.A. McIntyre. 1981. Trophoblast survival. *Transplantation* 32: In press.

Faulk, W.P., and H. Fox. 1981. In *Clinical Aspects of Immunology,* P. Lachmann and K. Peters, eds. 4th ed. London: Blackwell Scientific Publications.

Faulk, W.P., E. McCrady, B.-L. Hsi, P.J. Stevens, and J. Burgos. 1981. Thymus and amnion in epidermolysis bullosa letalis. Submitted for publication.

Feeney, J.G., L.A.C. Tovey, and J.S. Scott. 1977. Influence of previous blood transfusion on incidence of pre-eclampsia. *Lancet* i:874.

Feeney, J.G. 1980. Pre-eclampsia and changed paternity. In *Pregnancy Hypertension*. J. Bonnar, I. MacGillivray, and M. Symonds, eds. Lancaster: MTP Press, pp. 41-44.

Galbraith, G.M.P., R.M. Galbraith, and W.P. Faulk. 1980. Transferrin binding by human lymphoblastoid cell lines and other transformed cells. *Cell. Immunol.* 49:215.

Gardner, R.L., and J. Rossant. 1976. In *Embryogenesis in Mammals*. K. Elliott and M. O'Connor, eds. Amsterdam: Elsevier/North-Holland, pp. 5-25.

Gill, T.J., and C.F. Repetti. 1979. Immunologic and genetic factors influencing reproduction. *Am. J. Path.* 95:465.

Goldring, O.L., J.R. Kusel, and S.R. Smithers. 1977. Schistosoma mansoni: origin *in vitro* of host-like surface antigens. *Exp. Parasitol.* 43:82.

Goodfellow, P.N., C.J. Barnstable, W.F. Bodmer, D. Snary, and M.J. Crumpton. 1976. Expression of HLA system antigens in placenta. *Transplantation* 22:595.

Hamilton, T.A., H.G. Wada, and H.H. Sussman. 1980. Expression of human placental cell surface antigens on peripheral blood lymphocytes and lymphoblastoid cell lines. *Scand. J. Immunol.* 11:195.

Hearn, J.P. 1980. The immunobiology of chorionic gonadotrophins. In *Immunological Aspects of Reproduction and Fertility Control*. J.P. Hearn, ed. Lancaster, England: MTP Press, pp. 229.

Hertig, A.T. 1968. *Human Trophoblast*. Springfield, Illinois: Charles C. Thomas.

Hildemann, W.H., I.S. Johnson, and P.L. Jokiel. 1979. Immunocompetence in the lowest metazoan phylum: transplantation immunity in sponges. *Science* 204:420.

Hoyes, A.D. 1975. Structure and function of the amnion. *Obstet. Gynecol. Ann.* 4:1.

Hsi, B.-L., C.-J.G. Yeh, and W.P. Faulk. 1981. Human amniochorion: tissue specific; markers, transferrin receptors and histocompatibility antigens. *Placenta* In press.

Ikle, F.A. 1964. Dissemination von syncytiotrophoblastzellen in mutterlichen blut wahrend der gravudutal. *Bull. Schweiz. Akad. Med. Wiss.* 20:62.

International Forum. 1978. What is the importance of HLA compatibility for clinical outcome of renal transplantation. *Vox Sang.* 34:171-188.

Johansen, K., and H. Festenstein. 1974. Maternal HLA antibodies and fetal sex. *Brit. Med. J.* 4:202.

Jeannet, M., C. Werner, E. Ramirez, P. Vassakki, and W.P. Faulk. 1977. Anti-HLA, anti-human "Ia-like" and MLC blocking activity of human placental IgG. *Transplant. Proc.* 9:1417.

Johnson, P.M., W.P. Faulk, and A.-C. Wang. 1976. Immunological studies of human placentae: subclass and fragment specificity of binding of aggregated IgG by placental endothelial cells. *Immunology* 31:659.

Johnson, P.M., J. Navig, U.A. Ystehede, and W.P. Faulk. 1977. Immunological studies of human placentae: an immunofluorescence study of the distribution and character of immunoglobulins in chorionic villi. *Clin. Exp. Immunol.* 30:145.

Kelly, S.J. 1977. Studies of the developmental potential of 4- and 8-cell stage mouse blastomeres. *J. Exp. Zool.* 200:365.

Klopper, A. 1980. The new placental proteins. A review. *Placenta* 1:77.

Konlos, L., R. Zamir, H.Joshua, and I. Halbrecht. 1977. Common HLA antigens in couples with repeated abortions. *Clin. Immunol. Immunopathol.* 7:330.

Le Douarin, N.M., and F.V. Jotereau. 1980. Homing of lymphoid stem cells to the thymus and bursa of Fabricius studied in avian embryo chimaeras. In *Progress in Immunology*. Vol. 4. M. Fougereau and J. Dausset, eds. London: Academic Press, pp. 285-302.

Lindner, E. 1981. Binding of C1q and complement activation by vascular endothelium. *J. Immunol.* 126:648.

Luckett, W.P. 1978. Origin and differentiation of the yolk sac and extra-embryonic mesoderm in presomite human and rhesus monkey embryos. *Am. J. Anat.* 152:59.

McCormick, J.N., W.P. Faulk, H. Fox, and H.H. Fudenberg. 1971. Immunohistological and elution studies of the human placenta. *J. Exp. Med.* 133:1.

McIntyre, J.A., and W.P. Faulk. 1978. Suppression of mixed lymphocyte cultures by antibodies against human trophoblast membrane antigens. *Transplant. Proc.* 10:919.

McIntyre, J.A., and W.P. Faulk. 1979a. Antigens of human trophoblast: Effects of heterologous anti-trophoblast sera on lymphocyte responses *in vitro. J. Exp. Med.* 149:824.

McIntyre, J.A., and W.P. Faulk. 1979b. Trophoblast modulation of maternal allogeneic recognition. *Proc. Nat. Acad. Sci. USA* 76:4029.

McIntyre, J.A., and W.P. Faulk. 1979c. Immunobiology of trophoblast membrane glycoproteins. *Transplant. Proc.* 11:1892.

McIntyre, J.A., and W.P. Faulk. 1979c. Maternal blocking factors in human pregnancy are found in plasma not serum. *Lancet* ii:821.

McIntyre, J.A., and W.P. Faulk. 1980. Cross-reactions between cell surface membrane antigens of human trophoblast and cancer cells. *Placenta* 3:197.

McIntyre, J.A., and W.P. Faulk. 1981. Polymorphic trophoblast-lymphocyte cross-reactive (TLX) antigens. *Human Immunol.* In press.

Melchers, F., and J. Abramzynk. 1980. Murine embryonic blood between day 10 and 13 of gestation as a source of immature precursor B-cells. *Eur. J. Immunol.* 10:763.

Miller, J.F., E. Williamson, J. Glue, Y.B. Gordon, J.G. Grudzinsk, and A. Sykes. 1980. Fetal loss after implantation. *Lancet* ii:554.

Moore, M.A.S., and J.J.T. Owen. 1967. Experimental studies on the development of the thymus. *J. Exp. Med.* 126:715.

Moore, K.L. 1977. *The Developing Human*, 2nd ed. Eastbourne: W.B. Company.

Nathenson, G., J.B. Schoor, and S.D. Litwin. 1971. Gm factor gamma globulin incompatibility. *Pediatric Res.* 5:2.

Nicholas, J.S., and B.V. Hall. 1934. Development of isolated blastomeres of the rat. *Anat. Rec.* Suppl. 58:83.

Norris, E.H. 1938. The morphogenesis and histogenesis of the thymus gland in man: in which the origin of the Hassall's corpuscles of the human thymus is discovered. *Carnegie Contrib. Embryol..* 166:191.

Ogbimi, A.O., P.M. Johnson. P.J. Brown, and H. Fox. 1979. Characterization of the soluble fraction of human syncytiotrophoblast microvillous plasma membrane-associated proteins. *J. Reprod. Immunol.* 1:127.

O'Sullivan, M.J., J.A. McIntyre, and W.P. Faulk. 1980. Biochemical characterization of human trophoblast cell membrane proteins. In *Aspects of Developmental and Comparative Immunology.* J.B. Solomon, ed. Oxford: Pergamon Press, pp. 353-359.

O'Sullivan, M.J., and W.P. Faulk. 1981. Cell surface antigens of the trophoblast. In *Biology of Trophoblast.* C. Loke and A. Whyte, eds. Amsterdam: Elsevier/North-Holland.

Park, W.W. 1965. *The Early Conceptus, Normal and Abnormal.* Edinburgh: Livingstone.

Ramsey, E.M., M.L. Houston, and J.W. Harris. 1976. Interactions of the trophoblast and maternal tissues in three closely related primate species. *Am. J. Obst. Gynecol.* 124:647.

Redman, C.W.G. 1980. Immunological aspects of eclampsia and preeclampsia. In *Immunological Aspects of Reproduction and Fertility Control.* J.P. Hearn, ed. Lancaster, England: MTP Press, pp. 83-103.

Robertson, W.B. 1976. Utero placental vasculature. *J. Clin. Pathol. (Royal College of Pathology)* Supplement, 10:9.

Rocklin, R.E., J.C. Kitzmiller, C.G. Carpenter, M.R. Garovoy, and J.R. David. 1976. Maternal-fetal relation: absence of an immunologic blocking factor from the serum of women with chronic abortions. *N. Engl. J. Med.* 295:1209.

Rocklin, R.E., J. Kitzmiller, and M. Kaye. 1979. Immunobiology of the maternal fetal relationship. *Ann. Rev. Med.* 30:375.

Rossant, J. 1975. Investigation of the determinative state of the mouse inner cell mass. II. The fate of isolated inner cell masses transferred to the oviduct. *J. Embryol. Exp. Morphol.* 33:991.

Sawyer, M.H., N.E. Nachlas, and S. Panem. 1978. C-type antigen expression in human placenta. *Nature* 275:62.

Shah, L.C.P., A.O. Ogbimi, and P.M. Johnson. 1980. A cell membrane antigen expressed by both human breast carcinoma cells and normal human trophoblast. *Placenta* 1:299.

Shaw, A.R.E., M.K. Dasgupta, T. Kovithavongs, K.V. Johny, J.C. LeRich, J.B. Dossetor, and T.A. McPherson. 1979. Humoral and cellular immunity to paternal antigens in trophoblastic neoplasia. *Int. J. Cancer* 24:587.

Smith, N.C., M.G. Brush, and S. Luckett. 1974. Preparation of human placental villous surface membranes. *Nature* 252:302.

Smith, C.H., P.M. Nelson, B.F. King, T.M. Donohue, S.M. Ruzycki, and L.K. Kelley. 1977. Characterization of a microvillous membrane preparation from human placental syncytiotrophoblast. A morphologic, biochemical and physiological study. *Am. J. Obstet. Gynecol.* 128:190.

Smith, N.C., and M. Brush. 1978. Preparation and characterization of human syncytiotrophoblastic plasma membrane. *Med. Biol.* 56:272.

Smithers, S.R., R.M. Terry, and D.J. Hockley. 1969. Host antigens in schistosomiasis. *Proc. R. Soc. Lond. (Biol.)* 171:483.

Snow, M.H.L. 1973. Abnormal development of pre-implantation mouse embryos *in vitro* with (^{3}H)-thymidine. *J. Embryol. Exp. Morphol.* 29:601.

Solomon, J.B., and J.D. Horton. 1977. *Developmental Immunobiology.* Amsterdam: Elsevier/North-Holland.

Spindle, A.I. 1978. Trophoblast regeneration by inner cell masses isolated from cultured mouse embryos. *J. Exp. Zool.* 203:483.

Stevens, V.C. 1980. The current status of anti-pregnancy vaccines based on synthetic fractions of hCG. In *Immunological Aspects of Reproduction and Fertility Control* J.P. Hearn, ed. Lancaster, England: MTP Press, pp. 203-216.

Stimson, W.H., A.F. Strachan, and A. Shepherd. 1979. Studies on the maternal immune response to placental antigens: absence of a blocking factor from the blood of abortion-prone women. *Brit. J. Obstet. Gynaecol.* 86:41.

Szulman, A.E. 1973. The A, B and H blood group antigens in human placentae. *N. Engl. J. Med.* 286:1028.

Taylor, C., and W.P. Faulk. 1981. Prevention of recurrent abortions with leukocyte transfusions. *Lancet* ii:68.

Thiry, L., F. Yane, S. Sprecher-Goldberger, R. Cappel, M. Bossens, and F. Neuray. 1981. Expression of retrovirus antigen in pregnancy. II. Cytotoxic and blocking specificities in immunoglobulin eluted m placenta. *J. Reprod. Immunol.* 2:323.

Tormey, D.C., R.C. Imrie, and G.C. Mueller. 1972. Identification of transferrin as a lymphocyte growth promoter in human serum.. *Exp. Cell Res.* 74:163.

Tormey, D.C., and G.C. Mueller. 1972. Biological effects of transferrin on human lymphocytes *in vitro. Exp. Cell Res.* 74:220.

Trowsdale, J., P. Travers, W. Bodmer, and R.A. Patillo. 1980. Expression of HLA-A, B, C and β2-microglobulin antigens in human choriocarcinoma cell lines. *J. Exp. Med.* 152:115.

Volpe, E.P., R. Tompkins, and D.C. Reinschmidt. 1981. Evolutionary modifications of nephrogenic mesoderm to establish the embryonic centres of hemopoiesis. In *Aspects of Developmental and Comparative Immunology.* J.B. Solomon, ed. Oxford: Pergamon Press, pp. 193.

Wass, M., K. McCann, and K.D. Bagshawe. 1978. Isolation of antibodies to hCG/LH from human sera. *Nature* 274:369.

Whyte, A., and Y.M. Loke. 1979. Antigens of the human trophoblast plasma membrane. *Clin. Exp. Immunol.* 37:359.

Williams, K.A., A. Ting, M.E. French, S. Oliver, and P.J. Morris. 1980. Pre-operative blood transfusions improve cadaveric renal-allograft survival in non-transfused recipients. *Lancet* i:1104.

Wood, G.W. 1980. Mononuclear phagocytes in the human placenta. *Placenta* 1:113.

Yamashita, K., N. Wake, T. Araki, K. Ichinoe, and K. Maoto. 1979. Human leukocyte antigen expression in hydatidiform mole. Androgenesis following fertilization by a haploid sperm. *Am. J. Obst. Gynecol.* 135:597.

Yeh, G.C., and W.P. Faulk. 1981. Cellular differentiation and transferrin receptors. Protides of the Biological Fluids. (in press).

Yeh, G.C., B.-L. Hsi, and W.P. Faulk. 1981. Transferrin receptors of human placenta, tumour and transformed cells. In *Aspects of Developmental and Comparative Immunology.* I.

J.B. Solomon, ed. Amsterdam: Elsevier/North-Holland, pp. 361-367.

Youtananukorn, V., and P. Matangkasombut. 1972. Human maternal cell mediated immune reaction to placental antigens. *Clin. Exp. Immunol.* 11:549.

Part III

REGULATION OF THE MATERNAL IMMUNE RESPONSE

INTRODUCTION

Having explored the embryo and the placenta from the point of view of immunology and genetics, the book now focuses in this third section, on the maternal immune response and its regulation during pregnancy. One of the most important themes in modern immunology is that of immunoregulation, and the introduction of the placenta complicates this already prolific field by the addition of unique tissues, such as the trophoblast, that may generate a number of substances that can influence the immune response. The main themes of this section concentrate on immunoregulatory substances produced during pregnancy and on the number and types of immunologically stimulated cells that appear during pregnancy. The dilemma surrounding immune responsiveness to the fetus is enhanced by observations that pregnant females can reject skin grafts and eliminate leukemic cells bearing paternal histocompatibility antigens while carrying the fetus normally to term. These as well as related observations focus attention on the local fetal-placental environment where either immune mechanisms or the mechanical barrier of the placenta are critical in protecting the fetus. One must always bear in mind that the immunological phenomena found systemically in the mother may not reflect the critical mechanisms protecting the fetus from immune attack.

The potential role of immunoregulatory substances has attracted attention for quite some time. A variety of such substances have been purified from pregnancy sera and tissues, and they have *in vitro* effects but have not yet been demonstrated to have a physiological role *in vivo*. The same criterion must be applied to lymphocytes which are proposed to protect the fetus locally by immune suppression. William Stimson critically evaluates the large variety of molecules that have been proposed as immunoregulatory agents and the mechanisms by which they are supposed to act. He stresses that there may be a significant immunoregulatory role for steroids, since estradiol-17β can be both immunostimulatory and immunosuppressive in physiological concentrations. He has isolated a promising new immunoregulatory substance, which he calls pregnancy-associated prostaglandin synthetase inhibitor (PAPSI), that occurs during pregnancy. All of the other, currently known pregnancy-associated molecules are dealt with in a balanced and critical fashion.

One of the most extensively studied and controversial of these substances is alpha-fetoprotein, and it is discussed in detail by Thomas Tomasi, in whose laboratory the original observations on immunosuppression were made. The controversial nature of the observations may be related to technical difficulties in the preparation of the alpha-fetoprotein, such as whether or not copper is present in the preparation. Recent studies indicate that

279

alpha-fetoprotein may act through the helper cells that see antigen in association with Class II molecules. If this is indeed substantiated, then alpha-fetoprotein may not play an important role in the placenta, since Class II (I-region) antigens appear not to be present at the maternal-fetal interface. Nonetheless, this substance may still have a role in the neonatal environment.

The second major area of maternal immunoregulation is that involving cellular immunoregulation during pregnancy. R.A. Murgita, A.B. Peck, and Hans Wigzell discuss the absence of natural killer cells in the neonatal environment. This is particularly important because natural killer cells have a predilection for reacting with "embryonic-type" antigens, and their absence may be an important aspect of fetal and neonatal immune regulation. These authors postulate that either alpha-fetoprotein or the deficiency of macrophage-mediated interferon production could lead to the absence, or a low level, of natural killer cells in newborn mice and that this may prevent damaging autoimmune reactions against embryonic stem cells. An alternative explanation is that the natural killer cells are overwhelmed by the embryonic-type antigens in the fetus and appear only when these antigens wane. David Clark, Renata Slapsys, Anne Croy, Janet Rossant, and Mark McDermott emphasize the importance of local active suppression at the maternal-fetal interface for the regulation of maternal cytotoxic cells reactive with paternal alloantigens. They have isolated a nonspecific suppressor cell from the lymph nodes draining the uterus during pregnancy and have also found a soluble suppressor activity in the draining lymph nodes from allogeneically pregnant animals. The relevance of these factors *in vivo* must now be demonstrated. Finally, Warren Jones, Catherine Hawes, and Andrew Kemp review the data on various immunological parameters during human pregnancy. Although some changes may be seen, the weight of evidence suggests that these changes are minimal and do not affect the survival of the fetal allograft. They emphasize the need to look locally for relevant immunoregulatory substances or cells that may prevent the mother from rejecting the fetus.

All of the studies in this section point to the complexity of the mechanisms that foster the survival of the fetal homograft and the paucity of our knowledge about them. They highlight the importance of differentiating between what may be systemic epiphenomena detected in the mother during pregnancy and what may be the crucial mechanisms at the local level which allow fetal survival. The problem of how to study in a convincing manner the physiologically important immune changes in the pregnant uterus and in the lymph nodes draining it remains unresolved. Clearly, the development of techniques to explore immune responsiveness and regulation in the immediate area of the embryo and fetus is one of the most important objectives of reproductive immunology.

Chapter 13

THE INFLUENCE OF PREGNANCY-ASSOCIATED SERUM PROTEINS AND STEROIDS ON THE MATERNAL IMMUNE RESPONSE

W. H. STIMSON

The failure of the mother to reject the fetal allograft remains one of the great enigmas of present-day immunology. Despite a great deal of study, the mechanisms responsible remain largely unknown. The theories advanced to explain the phenomenon include

1. the existence of an anatomical barrier within the placenta, which separates the maternal and fetal circulations,
2. the possibility that the uterus is an immunologically privileged site,
3. a lack of histocompatibility antigens on the trophoblast, and
4. inhibition of the maternal immune response (Gill and Repetti, 1979).

It has been observed that the mother is sensitized to fetoplacental antigens, and thus, although she has the potential to mount a response against the fetus, a blocking of the recognition or effector stages of the immune response appears to occur (Rocklin et al., 1976; Stimson et al., 1979b). An explanation for this blocking could be the presence of regulatory factors in the blood, and substances capable of specific and nonspecific immunoregulation have been reported in maternal serum. The following review presents evidence in support of the view that certain pregnancy-associated proteins and hormones are the immunoregulatory factors in pregnancy serum.

STEROIDS

It is recognized that steroids can exert a profound influence over immunological reactivity, and a number of investigations have attempted to correlate increased steroid levels in pregnancy with the regulation of the maternal immune response. There are indications, however, that the serum and placental levels of available steroids are too low to have significant direct effects on lymphocytes (Schiff et al., 1975; Tomoda et al., 1976; Pavia et al., 1979). For this reason, we suggested that steroids may affect the immmune response indirectly, by inducing the thymus to produce immunoregulatory factors or cells (Stimson and Hunter, 1976).

EFFECTS OF STEROIDS MEDIATED THROUGH THE THYMUS

We originally found that estradiol-17β treatment of rats resulted in the appearance of thymus-dependent factors in the serum. The factors reduced the spontaneous binding of sheep erythrocytes to human lymphocytes, suppressed antigen-induced inhibition of leukocyte migration, and partially blocked phytohemagglutinin-induced (PHA) transformation. They also enhanced the number of lymphocytes bearing C3b receptors and IgM, but were not polyclonal B-cell activators, although they did increase the number of direct plaque-forming cells following *in vitro* sensitization (Stimson and Hunter, 1976, 1980). One of the factors has now been characterized as a previously unknown lipid with a molecular weight of 750 daltons.

Support for the idea of indirect steroid immunoregulation has come from the identification of steroid receptors in the rat thymus (Table 13-1). High-affinity cytosol binding species having specificities for estrogens, androgens, and corticosteroids are present in the thymus. However, a specific progesterone receptor has not been identified in rat thymus, even after estrogen treatment of animals or culture of thymic tissue with estradiol-17β for up to two days—procedures known to induce progesterone receptors in other tissues. It was found that estrogen- and androgen-binding activity was totally located in the thymic matrix, which is mainly medullary tissue, whereas thymocytes were devoid of receptors. Corticosteroid receptors, however, were distributed equally in both tissues, although receptor affinity appeared to be higher in the matrix than in the thymocytes (Table 13-2) (Reichman and Villee, 1978; Grossman et al., 1979; Brodie et al., 1980; McCruden and Stimson, 1980, 1981; Stimson et al., 1980b). Thus the region of the thymus apparently involved in immunoregulation possesses high-affinity binding species for certain naturally occurring steroid classes.

In order to assess the functional role of these steroid receptors, we examined the effects of steroids on the secretion of immunoregulatory factors by thymic epithelial cell cultures (Stimson et al., 1980a; Stimson and Crilly, 1981). An example of such an experiment is shown in Figure 13-1. Rat epithelial cells were cultured for up to 40 days in the presence of steroids and the culture supernatants (TES) harvested and examined for their capacity to regulate PHA-induced thymocyte proliferation. At the steroid concentrations studied, TES from control cultures caused a three- to fourfold increase in thymocyte transformation, while TES from testosterone-treated cultures enhanced proliferation up to 11-fold. As expected, supernatants from progesterone-treated cultures did not differ significantly from controls, whereas TES from estradiol and corticosterone cultures actually caused mitogen stimulation to fall below control levels for much of the culture period. Further studies have shown that the phenomena are dose related and occur at

TABLE 13-1
Characteristics of Steroid Cytosol Receptors in the Thymus

		STEROID			
					CORTICOSTERONE
PARAMETER	ESTRADIOL-17β	5α-DIHYDROTES-TOSTERONE	PROGESTERONE	THYMOCYTES	MATRIX*
Dissociation constant (K_d M)	5.7 x 10^{-10}	4.1 x 10^{-10}	NA†	2.5 x 10^{-8}	6.4 x 10^{-9}
Size (S)	6.5	3.8 (8.0)	NA†	4.0	4.0
Translocation to nucleus	Yes	Yes	No	Yes	Yes

* Residual tissue after removal of thymocytes—mainly medulla.
† NA = not applicable

hormone levels found during pregnancy. TES also affect the responses of cells from the thymus, spleen, lymph nodes, and bone marrow to PHA, pokeweed mitogen (PWM), concanavalin A (Con A), and allogeneic cells (Stimson and Crilly, 1981). Also, steroid-treated epithelial cultures produce TES which vary in their capacity to impart corticosteroid resistance to thymocytes and to decrease the display of Thy 1 antigen (Kruisbeek, 1979). It was apparent that the TES were multifactorial in nature, and fractionation by ultrafiltration (Table 13-3) indicated that the major inhibitory and enhancing factors had molecular weights of < 1000 and > 30000, respectively. Thus, the thymus is a steroid-responsive tissue, and the changes in steroid blood levels during pregnancy could alter the release of thymic humoral factors and the differentiation of precursor T cells in the maternal and fetal thymus.

Direct effects of steroids on leukocytes: In order to investigate whether peripheral blood leukocytes (PBL) have the capacity to respond to the sex steroids, PBL subpopulations were examined for high-affinity binding using whole cell

TABLE 13-2
Capacity and Distribution of Steroid Cytosol
Receptors in Thymic Tissue

	SPECIFIC STEROID BINDING (fmol/mg cytosol protein)	
STEROID	THYMOCYTES	MATRIX*
Estradiol-17β	< 1	15
5α-dihydrotestosterone	< 1	18(F), 8(M)
Progesterone	< 1	< 1
Corticosterone	65	41

* Residual tissue after removal of thymocytes—mainly medulla.

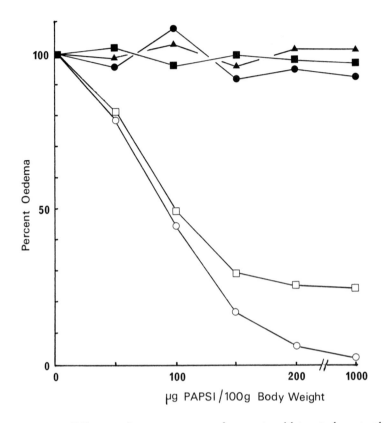

FIGURE 13-1. Effects of supernatants from steroid-treated rat thymic epithelial cell cultures on PHA-induced thymocyte proliferation.
Ratio TES:control = factor by which the normal thymocyte response to PHA is altered. Days = culture period. Steroid additions: ■, testosterone (0.001 μg/ml); ▲, progesterone (0.2 μg/ml); □, corticosterone (0.2 μg/ml); ♦, estradiol-17β (0.01 μg/ml); ●, control (no steroid). Supernatants were used at a 1:30 dilution. Results expressed as the mean (± S.D.) of triplicate determinations of four separate cultures. Arrow indicates time of steroid addition to cultures.

(Neifeld et al., 1977) and cytosol receptor (Stimson et al., 1980b) assays. Human granulocytes, monocytes, and T and B lymphocytes (Stimson, 1977) were assayed using estradiol-17β, 5α-dihydrotestosterone, progesterone, and [³H]-Org. 2058 (16α-ethyl-21-hydroxy-19-nor-4-pregnene-3,-2β-dione), a potent ligand for progesterone receptors, but we were totally unable to demonstrate specific binding of these steroids.

In contrast, we have shown the presence of glucocorticoid receptors in all leukocyte subpopulations, using the whole cell assay technique and [³H]-Org 7417 (11β-hydroxy-21-methyl-preg-4-ene-3,20-dione-17-butyrate) as ligand, thus confirming the studies of Lippman (1979). It is clear from Table

TABLE 13-3

Fractionation of Supernatants from Control and Steroid-treated Thymus Cultures: Effect on the Stimulation of Thymocytes with PHA

SUPERNATANT FRACTION (mol. wt)†	SUPERNATANT TREATMENT (RATIO)*			
	UNTREATED	ESTRADIOL	TESTOSTERONE	CORTICOSTERONE
Unfractionated	3.1	0.6	7.7	0.7
> 30 000	6.0	8.6	9.1	3.9
10 000–30 000	1.9	1.9	3.5	2.4
1000–10 000	1.4	4.3	1.1	2.8
< 1000	1.1	0.6	0.9	0.6

* Ratio = [Supernatant fraction–treated thymocytes + PHA]/Thymocytes + PHA.

Results expressed as the mean of four separate experiments. The results have been corrected for the effects caused by fractionation of the culture medium. Thymocytes alone = 426 dpm; thymocytes + PHA = 3323 dpm (S.I. = 7.8). Supernatants used were those obtained on day 30 of the cultures (see Figure 13-1).

† Fractionation by ultrafiltration.

13-4 that the dissociation constants and mean number of binding sites per cell are comparable in granulocytes and T and B cells. However, the equilibrium dissociation constant for macrophages/monocytes is significantly lower than that for the other cells, and the number of binding sites is two to three times higher, which could indicate a higher glucocorticoid sensitivity for this type of mononuclear cell.

The binding affinities of eleven natural and synthetic steroids for human leukocyte subpopulations were assessed relative to [^{3}H]-dexamethasone (Table 13-5). Similar patterns of competition were observed, and thus the cell types apparently show no differences in glucocorticoid specificity. It is of interest that progesterone and its associated synthetic steroids, Org. 2058 and R5020 (17,21-dimethyl,19-nor-4,9-pregnadiene-3,20-dione), showed a reasonable degree of competition for dexamethasone-binding to the receptors. Taking into account the apparent lack of specific progesterone-binding species in leukocytes, it appears likely that any direct inhibition of the immune reponse by progesterone (Pavia et al., 1979; Stites et al., 1979) is mediated through the leukocyte glucocorticoid receptors.

Finally, the possibility that steroids have certain effects on leukocytes not requiring the presence of high-affinity receptors is worthy of consideration. When phagocytic cells ingest microorganisms, they emit light. This chemiluminescence can be used as a measure of phagocytic cellular function (Easmon et al., 1980). We used this phenomenon to study the effects of steroids on the phagocytosis of zymosan by rat peritoneal macrophages (> 98 per cent pure). Although corticosterone and progesterone modified the response at concentrations greater than 10^{-8}M, as might be expected based on their interaction with the glucocorticoid receptor, estradiol-17β and estriol caused a dose-dependent increase in chemiluminescence, with a maximum at 10^{-8}M, that could not have

287

TABLE 13-4
Glucocorticoid Receptors in Leukocyte Subpopulations*

SUBPOPULATION	DISSOCIATION CONSTANT† K_d x 10^9 M	BINDING SITES/CELL†
T lymphocytes	8.6 ± 2.2	3521 ± 1240
B lymphocytes	11.1 ± 3.5	2564 ± 1043
Macrophages	2.8 ± 1.3	8629 ± 1722
Granulocytes	9.3 ± 3.3	3831 ± 987

* [^{3}H]-Organon 7414 employed in whole cell assays.
  Org 7417 = 11β-hydroxy-21-methyl-preg-4-ene-3,20-dione-17-butyrate.
† Mean ± S.D.

been receptor mediated (Figure 13-2). Estrogens can significantly stimulate the reticulo-endothelial system, and increased phagocytic activity can be detected in rodents during the estrus cycle and in pregnancy at times of increased estrogen levels (Nicol et al., 1964). Recently, a new phenolic compound has been indentified in the blood and urine of rodents, monkeys, and humans—trans-(±)-3,-bis [(3-hydroxyphenyl) methyl] dihydro-2-(3H)-furanone (HPMF). Its level has been shown to increase by up to 10 fold during the midluteal phase of the menstrual cycle and also shows a marked surge in pregnancy, reaching a maximum between weeks 14 and 22 of gestation (Setchell et al., 1980; Stitch et al., 1980). HPMF does not compete for steroid receptors in the thymus or leukocytes, and it has no effect on lymphocyte transformation induced by PHA, Con A, or PWM at concentrations up to 10^{-4}M. In

TABLE 13-5
Relative Binding Affinities of Steroids for the Glucocorticoid Receptor in Leukocyte Subpopulations

	RELATIVE BINDING AFFINITY (%)*			
STEROID	T LYMPHOCYTES	B LYMPHOCYTES	MACROPHAGES	GRANULOCYTES
Org 7417	550	520	720	590
Betamethasone	280	320	340	310
Dexamethasone	100	100	100	100
Prednisolone	58	55	62	43
Cortisol	32	28	36	27
R5020†	5.3	6.8	8.5	7.3
Progesterone	4.8	2.1	3.7	5.1
Org 2058†	1.8	2.2	2.9	3.2
Aldosterone	0.8	0.2	0.5	0.7
5α-dihydrotestosterone	< 0.2	< 0.2	< 0.2	< 0.2
Estradiol-17β	< 0.2	< 0.2	< 0.2	< 0.2

* Binding relative to dexamethasone = 100% using [^{3}H]-dexamethasone in whole cell assays.
† R5020 = 17,21-dimethyl,19-nor-4,9-pregnadiene-3,20-dione. Org2058 = 16α-ethyl-21-hydroxy-19-nor-4-pregnene-3,2β-dione.

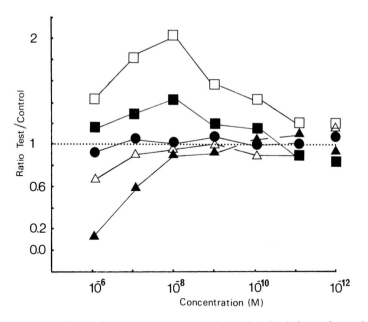

FIGURE 13-2. Effect of steroids on macrophage luminol-dependent chemiluminescence induced by zymosan.

▲, corticosterone; △, progesterone; ●, 5α-dihydrotestosterone; ■, estradiol-17β; □, estriol. Macrophages were preincubated with steroids for one hour before addition of zymosan. Control, chemiluminescence in the absence of steroids (ratio = 1).

experiments using the macrophage chemiluminescence model mentioned above, however, HPMF caused a six- to tenfold dose-dependent increase in light emission at 10^{-8} to 10^{-9}M, suggesting that estrogens and HPMF may directly alter macrophage function.

PREGNANCY-ASSOCIATED SERUM PROTEINS

Over the past decade, the amount of research into "pregnancy proteins" has increased tremendously, and approximately 30 of these substances have now been described. Unfortunately, little is known of their biological roles, but seven have been reported to possess immunosuppressive properties. Details of these proteins are shown in Table 13-6.

TABLE 13-6
Pregnancy-Associated Serum Proteins That Have Been Proposed as Immunoregulatory Factors

	NAME	ABBREVIATIONS	MOLECULAR WEIGHT
1.	Pregnancy-associated α_2-glycoprotein	α_2-PAG, PAG, PAM, SP3, PZP	360–380 000
2.	α-fetoprotein	AFP, αFP	64 000
3	Pregnancy-specific β_1-glycoprotein	PSβ_1G, SP1, PAPP-C, TBG	90 000(45 000)
4.	Human placental lactogen	HPL, HCS	21–23 000
5.	Human chorionic gonadotrophin	HCG	45 000
6.	Early pregnancy factor	EPF	240 000
7.	Pregnancy-associated plasma protein-A	PAPP-A	750 000

Pregnancy-associated α_2-glycoprotein (α_2-PAG): is a glycoprotein found in all normal sera at concentrations between 0.1 and 130 μg/ml. Males usually have a much lower level than females, and its levels tend to increase up to 60-fold during pregnancy and with estrogen treatment (Stimson, 1978). Variations in α_2-PAG blood levels have also been observed in cancer patients undergoing treatment and have been shown to correlate with the course of the disease in a significant number of cases (Stimson, 1975).

Alpha$_2$-PAG was first shown to have immunosuppressive properties when incorporated into mixed leukocyte cultures (Stimson, 1972) and has since been shown to significantly depress lymphocyte proliferation induced by PHA, Con A, mixed lymphocyte reaction (MLR) with allogeneic cells, and PPD (Stimson, 1976). It will also block BCG-induced inhibition of leukocyte migration, reduce spontaneous sheep erythrocyte binding to T cells (Stimson, 1976), and affect macrophage mobility in the electrophoretic test (Straube et al., 1975). However, this effect was much less when bacterial lipopolysaccharide (LPS) or anti-F(ab')$_2$ serum was used to selectively stimulate B cells, indicating that α_2-PAG may preferentially inhibit T-lymphocyte responses (Stimson, 1976). This suggestion is supported by the finding that α_2-PAG can be detected on some 25 per cent of T lymphocytes, but not on B cells (Stimson, 1977); however, some association with the latter has been indicated in one study (Horne et al., 1978). Although the glycoprotein will cause significant inhibition of lymphocyte transformation (up to 70 per cent) and of T-cell rosetting (25 per cent), total suppression is seldom achieved. The maximum effect was attained by approximately 400 μg/ml of α_2-PAG, a value below that normally found throughout pregnancy (Stimson, 1976). It is possible that only a limited number of cells have the capacity to bear α_2-PAG and to be inhibited by it. It is also possible that above relatively low blood levels, the binding sites on these cells (presumably a T cell subpopulation) are saturated with the protein. Alpha$_2$-PAG has also been identified on the surface of a large proportion of blood monocytes from pregnant women, cells which are capable of synthesizing it (Stimson et al., 1979a). Few *in vivo* studies have been carried out with α_2-PAG, possibly because of difficulties in obtaining enough pure material and because the legitimacy of crossing species barriers with proteins of high molecular weight is questionable. Nevertheless, intravenous administration of α_2-PAG can cause a limited, though significant, increase in the survival of heart transplants in mice (Svendsen et al., 1978).

Alpha-fetoprotein (AFP): has been the subject of numerous studies which have used the glycoprotein isolated from fetal, pregnant, and tumor-bearing animals and humans. Murgita and Tomasi (1975) demonstrated that AFP from mouse amniotic fluid depresses the primary immunoglobulin M (IgM) and secondary IgM, IgG, and IgA plaque-forming cell responses *in vitro* and that murine AFP inhibits the MLR and mitogen-induced lymphocyte transformation. Yachnin and Lester (1976) have shown that human AFP inhibits lymphocyte proliferation induced by allogeneic cells, phytomitogens, and antithymocyte serum. Human hepatoma-derived AFP has been shown not to affect B and T lymphocyte preparations, but rather to enhance "active" T rosetting and to depress PHA-induced transformation (Gupta and Good, 1977). Receptors for AFP have been demonstrated on a subpopulation of murine splenic T cells, and only those cells that bound AFP had reduced PHA responses (Keller and Tomasi, 1976).

AFP is not a general immunosuppressive factor. Murgita et al. (1978) found that AFP would not block murine T cell independent antibody responses or protein A-induced human B lymphocyte proliferation. Further, preincubation of murine spleen cells with AFP causes the activation of suppressor T cells (Murgita et al., 1977). Other studies have occasionally produced somewhat contradictory results. Sheppard et al. (1977) reported that murine AFP can inhibit the lymphocyte response to LPS but not to Con A and can depress the *in vitro* plaque-forming cell response while enhancing the induction of cytotoxic cells directed against P815 mastocytoma cells. Charpentier et al. (1977) and Goeken and Thompson (1977) failed to demonstrate significant suppression by the protein and also claimed that the activity of AFP depends on the method of purification. Variation in immunosuppressive activities of different preparations of AFP has also been described by Lester et al. (1976) and Murgita et al. (1978). Thus, AFP is not consistently immunosuppressive, and although five different forms of the glycoprotein have been defined in terms of electrophoretic mobility and sialic acid content, only one has inhibitory activity (Zimmerman et al., 1977). Keller et al. (1976) and Adinolfi (1978) suggest that AFP may have a suppressive role as a result of its carrier function as a plasma protein, mediated by hormones or plasma peptides bound to it.

It is still not clear whether AFP is an effective immunoregulatory factor. It is possible that it functions as a carrier for other compounds which are the true inhibitors. Finally, although fetal blood normally contains levels of AFP within the accepted immunosuppressive range, the quantity present in the maternal circulation (maximum 550 ng/ml) appears to be inadequate to contribute significantly to the inhibitory properties of pregnancy serum.

Pregnancy-specific β_1-glycoprotein (PSβ_1G): is currently receiving much attention because of an attempt to correlate its blood levels with abnormal pregnancy. It is synthesized by the syncytiotrophoblast and reaches a maximum concentration of 200 µg/ml in late pregnancy. PSβ_1G has been investigated for immunosuppressive properties only in lymphocyte transformation tests and inhibits PHA-induced proliferation by 75 per cent at 250 µg/ml but has no effect on the Con A response (Horne et al., 1976). Another study indicated that PSβ_1G caused only a small decrease in PHA responsiveness and that this decrease was not dose related (Cerni et al., 1977). Further, the 50 per cent inhibition point of mixed lymphocyte culture with PSβ_1G was found to be greater than 1000 µg/ml (Johannsen et al., 1976). Thus, on balance, it would appear that PSβ_1G does not have a realistic inhibitory profile and that maternal serum does not contain enough of the protein to be immunosuppressive.

Human placental lactogen (HPL): The inhibitory properties of human placental lactogen were first reported by Contractor and Davies (1973), who indicated that HPL suppresses lymphocyte proliferation at concentrations between 25 and 50 µg/ml, especially following preincubation of cells with the protein. A later study found that the simultaneous incubation of lymphocytes with up to 400 µg/ml of HPL had no effect on PHA-induced transformation, and it confirmed that preincubation resulted in a decrease in [³H]-thymidine incorporation, reaching a maximum at 200 µg/ml (Cerni et al., 1977). The amount required to inhibit mixed lymphocyte cultures by 50 per cent has been shown to be 100 to 1000 µg/ml (Johannsen et al., 1976), but Morse (1976) found that, after extensive purification, HPL suppressed PHA stimulation and the MLR only at concentrations greater than 1000 µg/ml. These results suggest that a contaminant might have been responsible for the inhibitory effects shown by the protein in other studies. On this basis, the highest values noted in pregnancy serum (10 µg/ml) and the placenta (200 µg/g of tissue) are inadequate to depress immune reactivity.

Human chorionic gonadotropin (hCG): Human chorionic gonadotropin is a placental glycoprotein, the blood level of which rises sharply in the first trimester to 160 IU/ml and then falls to under 80 IU/ml during the rest of the pregnancy. Early studies implied that crude hCG preparations ($<$ 4000 IU/mg) isolated from pregnancy urine possessed immunosuppressive activity. Daily injections were shown to prolong skin allograft survival (Pearse and Kaiman, 1967), to decrease graft-versus-host disease (Slater et al., 1977) and to inhibit the *in vitro* proliferative response of human lymphocytes to allogeneic cells, PHA, and varidase (Contractor and Davies, 1973; Han, 1974). However, highly purified hCG (> 12000 IU/mg) has since been found to be almost devoid of activity (Caldwell et al., 1975; Morse et al., 1976; Maes and Claverie, 1977). These investigations showed that the hormonal and immunosuppressive properties can be separated chromatographically, indicating that the active factor is a substance other than hCG. We have purified this substance and found, using sodium dodecyl sulfate/polyacrylamide gel electrophoresis, that it exhibits molecular weights of 24 000 and 49000. It has the capacity to inhibit responses induced by a variety of antigens and mitogens (PHA, PWM, Con A, MLR, PPD, LPS, dextran sulfate) and to suppress antibody production and delayed hypersensitivity reactions *in vitro*. Exposure of lymphocytes to the factor, followed by washing, still completely blocked mitogen-induced transformation, as did addition of the factor up to 24 hours following initiation of the cultures. This pregnancy protein appears to exhibit many characteristics of a lymphocyte chalone (Attalah, 1979). Although the inhibitory properties of hCG must now be viewed with caution, highly purified preparations have been shown to increase DNA synthesis in unstimulated cells, and hCG has been proposed as a B cell activator (Morse et al., 1976; Maes and Claverie, 1977).

Early pregnancy factor: Early pregnancy factor (EPF) was first demonstrated in pregnancy serum using the rosette inhibition technique (Morton et al., 1976). It has been detected in the sera of mice, humans, and sheep 6 to 24 hours after fertilization and has been shown to be present only during the first two trimesters. EPF has since been found to have a molecular weight of approximately 240000, to suppress the adoptive transfer of contact sensitivity to dinitrochlorobenzene (Noonan et al., 1979), and to decrease in cases of abortion (Morton et al., 1977).

The existence of such a factor is naturally desirable, given its potential in pregnancy detection and in early fetoplacental protection. The capacity of the rosette inhibition test to detect the proposed factor is doubtful, however (Cooper and Aitken, 1980), and thus further research is required to verify the role of EPF.

Pregnancy-associated plasma protein A: Pregnancy-associated plasma protein A (PAPP-A) is an incompletely characterized α_2-macroglobulin synthesized by the placental trophoblast. Blood levels have been shown to rise steadily throughout gestation but particularly during the third trimester. The concentration decreases sharply following delivery with a half-life of three to four days, and becomes undetectable after four to six weeks (Halbert and Lin, 1979).

It has been suggested that PAPP-A could possess immunosuppressive properties (Klopper et al., 1979), and indeed a recent study (Bischof et al., 1981) has indicated that it has the capacity to significantly suppress mitogen-induced lymphocyte proliferation. In order to assess whether PAPP-A and other pregnancy proteins (α_2-PAG, PSβ_1G, AFP, HCG, HPL) contribute to the ability of pregnancy serum to depress *in vitro* lymphocyte proliferation (Gill and Repetti, 1979), a pool was prepared from a large number of pregnancy sera, so as to contain high levels of the proteins. Each protein was then specifically removed by antibody-affinity chromatography, and the remaining material was examined for suppressive activity (Stimson, 1980, 1981). Apart from a limited effect following α_2-PAG absorption, removal of the other pregnancy-associated proteins did not influence the inhibitory properties of pregnancy serum. Thus, other as yet unidentified substances appear responsible for this phenomenon. Nevertheless, the possibility remains that one of the factors produced by the placenta could reach a sufficiently high local concentration to cause immunoregulation. However, Anderson (1978) was unable to find significant alterations in the mitogenic responses of lymphocytes from lymph nodes draining the fetal implantation site, as might be expected, based on the close proximity of these cells *in vivo* to the source of the putative immunosuppressive placental products.

OTHER NEW PREGNANCY-ASSOCIATED PROTEINS

We have recently identified two new pregnancy proteins whose postulated roles allow different aspects of maternal immunoregulation to be considered apart from the usual direct effects of these substances on lymphocytes.

Pregnancy-associated prostaglandin synthetase inhibitor (PAPSI): Pregnancy serum has been found to possess anti-inflammatory properties, as shown by its capacity to stabilize isolated leukocyte lysosomes (Hempel et al., 1970), inhibit carrageenin-induced edema (Persellin et al., 1974), modify neutrophil function (Persellin and Leibfarth, 1978), and to suppress the release of lysosomal enzymes from macrophages (Stimson et al., 1977). We have investigated this phenomenon and identified a new pregnancy-associated protein, PAPSI, which is probably responsible for at least part of the anti-inflammatory activity of pregnancy serum and could be involved in the regulation of a variety of inflammatory diseases during pregnancy (Persellin, 1977).

PAPSI has also been detected in early pregnancy urine but is absent from fetal serum and the serum and urine of normal males and nonpregnant females. Partial purification of PAPSI has been achieved using ultracentrifugation, gel chromatography, and SDS-PAGE. Its approximate molecular weight in serum is 1×10^6, whereas in urine it is 5×10^6 (aggregate). The molecular weight of its subunits is 8×10^3, and its isoelectric point is approximately pH 4.0. Administered intraperitoneally, PAPSI inhibits carrageenin-induced edema in the feet of rodents by up to 80 per cent at a dose of 1.5 mg/Kg and its effects on edema caused by pure mediators of inflammation (Cottney et al., 1976; Williams and Peck, 1977) are shown in Figure 13-3. PAPSI suppressed only edema induced by arachidonate and arachidonate-potentiated bradykinin. Furthermore, although PAPSI effectively inhibited the cyclo-oxygenase stage of the *in vitro* conversion of arachidonic acid to prostaglandins E_2 and $F_2\alpha$ (Kuehl et al., 1977), its ID_{50} was well below that of indomethacin (Table 13-7). However, the converse was true when synthetase activity was assessed *in vivo*, and it can be seen that PAPSI is a very effective inhibitor of prostaglandin synthesis. A specific antiserum to PAPSI has now been obtained, and, using immunoperoxidase staining, we have shown that the pregnancy protein is readily detected in the cytoplasm of the placental trophablast throughout gestation—except at term.

These studies indicate that pregnant women possess a potent inhibitor of prostaglandin synthesis. Secretion of PAPSI by the placenta would be expected to depress inflammatory responses which could potentially lead to rejection of the fetus, and its disappearance at the end of pregnancy suggests an involvement in the increase in prostaglandin levels at term (Filshie and Anstey, 1978).

Pregnancy-associated β_I-macroglobulin (β_I-PAM): It was shown initially that maternal lymphocytes would respond to crude placental antigen preparations and that there appeared to be a limited number of antigens involved. This suggested a high degree of crossreactivity between lymphocytes and antigen preparations. It was also found that maternal serum contained an IgG antibody that would specifically block this antigenic recognition (Stimson et al., 1979b). We therefore sought to characterize the antigens responsible for this phenomenon, including studies of immune complexes isolated from normal pregnancy serum, as these could be expected to contain the relevant antigens.

We reported (Stimson et al., 1981), as did others (Masson et al., 1977; Stirrat et al., 1978), that elevated levels of immune complexes may be found in the blood of pregnant women, and we have developed an enzyme-linked assay to assess these more efficiently (Stimson, et al., 1981). Immune complex preparations from primigravid women were dissociated at low pH and separated into their components by gel chromatography. A previously unknown antigen was identified which consistently recombines with the IgG fraction from the

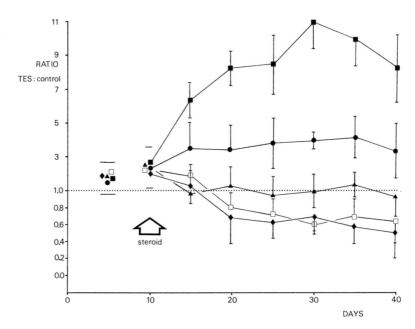

FIGURE 13-3. Effect of PAPSI on edema induced by the pure mediators of imflammation.

▲, histamine (1 μg); ■, 5-hydroxytryptamine (1μg); ●, bradykinin (500 ng); □, bradykinin + arachidonic acid; O, arachidonic acid (5 μg). Results expressed as the mean values (%) from five groups of six mice. Control value (100%) = foot edema induced by the pure mediators in the absence of PAPSI (administered i.p.).

TABLE 13-7
Effect of PAPSI on Prostaglandin
Synthetase Activity*

	ID_{50}†	
TEST	PAPSI	INDOMETHACIN
In vitro (μg/ml)	88 ± 33	0.045 ± 0.019
In vivo (mg/kg)‡	3.1 ± 2.2	27.6 ± 8.5

* Conversion of [^{3}H]-arachidonic acid to prostaglandins E_2 and $F_{2\alpha}$.
† ID_{50} = mean (± S.D.) 50% inhibition dose of five experiments carried out in triplicate.
‡ Substances administered i.p. 30 minutes prior to intradermal arachidonate injection.

dissociated complexes. These antigen-antibody aggregates were detectable with the immune-complex assay. This antigen was called β_1-PAM following further characterization (Table 13-8). A sandwich-type microenzyme immunoassay

TABLE 13-8
Physicochemical Properties of β_1-PAM from Pregnancy and Ovarian Cancer Serum

PARAMETER	β_1-PAMc*	β_1-PAMp†
Electrophoretic mobility‡	β_1	β_1
Ammonium sulfate fractionation range (M)	1.1 - 1.5	1.1–1.5
Isoelectric point (pH)	5.25	5.25
Molecular weight (x 10^{-6})		
Gel chromatography	1.41	(4.23, 1.38)**, 0.46
PAGE	1.54, (0.76)**	

* β_1-PAM from ovarian cancer patients.
† β_1-PAM from primigravid women
‡ Electrophoretic mobility in agar at pH 8.6 (0.1 M barbitone/HCl buffer).
** Values in parentheses represent the molecular weight of minor components
($< 22\%$ of the total β_1-PAM identified). These are probably related to the
subunit nature of β_1-PAM and its association with immune complexes.

was developed and used to assess β_1-PAM levels in sera from pregnant women, normal males and females, and cancer patients. As expected, β_1-PAM was present in the blood of pregnant women but it was not possible to detect it in sera from normal subjects and a number of cancer patients. However, the majority of women with ovarian cancer were also positive for β_1-PAM, often at levels far in excess of those found during pregnancy (Stimson and Farquharson, 1981).

Immunoperoxidase staining has shown β_1-PAM to be present in the placental trophoblast and in tissue from more than 95 per cent of ovarian tumors. In a preliminary trial, 26 ovarian cancer patients, with a variety of tumor types, were followed for up to two years and 4 to 15 serum samples obtained from each patient (Stimson and Shepherd, 1981). Of these women, 23 were β_1-PAM positive at the time of first operation and 24 had β_1-PAM levels that closely correlated with the stage and course of the disease, i.e., falling or remaining constant in static patients and rising significantly in those with progressive disease. Changes in levels showed a good "lead time" (greater than two months in 10 out of 11 patients from whom an adequate number of samples had been collected). A useful marker substance for ovarian cancer has so far eluded discovery; however, our initial results indicate that β_1-PAM may fulfill this role.

In a collaborative study with H.M. Dick, Royal Infirmary, Glasgow, it was shown that β_1-PAM prepared from single sources had the capacity to totally block the cytotoxic activity of a number of HLA typing sera (Table 13-9), but in no well-defined pattern. Thus, β_1-PAM may share antigenic specificities related to certain antigens of the major histocompatibility complex.

TABLE 13-9
β_1-PAM: Examples of Blocking Effect on
HLA Typing Sera

LYMPHOCYTE TYPE	NUMBER AND TYPE OF SERA BLOCKED*
A2, A11 ; B7, Bw44	3/7 B12 ; 3/8 B7
A2, A3 ; Bw35, Bw44	3/6 A3 ; 2/7 B12
A1, A28 ; B7, B27	2/3 A28 ; 2/8 B7
A2, A26 ; Bw44, Bw22	2/5 Bw22 ; 1/1 B7 + Bw22
A3, Aw24 ; B7, B17	4/6 A3 ; 3/3 Aw24
	2/9 B7 (+27 + 22)

Also blocked 1/4 Bw6 for all cells; no effect on DR sera
* Sera absorbed out with β_1-PAM (0.5 µg). 100% blocking obtained in
each case.

However, the glycoprotein has no apparent effect on lymphocyte proliferation induced by PHA, Con A, PWM, and allogeneic cells. Nevertheless, β_1-PAM may possess a function in specific immunoregulation by stimulating the production of certain maternal blocking antibodies.

ACKNOWLEDGMENTS

This research was supported by a grant from the Scottish Hospital Endowments Research Trust. I wish to express my gratitude to I.C. Hunter, D.M. Farquharson, A. McAdam, P.J. Crilly, and G. Gallaher for their many contributions to the studies described in this chapter.

REFERENCES

Adinolfi, M. 1978. Human alpha-fetoprotein 1956-1978. *Adv. Human Genetics* 9:165.
Anderson, D. J. 1978. The responsiveness of various maternal mouse lymphocyte populations to mitogenic stimulation *in vitro. Cell. Immunol.* 41:150.
Attalah, A.M. 1979. Lymphocyte chalone and lymphoid diseases. In *Naturally Occurring Biological Immunosuppressive Factors and Their Relationship to Disease.* R.H. Neubauer, ed. Boca Raton, Florida: C.R.C. Press, pp. 175-186.
Bischoff, P. et al. 1981. *J. Clin. Lab. Immunol.* In press.
Brodie, J.Y., I.C. Hunter, W.H. Stimson, and B. Green. 1980. Specific estradiol binding in cytosol from the thymus glands of normal and hormone-treated male rats. *Thymus* 1:337.
Caldwell, J.L., D.P. Stites, and H.H. Fudenberg. 1975. Human chorionic gonadotropin: effects of crude and purified preparations on lymphocyte responses to phytohemagglutinin and allogeneic stimulation. *J. Immunol.* 115:1249.

Cerni, C., G. Tatra, and H. Bohn. 1977. Immunosuppression by human placental lactogen (HPL) and the pregnancy-specific β_1 glycoprotein (SP-1). *Arch. Gynecol.* 223:1.

Charpentier, B., R.D. Guttman, J. Shuster, and P. Gold. 1977. Augmentation of proliferation of human mixed lymphocyte culture by human α-fetoprotein. *J. Immunol.* 119:897.

Contractor, S.F., and H. Davies. 1973. Effect of human chorionic somatomammotrophin and human chorionic gonadotrophin on phytohaemagglutinin-induced lymphocyte transformation. *Nature* 243:284.

Cooper, D.W., and R.J. Aitken. 1980. Failure to detect altered rosette inhibition titres in human pregnancy serum. *J. Reprod. Fertil.* 61:241.

Cottney, J., A.J. Lewis, and D.J. Nelson. 1976. Arachidonic acid-induced paw oedema in the rat. *Brit. J. Pharmacol.* 58:311P.

Easmon, C.S.F., P.J. Cole, A.J. Williams, and M. Hastings. 1980. The measurement of opsonic and phagocytic function by luminodependent chemiluminescence. *Immunology* 41:67.

Filshie, G.M., and M.D. Anstey. 1978. The distribution of arachidonic acid in plasma and tissues of patients near term undergoing elective or emergency caesarean section. *Brit. J. Obstet. Gynaecol.* 85:119.

Gill, T.J., and C.F. Repetti. 1979. Immunologic and genetic factors influencing reproduction. *Am. J. Pathol.* 95:465.

Goeken, N.E., and J.S. Thompson. 1977. Conditions affecting the immunosuppressive properties of human α-fetoprotein. *J. Immunol.* 119:139.

Grossman, C.J., L.J. Sholiton, and P. Nathan. 1979. Rat thymic estrogen receptor—I and II. *J. Steroid Biochem.* 11:1233 and 1241.

Gupta, S., and R.A. Good. 1977. α-Fetoprotein and human lymphocyte subpopulations. *J. Immunol.* 118:405.

Halbert, S.P., and T.-M. Lin. 1979. Pregnancy-associated plasma proteins: PAPP-A and PAPP-B. In *Placental Proteins.* A. Klopper and T. Chard, eds. Heidelberg: Springer-Verlag, pp. 89-103.

Han, T. 1974. Inhibitory effect of human chorionic gonadotrophin on lymphocyte blastogenic response to mitogen, antigen and allogeneic cells. *Clin. Exp. Immunol.* 18:529.

Hempel, K.H., L.A. Fernandez, and R.H. Persellin. 1970. Effect of pregnancy serum on isolated lysosomes. *Nature (New Biol.)* 225:955.

Horne, C.H.W., A.W. Thomson, C.M. Towler, F.K. MacMillan, and L.M. Gibb. 1978. Relationship of pregnancy-associated α_2-glycoprotein in peripheral blood leukocytes. *Scand. J. Immunol.* 8:75.

Horne, C.H.W., C.M. Towler, R.G.P. Pugh-Humphreys, A.W. Thomson, and H. Bohn. 1976. Pregnancy-specific β_1-glycoprotein—a product of the syncytiotrophoblast. *Experientia* 32:1197.

Johannsen, R., H. Haupt, H. Bohn, K. Heid, F.R. Seiler, and H.G. Schwick. 1976. Inhibition of the mixed lymphocyte culture (MLC) by proteins: Mechanism and specificity of the reaction. *A. Immunitat. A. Immunobiol.* 152:280.

Keller, R.H., N.J. Calvanico, and T.B. Tomasi. 1976. *Onco-Developmental Gene Expression.* W.H. Fishman and S. Sell, eds. New York: Academic Press, pp. 287-295.

Keller, R.H., and T.B. Tomasi. 1976. Alpha-fetoprotein synthesis by murine lymphoid cells in allogeneic reactions. *J. Exp. Med.* 143:1140.

Klopper, A., R. Smith, and I. Davidson. 1979. The measurement of trophoblastic proteins as a test of placental function. In *Placental Proteins.* A. Klopper and T. Chard, eds. Heidelberg: Springer-Verlag, pp. 23-42.

Kruisbeek, A.M. 1979. Thymic factors and T-cell maturation *in vitro*: a comparison of the effects of thymic epithelial cultures with thymic extracts and thymus dependent serum factors. *Thymus* 1:163.

Kuehl, F.A., J.L. Humes, R.W. Egan, E.A. Ham, G.C. Beveridge, and C.G. Van Arman. 1977. Role of prostaglandin endoperoxide PGG$_2$ in inflamatory processes. *Nature* 265:170.

Lester, E.P., J.B. Miller, and S. Yachnin. 1976. Human alpha-fetoprotein as a modulator of human lymphocyte transformation: correlation of biological potency with electrophoretic variants. *Proc. Nat. Acad. Sci. USA* 73:4645.

Lippman, M.E. 1979. Glucocorticoid receptors and effects in human lymphoid and leukemic cells. In *Monographs on Endocrinology.* Vol. 12, *Glucocorticoid Hormone Action.*

J.D. Baxter and G.G. Rousseau, eds. Heidelberg: Springer-Verlag, pp. 377-397.

Maes, R.F., and N. Claverie, 1977. The effect of preparations of human chorionic gonadotropin on lymphocyte stimulation and immune response. *Immunology* 33:351.

Masson, P.L., M. Delire, and C.L. Cambiaso. 1977. Circulating immune complexes in normal human pregnancy. *Nature* 266:542.

McCruden, A.B., and W.H. Stimson. 1980. Androgen and other sex steroid cytosol receptors in the rat thymus. *J. Endocrinol.* 85:47.

McCruden, A.B., and W.H. Stimson. 1981. Androgen binding cytosol receptors in the rat thymus: physicochemical properties, specificity and localisation. *Thymus* 3:105.

Morse, J.H. 1976. The effect of human chorionic gonadotrophin and placental lactogen on lymphocyte transformation *in vitro*. *Scand. J. Immunol.* 5:779.

Morse, J.H., G. Stearns, J. Arden, G.M. Agosto, and R.E. Canfield. 1976. The effects of crude and purified human gonadotropin on *in vitro* stimulated human lymphocyte cultures. *Cell. Immunol.* 25:178.

Morton, H., V. Hegh, and G.J.A. Clunie. 1976. Studies on the rosette inhibition test in pregnant mice: evidence of immunosuppression. *Proc. R. Soc. Lond. (Biol.)* 193:413.

Morton, H., B. Rolfe, G.J.A. Clunie, M.J. Anderson, and J. Morrison. 1977. An early pregnancy factor detected in human serum by the rosette inhibition test. *Lancet* i:394.

Murgita, R.A., and T.B. Tomasi. 1975. Suppression of the immune response by α-fetoprotein. I and II. *J. Exp. Med.* 141:269 and 440.

Murgita, R.A., E.A. Goidl, S. Kontiainen, and H. Wigzell. 1977. α-Fetoprotein induces suppressor T cells *in vitro*. *Nature* 267:257.

Murgita, R.A., I.C. Andersson, M.S. Sherman, H. Bennich, and J. Wigzell. 1978. Effects of human α-fetoprotein on human B and T lymphocyte proliferation *in vitro*. *Clin. Exp. Immunol.* 33:347.

Neifeld, J.P., M.E. Lippman, and D.C. Tormey. 1977. Steroid hormone receptors in normal human lymphocytes: induction of glucocorticoid receptor activity by phytohemagglutinin stimulation. *J. Biol. Chem.* 254:2972.

Nicol. T., D.L.J. Bilbey, L.M. Charles, J.L. Cordingley, and B. Vernon-Roberts. 1964. Oestrogen: the natural stimulant of body defence. *J. Endocrinol.* 30:277.

Noonan, F.P., W.J. Halliday, H. Morton, and G.J.A. Clunie. 1979. Early pregnancy factor is immunosuppressive. *Nature* 278:649.

Pavia, C., P.K. Siiteri, J.D. Perlman, and D.P. Stites. 1979. Suppression of murine allogeneic cell interactions by sex hormones. *J. Reprod. Immunol.* 1:33.

Pearse, W.H., and H. Kaiman. 1967. Human chorionic gonadotropin and skin allograft survival. *Am. J. Obstet. Gynecol.* 98:573.

Persellin, R.H., S.E. Vance, and A. Peery. 1974. Effect of pregnancy serum on experimental inflammation. *Brit. J. Exp. Pathol.* 55:26.

Persellin, R.H. 1977. The effect of pregnancy on rheumatoid arthritis. *Bull. Rheum. Dis.* 27:922.

Persellin, R.H., and J.K. Leibfarth. 1978. Studies on the effects of pregnancy serum on polymorphonuclear leucocyte functions. *Arth. Rheum.* 21:316.

Reichman, M.E., and C.A. Villee. 1978. Estradiol binding by rat thymus cytosol. *J. Steroid Biochem.* 9:637.

Rocklin, R.E., J.L. Kitzmiller, C.B. Carpenter, M.R. Garovoy, and J.R. David. 1976. Absence of an immunologic blocking factor from the serum of women with chronic abortions. *New. Engl. J. Med.* 295:1209.

Schiff, R.I., D. Mercier, and R.H. Buckley. 1975. Inability of gestational hormones to account for the inhibitory effects of pregnancy plasmas on lymphocyte responses *in vitro*. *Cell. Immunol.* 20:69.

Setchell, K.D.R., A.M. Lawson, F.L. Mitchell, H. Adlercreutz, D.N. Kirk, and M. Axelson. 1980. Lignans in man and in animal species. *Nature* 287:740.

Sheppard, H.W., S. Sell, P. Trefts, and R. Bahu. 1977. Effects of α-fetoprotein on murine immune responses. I. Studies in mice. *J. Immunol.* 119:91.

Slater, L.M., W. Bostick, and L. Fletcher. 1977. Decreased mortality of murine graft-versus-host disease by human chorionic gonadotropin. *Transplantation* 23:103.

Stimson, W.H. 1972. Transplantation—nature's success. *Lancet* i:684.

Stimson, W.H. 1975. Variations in the level of a pregnancy-associated α-macroglobulin in patients with cancer. *J. Clin. Pathol.* 28:868.

Stimson, W.H. 1976. Studies on the immunosuppressive properties of a pregnancy-associated α-macroglobulin. *Clin. Exp. Immunol.* 25:199.

Stimson, W.H., and I.C. Hunter. 1976. An investigation into the immunosuppressive properties of oestrogen. *J. Endocrinol.* 69:42.

Stimson, W.H. 1977. Identification of pregnancy-associated α-macroglobulin on the surface of peripheral blood leukocyte populations. *Clin. Exp. Immunol.* 28:445.

Stimson, W.H., I.C. Hunter, and C. Manos. 1977. Comparison of the effects of human pregnancy serum and anti-inflammatory compound on the release of lysosomal enzymes from macrophages. *Brit. J. Exp. Pathol.* 58:434.

Stimson, W.H. 1978. Pregnancy-associated α2-glycoprotein. pregnancy-specific β1-glycoprotein and pregnancy-associated plasma proteins A and B. *Bibl. Reprod.* 31:225.

Stimson, W.H., D.M. Farquharson, A. Shepherd, and J.M. Anderson. 1979a. Studies on the synthesis of pregnancy-associated α2-glycoprotein by the liver, placenta and peripheral blood leukocyte populations. *J. Clin. Lab. Immunol.* 2:235.

Stimson, W.H., A.F. Strachan, and A. Shepherd. 1979b. Studies on the maternal response to placental antigens: absence of a blocking factor from the blood of abortion-prone women. *Brit. J. Obstet. Gynaecol.* 86:41.

Stimson, W.H. 1980. Are pregnancy-associated serum proteins responsible for the inhibition of lymphocyte transformation by pregnancy serum? *Clin. Exp. Immunol.* 40:157.

Stimson, W.H., and I.C. Hunter. 1980. Oestrogen-induced immunoregulation mediated through the thymus. *J. Clin. Lab. Immunol.* 4:27.

Stimson, W.H., P.J. Crilly, and A.B. McCruden. 1980a. Effect of sex steroids on the synthesis of immunoregulatory factors by thymic epithelial cell cultures. *IRCS Med. Sci.* 8:263.

Stimson, W.H., McCruden, A.B., and P.J. Crilly. 1980b. The location of sex steroid cytosol receptors in rat thymus. *IRCS Med. Sci.* 8:341.

Stimson, W.H. 1981. Pregnancy-associated serum proteins and hormones: do they contribute to the regulation of the immune response in pregnancy? *Clin. Immunol. Newsletter* 2:7.

Stimson, W.H., and P.J. Crilly. 1981. Effects of steroids on the secretion of immunoregulatory factors by thymic epithelial cell cultures. *Immunology* 44:401.

Stimson, W.H., and D.M. Farquharson. 1981. Pregnancy-associated β1-macroglobulin (β1-PAM): a new serum protein associated with pregnancy serum-derived immune complexes and ovarian cancer. *J. Clin. Lab. Immunol.* 6:141.

Stimson, W.H., A. McAdam, and R.S. Hutchison. 1981. An assay for antigen-antibody complexes in human sera using Clq-enzyme conjugates. *J. Clin. Lab. Immunol.* 5:129.

Stimson, W.H., and A. Shepherd. 1981. To be published.

Stirrat, G.M., C.W.G. Redman, and R.J. Levinsky. 1978. Circulating immune complexes in pre-eclampsia. *Brit. Med. J.* 1:1450.

Stitch, S.R., J.K. Toumba, M.B. Groen, C.W. Funke, J. Leemhuis, J. Vink, and G.F. Woods. 1980. Excretion, isolation and structure of a new phenolic constituent of female urine. *Nature* 287:738.

Stites, D.P., C.S. Pavia, L.E. Clemens, R.W. Kuhn, and P.K. Siiteri. 1979. Immunologic regulation in pregnancy. *Arth. Rheum.* 22:1300.

Straube, W., B. Klausch, R. Hofman, H.L. Jensen, J. Gunther, and H. Kohler. 1975. Immunochemical investigations on the protein of the "pregnancy zone." *Arch. Gynecol.* 218:313.

Svendsen, P., T. Stigbrand, B. Teisner, J. Folkerson, M.G. Damber, B. von Schoultz, E. Kemp, and S.E. Svehag. 1978. Immunosuppressive effect of human pregnancy zone protein on H-2 incompatible mouse heart allografts. *Acta. Pathol. Microbiol. Scand. (C)* 86:199.

Tomoda, Y., M. Fuma, T. Miwa, N. Saka, and N. Ishizuka. 1976. Cell-mediated immunity in pregnant women. *Gynecol. Invest.* 7:280.

Williams, T.J., and M.J. Peck. 1977. Role of prostaglandin-mediated vasodilation in inflammation. *Nature* 270:530.

Yachin, S., and E. Lester. 1976. Inhibition of human lymphocyte transformation by human alpha fetoprotein (HAFP): comparison of foetal and hepatoma HAFP and kinetic studies of *in vitro* immunosuppression. *Clin. Exp. Immunol.* 26:484.

Zimmerman, E.F., M. Voorting-Hawkins, and J.G. Michael. 1977. Immunosuppression by mouse sialylated α-fetoprotein. *Nature* 265:354.

Chapter 14

MECHANISMS OF IMMUNOSUPPRESSION IN NEONATAL AND PREGNANT MICE

THOMAS B. TOMASI

Among the many theories that have been proposed to explain the immunological deficiencies noted in pregnancy and neonates is the presence of circulating, soluble suppressive factors. Suppression of immune reactions by pregnancy serum has been attributed to low-molecular-weight substances including corticosteroids, estrogens, progesterone, and circulating proteins, such as pregnancy-associated proteins, chorionic gonadotropin, and alpha-fetoprotein. Similarly, there appears to be a powerful inhibitor in neonatal serum that has been attributed to either a novel circulating protein or, more likely, a marked increase in the level of a normally occurring inhibitory factor. This area has been reviewed by the author (Tomasi, 1977) and will not be referred to in detail here. Rather, we choose to review the evidence for an inhibitory factor in neonatal fluids and the possible role of alpha-fetoprotein as a suppressive substance.

An oncofetal protein that has been implicated in immune suppression in neonatal and pregnant mice is alpha-fetoprotein (AFP). Several years ago, our laboratory reported that murine amniotic fluid (MAF) contained a suppressive factor for antibody formation to SRBC *in vivo* and subsequently presented data suggesting that apparently homogeneous preparations of AFP isolated from MAF were suppressive *in vitro* for antibody formation as well as for certain cell-mediated reactions (Tomasi et al., 1975). The concentrations of AFP used in these studies were low enough to suggest that this protein could play a physiological role in the neonate and/or pregnant animal (at the placental level) and possibly in those malignancies in which high levels of AFP were obtained.

Following these initial observations, a number of conflicting reports appeared concerning the immunosuppressive potency of AFP. Several laboratories have confirmed the immunosuppressive properties of both murine and human amniotic fluids and of AFP isolated from these fluids (Auer and Kress, 1977; Etlinger and Chiller, 1977; Gupta and Good, 1977; Murgita et al., 1977; Yachnin and Lester, 1977; Alpert et al., 1978; Gershwin et al., 1978). However, some investigators, although observing some degree of suppression in certain *in vitro* systems, have been unable to find significant inhibition with other assays (Sheppard et al., 1977a, 1977b). The possible reasons for these discrepancies have been reviewed in some detail by the author (Tomasi, 1978) and will not be repeated here. Suffice it to say, they remain largely unexplained, although it is our impression that technical factors are of key importance.

Since these early observations, several reports pertinent to the question of the immunosuppressive properties of AFP and its possible physiological role have appeared. These reports can be summarized as follows. There is little question that amniotic fluid (and neonatal serum) contain potent immune suppressive factor(s). It is also clear that the immune suppressive factor(s) is present in a higher concentration relative to the total amount of protein than in normal serum. The nature of the suppressive factor(s) has not been elucidated, although, as mentioned, evidence has been reported that AFP preparations are suppressive. We have investigated various methods of isolating murine AFP and have found that some of the techniques used by other workers yield only partially suppressive or nonsuppressive preparations. Also, the work of Auer and Kress (1977) has clearly shown the importance of the method of isolation of human AFP to its immunosuppressive activity. This may be related to the presence of a low-molecular-weight co-factor(s) complexed to AFP, which is necessary for suppression. The nature of the co-factor(s) is unknown, but AFP, like albumin with which it shares structural homology, binds a variety of low-molecular-weight components. Molecules that have been implicated include steroids, polyamines, and fatty acids. Recently, our laboratory (W. Anderson and T.B. Tomasi, submitted) has shown that AFP suppression of the immune response markedly increases when 1 mole of copper (Cu^{++}) is bound to the histidine residue at position 3 from the N terminus. Removal of Cu^{++} is accompanied by a loss of suppression. However, whether Cu binding is important physiologically is not clear. It is of interest, however, that MAF contains relatively high concentrations of copper (approximately 80 ng/mg of protein) and that the copper content of pregnancy serum is elevated.

An additional factor, which will be discussed in more detail below, is the demonstration that MAF contains a substance that will cause proliferation of lymphoid cells. Thus, it is entirely possible that the immunostimulatory properties that others (Charpentier et al., 1977; Soubiran et al., 1979) have seen with certain preparations of MAF and AFP may be due to a predominance of the stimulatory substances relative to the suppressive factor(s).

Finally, the possibility exists that the immunosuppressive factor, which has been found in apparently homogeneous AFP preparations, may well be a result of a contaminant that is not readily detected by available techniques of analysis. While it is, or course, extremely important to determine whether AFP or an unidentified contaminant is suppressive, an equally important question is whether the factor, whatever its nature, is a biologically important immunoregulatory substance or an artifact of our *in vitro* testing systems.

In this chapter, we present data on MAF with regard to suppression, focusing on two aspects: the mechanisms of suppression of certain *in vitro* immune responses and the question of whether the inhibitory factor(s) in MAF is able to suppress *in vivo* immune responses.

RESULTS AND DISCUSSION

Results obtained regarding the effect of amniotic fluid on a T-dependent lymph node proliferative assay are referred to in Table 14-1. The experimental protocol for this assay has been reported in detail (Suzuki and Tomasi, 1979). Briefly, regional lymph nodes (inguinal and para-aortic) from mice are sensitized with specific antigens by subcutaneous injection of Freund's complete adjuvant, removed eight days later, and re-exposed *in vitro* to optimal concentrations of the specific antigen. After four days in culture, proliferation is determined by tritiated thymidine incorporation. We have shown with certain antigens, such as ovalbumin (OVA), not only that this assay is T cell dependent, but also that the vast majority of proliferating cells are T cells. Previous data (see Suzuki and Tomasi, 1980) have shown that low concentrations of amniotic fluid may actually stimulate a proliferative reaction but that, as the concentration is increased, suppression occurs. Isolated AFP also shows the same biphasic effect.

Using the lymph node proliferative assay, we investigated the question of whether inhibition by macrophage activation factor (MAF) was mediated via antigen-presenting cells contained in macrophage (Mϕ) populations. In these experiments, we pulsed Mϕ for two hours at 37°C with antigen and used the pulsed adherent cells to present antigen to sensitized lymph node cells (SLNC) or T cells isolated from sensitized lymph node cells. The effect of preculturing with MAF either before or after antigen pulsing was tested. The data in Figure 14-1 show that MAF suppresses the presentation of OVA by Mϕ. Also, incubation of adherent peritoneal exudate cells (PECs)

TABLE 14-1
Inhibition by Three Different MAF Samples of the Lymph Node Proliferative Response Induced by Torpedo Acetylcholine Receptor (2.5 μg/ml)

ADDITION TO CULTURE	CPM	% INHIBITION
Control	32 857	0
MAF 6-30	1446	95.6
MAF 10-23	2580	91.6
MAF 12-10	7064	78.5

307

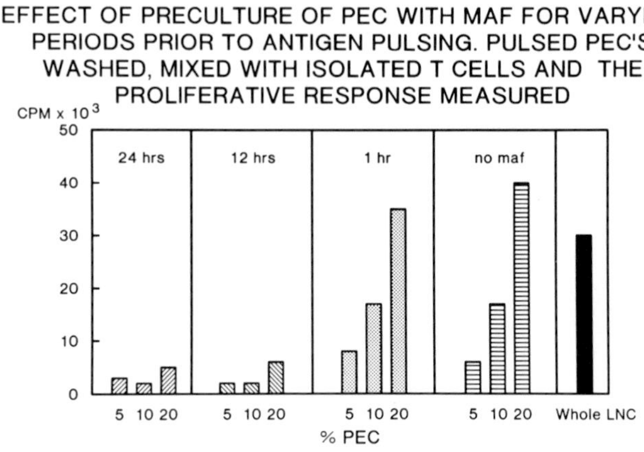

EFFECT OF PRECULTURE OF PEC WITH MAF FOR VARYING
PERIODS PRIOR TO ANTIGEN PULSING. PULSED PEC'S
WASHED, MIXED WITH ISOLATED T CELLS AND THE
PROLIFERATIVE RESPONSE MEASURED

FIGURE 14-1. Inhibition by three different MAF samples of the proliferation induced by Torpedo acetylcholine receptor (2.5 μg/ml).
For methods, see Suzuki and Tomasi, 1980.

for 24 hours in MAF *after* OVA-pulsing inhibited T cell proliferation, whereas incubation of the T cells from SLNC with MAF had no significant effect. Other experiments (data not shown) demonstrated that MAF did not inhibit the uptake of antigen by Mϕ. Moreover, when spleen cells are depleted of Mϕ and the response to PHA is partially maintained by the addition of β-mercaptoethanol (2-ME), MAF had no significant suppressive effect. On the other hand, the addition of PEC's to such cultures restored the ability of MAF to suppress. Therefore, we feel that a major, although perhaps not the sole, site of action of the suppressive factor(s) in MAF is at the level of the antigen-presenting cell. Since MAF was shown not to inhibit the uptake of antigen (Suzuki and Tomasi, 1980), it presumably inhibits processing of antigen or T-Mϕ interaction or the secretion of soluble factors by Mϕ or some combination of the three possibilities. MAF does not inhibit plasminogen activator secretion by Mϕ, which is induced by LPS or a T cell lymphokine (M. Tiku and T.B. Tomasi, unpublished observations). Whether suppression is a result of a direct action of MAF and AFP on the antigen-presenting cells (perhaps by affecting Ia antigens) or of indirect action via another cell type present in the PEC preparations was not established in this study.

   An important question that has been raised by the *in vitro* studies is whether one or more of the defects seen in the immune system of fetal, neonatal, and pregnant animals might be attributable to a soluble suppressive factor. Studies from other laboratories have suggested the possibility that the relative predominance of suppressor cells, characteristic of newborn mice, may possibly be due to the high concentration of AFP exerting an influence on the differentiation of precursor cells into active T suppressor cells (Murgita et al.,

1978). In view of the macrophage data reported here, we would postulate that some of the reported defects in Mϕ function in the fetus and newborn might be based on the effect of circulating factors, such as AFP. It is of interest that neonatal macrophages have been reported to be relatively ineffective in presenting antigens (Lu, 1979), and this is consistent with the data presented above. In order to test this further, we used a GAT-specific T cell clone and presented antigens with spleen cells from syngeneic (B6AF$_1$) mice of various ages. As shown in Table 14-2, one-day-old spleen cells are able to present antigens, although not as efficiently as adult cells, and there is a graded increase (with age) in the antigen-presenting capability of neonatal spleen. The origin of the T cell clones and techniques involved in the presentation experiments were reported by Kimoto and Fathman (1980). In other experiments (Tomasi et al., to be published) we have shown the following.

1.   Presentation of GAT by whole spleen cells approaches adult levels at around three to four weeks of age.
2.   The antigen-presenting cell is concentrated in the adherent population.
3.   Flotation (on BSA density of 1.081) of whole spleen cells, resulting in a separation into "floaters" (containing 10 per cent of the spleen cells) and "sinkers" demonstrates that the antigen-presenting cells are contained primarily within the floating population.
4.   Since antigen presentation in the neonate actually increases with graded decreases in cell numbers from 10^6 to 10^5, it seems likely that there is a suppressor population in the neonatal spleen.
5.   MLR-stimulating cells can be detected in two-day-old neonatal spleens through the use of cloned, alloreactive T cells. The development of the MLR-stimulating capacity follows a time course similar to that of antigen presentation, i.e., gradually reaches adult levels around three to four weeks of age. This may be of some significance since it has been speculated that the major MLR-stimulating cell and the accessory cell for antigen presentation may be one and the same cell (for review, see Steinman, 1981).
6.   Cells with dendritic morphology are present in low numbers in the newborn and gradually increase as immune reactivity appears. For example, if the floating population is adhered for two hours, approximately 35 to 40 per cent of the cells are dendritic cells (DC) in the adult spleen, whereas three-day neonatal spleens contain approximately 4 per cent DC. By 22 days, about 18 per cent of the adherent floaters are dendritic (Barr and Tomasi, to be published).

It is possible to reconcile the observations between laboratories regarding the cell type involved in MAF- and AFP-induced suppression if one assumes that, in the absence of effective presentation, free antigen directly

TABLE 14-2
GAT Presentation by Neonate Spleen Cells*

CELLS	MEDIA	GAT-PULSED	GAT PRESENT THROUGHOUT
Adult	289 ± 50	5177 ± 172	12 171 ± 569
1-Day	—	484 ± 75	1921 ± 614
4-Day	459 ± 35	467 ± 37	2483 ± 80
8-Day	419 ± 31	807 ± 94	2786 ± 304
14-Day	441 ± 16	1038 ± 39	5563 ± 227

* GAT-specific T cell clones (10^4/well) cultured with 1 x 10^6 irradiated (3200 R)
presenting cells from syngeneic (B6A)F_1 adult or neonatal spleen cells of various
ages. Spleen cells were either pulsed with GAT for 2 hours (GAT-pulsed) followed by washing (x3) or GAT was present throughout the three day culture
period. All results expressed as mean cpm ± SD.
Data from T.B. Tomasi, M. Kimoto, C.G. Fathman, and D. Czerwinski.

stimulates the generation of T suppressor cells. However, we have not been able to confirm the presence of suppressor cells of T lineage after incubating spleen cells *in vitro* with AFP, as reported by others (Murgita et al., 1977), although further studies are currently in progress in an attempt to clarify this issue.

Macrophages are one of the cell types thought to be involved in the production of tumor necrotic factor (TNF) by LPS in mice bearing certain tumors (Hoffmann et al., 1978; Berendt and North, 1980). We reasoned that, in view of the apparent functional deficiency of Mϕ (either immaturity and/or decreased numbers) in neonates, they may show a defect in their ability to reject tumors after the administration of LPS *in vivo*. Table 14-3 shows that neonatal B6AF$_1$ mice fail to reject Sarcoma I cells after 10 μg of LPS injected intravenously until they are about three weeks of age. Similar data have been obtained with the Meth A tumor. This time course is of some interest since, as discussed above, the antigen-presenting cells and the MLR-stimulating cells develop over a similar time span. Thus the failure to reject tumor may be due either to a deficiency in the cell (possibly Mϕ) producing TNF or to the inability of the animal to mount an effective immune response (T cell-mediated immunity), which has also been shown to be a requirement for tumor rejection (Berendt and North, 1980) and TNF production (Hoffmann et al., 1978). Further work on the origin of the defect in neonatal mice is in progress.

During the course of these investigations on the induction of TNF by LPS in neonatal animals, we noted an interesting phenomenon, although its biological significance is presently unclear. As shown in Table 14-4, when spleen cells taken at various periods following birth were examined for proliferative capacity in the presence of MAF, the neonatal cells were markedly stimulated up to about 20 days of age. These same MAF samples suppressed the response of adult cells (i.e., LPS, PHA, MLR) and did not induce proliferation of adult control cultures. Several fresh MAF samples containing no detectable

TABLE 14-3
Endotoxin-Induced Regression of Sarcoma I Tumors in Neonatal Mice

DAY AFTER BIRTH TUMOR GIVEN	DAY LPS GIVEN	TUMOR REJECTED/ NUMBER OF MICE
5	12	0/10
9	16	0/11
12	19	0/3
15	22	5/29
22	30	3/5
32	39	4/4
Adults (8–12 weeks)	7 days after tumor	30/33

Neonatal (B6A)F₁ mice injected with 1 x 10⁶ Sarcoma I cells at various ages after birth. LPS (10 μg) given i.v. 7 days later at which time tumor has reached a diameter of approximately 8 mm. Rejection defined as complete regression of tumor (usually within 7 to 14 days) with no reappearance of tumor after two months of observation.

LPS by the limulus assay were used with similar results. Stimulation of neonatal spleen cells was profound, and these cells proliferated to essentially the same magnitude as LPS-driven cultures. It appears therefore that neonatal cells are quite different from adult cells in their reactivity to MAF when cultured in 2.5 per cent normal human serum. A possibility being explored is that there are two factors in MAF, one stimulatory, the other suppressive, and that neonatal cells are relatively more reactive to the stimulatory factor. The proliferation is sensitive to radiation (800 r). Further characteristics of the proliferating cell(s) are currently being examined.

Whether the stimulatory effects of MAF observed at low concentrations on adult cells and in the neonatal period are related is unclear. It should also be noted that stimulatory effects have been reported for the PHA response

TABLE 14-4
Effect of Murine Amniotic Fluid (MAF) on the Proliferative Response of Neonatal Mouse Spleen Cells In Vitro*

SPLEEN AGE	PBS	MAF
1-Day	2121 ± 62	17 798 ± 709
7-Day	3231 ± 325	13 228 ± 439
11-Day	3442 ± 231	14 200 ± 1328
20-Day	970 ± 17	2422 ± 149
30-Day	659 ± 10	1559 ± 85
42-Day	853 ± 39	1674 ± 102

Mean proliferative response (cpm ± SD) in (B6A)F₁ mice (6-8 animals per group) in media + PBS, and in media + 200 μg/ml murine amniotic fluid (MAF).

with human AFP (Charpentier et al., 1977; Soubiran et al., 1979) and for certain murine allogeneic reactions (Krco et al., 1979). These latter studies show that, whereas allogeneic reactions mediated by I region differences are markedly suppressed by MAF and AFP, non-MHC MLR's are either unaffected or frequently stimulated. These results could conceivably be due to the "unopposed" activity of one or more stimulatory and/or growth-promoting substances in amniotic fluid. The murine allogeneic data also clearly show that suppression by MAF is not merely nonspecific inhibition of proliferative reactions in general, since genetic restrictions occur in inhibition by AFP, as first shown by Peck et al. (1978) (see Chapter 15).

Finally, a key question remains as to whether immunosuppression by MAF and AFP has any biological significance or whether it represents an *in vitro* artifact of little or no biological importance. Suggestive evidence for an *in vivo* effect was first obtained in our laboratory when we administered whole amniotic fluid to mice from birth to early adulthood and showed inhibition of the immune response to SRBC in young adults (Ogra et al., 1974). We have recently demonstrated that administration of high concentrations of MAF to adult animals will also suppress the appearance of direct PFC's to SRBC. The levels of AFP obtained in these adult mice receiving MAF are slightly higher than those seen in the pregnant animal, but only about one tenth of that generally found in the neonate. Suppression can be seen at several doses of SRBC, but is most evident when suboptimal doses are given, as shown in Table 14-5. It is interesting that supraoptimal doses of antigen (SRBC), which give approximately the same PFC response as suboptimal doses, are not as vigorously suppressed by MAF. Recent studies (Olinescu et al., 1978) have also shown *in vivo* suppression of the SRBC response by murine MAF and AFP, and these workers presented some evidence that this is due to an effect on Mϕ. However, these authors also found diminished capacity for phagocytosis by Mϕ cultured in the presence of MAF or AFP, an observation that we have not been able to consistently reproduce.

Studies by Gershwin et al. (1978) have demonstrated that isolated AFP but not albumin or transferrin (the other major constituents of MAF) administered *in vivo* accelerates tumor growth, delays regression, increases mortality, and lowers the threshold dose of virus necessary to induce tumors by the Maloney sarcoma virus. These authors also reported (Gershwin et al., 1980) accelerated plasmacytoma formation in AFP-treated mice. In addition, *in vivo* treatment with AFP appears to alter the appearance of thymic-dependent autoantibodies (such an anti-erythrocyte antibodies) in NZB mice (Gershwin et al., 1979). Our own studies in tumor systems (Tomasi et al., 1975) demonstrate that the intravenous injection of MAF into adult mice does promote the appearance of tumors when small cell inocula are used, and also increases the rate of tumor growth as well as mortality. Since purified AFP was

TABLE 14-5
In Vivo Suppression of SRBC PFC by Murine Amniotic Fluid

DOSE SRBC	PFC/10^6 Spleen Cells Mean ± SE		% SUPPRESSION PFC	% SUPPRESSION OF HA	SERUM AFP LEVELS DAY 5 (ng/ml)
	PBS	MAF			
Suboptimal (1 x 10^7)	106 ± 16	14 ± 6	87	96	190 200
Optimal (5 x 10^8)	662 ± 64	432 ± 115	35	40	168 000
Supraoptimal (1 x 10^{10})	147 ± 23	123 ± 18	17	—	217 500

One ml (2 OD units) of sterile endotoxin-negative MAF or NMS administered i.v. to CBA mice for three days before and four days after SRBC injection. PFC response (expressed as mean ± SE) and serum hemagglutinin (HA) titers on day five. Per cent suppression of response in MAF treated mice compared with those given PBS. Serum AFP levels determined by radioimmunoassay on day five. Statistically; suboptimal, $P < 0.01$; optimal, $P < 0.05$; supraoptimal, $P = $ not significant.

not used in these studies, no conclusions can be drawn regarding the nature of the tumor growth-promoting factor.

Recent work on experimental myasthenia gravis (Brenner et al., 1980) reports that AFP preparations inhibit the development of this syndrome. The authors propose that such inhibitory effects may explain remissions in myasthenia gravis during pregnancy and the rarity of neonatal myasthenia despite the placental passage of the IgG antiacetylcholine receptor antibody. In this regard, several authors have suggested that the so-called immunosuppression of pregnancy, as well as the inhibitory capacity of pregnancy sera on certain immune reactions, results from the elevations in AFP levels that accompany pregnancy (Murgita, 1976; Toder et al., 1979). However, the data concerning myasthenia are not entirely convincing since the AFP preparations used were impure and the mechanism of action proposed was the direct inhibition of antigen-antibody interactions by AFP. It is difficult to visualize how AFP would inhibit such interactions, and further work is certainly necessary in this area.

ACKNOWLEDGMENTS

This work was supported in part by National Institutes of Health grants AM-17554 and HD-09720. The Sarcoma I and Meth A tumors were kindly supplied by Robert North.

REFERENCES

Alpert, E., J.L. Dienstag, S. Sepersky, B. Littman, and R. Rocklin. 1978. Immunosuppressive characteristics of human AFP. Effect on tests of cell mediated immunity and induction of human suppressor cells. *Immunol. Comm.* 7:163.

Auer, I.O., and H.G. Kress. 1977. Suppression of the primary cell mediated immune response by human α-fetoprotein *in vitro. Cell. Immunol.* 30:172.

Berendt, M.J., and R.J. North. 1980. T-cell-mediated suppression of anti-tumor immunity. *J. Exp. Med.* 151:69.

Brenner, T., Y. Beyth, and O. Abramsky. 1980. Inhibitory effect of alpha-fetoprotein on the binding of myasthenia gravis antibody to acetylcholine receptor. *Proc. Nat. Acad. Sci. USA* 77:3635.

Charpentier, B., R.D. Guttmann, J. Shuster, and P. Gold. 1977. Augmentation of proliferation of human mixed lymphocyte culture by human alpha-fetoprotein. *J. Immunol.* 119:897.

Etlinger, H.M., and J.M. Chiller. 1977. Suppression of immunological activities by mouse amniotic fluid. *Scand. J. Immunol.* 6:1241.

Gershwin, M.E., J.J. Castles, A. Ahmed, and R. Makishima. 1978. The influence of alpha-fetoprotein on Moloney sarcoma virus oncogenesis. Evidence for generation of antigen nonspecific suppressor T cells. *J. Immunol.* 121:2292.

Gershwin, M.E., J.J. Castles, and R. Makishima. 1979. Alpha-fetoprotein associated alterations of New Zealand mouse immunopathology. *Immunopharmacol.* 1:331.

Gershwin, M.E., J.J. Castles, and R. Makishima. 1980. Accelerated plasmacytoma formation in mice treated with alpha-fetoprotein. *J. Nat. Cancer Inst.* 64:145.

Gupta, S., and R. Good. 1977. Inhibition of human lymphocyte transformation by human alpha-fetoprotein (HAFP): HAFP monomers and multimers and a resistant lymphocyte subpopulation. *J. Immunol.* 119:555.

Hoffmann, M.K., H.F. Oettgen, L.J. Old, R.S. Mittler, and U. Hämmerling. 1978. T-cell-mediated suppression of anti-tumor immunity. *J. Reticuloend. Soc.* 23:307.

Kimoto, M., and C.G. Fathman. 1980. Antigen-reactive T cell clones. *J. Exp. Med.* 152:759.

Krco, C.J., E. Johnson, C.S. David, and T.B. Tomasi. 1979. Differences in the susceptibility of MHC and non-MHC mixed lymphocyte reactions to suppression by murine amniotic fluid and its components. *J. Immunogenet.* 6:439.

Lu, C.Y. 1979. Neonatal murine macrophages are deficient in an H-2 gene regulated antigen presentation function. *Fed. Proc.* 38:1360.

Lu, C.Y., D.I. Beller, and E.R. Unanue. 1980. During ontogeny, Ia-bearing accessory cells are found early in the thymus but late in the spleen. *Proc. Nat. Acad. Sci. USA* 77:1597.

Murgita, R.A. 1976. The immunosuppressive role of alpha-fetoprotein during pregnancy. *Scand. J. Immunol.* 5:1003.

Murgita, R.A., E.A. Goidl, S. Kontiainen, and H. Wigzell. 1977. Alpha-fetoprotein induces suppressor T cells *in vitro. Nature* 267:257.

Murgita, R.A., S. Goidl, P.C. Beverley, S. Kontiainen, and H. Wigzell. 1978. Adult murine T cells activated *in vitro* by alpha-fetoprotein and naturally occurring T cells in newborn mice: identity in function and cell surface differentiation antigens. *Proc. Nat. Acad. Sci. USA* 75:2897.

Ogra, S.S., R.A. Murgita, and T.B. Tomasi. 1974. Immunosuppressive activity of mouse amniotic fluid. *Immunol. Comm.* 3:497.

Olinescu, A., D.E. Popescu, S. Babes, D. Ganea, and A. Dumitrescu. 1978. The effect of alpha-fetoprotein on the immune response. In *Carcinoembryonic Proteins.* B. Norgaard-Pedersen and N. Axelsen, eds. *Scand. J. Immunol.* Supplement. 8:391-401.

Peck, A.B., R.A. Murgita, and H. Wigzell. 1978. Cellular and genetic restrictions in the immunoregulatoy activity of alpha-fetoprotein. *J. Exp. Med.* 147:667.

Sheppard, H.W., S. Sell, P. Trefts, and R. Bahu. 1977a. Effects of alpha-fetoprotein on murine immune responses. I. Studies on mice. *J. Immunol.* 119:91.

Sheppard, H.W., S. Sell, P. Trefts, and R. Bahu. 1977b. Effects of alpha-fetoprotein on murine immune responses. II. Studies on Rats. *J. Immunol.* 119:98.

Soubiran, P., A.M. Mucchielli, J.-P. Kerckaert, B. Bayard, and R. Masseyeff. 1979. Stimulatory effect of human alpha fetroprotein and its molecular variants on *in vitro*-induced lymphocyte blastogenesis. *Scand. J. Immunol.* 10:179.

Steinman, R.M. 1981. Dendritic cells. *Transplantation* 31:151.

Suzuki, K., and T.B. Tomasi. 1979. Inhibition of antigen-induced lymph node cell proliferation by murine amniotic fluid and its components. *Immunology* 38:539.

Suzuki, K., and T.S. Tomasi. 1980. Mechanism of immune suppression by murine neonatal fluids. *J. Immunol.* 125:1806.

Toder, V., L. Blank, and L. Niebel. 1979. Immunosuppressive effect of alpha-fetoprotein at different stages of pregnancy in mice. *Develop. and Comparat. Immunol.* 3:537.

Tomasi, T.B. 1977. Serum factors which suppress the immune response. In *Regulatory Mechanisms in Lymphocyte Activation.* D.O. Lucas, ed. New York: Academic Press.

Tomasi, T.B. 1978. Suppressive factors in amniotic fluid and newborn serum. Is alpha-fetoprotein involved? *Cell. Immunol.* 37:459.

Tomasi, T.B., R.J. Dattwyler, R. A. Murgita, and R.H. Keller. 1975. Immunosuppression by alpha-fetoprotein. *Trans. Assoc. Am. Phys.* 88:292.

Tomasi, T.B., R.A. Murgita, R.L. 1977. Alpha-fetoprotein and the immune response during murine pregnancy and neonatal development. In *Development of Host Defenses.* D.H. Dayton, ed. New York: Raven Press, pp. 273-285.

Yachnin, S., and E.P. Lester. 1976. Inhibition of human lymphocyte transformation by human alpha-fetoprotein (HAFP): comparison of fetal and hepatoma HAFP and kinetic studies of *in vitro* immunosuppression. *Clin. Exp. Immunol.* 26:484.

Yachnin, S., and E.P. Lester. 1977. Inhibition of human lymphocyte transformation by human alpha-fetoprotein (HAFP): HAFP monomers and multimers and a resistant lymphocyte subpopulation. *J. Immunol.* 119:555.

Chapter 15

IMMUNOSUPPRESSIVE ELEMENTS IN THE FETAL AND NEONATAL ENVIRONMENT

R.A. MURGITA
A.B. PECK
HANS WIGZELL

The riddle of foreign tissue survival, such as in the case of the mammalian fetus, has long intrigued immunologists. In the present chapter, we would like to emphasize some of our own findings, together with those of others, on immunosuppressive elements found in the embryo and/or neonate. We have no intention of covering in detail every aspect of this highly complex field. For this reason, references to more complete discussions of specific topics are provided for the reader interested in studying related areas in more detail.

The major emphasis of this chapter is on the embryonic and neonatal periods of the murine system, providing comparisons to the human whenever data are available. Interest is focused on three major areas:

1. the existence of suppressor (or suppressor-inducing) cells in the fetal and the neonatal mouse;
2. the selective action of alpha-fetoprotein that inhibits certain T cell subsets from functioning, yet permits other T cell subsets to become activated when confronted with immunogen; and
3. the regulation of natural killer cells in the newborn in relation to the finding that such cells seemingly are able to react in a preferential manner against the "embryonic" stages of normal cells.

CELLS IN THE NEONATE WITH INHIBITORY CAPACITY FOR HUMORAL OR CELL-MEDIATED IMMUNE RESPONSES

Mice acquire the ability to produce antibodies to most T cell-independent antigens early in postnatal development (Mosier and Cohen, 1975; Mosier and Johnson, 1975; Rabinowitz, 1976; Hardy et al., 1976; Mosier et al., 1979; McKearn and Zuintano, 1979; Dekruyff et al., 1980). Responsiveness to T cell dependent antigens appears later (Takeya and Nomoto, 1967; Fidler et al., 1972; Spear and Edelman, 1974; Hardy et al., 1976; Rabinowitz, 1976). The initial interpretation was that B cells mature earlier during ontogeny than do T cells. Adult lymph node cells added to antigen-stimulated neonatal spleen cell cultures failed, however, to restore the poor responsiveness of newborn spleen. Removal of T cells from neonatal spleen with anti-Thy 1 and complement, however, allowed neonatal B cells to respond *in vitro* when adult T cells were added. Furthermore, small numbers of neonatal spleen T cells suppressed antibody responses of adult spleen cells to T-dependent (TD), as well as to T-independent (TI), antigens. It was concluded that poor antibody responsiveness in newborn animals is due, at least in part, to excess suppressor T cells interfering with helper T cells and perhaps also negatively influencing T-independent B cell functions (Mosier and Johnson, 1975). Subsequent studies showed that neonatal thymocytes constitute a rich source of suppressor cells for adult antibody synthesis (Mosier et al., 1977). It was inferred in these studies that suppressor cells residing within outer cortical thymocytes emigrate to the neonatal spleen, where similar immunosuppressive activity was detectable.

The thymic origin of nonspecific suppressor cells appearing early in ontogeny was further substantiated by studies showing that unprimed murine fetal thymocytes suppress T-dependent primary antibody responses of adult syngeneic spleen cells (Luckenbach et al., 1978). In the rat, antibody responsiveness during the neonatal period is regulated by a thymus-derived T cell suppressor mechanism (McCullagh, 1975a, 1975b). The impaired capacity of newborn rats to mount an anti-SRBC response can be restored by adoptive transfer of adult thoracic duct lymphocytes if the neonatal recipients have been thymectomized (McCullagh, 1975a) or exposed to low-dose X-irradiation (McCullagh, 1975b).

Little information exists concerning the mechanism, or mechanisms, of action of fetal and neonatal cells which inhibit antibody synthesis. Neonatal mice have a well-developed feedback inhibitory circuit (Murgita et al., 1978; Ptak et al., 1979). Lymphocytes from neonatal mice suppress antibody responses of adult lymphocytes to phosphorylcholine via auto-anti-idiotype suppression (Strayer and Köhler, 1976). The question of whether the primary target for neonatal T suppressors is a B cell or another T cell has not been resolved. Some suggest (Mosier et al., 1977) that neonatal T cells can exert a negative regulatory influence directly on B cells responding to the T cell-independent antigen TNP-Ficoll. Neonatal T cells have also been shown to suppress other T-independent antigens (Morese et al., 1976; Hardy and Mozes, 1978; Dekruyff et al., 1980). However, T-dependent antibody synthesis is much more sensitive to suppression by fetal or neonatal cells than is T-independent synthesis (Murgita et al., 1978; Luckenbach et al., 1978). Fetal thymocytes do suppress the *in vitro* antibody response of adult spleen against SRBC when present in cell ratios of 1:5 down to 1:500, whereas no suppression was observed on the response to the TI antigen NIP-POL even at a cell ratio of 1:5.

We have analyzed the immunoregulatory effect of purified spleen T lymphocytes from newborn mice on *in vitro* antibody synthesis to the thymus-dependent antigens SRBC and DNP-KLH and to the thymus- independent antigens DNP-Ficoll and DNP-POL (Murgita et al., 1978). The addition of newborn spleen T cells to yield as little as 0.05 per cent of the total cell population in the assay culture caused strong inhibition of the antibody synthesis to the two TD antigens, SRBC and DNP-KLH. Significant inhibition of TI antibody synthesis was observed in two out of seven experiments, then requiring a ratio of newborn T to adult spleen cells that was 1:20 or higher. It is notable that the suppression of TI responses by newborn lymphocytes *in vitro* reported by others (Mosier et al., 1977; Dekruyff et al., 1980) occurred only at ratios of newborn to adult cells higher than 1:5. It is our experience that even normal adult T cells begin to exert inhibiting effects when added in excess of 20 per cent of the total cell population in the present culture system. Nevertheless, newborn T cells, or an as yet undefined contaminating cell subpopulation, may exert weak inhibitory effects on TI antibody responses. The prevailing inhibitory effect mediated by newborn T lymphocytes on antibody synthesis *in vitro* is accordingly directed toward TD responses. This could mean that newborn inhibitory T cells directly suppress TD B cells more efficiently than they do TI B cells (Lewis et al., 1976). Alternatively, helper T cells required for TD antibody synthesis could be targets for suppression. Our preliminary evidence tends to favor the latter possibility. Thus, newborn T cells have been shown to exert helper effects for TD antibody synthesis when added to adult T cell-depleted spleen B cell cultures, while causing strong suppression of antibody production in parallel cultures of whole adult spleen cells (Stegagno et al., unpublished observations).

Negative selection experiments were performed to determine the cell surface antigen phenotype of the newborn inhibitory T cell. Rabbit antimouse brain (anti-T), as well as anti-Thy 1, anti-Lyt 1, and anti-Lyt 2 sera plus rabbit complement were used. Following selective lysis with these antisera, the remaining cells were tested for their ability to induce suppression of TD anti-SRBC resposes. The ability to cause suppression was eliminated by cytotoxic pretreatment of newborn spleen T cells from CBA mice with anti-T, anti-Thy 1, and anti-Lyt 1, but not anti-Lyt 2, sera (Murgita et al., 1978). Thus, surprisingly, lymphocytes capable of inhibiting TD antibody responses and residing in the spleen of newborn mice are T cells with an Lyt 1^+2^-phenotype (Murgita et al., 1978; Goidl et al., 1979). We further analyzed the Ia antigen phenotype of these newborn inhibitory T cells, as such markers have been found to be of importance in the delineation of various T cell subsets in other systems (Tada et al., 1978). Using spleen T cells from two- to four-day-old mice, we could show that the inhibitory activity of newborn T cells is effectively abrogated by anti-I-J^k sera (Hooper et al.,1979).

The presence of an I-J marker is known to be associated with T cells possessing suppressor properties. However, expression of Lyt 1^+2^- phenotype on newborn inhibitory T cells was incompatible with conventional suppressor effector T cells, which are Lyt 1^-2^+ in phenotype. It is possible that the Lyt 1^+2^- newborn inhibitory T cells represent "suppressor inducer" cells, rather than nonconventional effector suppressor cells. This premise is also supported by evidence (Ptak et al., 1979) that newborn mice contain considerable "latent" helper cells if overriding suppressor mechanisms are deliberately interrupted. Moreover, this study revealed distinct similarities between neonatal helper cells and a subpopulation of adult helper T cells which at the same time had powerful suppressor-inducing abilities. These findings may indicate that the neonatal Lyt 1^+2^- I-J$^+$ regulatory T cells are comparable to certain adult Lyt 1 cells which are efficient inducers of feedback inhibition (Cantor and Gershon, 1979). It is interesting here to note that whereas young NZB mice show an unusually early appearance of immunocompetence (Evans et al., 1968; Playfair, 1978) which may be associated with an absent Lyt 123$^+$ cell feedback suppression circuit (Cantor et al., 1978), newborn NZB spleen cells still retain active inhibitory cell activity as measured *in vitro* (Lebman and Calkins, 1980).

In conclusion, substantial evidence exist for the presence in newborn mice of an unusual and active population of T cells with efficient ability to suppress T cell-dependent antibody synthesis *in vitro*. The reason or reasons for the presence of active inhibitory cells in the neonate is still a matter of speculation. In addition, we feel that the presence of other non-T suppressor cells has not been fully excluded by available evidence published so far.

As for inhibitory cells of cell-mediated immune reactions, lymphocytes from cord blood of humans are known to inhibit division of the mother's lymphocytes *in vitro* (Olding and Oldstone, 1974; Olding et al., 1974). Examination of mother-baby mixed cultures for occurrence of male and female metaphases, using the fluorescent Y chromosome technique, revealed that the lymphocytes in mitosis were of the newborn type (Olding and Oldstone, 1974; Olding et al., 1974). Likewise, strong mitotic inhibition occurred when lymphocytes from newborn human males were mixed with lymphocytes from nonrelated, recently delivered mothers or from nonrelated, nonpregnant women. Others (Lawler et al., 1975) reported a similar inhibitory effect by male newborn lymphocytes on parental cells in two-way mixed cultures. Cell fractionation showed the suppressor activity in cord blood to be associated with the T lymphocyte population (Olding and Oldstone, 1976).

Further studies showed that PHA-stimulated cells from newborns could suppress the proliferation of lymphocytes from recently delivered mothers even when separated by a dialysis membrane (Olding et al., 1977). PHA-stimulated lymphocytes from nonrelated adults did not inhibit division of maternal lymphocytes when separated in double chamber cultures. Lymphocytes from one newborn failed to inhibit the proliferation of lymphocytes from another newborn. This indicated that proliferating newborn lymphocytes induce mitosis inhibition of adult cells via a low-molecular-weight factor or factors. Subsequent studies have confirmed that cord blood cells secrete factors that inhibit mitogen- and alloantigen-induced proliferation of adult cells

(Williams and Korsmeyer, 1978). The inhibitory cells in cord blood were characterized as T lymphocytes with Fc receptors for IgG (T_G) (Oldstone et al., 1977). This is consistent with findings that adult T lymphocytes with IgG receptors are efficient suppressors (Moretta et al., 1977). The mean proportion of T_G cells in cord blood is about three times the adult mean (Oldstone et al., 1977), but more recent studies (Durandy et al., 1979) indicate that the suppressor pathways in the human newborn may be more complex than just a matter of quantities. It is thus plausible that one of the inhibitory T cell subsets present in cord blood is analogous to the Lyt 1^+ suppressor inducer cell in the murine feedback inhibition circuit (Cantor and Gershon, 1979).

The embryonic liver is an early site of immunocompetence during murine ontogeny. Does the potential to suppress immune reactivity arise concomitantly with the appearance of immunocompetence in the liver? Mouse embryonic or neonatal liver cells strongly suppressed adult mixed lymphocyte reactivity and generation of cell-mediated lysis (Globerson et al., 1975). Suppression occurred regardless of whether stimulator cells were syngeneic or allogeneic to the liver donor. Newborn liver cells also suppressed GVH-induced mortality in sublethally irradiated F_1 mice challenged with parental spleen cells mixed with parental liver cells (Globerson and Umiel, 1978). Here, interference with GVH responses was seen only in combinations where injected spleen cells were syngeneic to the liver cells. Mouse embryonic liver suppressor cells belong to a subpopulation of cells bearing receptors for the lectin peanut agglutinin (Rabinovich et al., 1979).

Fetal spleen, but not thymus, cells reduce the ability of parental adult spleen cells to cause local GVH reactions in F_1 mice (Skowron-Cendrzak and Ptak, 1976). Suppression was observed with spleen cells from mice less than 24 hours old, but disappeared completely by five days of age. There was no requirement for histocompatibility between reacting and suppressor cells. Inhibitory activity was eliminated by pretreatment with anti-Thy 1 serum (Skowron-Cendrzak and Ptak, 1976). Further studies (Ptak and Skowron-Cendrzak, 1977) showed that newborn mice sensitized to picryl chloride (PCL) within 24 hours of birth fail to develop contact sensitivity when tested four weeks later,in contrast to animals sensitized at two days of age. Thus, painting with PCL at birth may generate specific suppressor cells which after four weeks still interfere with the ability of the animals to respond to the original skin sensitizer (Ptak and Skowron-Cendrzak, 1977). Likewise, spleen lymphocytes of fetal or one-day-old mice suppress passive transfer of contact sensitivity if injected along with sensitized lymphocytes. Moreover, fetal spleen cells were able to suppress the local GVH reaction elicited by immunized parental cells in F_1 recipients and to reduce the severity and mortality rate of GVH in cyclophosphamide-treated F_1 recipients. Thus, antigen given in the perinatal period may not cause a specific sensitization, but rather may trigger efficient, specific suppressor mechanisms.

In other experiments, stimulation of adult parental lymphocytes by newborn F_1 spleen lymphocytes was much lower than with adult F_1 cells (Bassett et al., 1977). This was not due merely to inadequate expression of MLR-activating antigens on the surface of newborn cells. Thus, mixtures of newborn and adult F_1 stimulator cells resulted in a reduction of proliferation of adult parental cells comparable to that observed with newborn stimulators alone, demonstrating newborn mouse suppressor cells which can inhibit adult responder cells in the MLR. Newborn F_1 spleen cells cultured alone produce a factor capable of suppressing an adult parental anti-F_1 MLR (Bassett et al., 1977). The newborn spleen suppressor activity for the semiallogeneic MLR persisted for the first 2½ weeks after birth and decayed quickly thereafter. Newborn spleen cells which suppress MLR were not able to interfere with the *in vitro* generation of cytotoxic lymphocytes. Thus, newborn suppressor cells may selectively inhibit lymphocyte subpopulations involved in proliferative, but not cytotoxic, reactions. Evidence for the T cell nature of the MLR-inhibitory cell was obtained with heterologous anti-T cell sera (Golub, 1971).

T helper cells present in adult but not newborn spleen are effective in allowing *in vitro* proliferation of low numbers of adult mouse thymus cells to allogeneic spleen cells *in vitro* (Argyris, 1978). Helper activity was absent in six- to eight-day-spleen cells, partially present in 20-day spleen cells, and reached adult levels in the spleen cells of 40-day-old mice. When neonatal and adult spleen cells were mixed, the helper activity for thymus responding cells in the MLR was eliminated. Thus, lack of helper activity in newborn spleen is due to active suppression. As with neonatal suppressors of GVH (Skowron-Cendrzak and Ptak, 1976) and contact sensitivity reactions (Ptak and Skowron-Cendrzak, 1977), the inhibitory cells for MLR were found in the newborn spleen but not in the thymus (Argyris, 1978). Inhibitory activity in neonatal spleen remained after removal of glass-adherent cells but was eliminated by removal of Thy 1^+ cells. Later studies showed that, in contrast to earlier studies (Pavia and Stites, 1979), murine neonatal spleen suppressor cells do inhibit both proliferation and generation of cytotoxic cells in mixed lymphocyte cultures (Argyris, 1979). Suppressor activity was inversely correlated with MLC reactivity in spleen cells of young animals. Thus, spleens from mice up to eight days of age had a low intrinsic MLR capacity but strong suppressor activity. From 11 to 16 days of age, spleens still have relatively low MLR, but they also have low suppressor effects. By 19 days of age, there is adult-like MLR and no suppressor activity is detectable.

A similar parallelism between the maturation of culture-generated suppressor functions in ontogeny and the immune functions they regulate *in vitro* has also been found by others (Rollwagen and Stutman, 1979a, 1979b).

We can conclude that newborn inhibitory cells of both antigen-specific and naturally occurring "nonspecific" types in many systems have been classified as T lymphocytes. Newborn spleens may contain, in addition to T cells, other suppressor cells of non-T type (Rodriquez et al., 1979; Rollwagen and Stutman, 1979; Hooper and Murgita, 1980). Recent studies (Hooper and Murgita, 1980) have defined a newborn spleen inhibitory cell for primary MLR on the basis of:

1. adherence to Ig-anti-Ig columns,
2. nonadherence to plastic dishes or Sephadex G-10,
3. insensitivity to cytotoxic treatment with anti-T cell reagents, and
4. differential agglutination with lectins known to selectively interact with T or non-T lymphocytes.

Other studies have reported on the existence of non-T newborn suppressors with macrophage-like properties (Rollwagen and Stutman, 1979) or with characteristics distinct from classical cell types (Rodriquez et al., 1979).

In conclusion, as seen in the antibody-producing system, there is convincing evidence of efficient inhibitory T cells in the neonatal mouse for various cell-mediated reactions. Whether these inhibitory T cells act or are generated in an independent manner or interact via other cells in the formation of regulatory circuits is unclear.

HUMORAL MEDIATORS OF FETAL AND NEWBORN IMMUNOSUPPRESSION

The number of humoral immunoregulatory factors described in recent years is enormous (for reviews, see Cooperband and Badger, 1979). We shall select and consider literature pertinent to immune regulation in the fetus and newborn. Some emphasis will be on our own contributions to this area, namely, the immunoregulatory role of alpha-fetoprotein (Murgita and Wigzell, 1979).

Fetal calf serum (FCS) is known to be capable of significantly altering lymphocyte reactivity *in vitro* (Mendelsohn et al., 1971; Irie et al., 1974; Forni and Green, 1976; Kerbel and Blakeslee, 1976; Shustic and Cohen, 1976; Sulit et al., 1976; Zielske and Golub, 1976; Opitz et al., 1977; Peck et al., 1977; Tomasi et al., 1977; Kedar et al., 1978; Kedar and Schwartzbach, 1979; Rollwagen and Stutman, 1979). Nonspecific suppressor cell activity is induced by the culturing of mouse lymphocytes in the presence of fetal calf serum for three to five days (Burns et al., 1975; Janeway et al., 1975; Nadler and Hodes, 1977; Kedar et al., 1978; Rollwagen and Stutman, 1979). Such suppressors affect antibody synthesis to thymus-dependent and independent antigens (Burns et al., 1975), mixed lymphocyte reactivity (Nadler and

Hodes, 1977; Rollwagen and Stutman, 1979), and cell mediated cytotoxic responses (Nadler and Hodes, 1977; Kedar et al., 1978). The inhibitory cells appear to be T lymphocytes in many of the systems (Burns et al., 1975; Janeway et al., 1975). Immunosuppressive effects have also been observed *in vitro* in experiments using fetal sera of human (Ayoub and Kasakura, 1971), bovine (Kedar et al., 1978), and murine (Murgita, 1976) origin. Amniotic fluid also has inhibitory properties *in vitro* (Lindahl-Kiessling et al., 1970; Murgita and Wigzell, 1976; Tyan, 1976; Etlinger and Chiller, 1977; Isakov et al., 1978) and *in vivo* (Ogra et al., 1974; Slades and Forrest, 1977; Toder et al., 1978). Mouse amniotic fluid (MAF) suppresses T-dependent and independent antibody responses, in addition to lipopolysaccharide-induced polyclonal antibody synthesis (Murgita and Wigzell,1976; Etlinger and Chiller, 1977). Sera from newborn mice (Jennings and Rittenberg, 1976; Murgita, 1976) and rats (Weiss, 1977) inhibit *in vitro* T cell-dependent antibody synthesis (Murgita, 1976; Labib and Tomasi, 1978), mixed lymphocyte reactivity (Weiss, 1977; Labib and Tomasi, 1978), and mitogen responses (Weiss, 1977; Labib and Tomasi, 1978). It is likely that fetal and newborn sera and amniotic fluids contain several substances capable of altering lymphocyte function *in vitro* in both positive and negative manners. It has been observed (Murgita and Wigzell, 1976; Etlinger and Chiller, 1977) that MAF exerts suppressive effects on both TD and TI antibody responses *in vitro*. However, the component that suppresses TI antibody synthesis is dialyzable, whereas a higher-molecular-weight moiety is responsible for the MAF-mediated inhibitory effect on TD responses.

Fetuin, a major constituent of fetal calf serum, has been reported to have both stimulatory (Puck et al., 1968; Hsu et al., 1973; Gupta et al., 1976) and suppressive effects on lymphocytes *in vitro*. Fetuin has also been shown to stimulate the proliferation of hematopoietic stem cells in mice (Knospe et al., 1971) and to induce blast transformation and DNA synthesis of human peripheral blood lymphocytes in culture (Hsu et al., 1973). It would thus appear that fetuin is yet another serum component possessing significant growth-promoting properties for mammalian cells *in vitro*. Fetuin has also been reported to have immunosuppressive properties (Mendelsohn et al., 1971; Yachnin, 1975). For example, fetuin inhibits the isomitogenic mixture called H-PHA to induced lymphocyte transformation in a manner that can be reversed by an increase in the dosage level of mitogen (Mendelsohn et al., 1971). Since H-PHA can interact with fetuin, it is assumed that fetuin-mediated inhibition of H-PHA responses can be attributed to competitive inhibition of the mitogen. However, fetuin suppresses lymphocyte responses to L-PHA, an isomitogen that does not interact with fetuin (Yachnin, 1975). Other results describing the fetuin-mediated suppression of one-way mixed lymphocyte reactivity (Yachnin, 1975) imply that fetuin exerts a direct inhibitory effect on activated lymphocytes *in vitro*. It is notable, however, that only high concentrations, ranging from 1.25 to 2.50 mg/ml of commercially prepared fetuins, were shown to be suppressive (Yachnin, 1975), whereas more purified fetuin was not significantly inhibitory at or below these concentrations (Tomasi et al., 1977). Furthermore, commercial preparations of fetuin can be shown to be contaminated with AFP, a distinct embryonic substance (Kithier and Poulik, 1972). This would tend to leave some doubt as to whether fetuin itself possesses immunosuppressive properties *in vitro*.

Mammalian alpha-fetoprotein, a glycoprotein of primarily hepatic origin (van Furth and Adinolfi, 1969; Gitlin and Perricelli, 1970; Abelev, 1971), is a major constituent of serum and amniotic fluids during embryonic life (Gitlin, 1975; Masseyeff et al., 1975). In the murine species, it is present in significant quantities in the newborn (Olsson et al., 1977). A number of different molecular subspecies of AFP have been detected by conventional polyacrylamide gel electrophoresis (Gustine and Zimmerman, 1973), extended agarose gel electrophoresis (Lester et al., 1976), and lectin (Bayard and Kerchaert, 1977; Smith and Kelleher, 1980) and estradiol (Uriel and Bouillon, 1975) affinity chromatography.

AFP is a well-studied "oncofetal" substance with regard to physicochemical properties and to normal and pathological distribution. Until recently, little was known about the biological functions of AFP. It may function as a estrogen-binding protein protecting fetal tissues against circulating maternal estrogens (Uriel et al., 1976). However, strong estrogen-binding capacity appears restricted to AFPs of certain animal species (Nunez et al., 1974). It is likely that AFP has important functions in addition to its hormone-binding capacity. We have been exploring the premise that AFP may play an immunoregulatory role during ontogeny. Thus, in the murine system, the physiological postnatal decline of serum AFP levels (Olsson et al., 1977) is seen to correlate with the onset of adult-like immune reactivity.

That highly purified AFP could suppress certain immune functions was initially demonstrated in the murine system (Murgita and Tomasi, 1975a, 1975b). AFP derived from mouse fetuses was shown to exert a potent inhibitory effect on primary and secondary anti-SRBC PFC responses *in vitro* (Murgita and Tomasi, 1975a). Significant inhibition of TD primary antibody synthesis occurred with concentrations of AFP as low as 1.0 $\mu g/ml$. IgA was most sensitive to suppression, IgG being intermediate and IgM responses the least susceptible (Murgita and Tomasi, 1975a). Lymphocyte proliferation induced by mitogens and alloantigens was also shown to be prone to AFP-mediated suppression (Murgita and Tomasi, 1975b). Zimmerman et al. (1977) confirmed that murine AFP from amniotic fluid or fetal plasma suppresses *in vitro* anti-SRBC antibody synthesis. They further showed that the sialic acid in the AFP molecule was necessary for immunosuppression. AFP isolated from hepatoma-bearing mice diminished primary *in vitro* antibody synthesis and reduced secondary antibody responses (Sheppard et al., 1977). Rat AFP of fetal (Sell et al., 1977), but not tumor (Parmely and Thompson,

1976; Sell et al., 1977), origin suppressed certain T cell reactions *in vitro*. It would appear that the suppressive ability of AFP is preferentially on T cell-dependent immune responses. Thus, IgA antibody synthesis, considered highly T dependent, is most sensitive to AFP-induced suppression, and IgM responses, thought to be relatively T-cell independent, are least affected by AFP. We compared the inhibitory effects of AFP on *in vitro* antibody responses to the TD antigens, SRBC and DNP-KLH, and the TI antigens, DNP-POL and DNP-Ficoll (Murgita et al., 1976; Murgita and Wigzell, 1979). AFP suppressed specific antibody responses to TD antigens, but failed to affect TI antigens. Polyclonal B cell antibody synthesis by LPS was also unaffected by AFP (Murgita and Wigzell, 1976). The additional finding (Murgita and Wigzell, 1976) that PHA responses were more suppressed by AFP than were Con A responses offered presumptive evidence for preferential action on certain T cell subpopulations (Stobo and Paul, 1973).

The selective immunoregulatory effects of AFP were emphasized by testing the impact of AFP on T cell proliferation toward defined histocompatibility antigens (Murgita et al., 1978; Peck et al., 1978; Wigzell et al., 1979; The effect of AFP was tested on four major genetic systems known to induce strong MLR activation in the mouse (Peck et al., 1977): the Ia (Class II) antigens, the serologically defined (Class I) products of the major histocompatibility complex (MHC) K or D regions, the Mls locus products, and the products of an as yet undefined non-MHC system associated with a limited number of strains (i.e., DBA/2, B10.D2). Addition of AFP to mixed lymphocyte cultures at initiation almost completely eliminated cell proliferation against Ia and/or M locus structures. Normal numbers of T blasts were observed when stimulations were carried out aginst SD or non-MHC loci determinants (Peck et al., 1978). As Mls is recognized in the context of Ia antigens, AFP may exert its suppressive activity on MLR through a selective interference with I region triggering systems. The genetic relationship between responding and stimulating cells also largely predetermines whether or not AFP will inhibit the generation of CTL's (Murgita et al., 1976; Peck et al., 1978). Thus, effector cells generated with certain stimulator strains including DBA/2 and B10.G, followed by testing for cytolysis on LPS-stimulated blasts (Peck et al., 1978), were found insensitive to AFP-mediated suppression. Also, AFP did exert strong suppressive effects on CTL generation when combinations with genetic differences in individual Class I regions of the MHC were used (Peck et al., 1978). Here, normal numbers of T cells responded by cell division when confronted with Class I antigens in the presence of AFP, but there was no generation of cytolytic T cells (Peck et al., 1978). Since there is evidence for a necessary T-T cell collaboration between Lyt 1 helper cells responding against Class II determinants and Lyt 2 CTL precursors reacting to Class I determinants for effective generation of CTL's, it seems likely that AFP is blocking at the Lyt 1 helper cell level. If alloantigenic differences exist with regard to both Class I and non-MHC antigens, then there will be a normal generation of killer T cells against Class I determinants in the presence of AFP (Peck et al., 1978). These findings indicate that lymphocyte-activating, non-MHC determinants circumvent AFP-induced inhibition, presumably by supplying a nonsuppressible helper T cell signal to the "poised CTL" reacting against Class I antigens.

Thus, AFP will not serve as a pandemic inhibitory agent for all immunological reactions. On the contrary, AFP appears to regulate TD antibody synthesis and certain cell-mediated reactions in a highly selective manner. A primary target for AFP-mediated suppression would appear to be an Lyt 1 T helper cell required for antibody responses to TD antigens and for collaboration with prekiller cells in the events leading to CTL generation (Bach et al., 1976; Alter and Bach, 1979). Data that fail to show significant inhibition of certain murine (Sheppard et al., 1977; Hassoux et al., 1978) or rat (Parmely and Thompson, 1976; Sell et al., 1977) immune reactions in the presence of AFP are more easily understood when the above knowledge of cellular and genetic restrictions is taken into account.

AFP is capable of suppressing certain types of human lymphocyte functions *in vitro* (Lester et al., 1976; Yachnin and Lester, 1976; Auer and Kress, 1977; Lester et al., 1977; Gupta and Good, 1977; Murgita et al., 1978). Here, AFP isolated from human fetal liver is a more potent suppressor of mitogen- and alloantigen-induced lymphocyte transformation than is tumor-derived AFP (Yachnin and Lester, 1976). Lester et al. (1976) showed a correlation between the relative amount of a particular negatively charged molecular variant of AFP and immunosuppressive strength. Thus, AFP derived from fetal liver (with a high content of electronegative species) is strongly immunosuppressive, whereas the more weakly inhibitory tumor AFP isolates contain fewer of these molecules. The effects of AFP derived from human fetuses were compared on lymphocyte activation induced by the B cell mitogen *Staphylococcus aureus* strain Cowan I (Forsgren et al., 1976), by the T cell mitogen PHA, and by irradiated allogeneic lymphocytes in the one-way mixed lymphocyte reaction (Murgita et al., 1978). Mitogenic responses to PHA were strongly inhibited in a dose-dependent manner over an AFP concentration range of 300 to 18 $\mu g/ml$ by a mechanism other than binding interference or competition with mitogen for cell surface receptors. In contrast, the proliferative response to *S. aureus* was normal or slightly enhanced by AFP. Strong suppression of the MLR required higher concentrations of AFP than were necessary for the inhibition of PHA responses. Other studies (Gupta and Good, 1977; Littman et al., 1977) also suggest that human AFP is selective in its suppressive effects on *in vitro* cellular immune responses. At the same time, there are reports of AFP-mediated augmentation of mitogen- and alloantigen-stimulated lymphocyte proliferation (Charpentier et al., 1977; Murgita et al., 1978; Soubiran et al., 1979). It should be noted that, unlike our experience with murine AFP, we (Murgita et al., 1978) and others (Yachnin and Lester, 1976) have observed a variation in the suppressive effects among individual isolates of human AFP. The reason for this is not clear, but differences in the methods used to isolate (Goeken and Thompson, 1977) the proteins, the fetal versus tumor origin (Yachnin and Lester, 1976), the existence of unique immunosuppressive subspecies (Lester et al., 1977), and loss during purification of active moieties which are normally bound to AFP in the native state (Keller et al., 1976) are possible factors.

We further delineated the mechanism of AFP-mediated suppression on *in vitro* murine antibody synthesis (Murgita et al., 1977; Goidl et al., 1979). Experiments were performed to determine whether *in vitro* exposure of adult spleen cells to AFP would result in the activation of "suppressor cells." Adult spleen cells were cultured in the presence of AFP or normal mouse serum (NMS) at a final concentration of 200 μg/ml for four days. The cells were then washed, and 1 x 10^6 cells were transferred to secondary assay cultures consisting of 20 x 10^6 fresh, normal, syngeneic spleen cells along with optimal immunizing doses of TD antigen. After an additional four days of incubation, the assay cultures were harvested and IgM antibody synthesis was measured. Spleen cells precultured in AFP effectively suppressed primary IgM antibody responses, but cells precultured in NMS did not (Murgita et al., 1977). Suppressor activity was significantly enriched in precultured purified spleen T cells in comparison with whole spleen precultures, and that suppressor activity was absent in precultures of nude mouse spleen (Murgita et al., 1977). Interestingly, the adult AFP-activated inhibitory T cells showed many similarities with naturally occurring newborn suppressor cells. Like newborn inhibitory T cells, adult AFP-activated T lymphocytes suppressed TD, but not TI, antibody synthesis (Murgita et al., 1978). Dilution analysis showed that as few as 10^4 adult inhibitory T cells, representing 0.05 per cent of the total assay culture cell population, markedly inhibited anti-SRBC responses (Murgita et al., 1977). Cytotoxic pretreatment of AFP-activated adult inhibitory T cells with heterologous anti-T, anti-Thy 1, or anti-Lyt 1 antibodies and complement eliminated the ability to initiate suppression. Cytotoxic pretreatment with anti-Lyt 2 failed to impair the inhibitory effect. Thus, like naturally occurring newborn suppressors, the AFP-induced inhibitory cells in adult spleens belong to the Lyt 1^+2^- subset of T lymphocytes. Additional studies determined the cell surface Ia antigen phenotype of adult AFP-activated inhibitory T cells (Hooper et al., 1979). Only sera with anti-I-J^k reactivity were effective in removing the suppressing ability. Thus, adult spleen T cells activated by a four-day preculture in the presence of AFP display the same functional properties and cell surface antigen phenotype (Thy 1^+, Lyt 1^+2^-, I-J$^+$) as naturally occurring newborn suppressor cells. Disappearance of inhibitory T cells from the spleens of newborn mice with increasing age is closely paralleled by a decrease in the levels of serum AFP (Murgita et al., 1978). Endogenous AFP may thus play an important immunoregulatory role in the developing fetus and newborn via induction of inhibitory T lymphocytes. In support of this, both newborn and adult AFP- activated inhibitory cells suppress IgG and IgA antibody synthesis *in vivo* (Hooper et al., 1979), and adult mice injected with AFP have heightened susceptibility to Moloney sarcoma virus oncogenesis as a result of AFP-induced T lymphocytes (Gershwin et al., 1978). In the human system, purified AFP activates *in vitro* cultured T cells from normal peripheral blood to become inhibitors of mixed lymphocyte reactivity (Murgita et al., 1977; Alpert et al., 1978).

REGULATION OF NATURAL KILLER CELL ACTIVITY IN THE NEWBORN

Natural killer (NK) cells constitute a distinct group of cells not belonging to any of the "classical" types of cells in the immune system (Herberman et al., 1975; Kiessling et al., 1975). They have attracted great interest because of their lytic behavior toward certain malignant cells *in vitro*. There is also a positive correlation between the level of NK activity in the murine system and the *in vivo* resistance to certain tumor types (Kiessling and Wigzell, 1979). In addition to their capacity to kill tumor cells, NK cells have lytic reactivity toward bone marrow stem cells (Hansson-Kiessling, 1981) and immature, cortical thymocytes (Hansson et al., 1979). NK cells would therefore seem to be endowed with a special ability to kill certain cells expressing embryonic or "primitive" features (Stern et al., 1980). As targets for NK cells include stem cells in the bone marrow, as well as immature thymocytes, it is likely that NK cells would also express some important features in the neonatal period. Since this is clearly true for rodents, we include here a summary of the results of NK activity in relation to age.

A characteristic feature of the NK system in mice and rats is the rapid rise and fall in NK activity with age (Herberman et al., 1975; Kiessling et al., 1975). Thus, NK reactivity is largely absent in fetal liver cells and in the spleens of newborn mice, but it shows a rapid rate of appearance at around three weeks of age. The activity peaks at six to eight weeks of age and declines thereafter. A similar although not as abrupt pattern is displayed in the rat. Human NK activity shows a less impressive correlation with age, although an age effect is clearly noted (Campbell et al., 1974). There may be several underlying reasons why NK cells seem to be comparatively more completely "suppressed" in the newborn mouse than several other cellular types. One possibility would be that these cells are absent in the very young animal. In support of such a view are the findings that spleen cells from young mice (two weeks old) possess only about half the number of NK cells, measured in a target binding assay (J. Roder, personal communication), as do those present in young adult mice. However, considering the relatively high numbers of erythropoietic cells in newborn spleen compared with the adult (Weissman, 1977), dilution effects must be seriously considered. Another plausible reason for the low level of NK activity noted in newborn mice may reside at the level of regulatory "suppressing" cells. Thus, it has been known for some time that in adult mice the organ distribution of NK activity is inversely related to the actual presence of "natural" targets for NK activity in the same organ (Kiessling and Wigzell, 1979). Thymus and bone marrow, which constitute two organ systems quite low in NK activity, are also the two organs with the highest numbers of cells that can be lysed *in vitro* by NK cells (Hansson et al., 1979a, 1979b).

The frequency of such naturally occurring NK targets is significantly increased in the fetal and newborn periods, in both murine and human systems. These target cells can be shown to be able to function as efficient "cold target" inhibitors in NK cytolytic assays. It would therefore follow that the presence of such potential target cells in a given organ would serve to reduce (i.e. suppress) the actual available NK cells in the same population. As this cold target inhibition would be expected to also involve conjugate formation between NK cells and target cells *in vivo*, this would automatically lower the number of NK cells available for the conventional target conjugate assay *in vitro* and thus yield falsely low values.

Finally, a third possible reason newborn mice have close to undetectable levels of NK cells may be at the level of endogenous regulation of interferon levels (Senik et al.,1979; Minator et al., 1980). The newborn mouse does seem to have somewhat altered macrophage activity (Lu et al., 1979). Polyinosinic-polycytidylic acid (poly I:C), an agent known to function via macrophages to induce NK cells through interferon release, has been found by us to be an inefficient inducer when spleen cells from newborn mice are used (Orn et al., 1980a, 1980b). On the other hand, the direct addition of interferon to newborn mouse spleen cells could induce a low but significant degree of NK activity. As interferon is know to be the major regulator of NK activity (Gidlund et al., 1978, 1979), we consider it likely that the newborn mouse may be particularly deficient with regard to macrophage-mediated interferon production. If this mode of interferon release is an important factor in the maintenance of normal levels of NK activity, for which there is some precedent (Clark et al., 1979), this may well add to a milieu inefficient for the generation of sizable numbers of functioning NK cells in newborn mice. It is interesting to note that serum from newborn (but not adult) mice inhibits *in vitro* NK activity (Rocklin et al., 1979). Our findings suggest that fetally derived AFP may play a role in this regard, as poly I-C will fail to induce efficient NK activation in adult spleen cells if AFP is present in physiological amounts

(Örn et al., 1980a, 1980b). Trivial reasons for this suppressive activity of AFP could be excluded, and interferon added to adult spleen cells in the presence of AFP did yield, as expected, a significant increase in NK activity. We would thus conclude that the very low NK activities observed, particularly in newborn rodents, can be explained as a result of several factors, all interacting to yield an "NK-suppressing" milieu.

SUMMARY AND CONCLUSIONS

The humoral and cellular milieu of the embryo and neonate represents, from the point of view of the immunologist, a bewildering and highly dynamic situation. A variety of elements have been depicted to have immunoregulatory power within this time period in the young, developing individual, mostly with immunosuppressive consequences. Whether these elements represent trivial side reactions and artifacts studied by ambitious scientists or whether they play relevant roles in the prevention of rejection disorders by the mother against its "foreign" fetus is yet to be resolved. It is likely that model genetic systems in which the survival of the fetus is different when normal and immunodeficent mothers (e.g., nude mice) are compared will be required to delineate the border line between biologically relevant factors and triviality.

We are convinced that many factors exist in the embryo and neonate with the potential to cause T cell reactions—in particular, to become either "suppressive" in nature or very selective in allowing only particular histocompatibility barriers to function as immunogens during this time period. We consider the finding that alpha-fetoprotein can function as a selective agent in this latter context to be an indication that this may in fact be one of the roles this protein has to fulfill in the embryonic and post-natal periods.

Whether the finding that natural killer cells frequently have preferential reactivity toward "embryonic" cells is of any relevance at all for the embryonal-maternal situation remains to be established but represents an area worth further analysis. However, it may well be that studies of immunosuppressive elements in the embryo and neonate will yield significant information as to the actual buildup and function of the normal immune system, while at the same time proving that these elements play no relevant role whatsoever for the survival of the semiallogeneic mammalian fetus.

ACKNOWLEDGMENTS

This work was supported by the Swedish Cancer Society.

REFERENCES

Abelev, G.I. 1971. Alpha-fetoprotein in ontogenesis and its association with malignant tumors. *Adv. Cancer Res.* 14:295.

Alpert, E., J.L. Dienstag, S. Sepersky, B. Littman, and R. Rocklin. 1978. Immunosuppressive characteristics of human AFP: effect on tests of cell mediated immunity and induction of human suppressor cells. *Immunol. Comm.* 7:163.

Alter, B.J., and F.H. Bach. 1979. Speculations on alternative pathways of T-lymphocyte response. *Scand. J. Immunol.* 10:87.

Argyris, B.F. 1978. Suppressor activity in the spleen of neonatal mice. *Cell. Immunol.* 36:354.

Argyris, B.F. 1979. Further studies on suppressor cell activity in the spleen of neonatal mice. *Cell. Immunol.* 48:398.

Auer, I.O., and H.G. Kress. 1977. Suppression of the primary cell-mediated immune response by human alpha-fetoprotein *in vitro. Cell. Immunol.* 30:173.

Ayoub, J., and S. Kasakura. 1971. *In vitro* response of fetal lymphocytes to PHA, and a plasma factor which suppresses the PHA response of adult lymphocytes. *Clin. Exp. Immunol.* 8:427.

Bach, F.H., M.L. Bach, and P.M. Sondel. 1976. Differential function of major histocompatibility complex antigens in T-lymphocyte activation. *Nature* 259:273.

Bassett, M., T.A. Coons, W. Wallis, E.H. Goldberg, and R.C. Williams. 1977. Suppression of stimulation in mixed leucocyte culture by newborn splenic lymphocytes in the mouse. *J. Immunol.* 119:1855.

Bayard, B., and J.-P. Kerckaert. 1977. Characterization and isolation of nine rat alpha-fetoprotein variants by gel electrophoresis and lectin affinity chromatography. *Biochem. Biophys. Res. Comm.* 77:489.

Burns, F.D., P.C. Marrack, J.W. Kappler, and C.A. Janeway Jr. 1975. Functional heterogeneity among the T-derived lymphocytes of the mouse. IV. Nature of spontaneously induced suppressor cells. *J. Immunol.* 114:1345.

Campbell, A.C., C. Waller, J. Wood, A. Aynsley-Green, and V. Yu. 1974. Lymphocyte subpopulations in the blood of newborn infants. *Clin. Exp. Immunol.* 18:469.

Cantor, H., and R.K. Gershon. 1979. Immunological circuits: cellular composition. *Fed. Proc.* 38:2958.

Cantor, H., L. McVay-Boudreau, J. Hugenberger, K. Naidorf, F.W. Shen, and R.K. Gershon. 1978. Immunoregulatory circuits among T cell sets. II. Physiologic role of feedback inhibition *in vivo*: absence in NZB mice. *J. Exp. Med.* 147:1116.

Charpentier, B., R.D. Guttman, J. Shuster, and P. Gold. 1977. Augmentation of proliferation of human mixed lymphocyte culture by human alpha-fetoprotein. *J. Immunol.* 119:897.

Clark, E.A., P.H. Russell, M. Egghart, and M.A. Horton. 1979. Characteristics and genetic control of NK cell-mediated cytotoxicity activated by naturally acquired infection in the mouse. *Int. J. Cancer* 24:688.

Cooperband, S.R., and A.M. Badger. 1979. Suppressor factors in serum and plasma. In *Naturally Occurring Biological Immunosuppressive Factors and Their Relationship to Disease.* R.H. Neubauer, ed. Boca Raton, Florida: CRC Press, pp. 115-138.

Dekruyff, R.H., Y.T. Kim, G.W. Siskind, and M.E. Weksler. 1980. Age related changes in the *in vitro* immune response: increased suppressor activity in immature and aged mice. *J. Immunol.* 125:142.

Durandy, A., A. Fischer, and C. Griscelli. 1979. Active suppression of B lymphocyte maturation by two different newborn T lymphocyte subsets. *J. Immunol.* 123:2655.

Durandy, A., A. Fischer, and C. Griscelli. 1979. Active suppression of B lymphocyte maturation by two different newborn T lymphocyte subsets. *J. Immunol.* 123:2644.

Etlinger, H.M., and J.M. Chiller. 1977. Suppression of immunological activities by mouse amniotic fluid. *Scand. J. Immunol.* 6:1241.

Evans, M.M., W.G. Williamson, and W.J. Irvine. 1968. The appearance of immunological competence at an early age in New Zealand black mice. *Clin. Exp. Immunol.* 3:375.

Fidler, J.M., M.O. Chiscon, and E.S. Golub. 1972. Functional development of the interacting cells in the immune response. II. Development of immunocompetence to heterologous erythrocytes *in vitro. J. Immunol.* 109:136.

Forni, G., and I. Green. 1976. Heterologous sera: a target for *in vitro* cell-mediated cytotoxicity. *J. Immunol.* 116:1561.

Forsgren, A., A. Svedjelund, and H. Wigzell. 1976. Lymphocyte stimulation by protein A of *Staphylococcus aureus. Eur. J. Immunol.* 6:207.

Gershwin, M.E., J.J. Castles, A. Ahmed, and R. Makishima. 1978. The influence of alpha-fetoprotein on Moloney sarcoma virus oncogenesis: evidence for generation of antigen nonspecific suppressor T cells. *J. Immunol.* 121:2292.

Gidlund, M., A. Orn, H. Wigzell, A. Senik, and I. Gresser. 1978. Enhanced NK cell activity in mice injected with interferon and interferon inducers. *Nature* 273:759.

Gidlund, M., E.A. Ojo, A. Orn, H. Wigzell, and R.A. Murgita. 1979. Severe suppression of the B-cell system has no impact on the maturation of natural killer cells in mice. *Scand. J. Immunol.* 9:167.

Gitlin, D. 1975. Normal biology of alpha-fetoprotein. *Ann. N.Y. Acad. Sci.* 259:7.

Gitlin, D., and A. Perricelli. 1970. Synthesis of serum albumin, prealbumin, alpha-fetoprotein, alpha-antitrypsin and transferrin by the hyman yolk sac. *Nature* 228:995.

Globerson, A., and T. Umiel. 1978. Ontogeny of suppressor cells. II. Suppression of graft-versus-host and mixed leucocyte culture responses by embryonic cells. *Transplantation* 26:438.

Globerson, A., R.M. Zinkernagel, and T. Umiel. 1975. Immunosuppression by embryonic liver cells. *Transplantation* 20:480.

Goeken, N.E., and J.S. Thompson. 1977. Conditions affecting the immunosuppressive properties of human alpha-fetoprotein. *J. Immunol.* 119:139.

Goidl E.A., H. Wigzell, and R.A. Murgita. 1979. Studies on the mechanisms of alpha-fetoprotein induction of immune suppressive activity. In *Developmental Immunobiology*, Proceedings of the Fifth Irwin Strasburger Memorial Seminar on Immunology. G.W. Siskind, S.D. Litwin, and M.E. Weksler, eds. New York: Grune and Stratton, pp. 35-55.

Golub, E.S. 1971. Brain-associated theta antigen: reactivity of rabbit anti-mouse brain with mouse lymphoid cells. *Cell. Immunol.* 2:353.

Gupta, S., and R.A. Good. 1977. Alpha-fetoprotein and human lymphocyte subpopulations. *J. Immunol.* 118:405.

Gupta, S., Z. Goel, and M.H. Grieco. 1976. Fetuin. *In vitro* effect on sheep erythrocyte rosette formation with human T lymphocytes. *Int. Arch. Allergy Appl. Immunol.* 52:273.

Gustine, D.L., and E.F. Zimmerman. 1973. Developmental changes in microheterogeneity of fetal plasma glycoproteins. *Biochem. J.* 132:541.

Hansson, M., K. Kärre, R. Kiessling, J.C. Roder, B. Andersson, and P. Häyry. 1979a. Natural NK cell targets in the mouse thymus: characteristics of the sensitive cell population. *J. Immunol.* 123:765.

Hansson, M., R. Kiessling, B. Andersson, K. Kärre, and J.C. Roder. 1979b. NK-cell sensitive T-cell subpopulation in thymus: inverse correlation to host NK activity. *Nature* 278:174.

Hardy, B., E. Mozes, and D. Danon. 1976. Comparison of the immune response potential of newborn mice to T-dependent and T-independent synthetic polypeptides. *Immunology* 30:261.

Hardy, B., and E. Mozes. 1978. Expression of T cell suppressor activity in the immune responses of newborn mice to a T-independent synthetic polypeptide. *Immunology* 35:757.

Hassoux, R., M.F. Poupon, and H. Uriel. 1978. Lack of inhibition by mouse alpha-fetoprotein (AFP) of *in vitro* induced lymphocyte blastogenesis. *Ann. Immunol.* (Inst. Pasteur) 129C:275.

Herberman, R.B., N.E. Nunn, and D.H. Lavrin. 1975. Natural cytotoxic reactivity of mouse lymphoid cells against syngeneic and allogeneic tumors. I. Distribution of reactivity and specificity. *Int. J. Cancer* 16:216.

Herberman, R.B., M.E. Nunn, H.T. Holden, and D.H. Lavrin. 1975. Natural cytotoxic reactivity of mouse lymphoid cells against syngeneic and allogeneic tumors. II. Characterization of effector cells. *Int. J. Cancer* 16:230.

Hooper, D.C., and R.A. Murgita. 1980. Evidence for two distinct neonatal suppressor cell populations. *Fed. Proc.* 39:354.

Hooper, D.C., T.L. Delovitch, H. Wigzell, and R.A. Murgita. 1979. Analysis of an immunoregulatory suppressor T cell system in newborn mice. In *The Molecular Basis of Immune Cell Function.* J.G. Kaplan, ed. Amsterdam: Elsevier/North-Holland Biomedical Press, p. 676.

Hsu, C.C.S., W.I. Waithe, P. Hathaway, and K. Hirschhorn. 1973. The effects of foetuin on lymphocytes: lymphocyte-stimulating property. *Clin. Exp. Immunol.* 15:427.

Irie, R.F., K. Irie, and D.L. Morton. 1974. Natural antibody in human serum to a neoantigen in human cultured cells grown in fetal bovine serum. *J. Nat. Cancer Inst.* 4:1051.

Janeway, C.A., S.O. Sharrow, and E. Simpson. 1975. T cell populations with different function. *Nature* 253:544.

Jennings, J.J., and M.B. Rittenberg. 1976. Evidence for separate subpopulations of B cells responding to T-independent and T-dependent antigens. *J. Immunol.* 117:1749.

Kedar, E., M. Schwartzback, E. Unger, and T. Lupu. 1978. Characteristics of suppressor cells induced by fetal bovine serum in murine lymphoid cell cultures. *Transplantation* 26:63.

Kedar, E., and M. Schwartzbach. 1979. Further characterization of suppressor lymphocytes induced by fetal calf serum in murine lymphoid cell cultures: comparison with *in vitro*-generated cytotoxic lymphocytes. *Cell. Immunol.* 43:326.

Keller, R.H., N.J. Calvanico, and T.B. Tomasi. 1976. Immunosuppressive properties of AFP: role of estrogens. In *Onco-Developmental Gene Expression.* W.H. Fishman and S. Sell, eds. New York: Academic Press, p. 287-295.

Kerbel, R.S., and D. Blakesee. 1976. Rapid adsorption of a fetal calf serum component by mammalian cells in culture. A potential source of artifacts in studies of antisera to cell-specific antigens. *Immunology* 31:881.

Kiessling, R., and H. Wigzell. 1979. An analysis of the murine NK cell as to structure, function, and biological relevance. *Immunol. Rev.* 44:165.

Kiessling, R., E. Klein, and H. Wigzell. 1975a. Natural killer cells in the mouse. I. Cytotoxic cells with specificity for mouse Moloney leukemia cells. Specificity and distribution according to genotype. *Eur. J. Immunol.* 5:112.

Kiessling, R., E. Klein, H. Pross, and H. Wigzell. 1975b. Natural killer cells in the mouse. II. Cytotoxic cells with specificity for mouse Moloney leukemia cells. Characteristics of the killer cell. *Eur. J. Immunol.* 5:117.

Kithier, K., and M.D. Poulik. 1972. Comparative studies of bovine alpha-fetoprotein and fetuin. *Biochim. Biophys. Acta* 278:505.

Knospe, W.H., P.H. Ward, R. Dors, W. Fried, R.J. Sassetti, and F.E. Trobaugh. 1971. Fetuin stimulation of hematopoietic stem cells (HSC). *Blood* 38:818.

Labib, R.S., and T.B. Tomasi. 1978. Immunosuppressive factors in mouse amniotic fluid and neonate serum. *Immunol. Comm.* 7:323.

Lawler, S.D., B.R. Reaves, and E.O. Ukaejiofo. 1975. Interaction of maternal and neonatal cells in mixed lymphocyte cultures. *Lancet* ii:1185.

Lebman, D.A., and C.E. Calkins. 1980. Demonstration of active suppressor cells in spleens of young NZB mice. *Cell. Immunol.* 51:419.

Lester, E.P., J.B. Miller, and S. Yachnin. 1976. Human alpha-fetoprotein as a modulator of human lymphocyte transformation: correlation of biological potency with electrophoretic variants. *Proc. Nat. Acad. Sci. USA* 73:4645.

Lester, E.P., J.B. Miller, and S. Yachnin. 1977. A postsynthetic modification of human alpha-fetoprotein controls and its immunosuppressive potency. *Proc. Nat. Acad. Sci. USA* 74:3988.

Lewis, G.K., R. Ranken, D.E. Nitecki, and J.W. Goodman. 1976. Murine B cell subpopulations responsive to T-dependent and T-independent antigens. *J. Exp. Med.* 144:382.

Lindahl-Kiessling, K., A. Mattsson, and V. Skoog. 1970. A factor in amniotic fluid inhibiting phytohemagglutinin (PHA) induced lymphocyte growth. *Life Sciences* 9:1427.

Littman, B.H., E. Alpert, and R. Rocklin. 1977. The effect of purified alpha-fetoprotein on *in vitro* assays of cell-mediated immunity. *Cell. Immunol.* 30:35.

Lu, C.Y., E.G. Calamai, and E.R. Unanue. 1979. A defect in the antigen-presenting function of macrophages from neonatal mice. *Nature* 282:327.

Luckenback, G.A., M.M. Kennedy, A. Kelly, and T.E. Mandel. 1978. Suppression of an *in vitro* humoral immune response by cultured fetal thymus cells. *Eur. J. Immunol.* 8:8.

Masseyeff, R., J. Gilli, B. Krebs, A. Callvand, and C. Bonet. 1975. Evolution of alpha-fetoprotein serum levels throughout life in humans and rats, and during pregnancy in rats. *Ann. N.Y. Acad. Sci.* 259:17.

McCullagh, P. 1975a. Role of the thymus in suppression of the immune responses in newborn rats. *Aust. J. Exp. Biol. Med. Sci.* 53:413.

McCullagh, P. 1975b. Radiosensitivity of suppressor cells in newborn rats. *Aust. J. Exp. Biol. Med. Sci.* 53:399.

McKearn, J.P., and J. Quintano. 1979. Ontogeny of murine B-cell responses to thymus-independent trinitrophenyl antigens. *Cell. Immunol.* 44:367.

Mendelsohn, J., A. Skinner, and S. Kornfeld. 1971. The rapid induction by phytohemagglutinin of increased aminiosobutyric acid uptake by lymphocytes. *J. Clin. Invest.* 50:818.

Minato, N., L. Reid, H. Cantro, P. Lengyel, and B.R. Bloom. 1980. Mode of regulation of natural killer cell activity by interferon. *J. Exp. Med.* 152:124.

Moretta, L., S.R. Webb, C.E. Grossi, P.M. Lydyard, and M.D. Cooper. 1977. Functional analysis of two human T cell subpopulations: help and suppression of B-cell responses by T cells bearing receptors for IgM or IgG. *J. Exp. Med.* 146:184.

Morse, H.C., B. Prescott, S.S. Cross, P.W. Stashak, and P.J. Baker. 1976. Regulation of the antibody response to type III pneumococcal polysaccharide. V. Ontogeny of factors influencing the magnitude of the plaque-forming cell response. *J. Immunol.* 116:279.

Mosier, D.E., and P.L. Cohen. 1975. Ontogeny of mouse T-lymphocyte function. *Fed. Proc.* 34:137.

Mosier, D.E., and B.M. Johnson. 1975. Ontogeny of mouse lymphocyte function. II. Development of the ability to produce antibody is modulated by T lymphocytes. *J. Exp. Med.* 141:216.

Mosier, D.E., I.M. Zitron, J.J. Mond, and W.E. Paul. 1979. Requirements for the induction of antibody formation in the newborn mouse. In *Developmental Immunobiology.* Proceedings of the Fifth Irwin Strasburger Memorial Seminar on Immunology. G.W. Siskind, S.D. Litwin, and M.E. Weksler, eds. New York: Grune and Stratton, p. 25.

Mosier, D.E., B.J. Mathieson, and P.S. Campbell. 1977. Lyphenotype and mechanism of action of mouse neonatal suppressor T cells. *J. Exp. Med.* 146:59.

Murgita, R.A., and H. Wigzell. 1979. Selective immunoregulatory properties of alpha-fetoprotein. *La Ricerca in Clinica e in Laboratorio.* 9:327.

Murgita, R.A., L.C. Andersson, E. Rouslahti, A. Kimura, and H. Wigzell. 1976. Effects of human and mouse alpha-fetoprotein on immune responses. *Scand. J. Immunol.* 6:731.

Murgita, R.A. 1976. The immunosuppressive role of alpha-fetoprotein during pregnancy. *Scand. J. Immunol.* 5:1003.

Murgita, R.A., and H. Wigzell. 1976. The effects of mouse alpha-fetoprotein on T cell-dependent and T cell-independent immune responses *in vitro. Scand. J. Immunol.* 5:1215.

Murgita, R.A., and T.B. Tomasi. 1975a. Suppression of the immune response by alpha-fetoprotein. I. The effect of mouse alpha-fetoprotein on the primary and secondary antibody response. *J. Exp. Med.* 141:269.

Murgita, R.A., and T.B. Tomasi. 1975b. Suppression of the immune response by alpha-fetoprotein. II. The effects of mouse alpha-fetoprotein on mixed lymphocyte reactivity and mitogen-induced lymphocyte transformation. *J. Exp. Med.* 114:440.

Murgita, R.A., E.A. Goidl, S. Kontiainen, P.C. Beverley, and H. Wigzell. 1978. Adult murine T cells activated *in vitro* by alpha-fetoprotein and naturally occurring T cells in newborn mice. Identity in function and cell surface differentiation antigens. *Proc. Nat. Acad. Sci. USA* 75:2897.

Murgita, R.A., D.C. Hooper, T.L. Delovitich, and H. Wigzell. Newborn and adult alpha-fetoprotein-activated inhibitory T lymphocytes express identical cell surface Ia determinants. In preparation.

Murgita, R.A., A.B. Peck, and H. Wigzell. 1978. Selective immunoregulatory properties of alpha-fetoprotein. *Fed. Proc.* 37: 1594.

Murgita, R.A., L.C. Andersson, M.S. Sherman, H. Bennich, and H. Wigzell. 1978. Effects of human alpha-fetoprotein on human B and T lymphocyte proliferation *in vitro. Clin. Exp. Immunol.* 33:347.

Murgita, R.A., E.A. Goidl, S Kontiainen, and H. Wigzell. 1977. Alpha-fetoprotein induces suppressor T cells *in vitro. Nature* 267:257.

Nadler, L.M., and R.J. Hodes. 1977. Regulatory mechanisms in cell-mediated immune responses. II. Comparison of culture-induced and alloantigen-induced suppressor cells in MLR and CML. *J. Immunol.* 118:1886.

Nunez, E., G. Vallette, C. Benassayag, and M. Jayle. 1974. Comparative study on the binding of estrogens by human and rat serum proteins in development. *Biochem. Biophys. Res. Comm.* 57:126.

Ogra, S.S., R.A. Murgita, and T.B. Tomasi. 1974. Immunosuppressive activity of mouse amniotic fluid. *Immunol. Comm.* 3:497.

Olding, L.B., and M.B.A. Oldstone. 1974. Lymphocytes from human newborns abrogate mitosis of their mother's lymphocytes. *Nature* 249:161.

Olding, L.B., and M.B.A. Oldstone. 1976. Thymus-derived peripheral lymphocytes from human newborns inhibit division of their mother's lymphocytes. *J. Immunol.* 116:682.

Olding, L.B., K. Benirschke, and M.B.A. Oldstone. 1974. Inhibition of mitosis of lymphocytes from human adults by lymphocytes from human newborns. *Clin. Immunol. Immunopathol.* 3:79.

Olding, L.B., R.A. Murgita, and H. Wigzell. 1977. Mitogen-stimulated lymphoid cells from human newborns suppress the proliferation of maternal lymphocytes across a cell-impermeable membrane. *J. Immunol.* 119:1109.

Oldstone, M.B.A., A. Tishon, and L. Moretta. 1977. Active thymus derived suppressor lymphocytes in human cord blood. *Nature* 269:333.

Olsson, M., G. Lundahl, and E. Ruoslahti. 1977. Genetic control of alpha-fetoprotein synthesis in the mouse. *J. Exp. Med.* 145:819.

Opitz, H.-G., U. Opitz, H. Lenke, G. Hewlett, W. Schreml, and H.D. Flad. 1977. The role of fetal calf serum in the primary immune response *in vitro. J. Exp. Med.* 145:1029.

Örn, A., M. Gidlund, E. Ojo, K.-O. Grönvik, J. Andersson, H. Wigzell, R.A. Murgita, A.Senik, and I. Gresser. 1980a. Factors controlling the augmentation of natural killer cells. In *Natural Cell-mediated Immunity Against Tumors.* R.B. Herberman, ed. New York: Academic Press, pp. 581-607.

Örn, A., M. Gidlund, H. Wigzell, and R.A. Murgita. 1980b. AFP can block the enhancement of natural killer cell activity mediated by interferon inducers. In *Mechanism of Natural Immunity.* Abstracts of the Fourth International Congress of Immunology. Paris: Academic Press. Abstract 11.6.15.

Parmely, M.J., and J.S. Thompson. 1976. Effect of alpha-fetoprotein and other serum factors derived from hepatoma-bearing rats on the mixed lymphocyte response. *J. Immunol.* 117:1832.

Pavia, C.S., and D.P. Stites. 1979. Immunosuppressive activity of murine newborn spleen cells. I. Selective inhibition of *in vivo* lymphocyte activation. *Cell. Immunol.* 42:48.

Peck, A.B., L.C. Andersson, and H. Wigzell. 1977. Secondary *in vitro* responses of T lymphocytes to non-H-2 alloantigen. Self-H-2 restricted responses induced in heterologous serum are not dependent on primary-stimulating non-H-2 alloantigens. *J. Exp. Med.* 145:802.

Peck, A.B., R.A. Murgita, and H. Wigzell. 1978. Cellular and genetic restrictions in the immunoregulatory activity of alpha-fetoprotein. I. Selective inhibition of anti-Ia associated proliferative reactions. *J. Exp. Med.* 148:667.

Peck, A.B., C.A. Janeway, H. Wigzell, and L.C. Andersson. 1977. Environmental and genetic control of T cell activation. *Transplant. Rev.* 35:146.

Peck, A.B., R.A. Murgita, and H. Wigzell. 1978. Cellular and genetic restrictions in the immunoregulatory activity of alpha-fetoprotein. II. Alpha-fetoprotein-induced suppression of cytotoxic T lymphocyte development. *J. Exp. Med.* 148:360.

Playfair, J.H.L., 1978. Strain differences in the immune response of mice. I. The neonatal response to sheep red blood cells. *Immunology* 15:35.

Ptak, W., and A. Skowron-Cendrzak. 1977. Fetal suppressor cells. Their influence on the cell-mediated immune responses. *Transplantation* 24:45.

Ptak, W., K.F. Naidorf, J. Strzyzewska, and R.K. Gershon. 1979. Ontogeny of cells involved in the suppressor circuit of the immune response. *Eur. J. Immunol.* 9:495.

Puck, T.T., C.A. Waldren, and C. Jones. 1968. Mammalian cell growth proteins. I. Growth stimulation by fetuin. *Proc. Nat. Acad. Sci. USA* 59:192.

Rabinovich, H., T. Umiel, Y. Reisner, N. Sharon, and A. Globerson. 1979. Characterizaion of embryonic liver suppressor cells by peanut agglutinin. *Cell. Immunol.* 47:347.

Rabinowitz, S.G. 1976. Measurement and comparison of the proliferative and antibody responses of neonatal, immature and adult murine spleen cells to T-dependent and T-independent antigens. *Cell. Immunol.* 21:201.

Rodriguez, G., G. Anderson, H. Wigzell, and A.B. Peck. 1979. Non-T cell nature of the naturally occurring, spleen-associated suppressor cells present in the newborn mouse. *Eur. J. Immunol.* 9:737.

Rocklin, R.E., J.L. Kitzmiller, and M.D. Kaye. 1979. Immunobiology of the maternal-fetal relationship. *Ann. Rev. Med.* 30:375.

Rollwagen, F.M., and O. Stutman. 1979a. Ontogeny of regulatory cells. *Fed. Proc.* 38:2075.

Rollwagen, F.M., and O. Stutman. 1979b. Ontogeny of culture-generated suppressor cells. *J. Exp. Med.* 150:1359.

Sell, S., H.W. Sheppard Jr., and M. Poler. 1977. Effects of alpha-fetoprotein on murine immune responses. II. Studies on rats. *J. Immunol.* 119:98.

Senik, A., I. Gresser, C. Maury, M. Gidlund, A. Orn, and H. Wigzell. 1979. Enhancement by interferon of natural killer cell activity in mice. *Cell. Immunol.* 44:186.

Sheppard, H.W. Jr., S. Sell, P. Trefts, and R. Bahu. 1977. Effects of alpha-fetoprotein on murine immune responses. I. Studies in mice. *J. Immunol.* 119:91.

Shustic, C., I.R. Cohen, R.S. Schwartz, E. Latham-Griffin, and S.D. Waksal. 1976. T lymphocytes with promiscuous cytotoxicity. *Nature* 263:699.

Skowron-Cendrzak, A., and W. Ptak. 1976. Suppression of local graft-versus-host reactions by mouse fetal and newborn spleen cells. *Eur. J. Immunol.* 6:451.

Slades, B., and J. Forrest. 1977. Depression of antibody production in rats by homologous amniotic fluid. *IRCS Med. Sci.* 5:319.

Smith, C.J.P., and P.C. Kelleher. 1980. Alpha-fetoprotein molecular heterogeneity. Physiological correlations with normal growth carcinogenesis, and tumor growth. *Biochim. Biophys. Acta* 605:1.

Soubiran, P., A. Mucchielli, J.P. Kerckaert, B. Bayard, and R. Masseyeff. 1979. Stimulatory effect of human alpha-fetoprotein and its molecular variants on *in vitro* induced lymphocyte blastogenesis. *Scand. J. Immunol.* 10:179.

Spear, P.G., and G.M. Edelman. 1974. Maturation of the humoral immune response in mice. *J. Exp. Med.* 139:249.

Stern, P., M. Gidlund, A. Orn, and H. Wigzell. 1980. Natural killer cells mediate efficient lysis of MHC lacking embryonal carcinoma cells. *Nature* 283:341.

Stobo, J.B., and W.F. Paul. 1973. Differential responsiveness of T cells to phytohemagglutinin and concanavalin A as a probe for T cell subsets. *J. Immunol.* 110:362.

Strayer, D.S., and H. Köhler. 1976. Immune response to phosphorylcholine. II. Natural "auto"-anti-receptor antibody in neonatal Balb/c mice. *Cell. Immunol.* 25:294.

Sulit, H.L., S.H. Golub, R.F. Irie, R.K. Gupta, G.A Grooms, and L.D. Morton. 1976. Human tumor cells grown in fetal calf serum and human serum: influences on the tests for lymphocyte cytotoxicity, serum blocking and serum arming effects. *Int. J. Cancer* 17:461.

Tada, T., T. Takemori, K. Okumera, M. Nonaka, and T. Tokushisa. 1978. Two distinct types of helper T cells involved in the secondary antibody response. Independent and synergistic effects of Ia⁻ and Ia⁺ helper T cells. *J. Exp. Med.* 147:446.

Takeya, K., and K. Nomoto. 1967. Characteristics of antibody response in young or thymectomized mice. *J. Immunol.* 99:831.

Toder, V., M. Blank, and L. Nebel. 1979. Immunosuppressive effect of alpha-fetoprotein at different stages of pregnancy in mice. *Dev. Comp. Immunol. 3:537.*

Tomasi, T.B., R.A. Murgita, R.L. Thompson, N.J. Calvanico, and R.J. Dattwyler. 1977. Alpha-fetoprotein and the immune response during murine pregnancy and neonatal development. In Development of Host Defences. M.D. Cooper and D.H. Dayton, eds. New York: Raven Press, pp. 273-285.

Tyan, M.L. 1976. Immunosuppressive properties of mouse amniotic fluid. *Proc. Soc. Exp. Biol. Med.* 151:343.

Uriel, J., and D. Bouillon. 1975. Affinity chromatography of human, rat, and mouse alpha-fetoprotein on estradiol-sepharose absorbants. *FEBS Letters* 53:305.

Uriel, J., D. Bouillon, C. Aussel, and M. Dupiers. 1976. Alpha-fetoprotein: the major high-affinity estrogen binder in rat uterine cytosols. *Proc. Nat. Acad. Sci. USA* 73:1452.

van Furth, R. van, and M. Adinolfi. 1969. *In vitro* synthesis of the foetal alpha-globulin in man. *Nature* 222:1296.

Weiss, A. 1977. Suppression of adult rat lymphoid proliferative responses by homologous neonatal serum. *J. Immunol.* 118:1121.

Weissman, I.L. 1977. T cell maturation and the ontogeny of splenic lymphoid architecture. In *Immuno-aspects of the Spleen.* J.R. Battisto and J.W. Streilein, eds. Amsterdam: Elsevier/North Holland Biomedical Press, p. 77-87.

Wigzell, H., R.A. Murgita, A.B. Peck, H. Binz, H. Frischknecht, P. Peterson, and K. Sege. 1979. Induction and inhibition of lymphocyte functions: a note on complexity and consequences. In *Escape from Immune Surveillance: An Interpretive Survey of the Interface Between Immune Mechanism and Disease.* B. Janicki, ed. New York: Academic Press, p.13.

Williams, R.C., and S.J. Korsmeyer. 1978. Studies of human lymphocyte interactions with emphasis on soluble suppressor activity. *Clin. Immunol. Immunopathol.* 9:335.

Yachnin, S. 1975. Fetuin, an inhibitor of lymphocyte transformation. The interaction of fetuin with phytomitogens and a possible role for fetuin in fetal development. *J. Exp. Med.* 141:242.

Yachnin, S., and E. Lester. 1976. Inhibition of human lymphocyte transformation by human alpha-fetoprotein (HAFP); comparison of fetal and hepatoma HAFP and kinetic studies on *in vitro* immunosuppression. *Clin. Exp. Immunol.* 26:484.

Zielske, J.U., and S.H. Golub. 1976. Fetal calf serum-induced blastogenic and cytotoxic responses of human lymphocytes. *Cancer Res.* 36:3842.

Zimmerman, E.F., M. Voorting-Hawking, and J.G. Michael. 1977. Immunosuppression by mouse sialylated alpha-fetoprotein. *Nature* 265:354.

Chapter 16

REGULATION OF CYTOTOXIC T CELLS
IN PREGNANT MICE

DAVID A. CLARK
RENATA SLAPSYS
ANNE CROY
JANET ROSSANT
MARK McDERMOTT

THE PROBLEM OF THE FETUS AS AN ALLOGRAFT

The fetus that results following mating in an outbred population bears a wide variety of paternally derived antigens foreign to the mother, and some of these represent transplantation or histocompatibility (H) antigens (Johnson and Calarco, 1980). Even in syngeneic matings that occur in inbred strains of mice where there is minimal histoincompatibility between the parents, the fetus expresses embryonic antigens against which the mother can react. Unlike grafts of other types of paternal tissue, however, the fetus seems exempt from immune rejection by the mother. Two contrasting models have been proposed to account for this paradox:

1. Fetal "trophoblast" cells that line the maternal-fetal interface may lack transplantation-type antigens, and this enables these cells to form an inert barrier between the fetus and the mother.
2. Local uterine or systemic immunosuppressor mechanisms may prevent an effective immune reaction.

Experimental evidence supporting the first model (discussed in detail in Clark, 1982) has been recently challenged by demonstration of Class I MHC antigens on murine placental trophoblast (Chatterjee-Hasrouni and Lala, 1979; Jenkinson and Owen, 1981; see Chapters 10 and 11), by demonstration of trophoblast specific antigens (Beer et al., 1972; Faulk et al., 1977; Sellens et al., 1978), by demonstration of a bidirectional fetal-maternal lymphoid cell traffic (Tillikainen et al., 1974; Gill, 1977; Herzenberg et al., 1979; Collins et al., 1980; Liegeois et al., 1981; Pollack et al., 1981), and by the spontaneous occurrence of both humoral and cellular immune responses to antigens of the fetus (Goodlin and Herzenberg, 1964; Nymand et al., 1971; Youtananukorn et al., 1974; Hamilton et al., 1976; Rocklin et al., 1976; Salinas et al., 1978). Appropriate immunization against paternal or embryonal antigens can sometimes compromise fetal survival (Beer et al., 1972; Nista et al., 1973; Parmiani and Della Porta, 1973; Milgrom et al., 1977; Hamilton et al., 1979), and there is evidence that similar phenomena can occur spontaneously. Palm (1974) reported that male progeny of syngeneic matings in certain strains of rats developed a graft-versus-host-like disease, and Brehsihann et al. (1977)

associated the binding of autoantibody to placenta with fetal wastage. However, the natural immunization that occurs during allogeneic pregnancy in humans seldom harms the fetus. Indeed, the type of immunity stimulated by allopregnancy appears to be incapable of mediating allograft rejection (Wegmann et al., 1978), and repeated exposure to paternal histocompatibility antigens in pregnancy tends to produce a state of tolerance (Prehn, 1960; Breyere and Barrett, 1961; Smith and Powell, 1977) in spite of the concomitant presence of sensitized maternal T cells (Maroni and Parrott, 1973; Youtananukorn et al., 1974; Rocklin et al., 1976).

The precise mechanism by which allografts are rejected is not completely understood. Rejection of solid tissue grafts, such as skin, solid tumor, heart, and kidney grafts, appears to be dependent on the participation of functional thymus-derived lymphocytes (T cells) and, in many instances, by infiltration of the graft by cytotoxic cells, such as cytotoxic T lymphocytes (CTL) (De Lustro and Haskill, 1978; Hayry et al., 1979). Some transplantation immunologists have attributed the survival of antigenic grafts in immunized recipients to the action of "blocking antibodies" that have been thought to prevent contact between cells of the graft and circulating "sensitized" cytotoxic cells in pregnant mice (Hellström and Hellström, 1970). Maternal IgG molecules appear to bind nonspecifically to blastocysts prior to implantation, and IgG molecules are also avidly taken up by the trophoblast layer of the placenta (Faulk et al., 1974; Bernard, 1977). Moreover, IgG molecules eluted from the placenta can nonspecifically block the stimulation of maternal lymphocytes by paternal cells and can enhance the growth of paternal tumors in an immunologically specific manner in otherwise immunocompetent histoincompatible hosts (Jeanette et al., 1974; Chaouat et al., 1979). IgG molecules obtained from the blood of women undergoing successful pregnancy can also block the secretion of the lymphokine MIF (macrophage migration inhibition factor) by sensitized maternal lymphocytes, and these antibodies are absent in women who abort their fetuses (Rocklin et al., 1976). Indeed, it is possible that the difficulty that many investigators have encountered in compromising fertility by immunizing the female against the histocompatibility antigens of her mate may be attributable to the stimulation of blocking antibodies by which a hypoantigenic fetal graft may be selectively favored or enhanced (Mitchison, 1953; Prehn and Lape, 1971; Segal et al., 1979). There are several problems with this explanation of survival protection of the antigenic fetoplacental unit during pregnancy. First, much of the IgG bound to placenta may represent a nonspecific uptake rather than paternal H-antigen-specific antibody that would block recognition by maternal lymphocytes (Faulk et al., 1974). Second, the presence of antibody in the placenta would not be expected to provide protection against sensitized maternal cells that might cross the placenta, enter the fetus, and cause graft-versus-host disease. Third, successful pregnancy has been reported in agammaglobulinemic females, where no blocking antibody should be synthesized (Holland and Holland, 1966; Kobayashi et al., 1980). Bell and Billington (1981) have recently reported that only H-2^b strain mice produce anti-H-2 antibody as a result of allopregnancy. Fourth, it is far from clear that MIF-producing T cells can effect allograft rejection. Nomoto et al. (1975) found that noncytotoxic sensitized cells producing MIF were ineffective in rejecting an antigenic tumor, whereas inoculation of sensitized cells that were cytotoxic but did not produce MIF was effective in causing rejection. Inhibition of target lysis by sensitized CTL, however, is only transiently inhibited by coating of the targets with alloantibody (Faanes et al., 1973). Thus, blocking factors would not be expected to provide adequate protection for the fetal allograft against CTL-type effector cells. Thomas and Shevach (1978) have shown that antibody fails to block T cell recognition of antigen on the surface of macrophages when the density of antigen is too low to allow antibody binding. Thus, a low antigen density on trophoblast would afford protection against antibody but not necessarily against sensitized T cells. Taken together, the above data suggest that the most likely mechanism explaining protection of the fetoplacental unit would be *either* the prevention of the generation of cytotoxic cells reactive with antigens expressed by trophoblast cells *or* a localized suppression of the accumulation, activation, or action of cytotoxic cells at the maternal-fetoplacental interface.

One should recall that *repeated* allogeneic pregnancies seem able to generate a state of tolerance to paternal H antigens which is systemic in nature (Prehn, 1960; Breyere and Barrett, 1961; Smith and Powell, 1977). Systemic tolerance to H-2 alloantigens that has been deliberately induced in mice by treatment with antigen, adjuvant, and immunosuppressant treatment is associated with a suppression of the ability to generate CTL *in vivo* in response to alloantigen challenge (Brooks et al., 1975). With these facts in mind, we set out to determine which types of mechanisms are operative in protecting the fetal allograft from maternal immune rejection during first pregnancy, and we decided to devote most of our attention to the regulation of CTL in pregnant inbred C3H/HeJ and CBA mice.

CTL IN SPONTANEOUS RESORPTIONS OF XENOGENEIC MOUSE FETUSES

Allen (1977) reported that female horses mated with donkeys mount a spontaneous immunologic attack on the hybrid trophoblast. Both antibodies and sensitized cytotoxic T lymphocytes seem to arise in response to xenografts (Hamilton and Gaugas, 1972; Simpson et al., 1973; Dennert, 1974). Rejection of skin xenografts seems to be mediated primarily by antibody

(Hamilton and Gaugas, 1972), whereas rejection of xenogeneic tumor requires sensitized T cells (Simpson et al., 1973). In the case of a fetal xenograft, Allen has observed that 90 per cent of mares undergoing (horse x horse) matings produce antibodies against the antigens of their foals during the first pregnancy, whereas in (horse x donkey) matings, where there is maternal attack of the hybrid trophoblast, antibodies were detectable in only 33 per cent of the mares, the antibody titer was low, and the appearance of antibody was delayed relative to the time of appearance in (horse x horse) pregnancies. (Goat x sheep) matings provide another example in which the female goat rejects her hybrid embryo with a high frequency, and here the mechanisms also appear to be immunological in nature (Tucker et al., 1971; McGovern, 1973). Rejection of the hybrid embryos did not appear to be attributable to hemolytic antibody production by the mothers (Tucker, 1971).

Some studies of xenopregnancy have also been conducted in rodents. C3H trophoblast appears to be rejected when transplanted into presensitized rat hosts (Simmons and Russell, 1967a, 1967b). Similar observations were made when rat blastocysts were transferred into mouse uteri (Håkansson et al., 1978), but survival is prolonged if xenogeneic blastocysts are implanted in immunodeficient congenitally athymic or irradiated animals.

Mus caroli, a species of wild mouse from Southeast Asia, does not normally interbreed with *Mus musculus*, but hybrid embryos can be produced by artificial insemination (West et al., 1977). Only about 1 to 1.5 per cent of the hybrid embryos survive to term. The embryos seem to implant normally but then die at midterm, suggesting the possibility that death represents a maternal rejection (Frels et al., 1980). To investigate this idea, we have studied *Mus caroli* blastocysts transplanted into pseudopregnant ICR and C3H/HeJ mice. These blastocysts implant on day 4 of pregnancy, but begin to undergo resorption by day 9, and 80 per cent die by day 11. C3H/HeJ mice generate cytotoxic cells against *Mus caroli* target cells in MLC reactions *in vitro*, and this reaction is augmented in the lymph nodes draining the uterus for several weeks following resorption. We have also recovered cytotoxic cells from embryos undergoing resorption, and these cells have the physical properties of cytotoxic T cells on velocity sedimentation analysis and are destroyed by anti-T cell serum and complement, as well as by anti-Lyt 1 and anti-Lyt 2 plus complement (Croy et al., 1981 and submitted). Although C3H/HeJ mice do develop cytotoxic serum antibodies against *Mus caroli* antigens, one of our mice has resorbed the incompatible fetuses without ever developing detectable antibodies against concanavalin A-transformed *Mus caroli* lymphoblasts. Taken together with previous data on interspecies pregnancies, these observations lead to the conclusion that the antigenic fetoplacental unit may be destroyed by immunologic mechanisms and that CTL may play an important role in this process. The regulation of CTL in pregnancy may therefore be quite important. Of course, it is also theoretically possible that the cytotoxic cells are, in this case, a consequence rather than a cause of embryonic death.

LOCAL REGULATION OF THE CTL RESPONSE IN ALLOPREGNANT MICE

One manifestation of successful allogeneic pregnancy that has been proposed to represent a maternal reaction to the antigens of the conceptus is enlargement of the lymph nodes draining the uterus (DLN) in rats (Beer and Billingham, 1974; Beer et al., 1975; Forster et al., 1979; McLean et al., 1980) in mice (Maroni and de Sousa, 1973; Baines et al., 1977; Ansell et al., 1978), and in certain strains of hamsters (Head et al., 1978). DLN enlargement has been reported to be greater in allogeneic than in syngeneic matings (Maroni and de Sousa, 1973) and is abolished if the mother is rendered tolerant to paternal antigens prior to mating (Beer et al., 1975). The explanation for DLN enlargement is not clear. Ansell et al. (1978) have suggested that one mechanism may be enhanced trapping of circulating T lymphocytes. On the other hand, B lymphocyte proliferation has been reported to occur during the first 48 hours following mating (McLean et al., 1980). Taken together, the above data suggest that the DLN enlargement occurring during allogeneic pregnancy may represent an immunologic reaction to the paternal alloantigens of the fetus.

Not all investigators have found the anticipated DLN enlargement in allopregnant mice. Chambers and Clarke (1979) observed as much enlargement in syngeneically mated mice, and Heatherington and Humber (1977) could not correlate the degree of DLN weight gain with the degree of histoincompatibility between mother and fetus. However, it is difficult to predict the strength of an adult histocompatibility antigen in a fetal tissue, and one must also take into consideration the presence of immunogenic embryonal antigens on the fetus (Hamilton et al., 1976; Salinas et al., 1978; Johnson and Calarco, 1980). Furthermore, the time during pregnancy that DLN weight is assessed may be important (Maroni and de Sousa, 1973, Chambers and Clarke, 1979; McLean et al., 1980; Shaya et al., 1981). In some strain combinations, enlargement may occur only as either an early or a late phenomenon. We have carefully studied the cell recovery from the DLN of C3H/HeJ and CBA/J female mice following mating to DBA/2J males (Clark, 1982). Following mating, there is a rapid but transient enlargement comparable to that described by McLean et al. (1980) and then atrophy sets in, reaching a nadir at about the time of implantation. At midpregnancy, a second burst of hypertrophy occurs and is followed by a second phase of atrophy. Similar atrophic changes have been reported in the study of DLN in women reaching the end of a normal pregnancy (Nelson et al., 1973). A third burst of hypertrophy in our allopregnant mice seems to develop just prior to or at term.

In order to evaluate the functional changes in the DLN from allopregnant C3H mice, we placed particular emphasis on *in vivo* methods that would be free of potential artifacts and pitfalls that may complicate the use of *in vitro* methods. To determine if pregnancy impaired the ability of maternal T cells in the DLN to attack and destroy the fetus, we isolated the DLN lymphocytes from 14½-day pregnant mice and inoculated them into newborn C3D2F$_1$ mice. It was found that the mortality of the recipients was greater when DLN cells from virgin donors were injected (Clark and McDermott, 1978). Furthermore, the reduced mortality in recipients of DLN from pregnant donors was associated with a decrease in the generation in the recipients' spleens of cells cytotoxic against target cells bearing the paternal H-2^d alloantigen as measured in a short-term *in vitro* ^{51}Cr-release assay. We also noted that the magnitude of the GVH as assessed by the spleen/body weight ratio method seemed to show an impairment when DLN cells were studied (0.00489 ± 0.00036 mean ± SEM uninjected neonates [N = 8], 0.00575 ± 0.00017 when injected with 2 x 10^6 F$_1$ DLN cells [N = 6], 0.00979 ± 0.00056 when injected with 2 x 10^6 virgin C3H DLN [N = 8], and 0.00502 ± 0.00046 when injected with 2 x 10^6 DLN cells from pregnant donors [N = 8]). These results contrast sharply with a report of *increased* GVH reactivity when mesenteric lymph node cells from pregnant mice were studied in an analogue of the splenomegaly assay (Barg et al., 1978). This outcome could be explained by differences between the strains of mice used or, more interestingly, by an anatomical restriction of suppression to the DLN.

Since it was possible that the impaired GVH reactivity we observed was caused by an increased sensitivity to the trauma of cell isolation in pregnant DLN, we tested CTL generation directly *in situ*. Using this method, we found a significant reduction in the CTL generation following alloantigen challenge of the DLN of allopregnant mice (Clark and McDermott, 1978; Clark et al., 1980). In contrast, the CTL response in the popliteal and inguinal nodes following footpad challenge was not impaired and may even have been enhanced (Clark et al., 1980). These data support the idea of a localized suppression of CTL generation in the DLN. In addition, it was found that the antibody response to intravaginal inoculation of sheep red cells on day 7½ of pregnancy as measured by the plaque assay (IgM + IgG + IgA plaques) was not impaired in the DLN (Clark et al., 1980). Thus, suppression in the DLN in early pregnancy seemed to selectively suppress the CTL response. We have not studied IgG subclasses, however, and it should be noted that production of cytotoxic antibody may be impaired during pregnancy (Bell and Billington, 1980).

IN VITRO ANALYSIS OF THE MECHANISM OF CTL SUPPRESSION DURING FIRST ALLOGENEIC PREGNANCY

We analyzed the mechanisms regulating CTL in allopregnant mice with two types of *in vitro* culture systems (Clark et al., 1980). The frequency of the precursors of CTL (CTLp) was measured in limiting dilution microcultures (Teh et al., 1977), and the proliferation of these CTLp into CTL was tested in mass MLC culture in 3 ml of medium in 17 x 100 mm tubes—a procedure that allows regulatory cell-cell interactions to be studied. We found that the frequency of CTLp was not significantly altered in the DLN, but tube cultures produced significantly less cytotoxic activity when DLN from allopregnant mice were used. Some suppression was also found in suspensions of axillary, brachial, and inguinal nodes that drain the mammary glands. Although the generation of CTL by spleen cells from pregnant mice showed an impairment in tube cultures, the defect in spleen could be attributed to a reduction in the frequency of CTLp due to dilution by proliferating hemopoietic cells, which account for most of the splenic enlargement during pregnancy (Clark et al., 1980). Indeed, we found that increased CTL generation per CTLp in the spleen is in agreement with the results obtained by Smith et al. (1978), who studied spleen cells from allopregnant mice. Taken together, these data suggest that there is a localized impairment in the generation of CTL from CTLp which occurs in the DLN and to some extent in the PLN of allopregnant mice, but not in the spleen.

We could also generate a soluble suppressive activity from the DLN of allopregnant mice, and the properties of this suppressive activity have been described elsewhere (Clark et al., 1980; Clark, 1981; Clark and McDermott, 1981):

1. Soluble suppressor activity behaves to a great extent as a localized activity, being present in DLN or PLN (which include the lymph nodes draining breast tissue). Activity was less often obtained from thymus and occasionally (two of seven experiments) from spleen.
2. As shown in Figure 16-1a, suppressive activity was elaborated optimally between 24 and 48 hours of incubation at 37°C. Figure 16-1b shows that, in this experiment, no suppressive activity could be obtained from spleen even after 72 hours of culturing.
3. Suppression is not H-2 specific (Clark et al., 1980; Clark and McDermott, 1981). Soluble suppressor activity from syngeneically

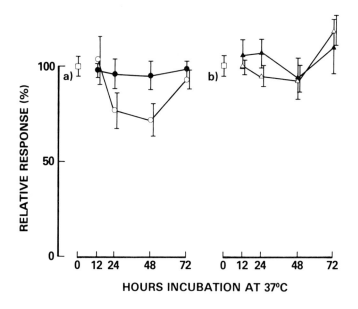

HOURS INCUBATION AT 37°C

FIGURE 16-1. Kinetics of elaboration of soluble suppressor activity *in vitro*. (a) Supernatant was obtained by incubation of a mixture of PLN and DLN from virgin (●) and 12- to 16-day-pregnant (O) mice for the length of time indicated. The CTL activity produced by test cultures to which the supernatant was added is compared with the result of cultures containing medium alone (■). (b) The effect of supernatants derived from the spleen cells of virgin (▲) and pregnant (Δ) mice.

        pregnant C3H DLN was able to suppress the CTL response to the $H-2^d$ alloantigen.

4.   Suppressive activity does not pass through an Amicon ultrafiltration membrane (25 000 dalton cutoff). This observation suggested that molecules with suppressive activity, such as thymidine, were not responsible for suppression (Clark et al., 1980).

5.   Soluble suppressor activity appears to inhibit cellular proliferation in the CTL response (Clark, 1982).

6.   Suppressive activity was absent in the supernatants derived from the DLN of CBA mice spontaneously resorbing their (CBA X DBA) F_1 embryos (Clark et al., 1980).

7.   Soluble suppressor activity during pregnancy appeared to parallel suppressor cell activity that was detectable by addition of untreated or mitomycin C-treated DLN or PLN to virgin PLN in the tube culture system. Both soluble and cell-associated suppression could be detected in DLN within 48 hours of mating and reached a maximum at the time of implantation. Suppression then decreased in intensity and was

minimal at approximately 9½ to 10½ days of pregnancy. Suppression subsequently recurred in the second half of pregnancy and disappeared at, or shortly after, parturition (Clark and McDermott, 1981).

8. Soluble suppressor activity was obtained from cells that were the same size as suppressor cells. DLN and PLN were separated by velocity sedimentation. It was found that the suppressor cell in DLN was predominantly a small lymphocyte sedimenting at 3 mm per hour. Soluble suppressor activity was recovered from a similar-sized cell population in PLN (Clark and McDermott, 1981).

Taken together, the above data indicate that suppression of CTL generation in successful allogeneic pregnancy can be attributed to a soluble non-MHC-specific suppressor factor(s) that decreases cellular proliferation. Indeed, we have found that the *peaks* of suppressor activity correlate rather closely with phases of DLN atrophy that follow episodes of increased DLN cellularity (Clark, 1982; Clark and McDermott, 1981).

We tested our suppressor to determine if it was a suppressor T cell (Kolsch et al., 1975; Truitt et al., 1978), but found that treatment of pregnant DLN or the fraction sedimenting between 2 and 4 mm per hour with heterologous anti-T cell serum (rabbit anti-mouse brain serum) and complement failed to eliminate suppressor cell activity (Clark and McDermott, 1981). Figure 16-2 illustrates some titration curves comparing the activity of mitomycin C-treated cells with the activity of those treated with RAMB plus complement plus mitomycin C. It can be seen that the suppressor activity of treated cells is increased by this treatment. Treatment with anti-Thy 1.2, Lyt 1.1, or Lyt 2.1 and complement did not affect suppression either, and treatment of virgin lymph node cells with these antisera did not render them suppressive (unpublished data). Thus, we have concluded that the suppressor cell in mice undergoing *first* allogeneic pregnancy cannot be shown in our laboratory to represent a classical suppressor T cell. A similar conclusion was reached elsewhere by Smith (1981), who studied suppressor cells in the DLN of multiparous allopregnant mice.

The resistance of our suppressor cell to anti-T cell serum and complement suggested the possibility of a suppressor B cell or null cell. Chatterjee-Hasrouni et al. (1980) described an increase in null cells in the DLN of allopregnant mice, and Roder et al. (1977) have described an age-associated splenic suppressor null cell with physical properties similar to our DLN suppressor cell. Suppressor "null" cells have also been identified in the spleens of neonatal mice and appear to precede the development of suppressor T cells in neonates (Rodriguez et al., 1979).

There have also been described (Calkins et al., 1976; Zembala et al., 1976; Ninneman, 1978; L'Age-Stehr et al., 1980) suppressor B cells that may enhance allograft survival (Lauchart et al., 1980). In this regard, it is

351

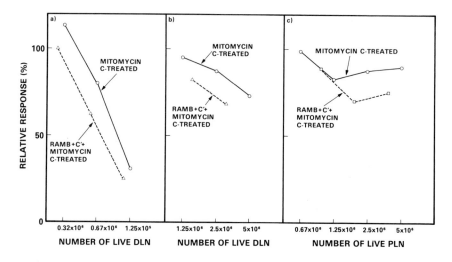

FIGURE 16-2. Suppressive effect of cells from pregnant mice is not eliminated by treatment with anti-T-cell serum and complement.
The cells obtained from mice 12 to 16 days after mating were treated with mitomycin C alone or with rabbit antimouse brain serum followed by complement (rabbit serum) containing mitomycin C (50 μg/ml). The CTL activity of cultures to which different numbers of treated cells were added was expressed as a percentage of the CTL yield from cultures to which no cells were added. (a) Experiment 1, DLN cells. (b) Experiment 2, DLN cells. (c) PLN cells from the DLN donors of Experiment 2.

intriguing that the localization of IgA-containing lymphoblasts in genital mucosal tissue, such as the uterus, appears to be regulated in part by hormones (McDermott et al., 1980; Wira et al., 1980). Indeed, the rapid appearance of suppressor activity we found following mating and temporary remission at the midpoint in pregnancy mimics the blood progesterone level that has been described in mice (Murr et al., 1975).

Progesterone has been proposed to represent "Nature's immunosuppressant" (Siiteri et al., 1977) and can block skin allograft rejection without seriously compromising the humoral immune response to the graft (Beer and Billingham, 1979)—a pattern of immunosuppression similar to that which we observed in the DLN (Clark et al., 1980). Furthermore, we found that CTL generation in the DLN was as impaired by pseudopregnancy as by allogeneic pregnancy (Clark and McDermott, 1981), which further indicates an important role for hormone in suppression. There are several arguments against a direct immunosuppressive role for progesterone in the immunosuppression of pregnancy. The concentration of progesterone that must be added *in vitro* to suppress human lymphocyte responses (Clemens et al., 1978) is greater than that present in the human placenta or syncytiotrophoblast membrane (N.C. Smith and Brush, 1978), and concentrations of progesterone used *in vivo* to prevent skin graft rejection produce serum progesterone levels considerably greater than those encountered during normal pregnancy (Sanyal, 1978; Beer and Billingham, 1979). We found that the concentration of progesterone in our suppressive supernatants was also too low to explain suppression (Clark and McDermott, 1981). Furthermore, we observed that the CTL generation in the DLN of pseudopregnant mice was suppressed *after* the time that the elevated progesterone levels of pseudopregnancy are reported to return to normal (Chambers and Clarke, 1979). Taken together, the above data suggest an indirect hormonal basis for suppression during pregnancy.

There is some evidence for the existence of a hormone-dependent suppression of rejection of grafts in the uterus. Progesterone treatment, pseudopregnancy, and normal pregnancy all prolong the survival of intrauterine grafts of allogeneic skin or tumor cells (Moriyana and Sugawa, 1972; Beer and Billingham, 1974), whereas pregnant rodents promptly reject first-set skin grafts placed at other sites (Beer and Billingham, 1974). Although intrauterine skin allografts are promptly rejected in a second-set manner in immune recipients (Beer and Billingham, 1974), this rejection is histologically characterized by infiltration with polymorphonuclear leukocytes (Dodd et al., 1980), suggesting the possibility that humoral antibody, leading to antibody-dependent cell-mediated cytotoxicity rather than CTL-mediated rejection, may be the primary effector of skin graft rejection in this situation (Hamilton and Gaugas, 1972; Jooste et al., 1973). Allogeneic blastocysts that are rejected if placed under the kidney capsule of suitably immunized hosts survive if allowed to implant in the uterus of immunized animals (Kirby et al., 1966). Fuchs et al. (1977) have reported local suppression of the response of maternal lymphocytes in the human placenta. It is possible that hormonal changes in pregnancy could cause such suppression by stimulation of the production of suppressive factors by endometrium, decidua, or trophoblast (Denari et al., 1976; Barg et al., 1978; Bell, 1979; McIntyre and Faulk, 1979). On the other hand, Bernard et al. (1978) reported the presence of a population of 7 μm-diameter "null" cells in the decidua of pregnant mice. Cells of this diameter usually sediment at 3 mm per hour, where we find our suppressor cell activity. We therefore conducted some experiments to test for a suppressor cell in the genital tract of allopregnant C3H/HeJ mice.

Table 16-1 illustrates an experiment in which the lymphocytes isolated from uterine blood and decidua were compared with lymphocytes obtained from DLN and cardiac blood. It can readily be seen that the uterine and decidual lymphocytes were potent suppressors. Decidual cells were suppressive in three of four tests, and uterine lymphocytes were suppressive in six of seven tests at 10^5 cells per culture; if decidua and uterine blood were considered together, seven of seven experiments showed suppression. This suppression was equal to or greater than the suppressive activity of DLN in six of seven studies. These observations suggest that the most potent suppression of CTL generation

TABLE 16-1
Local Suppressor Cells in the Genital Tract of Allopregnant C3H/HeJ Mice

	RELATIVE CYTOTOXIC ACTIVITY PER CULTURE (%)	
TEST CELLS ADDED	1 x 10^6 TEST CELLS	1 x 10^5 TEST CELLS
3 x 10^6 virgin PLN + virgin PLN	100 ± 8	100 ± 9
3 x 10^6 virgin PLN + virgin DLN	117 ± 11	129 ± 22
3 x 10^6 virgin PLN + pregnant DLN*	55 ± 8	116 ± 20
3 x 10^6 virgin PLN + virgin cardiac LC†	115 ± 10	102 ± 14
3 x 10^6 virgin PLN + pregnant cardiac LC	0§	67 ± 9§
3 x 10^6 virgin PLN + pregnant uterine LC	0§	36 ± 8§
3 x 10^6 virgin PLN + decidual LC‡	ND	10 ± 9§

* Draining lymph node cells (DLN) were removed from C3H/HeJ mice 14.5 to 19.5 days pregnant by DBA/2 males. The DLN were added to the 3 x 10^6 test pregnant lymph node cells (PLN) at the indicated number together with irradiated (C3D2)F₁ stimulators. No mytomycin C treatment of the test cells was used. The resulting cytotoxic activity per culture was quantified as described in detail elsewhere (Miller and Dunkley, 1974; Clark et al., 1977) and is expressed as a percentage of the control cultures containing virgin PLN alone. The 4 x 10^6 virgin PLN produced 179 units of activity and 3.1 x 10^6, 125 units.

† Cardiac blood was aspirated into a heparinized plastic syringe, the blood diluted, and the lymphocytes obtained by centrifugation over Lympholyte M (Cedarlane Laboratories, Hornby, Ontario, Canada). Uterine blood was obtained by puncture of the uterine veins, and 0.2 to 0.3 ml of blood aspirated.

‡ To obtain decidual lymphocytes, the fetoplacental unit was removed and the decidua scraped loose. Decidua pressed through a 60-mesh screen was resuspended and centrifuged over Lympholyte M.

§ Statistically significant suppression relative to virgin PLN.

may be associated with cells within the uterus and placenta. The nature of the suppressor cell, its kinetics, and the role of hormones in its generation and localization are currently under study. It is possible that the suppression we observe in PLN and occasionally in thymus and spleen may represent a distant effect of a more potent suppressor activity within the uterus. A localized antigen-nonspecific suppressor mechanism has considerable theoretical advantages in protecting the fetal allograft. Systemic suppression would be expected to seriously threaten the mother's survival since her susceptibility to infection would increase. Indeed, various investigators have reported an increased susceptibility to pneumococci, malarial parasites, and certain viruses, particularly herpes, where there exists a local increase in susceptibility within the genital tract (Young and Gomez, 1979; Harkness,1980; van Zon and Eling, 1980). Nevertheless, the magnitude of the increase in susceptibility to infections is slight relative to the effects of other nonspecific immunosuppressive agents that are given to combat rejection of kidney or heart allografts. We have shown that nonspecific suppression in the DLN occurs prior to implantation—a timing that is optimal to block initiation of an antigraft reaction. The magnitude of suppression in the DLN is comparable to that seen in the spleens of H-2 skin graft-tolerant and alloantigen-challenged recipients (Brooks et al., 1975). In an outbred population, each successive fetus may differ antigenically. Nonspecific suppression would not cause an undue selective pressure favoring a fetus of one antigenic type.

LOCAL NONSPECIFIC SUPPRESSOR AND ANTIGEN MAY ACTIVATE SYSTEMIC ANTIGEN-SPECIFIC SUPPRESSORS

One consequence of repeated allogeneic pregnancy that has been reported is the induction of antigen-specific transplantation tolerance to paternal tissues (Prehn, 1960; Breyere and Barrett, 1961; Smith and Powell, 1977). In the case of tolerance to the male H-Y antigen, tolerance can be adoptively transferred from multiparous females to virgin mice by suppressor T cells (Smith and Powell, 1977). Similar results have been obtained in mice for other paternal histocompatibility antigens (Chaouat et al., 1979, 1980) and in man (McMichael and Sasazuki, 1977).

Nonspecific suppressor cells, such as suppressor B cells and suppressor null cells, may act by inducing suppressor T cell activity (Ninneman, 1978; Caulfield and Cerney, 1980; L'Age-Stehr et al., 1980). Nonspecific suppression, whether achieved by natural genetic means (Kapp et al., 1976) or by administration of immunosuppressants (Brooks et al., 1975; Reiger et al., 1978; Simpson and Gozzo, 1979), has also been reported to favor the generation of suppressor T cells. We found that 1 to 3 allogeneic (but not syngeneic) pregnancies impaired CTL generation *in vivo* in the spleens of P-815 challenged C3H mice, and this impairment persisted beyond the time of weaning. We also found that inoculation of TNP-modified syngenic spleen cells into the vaginal wall of 7½-day pregnant mice markedly suppresssed the subsequent ability of PLN or spleen cells to generate TNP-specific CTL (Clark, 1982). Similarly treated virgin mice did not develop suppression.

Chaouat et al. (1981) have recently reported that injection of alloantigen together with a placental extract facilitates the generation of suppressor T cells. Sensitization by intrauterine skin allografts during pregnancy can, however, prime for systemic transplantation immunity (Rukavina et al., 1981). Further studies must therefore be done to clarify the requirements for the induction of tolerance rather than immunity to the fetal allograft or fetal xenograft.

SUMMARY AND CONCLUSIONS

Reproduction in an outbred population generates a fetus that bears paternal transplantaion antigens against which the mother mounts an immune response. Nevertheless, the fetus is usually not rejected. In a mouse mated xenogeneically, however, the fetus can certainly be attacked and destroyed. Following mating, a suppressor activity rapidly appears in the lymph nodes draining the uterus and is absent in the DLN of CBA mice spontaneously resorbing their fetuses. Suppression may be attributed to a small lymphoid cell from which a soluble suppressor activity can be obtained. Suppression is nonspecific, and the suppressor cell appears to be resistant to anti-T cell reagents. Suppression in early pregnancy may be localized *in vivo* to certain lymphoid organs, and evidence of systemic suppression activity may be seen in circulating blood lymphocytes in late pregnancy. The most potent source of suppression appears to be the uterine blood and decidual lymphocytes. This localized, nonspecific suppression may play a role in induction of tolerance and generation of suppressor T cells to alloantigens presented to the immune system by the genital route during pregnancy The precise relevance of nonspecific and specific suppressor mechanisms to the survival of the fetal graft remains to be defined.

ACKNOWLEDGMENTS

This work was supported in part by grants from the Medical Research Council of Canada, the World Health Organization, and the Ministry of Health of Ontario.

REFERENCES

Allen, W.R. 1979. Maternal recognition of Pregnancy and immunological implications of trophoblast-endometrium interactions in equids. In *Maternal Recognition of Pregnancy*. J. Whelan, ed. Ciba Foundation Symposium 64, New York: Excerpta Medica, pp. 323-352.

Ansell, J.D., C.M. McDougall, G. Speedy, and C.J. Inchley. 1978. Changes in lymphocyte accumulation and proliferation in lymph nodes. *Clin. Exp. Immunol.* 31:397.

Baines, M.G., H.F. Pross, and K.G. Miller. 1977. Effect of pregnancy on the maternal lymphoid system of mice. *Obstet. and Gynecol.* 50:457.

Barg, M., R.C. Burton, J.A. Smith, G.A. Luckenbach, J. Decker, and G.F. Mitchell. 1978.

Effects of placental tissue on immunological responses. *Clin. Exp. Immunol.* 34:441.

Bell. S.C. 1979. Immunochemical identity of decidualizationassociated protein and α2 acute-phase macroglobulin in the pregnant rat. *J. Reprod. Immunol.* 1:193.

Bell. S.C., and W.D. Billington. 1980. Major anti-paternal alloantibody induced by murine pregnancy is non-complement-fixing IgG1. *Nature* 288:387.

Bell,S.C., and W.D. Billington. 1981. Humoral immune response in murine pregnancy. I. Anti-paternal alloantibody levels in maternal serum. *J. Reprod. Immunol.* 3:3.

Beer, A.E., R.E. Billingham, and S.L. Yang. 1972. Further evidence concering the autoantigenic status of the trophoblast. *J. Exp. Med.* 35:1177.

Beer, A.E., and R.E. Billingham. 1974. Host responses to intrauterine tissue, cellular and fetal allografts. *J. Reprod. Fertil.* Suppl. 21:59.

Beer, A.E., J.S. Scott, and R.E. Billingham. 1975. Histocompatibility and maternal immunological status as determinants of fetoplacental weight and litter size in rodents. *J. Exp. Med.* 142:180.

Beer, A.E., and R.E. Billingham. 1979. Maternal immunological recognition mechansims during pregnancy. In *Maternal Recognition of Pregnancy.* J. Whelan, ed. Ciba Foundation Symposium 64, Amsterdam: Excerpta Medica, pp. 293-322.

Bernard, O. 1977. Possible protecting role of maternal immunoglobulins in embryonic development in mammals. *Immunogenetics* 5:1.

Bernard, O., M.P. Scheid, M.A. Ripoche, and D. Bennett. 1978. Immunologic studies of mouse decidual cells. I. Membrane markers of decidual cells in the days after implantation. *J. Exp. Med.* 148:580.

Bresnihan, B., R.R. Gregor, N. Oliver, R.M. Newkanio, R.E. Lovins, W.P. Faulk and G.R.V. Hughes. 1977. Spontaneous abortion in systemic lupus erythematosus: an association with trophoblast reactive lymphocytotoxic antibodies. *Lancet* ii:1205.

Breyere, E.J., and M.K. Barrett. 1961. Tolerance induced by parity in mice incompatible at the H-2 locus. *J. Nat. Cancer Inst.* 27:409.

Brooks, C.G., L. Brent, P.J. Kilshaw, R.R.C. New, and M. Pinto. 1975. Specific unresponsiveness to skin allografts in mice. IV. Immunologic reactivity of mice treated with liver extracts, *Bordatella pertussis,* and antilymphocyte serum. *Transplantation* 19:134.

Calkins, D.F., O. Orbach-Arbouys, O. Stutman, and R.K. Gershon. 1976. Cell interactions in the suppression of *in vitro* antibody response. *J. Exp. Med.* 143:1421.

Chaouat, G., and G.A. Voisin. 1979. Regulatory T cell subpopulations in pregnancy. I. Evidence for suppressive activity in the early phase of the MLR. *J. Immunol.* 122:1283.

Chaouat, G., and G.A. Voisin. 1980. Regulatory T-cell subpopulations in pregnancy. II. Evidence for suppressive activity of the late phase of MLR. *Immunol.* 39:239.

Chaouat, G., G.A. Voisin, D. Excalier, and P. Robert. 1979. Facilitaion reaction (enhancing antibodies and suppressor cells) from the mother to the paternal antigens of the conceptus. *Clin. Exp. Immunol.* 35:13.

Chambers, S.P., and A.G. Clarke. 1979. Measurement of thymus weight, lumbar node weight and progesterone levels in syngeneically pregnant, allogeneically pregnant, and pseudopregnant mice. *J. Reprod. Fertil.* 55:309.

Chatterjee-Hasrouni, S., and P.K. Lala. 1979. Localization of H-2 antigens in trophoblast cells. *J. Exp. Med.* 149:1238.

Chatterjee-Hasrouni, S., V. Santer, and P.K. Lala. 1980. Characterization of maternal small lymphocyte subsets during alogeneic pregnancy in the mouse. *Cell. Immunol.* 50:290.

Clark, D.A., R.A. Phillips, and R.G. Miller. 1977. Characterization of the cells that suppress the cytotoxic activity of T lymphocytes. II. Physical properties, specificity, and developmental kinetics of the inhibitor splenic non-T cell population. *Cell Immunol.* 34:25.

Clark, D.A., and M.R. McDermott. 1978. Impairment of host vs graft reaction in pregnant mice. I. Suppression of cytotoxic T cell generation in lymph nodes draining the uterus. *J. Immunol.* 121:1389.

Clark, D.A., M.R. McDermott, and M.R. Szewczuk. 1980. Impairment of host vs graft reaction in pregnant mice. II. Selective suppression of cytotoxic T-cell generation correlates with soluble suppressor activity and with successful allogeneic pregnancy. *Cell Immunol.* 52:106.

Clark, D.A., and M.R. McDermott. 1981. Active suppression of host vs graft reaction in pregnant mice. III. Developmental kinetics, properties, and mechanism of induction of suppressor cells during first pregnancy. *J. Immunol.* 127:1267.

Clark, D.A. 1982. Impairment of host immunity and the survival of the fetus. In *Immunobiology of Transplantation, Cancer and Pregnancy.* P.K. Ray, ed. New York: Pergamon Press. In press.

Clemens, L.E., P.K. Siiteri, and D.P. Stites. 1979. Mechanism of immunosuppression of progesterone on maternal lymphocyte activation during pregnancy. *J. Immunol.* 122:197.

Collins, C.D., F.J. Chrest, and W.H. Adler. 1980. Maternal cell traffic in allogeneic embryos. *J. Reprod. Immunol.* 2:163.

Croy, B.A., J. Rossant, and D.A. Clark. 1981. Is there maternal immune rejection of the fetus in failed murine interspecies pregnancy? *J. Reprod. Immunol.* Suppl. 32:S32.

De Lustro, F., and J.G. Haskill. 1978. *In situ* cytotoxic T cells in a methylchlolanthrene-induced tumor. *J. Immunol.* 121:1007.

Denari, J.H., N.J. Germino, and J.M. Rosner. 1976. Early synthesis of uterine proteins after a decidual stimulus in the pseudopregnant rat. *Biol. Reprod.* 15:1.

Dennert, G. 1974. Effector mechanisms of cell-mediated immunity to xenogeneic cell antigens. *J. Immunol.* 113:201.

Dodd, M., T.A. Andrew, and J.S. Cotes. 1980. Functional behaviour of skin allografts transplanted to rabbit deciduomata. *J. Anat.* 130:381.

Faanes, R.B., Y.S. Choi, and R.A. Good. 1973. Escape from isoantiserum inhibition of lymphocyte-mediated cytotoxicity. *J. Exp. Med.* 137:171.

Faulk, W.P., M. Jeannet, W.D. Creighton, A. Carbondara, and F. Hay. 1974. Studies of the human placenta. II. Characterization of immunoglobulins on the trophoblast basement membranes. *J. Reprod. Fert.* (Suppl.) 21:43.

Faulk, W.P., C. Yeager, J.A. McIntyre, and M. Ueda. 1979. Oncofetal antigens of human trophoblast. *Proc. Roy. Soc. Lond. B.* 206:163.

Forster, P.M., J.M. McLean, and A.C.C. Gibbs. 1979. Lymphoid responses to pregnancy and pseudopregnancy in the rat. *J. Anat.* 128:837.

Frels, W.I., J. Rossant, and V.M. Chapman. 1980. Intrinsic and extrinsic factors affecting the viability of *Mus caroli* x *M. musculus* hybrid embryos. *J. Reprod. Fert.* 59:387.

Fuchs, T., L. Hammarstrom, E. Smith, and J. Brundin. 1977. *In vivo* suppression of uterine lymphocytes during early human pregnancy. *Acta Obstet. Scand.* 56:151.

Gill, T.G. 1977. Chimerism in humans. *Transplant. Proc.* 9:1069.

Goodlin, R.C., and L.A. Herzenberg. 1964. Pregnancy induced hemagglutinins to paternal H-2 antigens in multiparous mice. *Transplantation* 2:57.

Håkansson, S., O. Lundkuist, and B.O. Nilsson. 1978. Survival of rat blastocysts transplanted into the uterus of hyperimmunized mice during delay of implantation. *Int. J. Fertil.* 23:148.

Hamilton, M.S., I. Hellström, and G. Van Belle. 1976. Cell-mediated immunity to embryonic antigens of syngeneically and allogeneically mated mice. *Transplantation* 21:261.

Hamilton, M.S., A.E. Beer, R.D. May, and E.S. Vitelta. 1979. The influence of immunization of female mice with F9 teratocarcinoma cells on their reproductive performance. *Transplant. Proc.* 11:1069.

Harkness, R.A. 1980. Estrogens and host resistance. *J. Royal Soc. Med.* 73:161.

Hayry, P., E. von Willebrand, and A. Soots. 1979. *In situ* effector mechanisms in rat kidney allograft rejection. III. Kinetics of the inflammatory response and generation of donor-directed killer cells. *Scand. J. Immunol.* 10:95.

Hamilton, D.N.H., and J.M. Gaugas. 1972. Humoral and cellular factors in xenogeneic rejection by mice. *Transplantation* 6:620.

Head, J.R., M.S. Hamilton, and A.E. Beer. 1978. Maternal hamster immune responses to alloantigens of the fetus. *Fed. Proc.* 37:2054.

Hellström, K.E., and I. Hellström. 1970. Immunologic enhancement as studied by cell culture techniques. *Ann. Rev. Microbiol.* 24:373.

Herzenberg, L.A., D.W. Bianchi, J. Schroeder, H.M. Cann, and G.M. Iverson. 1979. Fetal cells in the blood of pregnant women. Detection and enrichment by fluorescence-activated cell sorting. *Proc. Nat. Acad. Sci. USA* 76:1453.

Hetherington, C.M., and D.P. Humber. 1977. The effect of pregnancy on lymph node weight in the mouse. *J. Immunogenetics* 4:271.

Holland, N.H., and P. Holland. 1966. Immunologic maturation in an infant of an agammaglobulinemic mother. *Lancet* II:1152.

Jeannet, M., W.P. Faulk, W.D. Crighton, and K. Fournier. 1974. Blocking of mixed lymphocyte cultures by IgG eluted from human placenta. In *Proceedings of 8th Leucocyte Culture Conference*. K. Lindahl-Kiessling and D. Osoba, eds. New York: Academic Press, pp.243-248.

Jenkinson, E.J., and V. Owen. 1981. Ontogency and distribution of major histocompatibility complex (MHC) antigens on mouse placental trophoblast. *J. Reprod. Immunol.* 2:173.

Johnson, L.V., and P.G. Calarco. 1980. Mammalian preimplanta development: the cell surface. *Anat. Record* 196:201.

Jooste, S.V., H.J. Winn, and P.S. Russell. 1973. Destruction of rat skin by humoral antibody. *Tranplant. Proc.* 5:715.

Kapp, J.A., C.W. Pierce, F. de la Croix, and B. Benacerraf. 1976. Immunosuppressive factor(s) extracted from lymphoid cells of nonresponder mice primed with L-glutanic acid-L-alanine-L-tyrosine (GAT). I. Activity and specificity. *J. Immunol.* 116:305.

Kirby, D.R.S., W.D. Billington, and D.A. James. 1966. Transplantation of eggs to the kidney and uterus of immunized mice. *Transplantation* 4:713.

Kobayashi, R.H., C.J. Hyman, and R. Stiehm. 1980. Immunologic maturation in an infant born to a mother with agammaglobulinemia. *Amer. J. Dis. Children* 134:942.

Kolch, E., R. Stumpf, and G. Weber. 1975. Low zone tolerance and suppressor T cells. *Transplantation Rev.* 26:56.

L'Age-Stehr, J., J. Teichman, R.K. Gershon, and H. Carter. 1980. Stimulation of regulatory T cell circuits by Ia-associated structures on activated B cells. *Eur. J. Immunol.* 10:21.

Lauchart, W., B.J. Alkins. and D.A.L. Davies. 1980. Only B lymphocytes induce active enhancement of rat cardiac allografts. *Transplantation* 29:259.

Maroni, E.S., and M.A.B. de Sousa. 1973. The lymphoid organs during pregnancy in the mouse. A comparison between syngeneic and allogeneic mating. *Clin. Exp. Immunol.* 31:107.

Maroni, E.S., and D.M.V. Parrott. 1973. Progressive increase of immunity against paternal transplantation antigens after multiple pregnancies. *Clin. Exp. Immunol.* 13:253.

McDermott, M.R., D.A. Clark, and J. Bienenstock. 1980. Evidence for a common mucosal immunologic system. II. Influence of the estrous cycle on B immunoblast migration into genital and intestinal tissue. *J. Immunol.* 124:2536.

McGovern, P.T. 1973. The effect of maternal immunity on the survival of goat X sheep hybrid embryos. *J. Reprod. Fertil.* 34:215.

McIntyre, J.A. and W.P. Falk. 1979. Trophoblast modulation of maternal allogeneic recognition. *Proc. Nat. Acad. Sci. USA* 76:4029.

McLean, J.M., E.I. Shaya, and A.C.C. Gibbs. 1980. Immune response to first mating in the female rat. *J. Reprod. Immunol.* 1:285.

McMichael, A.J., and T. Sasazuki. 1977. A suppressor cell in the human mixed lymphocyte reaction. *J. Exp. Med.* 146:368.

Milgrom, F., E. Comimi-Andrada, and A.P. Chaudhry. 1977. Fetal and neonatal fatality in rat hybrids from mothers stimulated with paternal skin. *Transplant. Proc.* 9:1409.

Miller, R.G., and M. Dunkley. 1974. Quantitative analysis of the ^{51}Cr release cytotoxicity assay for cytotoxic lymphocytes *Cell. Immunol.* 14:284.

Mitchison, N.A. 1953. The effect on the offspring of maternal immunization in mice. *J. Genetics* 51:406.

Moriyana, I., and T. Sugawa. 1972. Progesterone facilitates implantation of xenogeneic cultured cells in hamster uterus. *Nature* (New Biol.) 236:150.

Murr, S.M., G.H. Stabenfeldt, G.E. Bradford, and I.I Geshwind. 1975. Plasma progesterone during pregnancy in the mouse. *Endocrinol.* 94:1209.

Nelson, J.H., T. Lu, J.E. Hall, S. Krown, J.M. Nelson, and C.W. Fox. 1973. The effect of trophoblast on the immune status of women. *Amer. J. Obstet. Gynecol.* 117:689.

Ninneman, J.L. 1978. Melanoma associated immunosuppression through B cell activation of suppressor T cells. *J. Immunol.* 120:1573.

Nista, A., M.L. Sezzi, and S. Belleli. 1973. Pregnancy rejection induced by neuraminidase-treated placental cells. *Oncology* 28:402.

Nomoto, K., M. Sato, Y. Yano, K. Taiguchi, and K. Takeya. 1975. Dissoccation between delayed hypersensitivity and cytotoxic activity against syngeneic or allogeneic tumour grafts. In *Host Defences, Against Cancer and its Potentiation*. D. Mizano, G. Chihara, F. Fukuoka, T. Yato, and Y. Yamamura, eds. Baltimore: University Park Press, pp. 55-65.

Nymond, G., I. Heron, K.G. Jensen, and A. Landsqaard. 1971. Cytotoxic antibodies in serum of pregnant women at delivery. *Acta Pathol. Microbiol. Scand.* (B) 79:595.

Palm, J. 1974. Maternal fetal histoincompatibility in rats: an escape from adversity. *Cancer Res.* 34:2061.

Parmiani, G., and G. Della-Porta. 1973. Effects of anti-tumor immunity on pregnancy in the mouse. *Nature* (New Biol.) 241:26.

Pollack, M.S., N.Kapoor, N. Sorell, Y. Morishinia, B. Dupont, and R.J. O'Reilly. 1981. Absence of demonstrable suppressor cell activity in a severe combined immuno-deficiency patient with sustained engraftment of DR-positive maternal T cells. *Transplant. Proc.* 13:270.

Prehn, R.T. 1960. Specific hormograft tolerance induced by successive matings and implications concerning choriocarcinoma. *J. Nat. Cancer Inst.* 25:883.

Prehn, R.T., and M.A. Lappe. 1971. An immunostimulation theory of tumour development. *Transplant. Rev.* 7:26.

Reiger, M., J. Gunther, H. Kristofova, and I. Hilgert. 1978. Evidence of suppressor cell-mechanism of allograft tolerance induced by spleen extract and hydrocortisone. *Folia Biol.* (Praha) 24:145.

Rocklin, R.R., J.L. Kitzmiller, C.B. Carpenter, M.R. Garvoy, and J.R. David. 1976. Maternal - fetal relation. Absence of an immunologic blocking factor from the serum of women with chronic abortions. *New Engl. J. Med.* 295:1209.

Roder, J.C., D.A. Bell, and S.K. Singhal. 1977. Regulation of the immune response in autoimmune NZB/NZW F1 mice. I. The spontaneous generation of splenic suppressor cells. *Cell. Immunol.* 29:272.

Rodriguez, G., B. Anderson, H. Wigzell, and A.B. Peck. 1979. Non-T cell nature of the naturally occurring spleen-associated suppressor cells present in the newborn mouse. *Eur. J. Immunol.* 9:737.

Rukvina, D., N. Matejcic, and M. Doric. 1981. Consequences to maternal and offspring reactivity of immunologic manoeuvres during pregnancy. *Abstracts of the 4th International Congress of Immunology*, Paris, July 1981, Abstract #16.4.15.

Salinas, F.A., H.K.B. Silver, K.M. Sheikh, and S.B. Chandera. 1978. Natural occurrence of human tumour-associated anti-fetal antibodies during normal pregnancy. *Cancer* 42:1653.

Sanyal, M.K. 1978. Secretion of progesterone during gestation in the rat. *J. Endocrinol.* 79:179.

Segal, S., T. Siegal, H.Altaraz, A. Lev-El, A. Nevo, L. Nevel, A. Katzenelsin, and M. Feldman. 1979. Fetal bone grafts do not elicit allograft rejection because of oprotecting anti-Ia alloantibodies. *Transplantation* 28:88.

Sellens, M.H., E.J. Jenkinson, and W.D. Billington. 1978. Major histocompatibility complex and non-major histocompatibility complex antigens on mouse ectoplacental cone and placental trophoblastic cells. *Transplant.* 25:173.

Shaya, E.I., J.M. McLean, and A.C.C. Gibbs. 1981. Accumulation and proliferation of lymphocytes in the lymph nodes of the female rat following first mating. *J. Anat.* 132:137.

Simmons, R.L., and P.S. Russell. 1967. Immunologic interactions between mother and fetus. *Adv. Obstet. Gynecol.* 1:38.

Simmons, R.L. and P.S. Russell. 1967b. Xenogeneic antigens in mouse trophoblast. *Transplantation* 5:85.

Simpson, E., P.C.L. Beverley, and V. Jones. 1973. Lymphocyte-dependent antibody in a tumor xenograft system. *Transplantation Proc.* 5:161.

Simpson, M.A., and J.J. Gozzo. 1979. Isolation of suppressor cells from immunosuppressed and antigen-challenged mice. *Transplant. Proc.* 11:452.

Sinclair, N.R. StC., and F.Y. Law. 1979. Antibody-mediated immunosuppression of a cytotoxic cell response not involving a simple antigen-masking mechanism. *J. Immunol.* 123:1439.

Siiteri, P.K., F. Febres, L.E. Clemens, R.J. Chong, B. Gondos, and D.P. Stites. 1977. Progesterone and the maintenance of pregnancy: is progesterone Nature's immunosuppressant? *Ann. N.Y. Acad. Sci.* 286:384

Smith, G. 1981. Maternal regulator cells during murine pregnancy. *Clin. Exp. Immunol.* 44:90.

Smith, J.A., R.C. Burton, M. Barg, and G.F. Mitchell. 1978. Maternal alloimmunization in pregnancy. *Transplantation* 25:216.

Smith, N.C., and M.G. Brush. 1978. Preparation and characterization of human syncytiotrophoblast plasma membrane. *Med. Biol.* 56:272.

Smith, R.N., and A.E. Powell. 1977. The adoptive transfer of pregnancy induced unresponsiveness to male skin grafts with thymus-dependent cells. *J. Exp. Med.* 146:899.

Teh, H.S., E. Harley, R.A. Phillips, and R.G. Miller. 1977. Quantitative studies on the precursors of cytotoxic lymphocytes. I. Characterization of a clonal assay and determination of the size of clones derived from single precursors. *J. Immunol.* 118:1049.

Thomas, D.W., and E.M. Shevach. 1978. Nature of the antigenic complex recognized by T lymphocytes VI. The effect of anti-TNP antibody on T cell responses to TNP-cojugated macrophages. *J. Immunol.* 121:1145.

Tiilikainen, A., J. Schroeder, and A. de la Chapelle. 1974. Fetal leukocytes in the maternal circulation after delivery. *Transplantation* 17:355.

Truitt, G.A., R.R. Rich, and S.S. Rich. 1978. Suppression of cytotoxic lymphocyte responses *in vitro* by soluble products of alloantigen-activated spleen cells. *J. Immunol.* 121:1045.

Tucker, E.M., P.T. McGovern, and J.L. Hancock. 1971. Serological investigations into the cause of death of goat X sheep hybrid fetuses. *J. Reprod. Fertil.* 27:417.

Van Zon, A.A.J.C., and W.M. Eling. 1980. Depressed malarial immunity in pregnant mice. *Infect. Immun.* 28:630.

Wegmann, T.G., B. Singh, G.A. Carlson. 1979a. Allogeneic placenta is a paternal strain antigen immunoabsorbant. *J. Immunol.* 122:270.

Wegmann, T.G., C.A. Waters, D.W. Drell, and G.A. Carlson. 1978. Pregnant mice are not primed but can be primed to fetal alloantigens. *Proc. Nat. Acad. Sci. USA* 76:2410.

Wegmann, T.G., J. Barrington Leigh, G.A. Carlson, T.R. Mosmann, R. Raghupathy, and B. Singh. 1980. Quantitation of the capacity of placenta to absorb monoclonal anti-fetal H-2K antibody. *J. Reprod. Immunol.* 2:53.

West, J.D., W.I. Frels, V.E. Papaioannou, J.P. Karr, and V.M Chapman. 1977. Development of interspecific hybrids of *Mus. J. Embryol. Exp. Morphol.* 41:233.

Wira, C.R., E. Hyde. C.P. Sandoe, D. Sulliva, and S. Spencer. 1980. Cellular aspects of the rat uterine IgA response to estradiol and progesterone. *J. Steroid Biochem.* 12:451.

Young, E.J., and C.I. Gomez. 1979. Enhancement of Herpes virus type 2 infection in pregnant mice. *Proc. Soc. Exp. Biol. Med.* 160:416.

Youtananukorn, V., P. Matankasombut, and V. Osthanondh. 1974. Onset of human cell-mediated immune reaction to placental antigens during first pregnancy. *Clin. Exp. Immunol.* 16:593.

Zembala, M., G.L. Asherson, J. Noworolski, and B. Mahew. 1976. Contact sensitivity to picryl chloride: the occurrence of B suppressor cells in the lymph nodes and spleen of immunized mice. *Cell. Immunol.* 25:266.

Chapter 17

STUDIES ON CELL-MEDIATED IMMUNITY IN HUMAN PREGNANCY

W. R. JONES
CATHERINE S. HAWES
ANDREW S. KEMP

Mammalian pregnancy, in which an antigenically foreign fetus is tolerated by the mother, represents a successful allograft. One of a number of hypotheses advanced to account for this success is that maternal immune responsiveness is depressed (Medawar, 1953; reviewed by Beer and Billingham, 1971; Billington, 1975; Gusdon, 1976; Rocklin et al., 1979).

It has been postulated that such an impairment of the maternal immune response would lead to increased susceptibility to, and/or a more severe outcome of, infection during pregnancy (Purtilo et al., 1972). The balance of evidence suggests, however, that this is not the case. Thus, an increased incidence of poliomyelitis in pregnant women found by Siegel and Greenberg (1955) was attributed to their parity and associated increase in exposure. Their mortality rate was decreased. Similarly, D'Cruz et al. (1968) found that an increased incidence of hepatitis in pregnant women in Bombay was associated with severe anemia and malnutrition. Mortality due to influenza (Eickhoff et al., 1961) and varicella pneumonia (Pickard, 1968) was higher in late pregnancy, which the latter author attributed to decreased respiratory function. De March (1975) found no evidence of relapse of tuberculosis during pregnancy provided that adequate treatment was maintained. Investigation of deaths from fungal infection in pregnant women showed that, with the exception of coccidioidomycosis, all patients were immunosuppressed as a result of other clinical conditions and treatment (Purtilo, 1975). Pregnancy does appear to greatly increase the risk of dissemination of infection by *Coccidioides immitis* (Purtilo, 1975) and malarial parasites (Bray and Anderson, 1979; Van Zon and Eling, 1980).

Early reports suggested that, during pregnancy, tuberculosis patients became anergic to tuberculin skin testing. These studies were considered inadequate by Lichtenstein (1942), who found no evidence of anergy when 82 pregnant patients were tested, although there was some depression of reactivity during the third trimester compared with nonpregnant women with the same extent of disease. Following a comparison of two groups of women, Finn et al. (1972) also suggested that tuberculin reactivity was reduced during pregnancy. However, comparisons made between different subjects overlook the problem of different histories of antigen exposure (Jenkins and Scott, 1972). This is overcome in longitudinal studies, in which the subjects serve as their own controls. Montgomery et al. (1968) tested 23 women during and after pregnancy and concluded that the minor fluctuation found in the degree of reactivity paralleled that seen in nonpregnant individuals. In a much larger series, Present and Comstock (1975) found no indication that tuberculin sensitivity was affected by pregnancy.

365

Studies of cell-mediated immunity (CMI) *in vitro* during pregnancy have been largely limited to the assay of lymphocyte transformation. Little attention has been focused on other components of CMI, such as monocyte activity and lymphokine production.

Animal studies have suggested a stimulation of the reticuloendothelial system during pregnancy. There was an increased blood clearance of injected particles in rats (Wexler and Kantor, 1966; Douglas and Grogan, 1970; Graham and Saba, 1973) and mice (Sljivic et al., 1975). However, reduced activity in pregnant mice has also been described (Nicklin and Billington, 1979). In contrast, the function of the monocyte-macrophage system has not been studied during human pregnancy. When morphological criteria were used, monocyte numbers were found to be either unchanged (Efrati et al., 1964) or slightly increased (Andrews and Bonsnes, 1951). When monocytes were identified by latex particle ingestion, a significant increase was found throughout the duration of pregnancy (Plum et al., 1978).

Differing results have also been found in studies of lymphokine production and lymphocyte transformation. Thus, normal lymphokine production throughout pregnancy was described by Youtananukorn et al. (1974), who used a PPD-induced macrophage inhibition factor assay. However, Smith et al. (1972) found that, with the macrophage electrophoretic migration assay for lymphokine detection, the response to PPD became lower in late pregnancy.

Phytohemagglutinin (PHA)-induced lymphocyte transformation during pregnancy has been the subject of a large number of investigations and has been concluded to be depressed (Finn et al., 1972; Purtilo et al., 1972; Nelson et al., 1973; Thong et al., 1973; Jha et al., 1974; Petrucco et al., 1976; Tomoda et al., 1976; Garewal et al., 1978), unchanged (Watkins, 1972; Kaye, 1973; Khoo et al., 1975; Need et al., 1976; Birkeland and Kristofferson, 1977; Poskitt et al., 1977; Covelli and Wilson, 1978; Plum et al., 1978; Lopatin et al., 1980), or increased (Metcalf and Metcalf, 1972; Carr et al., 1973). Some, but not all, of these differences can be attributed to differing methods (see under Discussion)—in particular, whether autologous plasma was present in the system studied, either on unwashed cells or in the culture medium.

There have been relatively few studies of antigen-induced lymphocyte transformation during human pregnancy. Using the single antigen PPD, Birkeland and Kristofferson (1977) and Covelli and Wilson (1978) found lower responses in the presence of autologous plasma in the later stages of pregnancy relative to nonpregnant controls. A comprehensive study by Lopatin et al. (1980) using a number of orally-associated bacterial antigens, including streptokinase-streptodornase and tetanus toxoid, also showed impaired responsiveness during pregnancy, both in the presence and in the absence of autologous plasma.

EXPERIMENTAL STUDIES

A longitudinal study of both *in vivo* and *in vitro* cell-mediated immune responsiveness was undertaken in a group of 14 pregnant women aged 20 to 32 years. CMI was examined once during pregnancy (five subjects in the second trimester; nine in the third) and subsequently in all subjects between 6 and 12 weeks after delivery. Skin test reactions to four antigens were measured, and at the same time cell function *in vitro* was assessed by studies on monocyte and neutrophil migration, production of monocyte chemotactic factor (MCF), and lymphocyte transformation induced by PHA and antigens. All the *in vitro* assays were performed in the absence of autologous sera. Further data on cell migration were obtained by studying an additional group of 41 women aged 19 to 37 years and ranging from 5 to 36 weeks' gestation. Of these, 14 were retested between 6 and 19 weeks postpartum. The control subjects were 28 nonpregnant women aged 20 to 38 years who were not taking oral contraceptives. Some of the data arising from these studies, together with the methods used, have been reported elsewhere (Hawes et al., 1979; Hawes et al., 1980a, 1980b).

DELAYED HYPERSENSITIVITY SKIN TESTING

Of the 14 women tested during pregnancy, 13 were positive to streptokinase-streptodornase (SKSD), 8 to *Candida albicans* extract, 7 to purified protein derivative (PPD), and 5 to tetanus toxoid. On postnatal retesting, two skin tests that had been positive during pregnancy were now negative. No test became positive postnatally if negative during pregnancy. The reaction levels obtained during pregnancy are compared with the subsequent postnatal reactions in Figure 17-1. While the mean value of the reaction size to each antigen during pregnancy was lower than that observed postnatally, comparison of the results by Mann-Whitney analysis showed that there was no significant difference ($p > 0.2$) in reactivity to any of the antigens used, either during or after pregnancy.

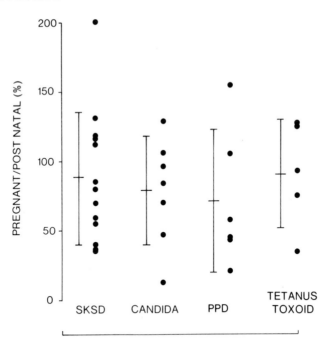

FIGURE 17-1. Delayed hypersensitivity skin reactivity during and after pregnancy.
The induration reactions (in μm) elicited during pregnancy are expressed as a percentage of that measured postpartum in each individual. The mean and S.D. of the comparative reactivities to each antigen are shown. One subject became nonreactive to *Candida* and one to PPD postpartum.

WHITE CELL NUMBERS DURING PREGNANCY

Total differential white cell counts were performed on blood samples obtained from 46 pregnant women aged 17 to 41 years, gestation range 7 to 36 weeks, and then compared with counts of samples from 25 control women (ages 20 to 42 years). A duplicate blood smear from each sample was stained for naphthyl esterase for monocyte indentification. There was a significant increase in the total white cell count throughout pregnancy (Table 17-1). This was largely due to an increased number of neutrophils, since the total mononuclear cell population, consisting of lymphocytes and monocytes, was similar to that in the controls in the first and second trimesters and slightly decreased in the third trimester. However, both the absolute number of monocytes and the proportion of monocytes in the mononuclear cell population increased throughout pregnancy.

TABLE 17-1
White Cell Numbers During Pregnancy

	CONTROL	1st TRIMESTER	2nd TRIMESTER	3rd TRIMESTER
Number of samples	25	13	6	26
Total WBC	6848 ± 1121	9953 ± 1646*	10417 ± 1201*	9946 ± 2244*
Total mononuclear cells	2544 ± 663	2346 ± 654‡	2379 ± 733‡	2155 ± 615†
Total monocytes	445 ± 136	600 ± 240§	563 ± 187‡	681 ± 284*
% Monocytes in mononuclear cell population	18 ± 5	27 ± 12*	26 ± 14†	32 ± 13*

Note: total cell counts are expressed per mm^3 (mean ± SD). The results in each trimester were compared with controls using Student's t-test.
Significance levels: *, $P < 0.001$; §, $P < 0.02$; †, $P < 0.05$; ‡, P = not significant.

MONOCYTE MIGRATION IN PREGNANCY

The random migration of monocytes was assessed in the absence of chemotactic stimulus. Random migration of monocytes obtained from pregnant women was unchanged during the first trimester, but increased significantly during the second and third trimesters, in comparison to controls (Table 17-2). No influence of gravidity was found when comparisons of random migration were made between primigravidae and multigravidae in each trimester ($p > 0.1$, Student's t-test).

In order to determine whether the random migration returned to normal levels in the postnatal period, 28 subjects tested during pregnancy (5 during the second trimester, 23 during the third) were retested between 6 and 19 weeks postpartum. They were divided into two groups according to whether their monocyte random migration during pregnancy was above or below the mean of the control individuals (Figure 17-2). In those subjects in whom the random migration during pregnancy had been above the mean of the controls, monocyte migration was significantly reduced postpartum, with a mean decrease of 11 μm ($N = 21$, $p < 0.001$). In contrast, no significant change in monocyte migration was found on retesting those in whom the previously determined random migration was below the control average. The mean decrease in this group was 1 μm ($N = 7$). These changes can be compared with the result that was obtained on repeating the assay in a group of 17 normal adults. The mean change in the second determination of migration in this group was an increase of 0.4 μm.

In contrast to the random migration, migration of monocytes from either pregnant or postnatal women toward the chemotactic stimulus of casein was not significantly different from that of the controls (Table 17-3).

369

TABLE 17-2
Random Monocyte Migration in Pregnancy

	CONTROL	1st TRIMESTER	2nd TRIMESTER	3rd TRIMESTER
Number of samples	28	14	16	25
Migration in μm (mean ± SD)	35 ± 11	38 ± 10‡	46 ± 9*	40 ± 9†

Note: comparison with the control group was by Student's t-test. Significance levels: *, P < 0.001; †, P < 0.05; ‡, P = not significant.

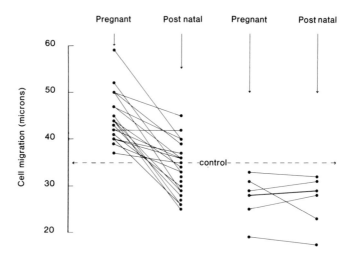

FIGURE 17-2. Random monocyte migration during and after pregnancy.
Random migration of monocytes from 28 women was determined during pregnancy and repeated postpartum. In 21 women, random migration during pregnancy was greater than the control mean (35 μm, N = 28). The mean decrease in migration postpartum was 11 μm (p < 0.001, Mann-Whitney U test). Migration decrease during pregnancy was below that of the controls in seven women, and no significant change was found on retesting (mean decrease of 1 μm).

MIGRATION OF NEUTROPHILS

The random migration of neutrophils from pregnant women (between 7 and 36 weeks' gestation) did not differ significantly from that of the controls. Nor were there any significant differences between subjects and controls in the chemotactic responses of neutrophils obtained during pregnancy (Table 17-4).

TABLE 17-3
Monocyte Chemotactic Response

	CONTROL	PREGNANT	POST-PARTUM
Number of samples	28	55	28
Migration in μm (mean ± SD)	91 ± 17	93 ± 20‡	93 ± 17‡

Note: Casein (5 mg/ml) was used as a chemotactic stimulus.
Comparison with the control group was by Student's t-test. ‡, P = not significant.

TABLE 17-4
Migration of Neutrophils

	CONTROL	2nd TRIMESTER	3rd TRIMESTER
Number of samples	20	12	26
Random	31 ± 5	29 ± 6‡	29 ± 5‡
Chemotactic	97 ± 17	96 ± 14‡	93 ± 16‡

Note: random migration was determined with HBSS below the filter, and chemotactic response using casein (5 mg/ml) as stimulus. Migration is expressed in μm (mean ± SD).
Comparison with the control group was by Student's t-test. ‡, P = not significant.

PRODUCTION OF MONOCYTE CHEMOTACTIC FACTOR DURING PREGNANCY

Production of monocyte chemotactic factor (MCF) was assessed during and after pregnancy in PHA- and SKSD-, *Candida albicans*-, and tetanus toxoid-stimulated cultures by comparing monocyte migration toward control and stimulated lymphocyte culture supernatants. Since high spontaneous activity prevents detection of MCF in stimulated cultures, only those assays in which migration to the control culture was less than 70 μm were considered.

Of the 14 women tested during and after pregnancy, significant PHA-induced MCF was detected in cultures obtained from 13 during pregnancy and from 9 postnatally. The increases in migration detected were 31 ± 11 μm (mean ± S.D.) and 28 ± 10 μm respectively. These were not significantly different (Mann-Whitney U test, p > 0.1). Thirteen subjects were skin test positive to SKSD. The increase in migration promoted by MCF-positive cultures obtained during pregnancy was 25 ± 9 μm (N = 10) which also was not significantly different from that detected in the postnatal period (23 ± 16 μm, n = 8).

Seven subjects were skin test positive to *Candida albicans* extract when tested both during and after pregnancy. Significant MCF was detected in three cultures obtained during pregnancy and in two during the postnatal

371

period. The mean increases in migration were comparable (20 μm and 17 μm respectively). Only four cultures were set up from skin test positive women; for these cultures, tetanus toxoid was used as antigen. The increase in migration obtained in MCF-positive cultures during pregnancy (11 μm, n = 2) was the same as that found postnatally (11 μm, n = 3).

LYMPHOCYTE TRANSFORMATION DURING PREGNANCY

Transformation of lymphocytes was assessed by culturing mononuclear cells with PHA, SKSD, and *Candida albicans* extract, and the responses obtained during and after pregnancy were compared in the study group. Cultures were performed in the absence of autologous sera. The ^{3}H-thymidine incorporation in response to dose ranges of each stimulant is shown in Figure 17-3. Antigen-induced transformation was compared in those individuals who were skin test positive to these antigens. PHA and antigen-induced transformation was higher during pregnancy than postpartum, attaining significance for at least one concentration of each stimulant (p < 0.01, Student's t-test). Spontaneous transformation was also increased during pregnancy; however, this was not significant (p > 0.2, Student's t-test)

DISCUSSION

The assessment of cell-mediated immunity in human pregnancy by delayed hypersensitivity skin testing and a range of *in vitro* assays showed that reactivity during pregnancy was either normal or enhanced.

The increased total white cell count during pregnancy, due largely to increased numbers of neutrophils, confirms the findings of others (Andrews and Bonsnes, 1951; Efrati et al., 1964). These authors found little alteration in monocyte numbers during pregnancy, but based their studies on morphological identification. However, functional or histochemical criteria appear to be necessary to distinguish monocytes from other mononuclear cells (Zucker-Franklin, 1974). The significant increase in the proportion of monocytes in the mononuclear cell population found in this study using esterase staining for identification is in agreement with the results of Plum et al. (1978), who identified monocytes by latex particle ingestion. In the nonpregnant control group, the proportion of esterase-positive cells in the mononuclear cell population was 18 per cent, which is similar to the finding of 19 per cent by Horwitz et al. (1977), although less than the 28 per cent proportion reported by Soman and Kaplow (1980).

The random migration of monocytes from pregnant women was significantly enhanced in the second and third trimesters, and migration decreased to control levels postpartum. As there is an increased proportion of monocytes in the mononuclear cell population during pregnancy, it is possible that the increased migration reflected the larger number of monocytes in the cell preparation, since a fixed total number of mononuclear cells was added to each migration chamber. However, a doubling of the mononuclear cell concentration from 5×10^6 cells/ml to 10×10^6 cells/ml did not significantly increase migration. Furthermore, random migration was not enhanced in the first trimester, although monocyte numbers were increased. Monocytes from pregnant subjects showed a normal chemotactic response to casein.

In contrast to the findings with monocytes, neutrophil random migration was not enhanced during pregnancy, although the neutrophil count was greatly increased. The normal migratory activity of neutrophils found during pregnancy is in contrast to the increased intracellular activity of several enzymes (reviewed by Fleming, 1975).

Lymphocyte function was assessed in vitro by assaying monocyte chemotactic factor production and transformation induced by the mitogen PHA and several antigens.

Comparison of production of MCF in cultures stimulated with PHA and SKSD, and in the smaller number of cultures stimulated with Candida albicans extract and tetanus toxoid, showed that the lymphocyte reactivity during pregnancy was equivalent to that postpartum. The amount of PHA-induced MCF detected in each group was similar to that found in a control adult population. Previous investigators of other lymphokines during pregnancy used the antigen PPD, and found either normal (Youtananukorn et al., 1974) or lowered (Smith et al., 1972) activity.

PHA-induced lymphocyte transformation has been variously reported (see Introduction) as depressed, unchanged, or increased as in the present study. Some of these differences can be explained by technical considerations. During pregnancy, the relative neutrophilia, which can interfere with transformation (Oppenheim et al., 1974), may have led to the depressed responses reported when whole blood or unfractionated leukocyte cultures (Finn et al., 1972; Purtilo et al., 1972; Jha et al., 1974; Petrucco et al., 1976; Garewal et al., 1978) were used. We overcame this problem in the present study by culturing mononuclear cells separated from neutrophils by Ficoll-Hypaque gradient centrifugation.

The depressed responses obtained in cultures in which autologous plasma or serum (Finn et al., 1972; Jha et al., 1974; Garewal et al., 1978) was used were presumably due to the influence of the plasma rather than to intrinsic lymphocyte hyporesponsiveness, since lowered PHA responses have been described in lymphocytes from both control and pregnant women in the

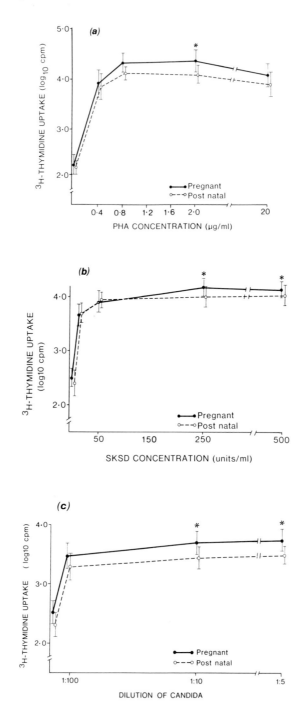

presence of pregnancy plasma when compared to control plasma (Leikin, 1972; St. Hill et al., 1973; Need et al., 1976; Tomoda et al., 1976; Figueredo et al., 1979). There are conflicting reports in this area, however, since others have not found any significant evidence of suppression of lymphocyte transformation by maternal plasma (Purtilo et al., 1972; Kaye, 1973; Petrucco et al., 1976; Poskitt et al., 1977). To overcome the influence of pregnancy serum or plasma on cellular activity, a single batch of pooled serum from healthy males was used in all cell cultures throughout the present study: In this way, the intrinsic reactivity of the lymphocytes could be assessed.

The necessity of using a range of mitogen concentration was demonstrated by Hosking et al. (1971), who showed that in immunodeficiency states, subnormal responses were more likely to be detected at suboptimal concentrations of PHA. Furthermore, Carr et al. (1973) found that comparative responses of Ficoll-separated mononuclear cells from pregnant women cultured in homologous plasma varied markedly according to PHA concentration; at lower doses, transformation was significantly greater than in controls, particularly in the third trimester, whereas at the highest concentration, reactivity was less than in the controls. A variable response according to dose was also described by Khoo et al. (1975). This influence of PHA concentration could explain why a lowered response was obtained by Thong et al. (1973), who used only one concentration of PHA.

The normal-to-enhanced PHA response of lymphocytes from pregnant women described here is in agreement with the majority of studies performed using separated mononuclear cells cultured in homologous sera (Carr et al., 1973; Khoo et al., 1975; Poskitt et al., 1977; Covelli and Wilson, 1978; Plum et al., 1978; Lopatin et al., 1980).

Significantly enhanced transformation to at least one concentration of the antigens SKSD and *Candida* extract was also found during pregnancy. Other authors have described impaired transformation in response to PPD in the latter part of pregnancy (Birkeland and Kristofferson, 1977; Covelli and Wilson, 1978). Impaired responses at one or more stages of pregnancy to a number of antigens, including SKSD, were found in a longitudinal study that used methods similar to those used in the present study (Lopatin et al., 1980). These discrepancies are difficult to explain at present, and it would seem that

---

FIGURE 17-3. Lymphocyte transformation during pregnancy.
(a) Log_{10} cpm (mean $\pm$ S.D.) incorporated by lymphocytes during and after pregnancy in response to PHA (N = 14). Transformation was significantly higher (p < 0.01, Student's t-test) during pregnancy at the point marked "*". (b) log_{10} cpm (mean $\pm$ S.D.) incorporated by lymphocytes obtained during and after pregnancy in response to SK/SD (12 skin test positive subjects). Transformation was significantly higher (p < 0.01, Student's t-test) during pregnancy at the points marked "*". (c) Log_{10} cpm (mean $\pm$ S.D.) incorporated by lymphocytes in response to *Candida albicans* extract (eight skin test positive subjects). Transformation was significantly higher (p < 0.01, Student's t-test) during pregnancy at the points marked "*"

further work is needed in this area. In the present study, no impairment of activity was found during pregnancy on examination of different aspects of lymphocyte reactivity, e.g., lymphokine production and transformation in response to antigens.

With cells from pregnant women, higher transformation activity was detected in control cultures as well as in mitogen- or antigen-stimulated cultures. Carr et al. (1973) suggested that the greater reactivity of cells from pregnant women reflected the immunological stimulus provided by fetal antigens. Another possible explanation is that the increased proportion of monocytes in mononuclear cell preparations from pregnant women enhances lymphocyte responsiveness. Thus, de Vries et al. (1979) demonstrated in reconstitution experiments that, if the proportion of monocytes in the mononuclear cell population is increased from 18 per cent to 32 per cent, the values observed in control and in third-trimester pregnant women, respectively, then lymphocyte mitogen responsiveness is greatly enhanced.

The increased number of monocytes and enhanced random migration observed during pregnancy are perhaps due to immunological or hormonal stimulation of the reticuloendothelial system. Antigenic stimulus can induce changes in the human blood monocyte population. Thus, peripheral blood monocytosis has been found during bacterial infection (Maldonado and Hanlon, 1965) and after *Corynebacterium parvum* injection (Gill and Waller, 1977; Hokland et al., 1980), and enhanced random migration of monocytes was observed during influenza infection (Kleinerman et al., 1975). Animal models suggest that pregnancy provides an immunological stimulus since histological changes of lymph nodes draining the pregnant uterus are similar to those seen during antigenic stimulation (Maroni and de Sousa, 1973; Beer and Billingham, 1974; McLean et al., 1974; Ansell et al., 1978; Bauminger and Peleg, 1978).

Hormonal changes may also affect the activity of the human monocyte-- macrophage system, since estrogen treatment significantly increased the clearance rate of intravenous albumin (Magarey and Baum, 1971), and increased numbers of peripheral blood monocytes have been described during the luteal phase of the menstrual cycle (Bain and England, 1975; Mathur et al., 1979). Animal experiments have indicated an increased hepatic phagocytic activity after estrogen treatment of normal rats (Nicol and Vernon-Roberts, 1965; Cordingly, 1969), an increased clearance rate of intravenously injected particles in pregnant rats (Wexler and Kantor, 1966; Douglas and Grogan, 1970; Graham and Saba, 1973), and either increased (Sljivic et al., 1975) or decreased (Nicklin and Billington, 1979) clearance of intravenously injected particles in pregnant mice. The increased numbers and enhanced random migration of monocytes found during human pregnancy indicate a stimulation of the reticuloendothelial system and "activation" of the circulating blood monocytes. This may be due either to an immunological or to a hormonal stimulus.

376

The *in vivo* test of CMI to four antigens showed no impairment of responses during pregnancy. As mentioned above, the use of a longitudinal study, in which subjects serve as their own controls, overcomes the problem of different histories of antigen exposure (Jenkins and Scott, 1972). The delayed skin test responses were measured by the same individual to minimize observer variation (Sokal, 1975). Slightly over half the reaction sizes were increased postnatally, and the remainder were reduced, two becoming negative. Longitudinal skin test studies by Montgomery et al. (1968) and Present and Comstock (1975), in which the single antigen PPD, was used, also showed no influence of pregnancy on the level of skin reactivity, with minor fluctuations paralleling those seen in control individuals. The normal skin reactivity found in these studies indicates a lack of influence of proposed nonspecific humoral immunosuppressive factors in pregnancy serum on CMI *in vivo*.

The *in vitro* assays of lymphocyte reactivity indicated normal to increased activity of the afferent limb of the cellular immune response during pregnancy. An additional *in vitro* parameter of cell-mediated immunity, monocyte migration, which is a function of the efferent limb of CMI, was enhanced during pregnancy. The *in vitro* findings correlate with the normal *in vivo* delayed hypersensitivity responses and provide no evidence for nonspecific depression of cell-mediated immunity during human pregnancy.

REFERENCES

Andrews, W.C. and R.W. Bonsnes. 1951. The leucoytes during pregnancy. *Am. J. Obstet. Gynecol.* 61:1129.

Ansell, J.D., C.M. McDougall, G. Speedy, and C.J. Inchley. 1978. Changes in lymphocyte accumulation and proliferation in the lymph nodes draining the pregnant uterus. *Clin. exp. Immunol.* 31:397.

Bain, B.J., and J.M. England. 1975. Variations in leukocyte count during menstrual cycle. *Brit. Med. J.* 2:473.

Bauminger, S., and S. Peleg. 1978. Changes in immunological activity of rat lymphoid organs during pregnancy. *Clin. exp. Immunol.* 32:179.

Beer, A.E., and R.E. Billingham. 1971. Immunobiology of mammalian reproduction. *Adv. Immunol.* 14:1.

Beer, A.E., and R.E. Billingham. 1974. Host response to intra-uterine tissue, cellular and fetal allografts. *J. Reprod. Fertil.* Suppl. 21:559.

Billington, W.D. 1975. Immunological aspects of normal and abnormal pregnancy. *Eur. J. Obst. Gyn. Reprod. Biol.* 5:147.

Birkeland, S.A., and K. Kristofferson. 1977. Cellular immunity in pregnancy: blast transformation and rosette formation of maternal T and B lymphocytes. *Clin. exp. Immunol.* 30:408.

Bray, R.S. and M.J. Anderson. 1979. *Falciparum* malaria and pregnancy. *Trans. R. Soc. Trop. Med. Hyg.* 73:427.

Carr, M.C., D.P. Stites, and H.H. Fudenberg. 1973. Cellular immune aspects of the human

377

fetal-maternal relationship. II. *In vitro* response of gravida lymphocytes to phytohae-magglutinin. *Cell. Immunol.* 8:448.

Cordingly, J.L. 1969. The mechanism of oestrogen stimulation of reticulendothelial activity. *J. Anat.* 104:190.

Covelli, H.D., and R.T. Wilson. 1978. Immunologic and medical considerations in tuberculin-sensitized pregnant patients. *Am. J. Obstet. Gynecol.* 132:256.

D'Cruz, I.A., S.G. Balani, and L.S. Lyer. 1968. Infectious hepatitis and pregnancy. *Obstet. Gynecol. (N.Y.)* 31:449.

de March, P.1975. Tuberculosis and pregnancy. *Chest* 68:800.

de Vries, J.E., A.P. Caviles, W.S. Bont and J. Mendelsohn. 1979. The role of monocytes in human lymphocyte activation by mitogens. *J. Immunol.* 122:1099.

Douglas, B.H. and J.B. Grogan. 1970. Effect of pregnancy and hypertension on reticuloendothelial activity. *Am. J. Obstet. Gynecol.* 107:44.

Eickhoff, T.C., I.L. Sherman, and R.E. Serfling. 1961. Observations on excess mortality associated with epidemic influenze. *J. Amer. Med. Ass.* 176:776.

Figueredo, M.A., P. Palomino, and F. Ortiz. 1979. Lymphocyte response to phytohaemagglutinin in the presence of serum from pregnant women: correlation with serum levels of alpha-fetoprotein. *Clin. exp. Immunol.* 37:140.

Finn, R., C.A. St. Hill, A.J. Govan, I.G. Ralfs, F.J. Gurney, and V. Denye. 1972. Immunological responses in pregnancy and survival of fetal homografts. *Brit. Med. J.* 3:150.

Fleming, A.F. 1975. Haematological changes in pregnancy. *Clin. Obstet. Gynaec.* 2:269.

Garewal, G., S. Sehgal, B.K. Aikat, and A.N. Gupta. 1978. Cell-mediated immunity in pregnant patients with and without a previous history of spontaneous abortion. *Brit. J.Obstet. Gynaecol.* 85:221.

Gill, P.G., and C.A. Waller. 1977. Quantitative aspects of human monocyte function and its measurement in cancer patients. In *The Macrophage and Cancer*. K. James, B. McBride, and A. Stuart, eds. Edinburgh: Proc. EURES Symposium, pp. 374-385.

Graham, C.W., and T.M. Saba. 1973. Development of humoral and cellular aspects of macrophage function in fostered and nonfostered neonates. *J. Reticuloendothel. Soc.* 13:7.

Gusdon, J.P. 196. In *Immunology of Human Reproduction.* J.S. Scott, and W.R. Jones, eds. New York: Academic Press, pp. 103-125.

Hawes, C.S., A.S. Kemp, and W.R. Jones. 1979. Random monocyte migration: an *in vitro* correlation with the delayed hypersensitivity skin reaction. *Clin. exp. Immunol.* 37:567.

Hawes, C.S., A.S. Kemp, ad W.R. Jones. 1980a. Enhanced monocyte migration during human pregnancy. *J. Reprod. Immunol.* 2:37.

Hawes, C.S., A.S. Kemp, and W.R. Jones. 1980b. Detection of a truly chemotactic lymphokine for human monocytes using millipore membranes. *J. Clin. Lab. Immunol.* 3:71.

Hokland, P., J. Ellegaard, and I. Heron. 1980. Immunomodulation by *Corynebacterium parvum* in normal humans. *J. Immunol.* 124:2180.

Horwitz, D.A., A.C. Allison, P. Ward, and N. Kight. 1977. Identification of human mononuclear leucocyte populations by esterase staining. *Clin. exp. Immunol.* 30:289.

Hosking, C.S., M.G. Fitzgerald, and M.J. Simons. 1971. Quantified deficiency of lymphocyte response to phytohaemagglutinin in immune deficiency diseases. *Clin. exp. Immunol.* 9:467.

Jenkins, D.M., and J.S. Scott. 1972. Immunological responses in pregnancy. *Brit. Med. J.* 3:528.

Jha, P., G.P. Talwar, and V. Hingorani. 1974. Depression of blast transformation of peripheral leukocytes by plasma from pregnant women. *Am. J. Obstet. Gynecol.* 122:965.

Kaye, M.D. 1973. Human lymphocyte responses during pregnancy. *J. Reprod. Fertil.* 32:333.

Khoo, S.K., S.V. Tillack, and E.V. Mackay. 1975. Cell-mediated immunity: effect of female genital tract cancer, pregnancy and immunosuppressive drugs. *Aust. N.Z. J. Obstet. Gynaecol.* 15:156.

Kleinerman, E.S., R. Snyderman, and C.A. Daniels. 1975. Depressed monocyte chemotaxis during acute influenze infection. *Lancet* ii:1063.

Leiken, S. 1972. Depressed maternal lymphocyte response to phytohaemagglutinin in pregnancy. *Lancet* ii:43.

Lichtenstein, M.R. 1942. Tuberculin reaction in tuberculosis during pregnancy. *Am. Rev. Tuberc.* 46:89.

Lopatin, D.E., K.S. Kornman, and W.J. Loesche. 1980. Modulation of immunoreactivity to periodontal disease-associated microorganisms during pregnancy. *Infect. Immunity* 28:713.

Maldono, J.E. and D.G. Hanlon. 1965. Monocytosis; a current appraisal. *Mayo Clinic Proc.* 40:248.

Maroni, E.S., and M.A.B. de Sousa. 1973. The lymphoid organs during pregnancy in the mouse. Comparison between a syngeneic and an allogeneic mating. *Clin. exp. Immunol.* 13:107.

Mathur, S., R.S. Mathur, J.M. Goust, H.O. Williamson, and H.H. Fudenberg. 1979. Cyclic variations in white cell subpopulations in the human menstrual cycle: correlations with progesterone and estradiol. *Clin. Immunol. Immunopathol.* 13:246.

Medawar, P.B. 1953. Some immunological and endocrinological problems raised by the evolution of viviparity in vertebrates. *Soc. Exp. Biology: Evolution* 7:320.

Metcalf, W.K., and N.F. Metcalf. 1972. Platelets, pregnancy and phytohaemagglutinin. *Am. J. Obstet. Gynecol.* 114:602.

Montgomery, W.P., R.C. Young, M.P. Allen, and K.A. Harden. 1968. The tuberculin test in pregnancy. *Am. J. Obstet. Gynecol.* 100:829.

McLean, J.M., J.G. Mosley, and A.C.C. Gibbs. 1974. Changes in the thymus, spleen and lymph nodes during pregnancy and lactation in the rat. *J. Anat.* 118:223.

Need, J.A., D.M. Jenkins, and J.S. Scott. 1976. The response of lymphocytes to phytohaemagglutinin in women with pre-eclampsia. *Brit. J. Obstet. Gynaecol.* 83:438.

Nelson, J.H., T. Lu, E.J. Hall, S. Krown, J.M. Nelson, and C.W. Fox. 1973. The effect of trophoblast on immune state of women. *Am. J. Obstet, Gynecol.* 117:689.

Nicol. T., and B. Vernon-Roberts. 1965. The influence of the oestrus cycle, pregnancy and ovariectomy on RES activity. *J. Reticuloendothel. Soc.* 2:15.

Micklin, S. and W.D. Billington. 1979. Macrophage activity in mouse pregnancy. *J. Reprod. Immunol.* 1:117.

Oppenheim, J.J., S. Dougherty, S.P. Chan, and J. Baker. 1974. Use of lymphocyte transformation to assess clinical disorders. In *Laboratory Diagnosis of Immunologic Disorders.* G.N. Vyas, D.P. Stites, and G. Brecher, eds. New York: Grune and Stratton, p. 87.

Petrucco. O.M., R.F. Seamak, K. Holmes, I.J. Forbes and R.G. Symons. 1976. Changes in lymphocyte function during pregnancy. *Brit. J. Obstet. Gynaecol.* 83:245.

Pickard, R.F. 1968. Varicella pneumonia in pregnancy. *Am. J. Obstet. Gynecol.* 101:504.

Plum, J., M. Thiery, and L. Sabbee. 1978. Distribution of mononuclear cells during pregnancy. *Clin. exp. Immunol.* 31:45.

Poskitt, P.K.F., E.A. Kurt, B.B. Paul, R.J. Selaraj, A.J. Sbarra, and G.W. Mitchell. 1977. Response to mitogen during pregnancy and the post partum period. *Obstet. Gynecol.* 50:319.

Present, P.A. and G.W. Comstock. 1975. Tuberculin sensitivity in pregnancy. *Am. Rev. Resp. Dis.* 112:413.

Purtilo, D.T. 1975. Opportunistic mycotic infections in pregnant women. *Am. J. Obstet. Gynecol.* 122:607.

Purtilo, D.T., H.M. Hallgren, and E.J. Yunis. 1972. Depressed maternal lymphocyte response to phytohaemagglutinin in human pregnancy. *Lancet* i:769.

Rocklin, R.E. J.L. Kitzmiller, and M.D. Kaye. 1979. Immunobiology of the maternal-fetal relationship. *Ann. Rev. Med.* 30:375.

Siegel, M., and M. Greenberg. 1955. Incidence of poliomyelitis in pregnancy. Its relationship to maternal age, parity and gestational period. *New. Engl. J. Med.* 253:841.

Sljivic, V.S., D.W. Clark, and G.W. Warr. 1975. Effects of oestrogens and pregnancy on the distribution of sheep erythrocytes and the antibody response in mice. *Clin. exp. Immunol.* 20:179.

Smith, J.K., E.A. Caspary, and E.J. Field. 1972. Lymphocyte reactivity to antigen in pregnancy. *Am. J. Obstet. Gynecol.* 113:602.

Sokal, J.E. 1975. Measurement of delayed skin-test responses. *New. Engl. J. Med.* 293:501.

Soman, S., and L.S. Kaplow. 1980. Monocyte contamination in Ficoll-Hypaque mononuclear cell concentrates. *J. Immunol. Methods* 32:215.

St. Hill, C.A., R. Finn, and V. Denye. 1973. Depression of cellular immunity in pregnancy due to a serum factor. *Brit. Med. J.* 3:513.

Thong, Y.H., R.W. Steele, M.M. Vincent, S.A. Hensen, and J.A. Bellanti, 1973. Impaired *in vitro* cell mediated immunity to rubella virus during pregnancy. *New Engl. J. Med.* 289:604.

Tomoda, Y., M. Fuma, T. Miwa, N. Saiki, and N. Ishizuka. 1976. Cell-mediated immunity in pregnant women. *Gynecol. Invest.* 7:280.

Van Zon, A.A.J.C., and W.M.C. Eling. 1980. Depressed malarial immunity in pregnant mice. *Infect. Immunity* 28:630.

Watkins, S. 1972. Immunological responses in pregnancy. *Brit. Med. J.* 3:353.

Wexler, W.M., and F.S. Kantor. 1966. Reticuloendothelial function in pregnancy. *Yale J. Biol. Med.* 38:315.

Youtananukorn, V., P. Matangkasombut, and V. Osathaondh. 1974. Onset of human maternal cell-mediated immune reaction to placental antigens during the first pregnancy. *Clin. exp. Immunol.* 16:593.

Zucker-Franklin, D. 1974. The percentage of monocytes among mononuclear cell fractions obtained from normal human blood. *J. Immunol.* 112:234.

Part IV

AUTOIMMUNITY AND INFERTILITY

INTRODUCTION

The final section of the book describes the consequences—both natural and induced—of the immune response to sperm, ova, and the pregnancy-associated molecules. The natural consequence is either abortion, as already discussed, or autoimmune diseases of the reproductive organs. The aim of induced immunity is to prevent fertility, either temporarily or permanently, without causing autoimmune diseases. As such, it has impressive potential for reproductive control in populations.

K. Tung discusses the various models of autoimmunity involving the testis and the spermatozoa and the immunological consequences of various attempts to vaccinate against sperm. The complex nature of the results obtained with a variety of different species makes it difficult to draw any general conclusions from studies in this area, but there does appear to be a genetically influenced immune responsiveness in the production of antisperm antibody following vasectomy. He raises the interesting possibility that vaccination against pregnancy in a population may endow those individuals who are low responders to the vaccine with enhanced reproductive capacity. This could be overcome, however, through the use of polyvalent vaccine or immunologically modified antigens, as pointed out by in chapter 20.

The point on autoimmunity after vasectomy is expanded in some detail by N. Alexander, who points out that it has been known for a long time that this common form of contraception leads to the production of antisperm antibodies. She discusses her finding that these antibodies can cause arteriosclerotic changes in a number of animal models, including cynomologous monkeys, as a result of the formation of antibody-antigen complexes. The situation in man has not yet been investigated adequately with long-term epidemiological studies, but clearly any antifertility vaccine for use in either males or females must avoid the problem of induced autoimmunity.

A great deal of effort has been put into developing a vaccine to human chorionic gonadotropin (hCG), which is a protein hormone necessary for the maintenance of pregnancy in its early stages. G.P. Talwar et al. review the clinical experiences with this vaccine to date and some of the problems associated with it. They do so in the context of other potential approaches to fertility control. V. Stevens, J. Powell, and A. Lee then present a detailed and elegant immunochemical study of hCG.

E. Goldberg, T. Wheat, and V. Gonzales-Prevatt have taken a different approach to the study of potential antifertility vaccines, namely investigation of a unique sperm antigen that is a variant of the enzyme lactic dehydrogenase, the testis-specific isozyme LDH-C_4 (LDH-X). Their work is an elegant study of the structure and immunochemistry of LDH-X and shows that vaccination against it in a variety of animals, including baboons, leads to a reduction in fertility. For antigens in the reproductive tract, it is necessary to study not only the chemistry of the antigen but also the way in which it is administered. Systemic immunization may reduce mucosal immunity and vice versa, and so we need to learn more about how to immunize against sperm antigens in the reproductive tract.

Finally, B. Dunbar presents studies on the chemistry and immunochemistry of antigens from the zona pellucida which may be used in a vaccine to reduce fertility. Immunization against crude zona preparations can lead to a reduction in fertility, and attempts are now directed at purifying and characterizing the zona proteins obtained by new preparative techniques so that the appropriate ones can be used for immunization. Once again, however, the problem of generating autoimmune responses must be addressed in exploring these moieties as potential antifertility vaccines. In any event, the isolation and characterization of these proteins will potentially provide a large amount of information about how the sperm attaches to specific receptors on the ovum and about the subsequent changes that prevent polyspermy.

Studies of the influence of specific immunization on fertility should provide new insights into the nature and function of the proteins of the reproductive tract. Hopefully, these studies will also lead to the development of a clinically useful vaccine to reduce fertility. At any rate, they will most certainly provide a clearer insight into the mechanism of fertilization.

Chapter 18

MODELS OF AUTOIMMUNITY TO SPERMATOZOA AND TESTIS

KENNETH S.K. TUNG

The study of experimental models of disease often helps to provide insight into the pathogenesis and etiology of human disease. In the case of immunologic diseases involving the testis, their existence has not been established. Thus, the finding that similar diseases occur *naturally* in experimental animals will provide support for their possible occurrence in man. On the other hand, there has been abundant circumstantial evidence for autoimmunity (in men) and isoimmunity (in women) to spermatozoa as a cause of idiopathic infertility (reviewed in Rumke, 1980). Understanding of the pathogenic mechanisms of these human diseases may come from studies on experimental models of antisperm immunity. Studies on experimental models have also generated useful information on the nature of biological and pathophysiologic processes. Thus, investigations of autoimmunity of sperm and the testis have helped to elucidate the blood-testis barrier as an immunologic barrier. Future studies are expected to help clarify the cell biology and the nature of the molecules involved in fertilization events, the status of immunologic tolerance and autoimmunity to sequestered tissue antigens, and the relationship between the endocrine system and the immune system. I have listed in Table 18-1 examples of clinical diseases involving the testis and sperm that may have an immunologic basis, as well as their likely experimental counterparts. This chapter provides a review of the experimental models.

MODELS OF AUTOIMMUNE DISEASES OF THE TESTIS

Experimental allergic orchitis (EAO) can be readily induced in guinea pig (Voisin et al., 1951; Freund et al., 1953), mouse (Bernard et al., 1978), and rabbit (Tung and Woodroffe, 1978) by injection of sperm or testis in complete Freund's adjuvant (CFA). EAO is an ideal model for the investigation of the immunopathology and pathogenesis of testicular autoimmunity. Many studies have investigated the immunologic mechanisms required for EAO induction, the influence of testicular physiology and anatomy on the disease process, and how EAO is initiated. Since these topics have been covered extensively in several recent reviews (Rumke, 1980; Tung, 1980; Tung et al., 1981a), only selected aspects of EAO will be summarized here.

In the past decade, several new models of testicular autoimmune diseases have been described. Vasectomy is known to result in autoimmune responses to sperm and testicular antigens and testicular autoimmune

TABLE 18-1
Examples of Possible Clinical Infertility with an Immunological Basis and Their Experimental Counterparts

POSSIBLE EXAMPLES OF CLINICAL IMMUNOLOGIC INFERTILITY	EXPERIMENTAL MODEL THAT MAY CORRESPOND WITH THE CLINICAL ENTITIES
Infertile men with oligospermia or aspermia and testicular pathology	Experimental allergic orchitis
	Naturally occurring orchitis and/or aspermatogenesis with possible immunologic mechanisms:
	"A" line beagle dogs
	T/t^{w18} backcross mouse
	Black mink
?	Postvasectomy orchitis in rabbit and guinea pig
Postinfectious (mumps) orchitis	Orchitis/aspermatogenesis in guinea pigs immunized with salivary gland in CFA (Nagano and Okumura, 1973)
Infertility in men with normal sperm count and autoantibodies to spermatozoa	Effects of antisperm autoantibodies on sperm behavior and fertilization *in vitro*
Infertility in men with normal sperm count and antisperm antibodies after vasovasostomy	As above
Infertility in women with antisperm antibodies in serum and cervical secretion	Reduction of fertility and/or fecundity Effects of antisperm autoantibodies or isoantibodies on sperm behavior or fertilization *in vitro*

diseases. However, because of space limitations, we shall not review studies on the immunopathology of vasectomy. Several recent reviews on this topic are available (Anderson and Alexander, 1979; Lepow and Crozier, 1979; Teuscher et al., 1981; see chapter 19). Studies of naturally occuring orchitis help to elucidate the etiology and pathogenic mechanisms of testicular autoimmunity (Fritz et al., 1976; Dooher et al., 1981b; Tung et al., 1981a, 1981b). Finally, in order to bring up to date the possible mechanisms of immunologic tolerance to sperm and testicular autoantigens, several recent studies will be discussed, including:

1. *in vitro* induction of autoreactive T lymphocytes against testicular antigens and of EAO in the rat,
2. induction of testicular disease in adults following neonatal thymectomy,
3. antigen-mediated prevention of EAO and
4. evidence for the existence of immunoregulatory processes within the testis.

EXPERIMENTAL ALLERGIC ORCHITIS

Although several aspermatogenic antigens have been purified (Hagopian et al., 1975; Jackson et al., 1975), their applications to EAO are limited. In studies to be described, several crude but biologically active antigenic preparations were used. Their isolation steps are summarized in Table 18-2.

Pathologic and immunologic features diagnostic of immunologic testicular diseases
As Rumke (1980) has pointed out, infertility in men with aspermia or oligospermia and whose testes showed aspermatogenesis may be the consequence of autoimmune testicular diseases. While aspermatogenesis per se is a nonspecific testicular response to a number of different injuries, the finding of the following five immunopathologic lesions provides strong evidence for immunologic testicular disease.

Monocytic orchitis (Figure 18-1A and B) is lesion often found near the rete (Johnson, 1973), but can also exist independently in peripheral seminiferous tubules. It consists of focal to diffuse interstitial and perivascular infiltrations of lymphocytes, monocytes, and eosinophils. Invasion of the

TABLE 18-2
Summary of Extraction Procedures for Obtaining Crude Aspermatogenic Antigenic Preparations Described in This Chapter

ANTIGENIC PREPARATION	ISOLATION PROCEDURES
ASPM (Freund et al., 1955)	Guinea pig testis homogenate extracted in acetic acid; supernatant in 30% ammonium sulfate (AS); precipitate in 70% AS
T (Toullet et al., 1973)	GP sperm in water; centrifuged (500 x g); supernatant centrifuged (80 000 x g); precipitate sonicated and washed
S (Toullet et al., 1973)	GP sperm in water; centrifuged (500 x g); supernatant centrifuged after 5% trichloroacetic acid (TCA); excluded volume in Sephadex G100
P (Toullet et al., 1973)	GP sperm in water; centrifuged (500 x g); supernatant centrifuged (80 000 x g); precipitate after 5% TCA; aqueous phase after chloroform-butanol; "included" volume in Sephadex G100.
TCAsup; (Jackson et al., 1975)	GP testis homogenized in chloroform-methanol-acetone; extracted with HCl, pH 3; precipitated in 80% AS; supernatant in 5% TCA
G75m (Teuscher et al., 1981)	TCAsup; second peak on Sephadex G75 (proteins with MW between 13 000 and 50 000 daltons)

From Tung et al., 1980, reprinted by permission.

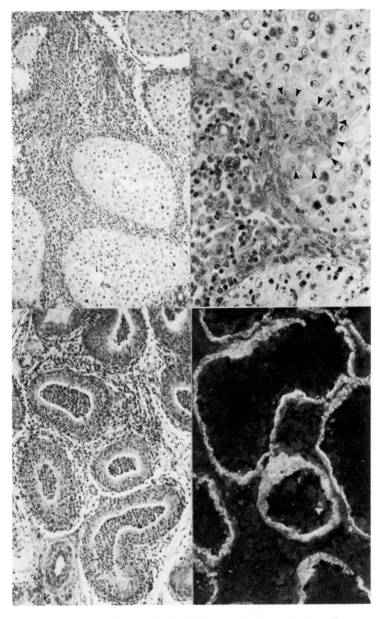

FIGURE 18-1. Findings characteristic of immunologic testicular diseases.
(A and B) Orchitis; (C) neutrophil-rich lesions in the ductus efferentes and/or the epididymides and immune complexes along the tubular basement membrane of seminiferous tubules detectable by immunofluorescence. In orchitis, multifocal invasion of seminiferous tubules by nodules of macrophages (arrowhead in B) is typical.

seminiferous tubules by monocyte-macrophages results in intratubular micro-granulomata (Waksman, 1959; Tung et al., 1970).

Inflammation and obstruction of the ductus efferentes (Figure 18-1C) is an early lesion of EAO in the guinea pig (Brown and Glynn, 1969), characterized by periductal monocytic and intraductal neutrophilic infiltration confined to the ductus efferentes. In severe lesions, the ducts may have no detectable lumen and are replaced by granulomatous inflammatory cells and periductal fibrosis. Reduction in sperm outflow may result in concomitant dilatation of the rete testis and the seminiferous tubules.

Sperm antigen-antibody complexes surrounding seminiferous tubules appear on direct immunofluorescence as focal to diffuse granular deposits of host IgG and/or complement C3 (Figure 18-1D) (Tung and Woodroffe, 1978). Sperm antigens in the immune complexes have been detected by IgG isolated from serum of the host. The antisperm antibody in the immune complexes can be determined by quantitative elution of the diseased organs. Thus, IgG eluted from the testis under conditions that dissociate antigen from antibodies (such as acid or alkaline pH, heat, strong ionic buffer, or chaotropes) contained much more antisperm antibody activity than did serum IgG.

Caudal epididymitis, in the absence of any experimental manipulation or infection, also indicates immunologic injury. Rupture of ducts may result in the formation of large sperm granulomata.

Vasitis has been observed in isolation in guinea pigs and mice with EAO. Neutrophilic infiltrates throughout the wall of the vas are associated with phagocytosis of sperm by neutrophils in the lumen.

Pathogenesis of EAO

The pathogenesis of EAO is complex (Tung, 1980). There is evidence that both T cell-mediated (or cell-mediated immune) mechanisms and mechanisms mediated by non-T cells may be operative in this disease.

There is evidence that a cell-mediated mechanism is involved in EAO.

1. A cell-mediated immune (CMI) response to aspermatogenic antigen precedes the development of EAO. In guinea pigs immunized with G75m antigens in complete Freund's adjuvant (CFA), the proliferative response of lymph node cells in the presence of G75m was first detected on day 4, reached maximal activity by day 7, and declined after day 15. Peritoneal exudate cells capable of secreting macrophage inhibitory factor (MIF) in the presence of G75m were detected on day 7, and reached plateau activity on day 10. On the other hand, EAO was not detected until day 10 and reached a plateau incidence on day 14 (Figure 18-2A, B, and C) (Meng and Tung, in preparation). Using delipidated acid extract of guinea pig

393

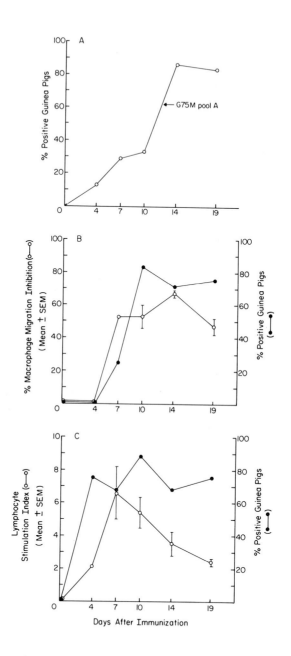

testis, Hojo et al. (1980) found a similar temporal relation between lymph node cell proliferative response, delayed-type hypersensitivity reaction, and EAO occurrence in the guinea pig. Furthermore, the CMI response was found, in both studies, to decline by day 18, only to be followed by a second peak of CMI activity on day 30 (Hojo et al., 1980). These findings provide indirect evidence for CMI involvement in EAO and point to the possible existence of suppressive or negative immunoregulation once EAO is fully developed. It is possible that the second peak of the biphasic CMI response represents immune responses to autoantigens released from the diseased testis.

2.   Orchitis and aspermatogenesis are adoptively transferred by lymph node cells, peritoneal exudate cells, and enriched T lymphocytes. Based on the technique of local adoptive transfer in syngeneic Strain 13 guinea pigs, lesions indistinguishable from EAO were obtained (Tung et al., 1971a; Kantor and Dixon, 1972; Tung et al., 1977). Over 90 per cent of the recipients of lymph node cells and 100 per cent of recipients of peritoneal exudate cells develop lesions in the seminiferous tubules and the rete testis. Lesions produced by local adoptive transfer were antigen-specific, since cells from donors immunized with CFA alone or CFA with several irrelevant antigens did not transfer disease to susceptible animals. The requirement of T lymphocytes in the adoptive transfer of EAO was determined by comparing the efficiency of transfer between highly enriched T cells and non-T cells from peritoneal exudate cells (Tung et al., 1977) and lymph node cells (Tung, 1980). In these experiments, guinea pig T cells were isolated either by rosette formation of these cells with rabbit erythrocytes (Stadecker et al., 1973) that were treated with 2-aminoethyl isothiouronium bromide hydrobromide (Kaplan and Clark, 1974; Tung et al., 1977) and then separated from the nonrosetted cells in a Hypaque-Ficoll gradient, or by depletion of macrophages and neutrophils on a nylon wool column (Julius et al., 1973). With both peritoneal exudate cells and lymph node cells, the enriched T cell fraction transferred orchitis typical of EAO. Furthermore, the number of T cells correlated positively with the frequency of adoptive transfer.

3.   EAO induction in the mouse is thymus-dependent. Hypothymic Balb/c nu/nu mice failed to develop EAO (Bernard et al., 1978). After these mice were reconstituted with thymus cells from litter mate

---

FIGURE 18-2. The temporal relation between time of onset and magnitude of: (A) testicular disease; (B) antigen-induced lymphokine-producing T lymphocytes in peritoneal exudates; and (C) antigen-induced proliferating lymph node cells following the induction of EAO with aspermatogenic antigen, G75m, in CFA.

Balb/c nu/+ mice, the hypothymic mice gained the capacity to develop EAO. This finding indicates the thymus dependency of EAO induction.

Findings that indicate the involvement of mechanisms other than CMI in the pathogenesis of EAO.

1. EAO can be adoptively transferred by lymph node cells depleted of T lymphocytes. T cell-depleted lymph node cells from guinea pigs sensitized with aspermatogenic antigens in CFA, which consisted mainly of B lymphocytes, macrophages, NK cells, and K cells, transferred EAO with the same efficiency as cell population enriched in T lymphocytes (Tung, 1980). Furthermore, the lesions transferred by the two cell populations were indistinguishable. A study that led to a similar conclusion was carried out earlier on the mouse (Bernard et al., 1978). A mixture of lymph node and spleen cells adoptively transferred EAO to syngeneic Balb/c recipients. When T lymphocytes were depleted by treatment of lymphoid cells with anti-Thy 1 antiserum and complement, transfer was only partially inhibited. More recently, it was discovered that murine spleen lymphocytes could be stimulated to proliferate when co-cultured with interstitial cells from syngeneic testis (Hurtenbach et al., 1980). In this *in vitro* model, the depletion of responder T cells with a monoclonal anti-Thy 1.2 antiserum and complement only partially suppressed, but did not abolish, the proliferative response.

2. Sperm antigen-antibody complexes are detectable in the testes of rabbits with EAO. This finding is an example of *in situ* immune complexes (Figure 18-1D) (Tung and Woodroffe, 1978). A similar event apparently occurs in glomerulonephritis in Balb/c mice treated with lipopolysaccharide (Isui et al., 1977), in experimental tubulointerstitial nephritis (Unanue et al., 1967; Klassen et al., 1971), in murine autoimmune thyroiditis (Clagett et al., 1974), in vasectomized rabbits (Bigazzi et al., 1976) and in the infertile black mink (Tung et al, 1981a, 1981b). However, it is uncertain as to whether these sperm immune complexes are of pathogenetic significance or the consequence of a breakdown of tissue barriers.

3. Aspermatogenesis and epididymitis are transferred by immune sera to CFA-treated guinea pigs. After transfer of immune sera, humoral antibodies reached the sperm in the rete (Tung et al., 1971b) and, perhaps by reflux into the seminiferous tubules, could induce two types of lesions of EAO (Toullet and Voisin, 1976). This possibility is supported by the finding that the blood-testis barrier at the region of the ductus efferentes and the rete is less complex than the barrier at the

seminiferous tubules (Suzuki and Nagano, 1978). However, the need to prime recipients with CFA in order for serum transfer of severe lesions of EAO to take place remains unexplained. A provocative idea that is worthy of study is whether the transfer of disease is due to transfer of anti-idiotype antibody. Anti-idiotype antibodies can stimulate T lymphocytes bearing the appropriate idiotype leading to T cell-mediated immunopathology (Sy et al., 1980; Thomas et al., 1981).

MODELS OF SPONTANEOUS MALE INFERTILITY ASSOCIATED WITH TESTICULAR AUTOIMMUNITY

The infertile "A" line beagle dog
The pure-bred beagle colony at the Division of Biological and Medical Research, Argonne National Laboratory, has been a closed breeding unit since 1960. The ancestral relationships of the animals in this colony have been well documented and include two partially inbred lines ("A" and "B"). The "A" line beagle dogs, derived from three sibling progenitors, have a high incidence of lymphocytic thyroiditis (Fritz et al., 1970) and increasing evidence of male infertility. Fritz et al. (1976) analyzed 69 "A" line dogs over one year of age for testicular size, semen analysis, testicular histopathology, and lymphocytic thyroiditis. These findings were then correlated with the "ancestral composition" as determined by calulation of the summation of ancestry (in this case the progenitors of the "A" and "B" inbred lines) contributing to the genotype of the animals.

Of the 69 dogs, 22 had different degrees of nodular lymphocytic orchitis involving the interstitium and/or the seminiferous tubules (Table 18-3). The presence of orchitis was associated with an increased frequency of aspermatogenesis or degenerative changes, "A" ancestral gene frequency, small testicles, and hypospermia or aspermia.

Lymphocytic thyroiditis and orchitis tended to coexist. Statistical analysis indicated that the occurrence of lesions in the two organs was related and both diseases were in turn related to the "A" ancestral composition. Thus, the incidence of both diseases increased with increased relatedness to the three sibling progenitors. Among dogs that derived 100 per cent of their genes from the three progenitors, 85 per cent had lymphocytic thyroiditis and 60 per cent had orchitis. The "A" line beagle dogs represent one of the best examples of genetically determined autoimmune disease involving multiple "endocrine" organs (Sotsiou et al., 1980).

TABLE 18-3
Effects of Testicular Germ Cells (Te II Cells) on Proliferative Responses of Spleen Cells *In Vitro*

EXPERIMENT	RESPONDER CELLS	STIMULATOR CELLS	MODULATOR CELLS	STIMULATION INDEX*
I	T/t^{w18} spleen cells	T/t^{w18} (sterile) Te II cells	—	7.0
	T/t^{w18} spleen cells	T/t^{w18} (fertile) Te II cells	—	0.8
	$+/+$ spleen cells	T/t^{w18} (sterile) Te II cells	—	5.3
	$+/+$ spleen cells	$+/+$ Te II cells	—	2.0
II	$+/+$ spleen cells	A/J spleen cells	—	9.4
	$+/+$ spleen cells	A/J spleen cells	$+/+$ Te II cells	2.8
	$+/+$ spleen cells	A/J spleen cells	T/t^{w18} Te II cells	8.7

From Dooher et al., 1980.

Stimulator cells were either 5 x 10^5 irradiated spleen cells or 2 x 10^5 Te II cells obtained by protease treatment of testis fragments. Responder cells were 5 x 10^5 spleen cells. Cells were cultured for four days at 37°C with 7% CO_2, in RPMI 1640 enriched with 1% normal mouse serum and 0.05 mM 2-mercaptoethanol.

* CPM experimental/CPM background.

The T/t^{w18} backcross mouse

In the course of raising a congenic mouse strain with the t^{w18} haplotype on the inbred BTBRTF/NeV (genotype $+/+$) background, repeated backcross matings were carried out. Although the original T^j/t^{w18} breeders were fertile, it was evident that the male (6 per cent) and female (14 per cent) backcross progeny with the tailless phenotype (T/t^{w18} genotype) were infertile (Dooher et al., 1981). Because of the nature of the breeding conditions by which these mice were generated, the infertility was considered to be a consequence of genetic factors located on chromosome 17. Of 20 infertile T/t^{w18} backcross males, 8 (40 per cent) had bilateral or unilateral orchitis. The basis of the infertility of the remaining 12 mice was not known; their testes were normal. The testicular lesions consisted of focal infiltrations of lymphocytes and plasma cells present mainly in the interstitial space. On electron microscopy, peritubular adventitial cells tended to separate the inflammatory mononuclear cells from the seminiferous tubules. Macrophages and neutrophils were essentially absent. Inside the seminiferous tubules, there were degenerative germ cells and phagocytosis of germ cells by Sertoli cells. The morphologic appearance of the testicular lesions was one of cell-mediated immunity. Furthermore, a preliminary study on T/t^{w18} mice with orchitis did not detect testicular immune complexes or serum antisperm antibodies by immunofluorescence (Tung, Arzt, Dooher, and Bennett, unpublished observations).

The study also investigated the immunologic properties of the testicular cells from the T/t^{w18} backcross mice (Dooher et al., 1981). The rationale of this study was based on an earlier observation (Hurtenbach et al., 1980) that germ cells of normal mouse testis (Te II cells) had the capacity to induce suppressor T cells. Among the T/t^{w18} backcross males, occasional animals were found to be fertile at the age of ten weeks, whereas by five to six months, all T/t^{w18} males were sterile. In this study, the Te II cells of the "fertile" T/t^{w18} males (ten weeks old) and of the "sterile" T/t^{w18} males (five to six months old), as well as the Te II cells of $+/+$ mice, were compared. Te II cells of $+/+$ mice and of the fertile T/t^{w18} mice stimulated significantly more proliferative responses of spleen lymphocytes of T/t^{w18} or $+/+$ mice than did the Te II cells of sterile T/t^{w18} mice (Table 18-4, experiment I). Furthermore, the proliferative response of $+/+$ spleen cells to allogeneic A/J spleen cells was significantly reduced in the presence of Te II cells of $+/+$ mice but not in the presence of Te II cells of T/t^{w18} mice (Table 18-4, experiment II). These findings were interpreted to suggest that spermatogenic cells from older T/t^{w18} males no longer had the suppressive capacity of normal germ cells. Although this change correlated with the appearance of orchitis, it is unclear whether the loss of the suppressor capacity of germ cells in the Te II cell population of the sterile T/t^{w18} males is the cause or consequence of orchitis.

The infertile black mink

The process of breeding mink for fine black fur has selected for the undesirable phenotype of male infertility. Some mink were infertile soon after puberty (primary infertility), and others became infertile after a period of proven fertility (secondary infertility). In April 1980, at the early regression phase of mink in Utah, we, in collaboration with LeGrand Ellis of Utah

TABLE 18-4
Suppression of Lymphoproliferative Response by Testicular Germ Cells:
Studies on Cellular Mechanisms of the Phenomenon

EXPERIMENT	RESPONDER CELLS	STIMULATOR CELLS	MODULATOR CELLS	STIMULATION INDEX*
I	B6 spleen cells	A/J spleen cells	–	37.0
	B6 spleen cells	A/J spleen cells	B6 Te II cells (2 x 10^5)	40.4
	B6 spleen cells	A/J spleen cells	B6 Te II cells (2 x 10^5)	3.4
II	B6 spleen cells	A/J spleen cells	–	44.2
	B6 spleen cells (anti-Ly 2.2 +C)	A/J spleen cells	–	1.2
	B6 spleen cells	A/J Te I cells	–	18.2
	B6 spleen cells (anti-Ly 2.2 +C)	A/J Te I cells	–	7.7
III	$+/+$ spleen cells	B6 spleen cells	–	24.1
	$+/+$ spleen cells	B6 spleen cells	$+/+$ Te II cells	2.9
	$+/+$ spleen cells	B6 spleen cells	W/W^v Te II cells	24.2

From Hurtenbach et al., 1980

* CPM experimental/CPM background.
Note: B6 = C57Bl/6 strain; Te I = testicular interstitial cells; Te II = testicular cells in the seminiferous tubules; C = complement. Stimulator cells were either 5 x 10^5 irradiated spleen cells or 2 x 10^5 testicular cells. Responder cells were 5 x 10^5 spleen cells. Cells were co-cultured for four days at 37°C with 7 % CO_2 in RPMI 1640 enriched with 1 % normal mouse serum and 0.05 mM 2-mercaptoethanol.

State University, studied 77 infertile and 9 fertile male black mink. In addition, many fertile male and female mink with pastel, opal and violet fur were studied for antisperm antibodies (Tung et al., 1981a, 1981b). The findings indicate that mink with primary infertility had low levels of antisperm antibodies and their main testicular abnormality was aspermatogenesis or underdeveloped spermatogenesis. Autoimmunity to sperm is thus not a significant cause of primary infertility. In contrast, mink with secondary infertility had higher levels of antisperm antibodies, detectable by indirect immunofluorescence. Many testes had severe orchitis and/or aspermatogenesis and epididymitis. In many of these testes, there was peritubular granular deposition of mink IgG and C3, presumptive evidence of immune complexes (Figure 18-1D). Although sperm antigen was not detected in the immune complexes, acid elution of testes with immune complexes recovered IgG that was enriched in antisperm antibody activity Ten times more antisperm antibody activity was recovered in the acid-eluted mink IgG than in the serum IgG of the same animal. Furthermore, phosphate-buffered saline failed to elute antisperm antibody from the same pool of testicular tissues. These findings are thus consistent with *in situ* deposition of sperm antigen-antibody complexes in the peritubular basement membrane of the testis.

While autoimmune testicular disease in the mink is clearly an important feature of the infertile state, it is unclear how the autoimmune disease evolves. There are at least two possibilities, and they are not mutually exclusive. First, the testicular autoimmunity may be the primary event that provides the etiologic basis for secondary infertility. If this is correct, it would be logical to extrapolate that genes that code for fur characteristics may be closely linked to the immune response genes for testicular autoantigens. As a second possibility, the primary defect may lie with some nonimmunologic mechanisms that determine the development and maintenance of normal testicular function. Thus, the pathogenesis of primary infertility may be related to some abnormalities in the hypophysial-pituitary-testicular axis. In consequence, the testis may fail to develop, to descend normally, or to respond appropriately to gonadotropin. In support of this hypothesis is the preliminary finding that both high and low levels of luteinizing hormones were detectable in groups of infertile mink (Dufau, Ellis, and Tung, unpublished observations). In addition, the failure to maintain the normal blood-testis barrier might lead to excessive release of testis-specific antigens during the seasonal regenerative phase of the testis. Autostimulation by testis-specific antigens could then lead to autoimmune testicular disease, characteristically found in mink with secondary infertility soon after the breeding season. The dissection of any relationship between the immunologic and nonimmunologic bases of the infertile state is the focus for future studies.

STUDIES RELEVANT TO MECHANISMS OF IMMUNOLOGIC TOLERANCE TO TESTICULAR AUTOANTIGENS

There is now ample evidence that testicular antigens, which are present in and on germ cells, are readily immunogenic to autologous hosts (Tung, 1980). As shown by the study on *in vitro* induction of EAO in the rat (Wekerle, 1978) (see below), testicular autoantigen-reactive T lymphocytes are also present. The blood-testis barrier is undoubtedly important in isolating the testicular antigens from the immune system (Johnson, 1973). However, we have argued recently that the blood-testis barrier may not be complete and that soluble testicular autoantigens may normally reach the lymphoid organs (Tung, 1977, 1980). In theory, leakage of testicular antigens may be advantageous if such antigens are tolerogenic. We shall now describe several studies that provide further support for active immunologic suppression of testicular autoimmunity; they suggest that these mechanisms may act systemically as well as within the testis.

The presence in rat of T lymphocytes that react with syngeneic testicular cells and can adoptively transfer EAO
Experiments by Wekerle and his colleagues (Wekerle et al., 1978; Wekerle and Begemann, 1978) showed that incubation of lymph nodes or spleen but not thymus lymphocytes of untreated rats with syngeneic dissociated testicular cells led to the formation of lymphocyte rosettes around a central testicular cell. After four to five days, the lymphocytes in the rosette transformed to lymphoblasts. The lymphoblasts, defined as T cells on the basis of undetectable C3b receptors, were cytotoxic to monolayers of testicular cells, and when adoptively transferred to the testis of syngeneic recipients, they induced an inflammatory lesion. Experiments using congenic strains of rats demonstrated that the cytotoxic T lymphocytes and the target testicular cells had to share the major histocompatibility antigens in order for cytotoxicity to occur. As a control for the specificity of the cytotoxicity, lymphocytes that rosetted with testicular cells were found to respond poorly against allogeneic lymphocytes *in vitro* when compared with lymphocytes that did not form rosettes. This study provides strong evidence for the existence in normal unimmunized animals of precursor effector T lymphocytes against testicular autoantigens.

Neonatal thymectomy is associated with testicular disease resembling EAO
Rats thymectomized within the first week after birth were often infertile when they became adults (Hattori and Brandon, 1977). Infertility was associated with testicular atrophy, interstitial edema, multinucleated giant

spermatids within aspermatogenic seminiferous tubules, and empty epididymides. Interstitial cells appeared normal. This finding was confirmed by Lipscomb et al. (1979), who also detected focal orchitis and serum antisperm antibodies. However, antibody presence did not correlate with testicular disease, and adoptive transfer of spleen cells from thymectomized Lewis rats to syngeneic recipients resulted in antisperm antibodies but not in testicular disease (Lipscomb et al., 1981). Rats of the Lewis strain were more susceptible to thymectomy-induced infertility than the Wistar and Fisher strains. Finally, serum levels of follicule-stimulating hormones, luteinizing hormone, and testosterone were normal in rats with neonatal thymectomy.

In the absence of any known or proven functional relation between the thymus and the gonads, the above findings can best be interpreted as the consequence of removal of the thymus as a central lymphoid organ. The thymus is the site for differentiation of T cells (Cantor and Boyse, 1977) and is the place where thymus hormones, including thymopoietin (Schlesinger et al., 1975), are produced. Since these thymus products are thought to maintain the physiologic levels of immunocompetence of the host (Katz, 1977), thymectomy at a critical period may preferentially remove subpopulations of T cells that subserve the immunosuppressive function (i.e., the suppressor T cells or their precursor cells). Unopposed effector T cells may spontaneously induce orchitis and aspermatogenesis. This hypothesis would predict that the disease could be avoided by reconstitution of neonatally thymectomized animals with different subpopulations of thymus cells or thymus hormones. Of course, the orchitis seen in neonatally thymectomized animals could also be secondary to viral infection. Further evidence of the existence of suppressor T lymphocytes in the prevention of testicular autoimmunity comes from studies on antigen-mediated prevention of EAO induction.

Antigen-mediated suppression of immune response to testis-specific autoantigens and of EAO induction
That the treatment of guinea pigs with testis antigen can lead to unresponsiveness to EAO induction is best illustrated by the so-called immune deviation phenomenon (Asherson, 1967). Brown et al. (1967) and Chutna and Rychlikova (1964) demonstrated that guinea pigs that were immunized with testis antigens in saline or in incomplete Freund's adjuvant became unresponsive to subsequent development of EAO when challenged with testis antigens in CFA. Later, Chutna (1970) demonstrated that the first injection of testis antigen in saline abrogated both the complement-dependent, cytotoxic antibody response and the delayed hypersensitivity reaction to sperm antigens, but not the anaphylactic antibody response. Furthermore, it was shown that the state of unresponsiveness could be reversed if guinea pigs were given cyclophosphamide at the time of the first injection. This finding is of interest because it has since been shown that the immune response seen in another

experimental autoimmune disease, allergic encephalomyelitis, can be adoptively transferred with lymphoid cells. Lymph node cells from rats immunized with encephalitogenic antigens in incomplete Freund's adjuvant, when adoptively transferred to syngeneic recipients, rendered the recipients unresponsive to the development of allergic encephalomyelitis when they were subsequently challenged with encephalitogenic antigen in CFA (Swierkosz and Swanborg, 1975). Furthermore, it has also been demonstrated that a subset of suppressor T cells is highly sensitive to cyclophosphamide treatment (Gill and Liew, 1978). While these studies indicate that injections of testis-specific antigens can lower the ability of the host to develop EAO, there is no available evidence that this situation occurs in the physiologic state.

Immunoregulatory mechanisms within the testis
The concept of immunoregulatory function within the testis is analogous to the regulatory role played by the placenta and placental hormones against immunologic rejection of the fetus as an allograft. The existence of the blood-testis barrier at the level of the seminiferous tubule is clearly an important immunologic barrier (Johnson, 1973). A study by Hurtenbach et al. (1980) suggests that cells within the seminiferous tubules can also impose an immunosuppressive influence through the induction of suppressor T lymphocytes. Although the physiologic significance of this mechanism remains to be established, the finding is of considerable interest.

The immunosuppressive property of germ cells is demonstrated by two observations. First, spleen lymphocytes were found to proliferate when stimulated by interstitial cells (designated Te I cells) but not by cells within the seminiferous tubules (Te II cells) of syngeneic testis. Since suppression decreased as the stimulator cell number was increased, active suppression was probably involved. Secondly, Te II cells but not Te I cells were found to suppress the proliferative response of spleen lymphocytes to allogeneic lymphocytes (Table 18-4, experiment I). Te II cells did not directly suppress the proliferative response of spleen lymphocytes. Instead, Te II cells apparently induced suppressor T cells within the responder splenic cell population that bore the Lyt 2 surface marker (Table 18-4, experiment II). Sertoli cells were tentatively excluded as the cells within the Te II cell population responsible for immunosuppression since the Te II cells from the W/W^v mutant mouse, whose seminiferous tubules contained only Sertoli cells, did not suppress the proliferative response of spleen lymphocytes (Table 18-4, experiment III). This study suggests that germ cells are able to induce suppressor T lymphocytes, and that the suppression is not antigen specific. However, the data excluding Sertoli cell involvement are less convincing, since the study did not directly identify the cell types in the cellular co-culture, and since the culture conditions (37°C with 1 per cent mouse serum) were not selective for Sertoli cell growth *in vitro*. Furthermore, recent data have shown that Leydig cells are also able to suppress lectin or antigen-induced T lymphocyte proliferative response *in vitro* (Born and Wekerle, 1981).

EXPERIMENTAL MODELS OF AUTOIMMUNITY AND ISOIMMUNITY TO SPERM

The discovery of autoimmunogenicity of sperm was followed by attempts to induce infertility in female and male animals with either active immunization with sperm or testis from the same species, or passive immunization with antiserum to these antigens (reviewed in Katsh, 1959a; Tyler, 1961). The ultimate goal was to develop an antifertility vaccine free of side effects and based on well-characterized sperm antigens. These studies also serve as models for the investigation of idiopathic infertility in human couples due possibly to autoimmune or isoimmune responses to sperm antigens (Table 18-1). Furthermore, through the use of antiserum of known specificity to block cellular events leading to fertilization, one may gain further insight into the molecular events involved in the prefertilization and fertilization processes. In this discussion, we shall cover studies published since the early 1960's, all of which can conveniently be divided into *in vivo* and *in vitro* studies.

While only the studies on the effect of sperm immunity will be reviewed, it should be stated parenthetically that this represents merely one aspect of the complex problem of immunologic infertility. Important host factors that will not be discussed but which undoubtedly have an impact on the development of sperm immunity include immunosuppressive properties of the seminal plasma (Stites and Erickson, 1975; Marcus et al., 1978; Prakash and Lang, 1980), the local secretory immune system in the male and female genital tracts (WHO workshop, 1976), cell-mediated immunity to sperm as a possible pathogenetic mechanism (Morton and McAnulty, 1979), and the hormone-related cyclical changes in the composition and immunoglobulin content of cervical mucus (Schumacher, 1980). However, a discussion of these aspects is beyond the scope of this chapter.

EFFECTS OF IMMUNIZATION WITH SPERM, TESTIS, OR THEIR ANTIGENS ON FEMALE REPRODUCTIVE PERFORMANCE *IN VIVO*

Of 150 papers written on the subject before 1960, about two-thirds of them claimed to have induced infertility (Tyler, 1961). Doubt was then expressed on the validity of this experimental model and the feasibility of sperm antigens as an antifertility vaccine. Contrary to these earlier reports, studies in the past 20 years have engendered a more optimistic view. Of 37 reports by 18 different authors (or laboratories) published since 1964, 33 (or 90 per cent) reported reduced fertility following immunization with sperm, testis, or teratocarcinoma antigens. Furthermore, infertility has been induced in many species, including rabbit, mouse, guinea pig, heifer, sheep, and baboon. These studies, summarized in Tables 18-5, 18-6, and 18-7, will be analyzed according to experimental design, mechanisms of infertility induction, nature of the immunizing antigen, and significance of antisperm antibody as cause of the infertile state.

Experimental design
One of the most likely explanations for the success in recent studies is the induction of an adequate immune response to sperm or testis antigens through the use of hyperimmunization regimes and powerful adjuvants. Thus, mice were frequently injected with about 1×10^6 sperm thrice weekly for 8 to 12 weeks. When antigens were injected in CFA, three to four injections were usually given.

Three different approaches were adopted. In the first, females that had been immunized with sperm, testis, or teratocarcinoma antigens were either mated or artificially inseminated following ovulation induction or were implanted with embryos after being made pseudopregnant. Control animals were immunized with adjuvant, saline, or seminal plasma with adjuvant. In the second approach, ejaculated or epididymal sperm were incubated with serum, cervical mucus, or their IgG, Fab (of IgG), or secretory IgA components from homologous or heterologous animals immunized with sperm or testis. The mixture was then artificially inseminated into unimmunized animals in which ovulation had been induced. In the third approach, antiserum to sperm antigens was transferred systemically into recipients at defined stages of pregnancy. This approach provides evidence that humoral antibodies are important pathogenic mediators of infertility. In the second and third approaches, control animals received preimmune sera, sera from animals immunized with adjuvant alone, or antisperm antisera that had been absorbed with sperm or testis. The second approach may be criticized as being less than physiologic since antibodies other than Fab may prevent sperm from reaching the eggs by simply agglutinating them. However, Munoz and Metz (1981) and Moore (1981) found sperm in oviducts after the sperm were injected along with bivalent antisperm antibodies. Furthermore, sperm were agglutinated by antiserum to seminal plasma but retained their fertilization capacity.

TABLE 18-5
Effect of Immunization of Female Rabbits with Spermatozoa on their Reproductive Performance

STUDY	IMMUNIZATION	RESULTS	CONTROL	OTHER FINDINGS	ANTIBODY STUDY
Behrman and Kakayama, 1965.	Female rabbits immunized with testis in CFA.	Mating did not lead to pregnancy during antibody peak.	Pregnancy ensued when antibody level fell.	None.	Passive hemagglutination.
Menge and Prottman, 1967.	Bull or GP immunized with rabbit testis or spermatozoa or semen in CFA.	Insemination of rabbit sperm incubated with heteroantisera led to reduced fertilization.	Heteroantisera absorbed with sperm abolished the effect. Heterologous antisera to seminal plasma had no effect.	50–70% of sperm motile at insemination. Seminal plasma immunization induced spermagglutinin.	Spermagglutination, gel diffusion, immunoelectrophoresis.
Menge, 1968.	Female rabbits immunized with rabbit semen or testis in CFA.	Reduction of fertilization rate and embryo survival as determined by embryo transfer.	Immunization with saline or seminal plasma in CFA had no effect.	General correlation between sperm agglutinin antibody and infertility.	Spermagglutination.
Bell, 1969a.	Female rabbits immunized with sperm in sodium alginate weekly for ten weeks.	At second mating, significant reduction in fertilization: 9% vs. 99% (control).	Saline injection had no effect on fertility.	Fertility inversely correlated with antibody to spermatozoa.	Passive hemagglutination.
Metz and Anika, 1970.	GP antiserum to rabbit sperm in CFA. Papain digestion separated into agglutinating and nonagglutinating antibodies.	Incubation of sperm with heteroantiserum IgG or Fab reduced fertilization rate (blastocyst/corpus luteum ratio) and passage of sperm into oviduct.	Preimmune sera had no effect.	None.	None.
Menge, 1970a.	Female rabbits immunized with semen and poly A-U.	Insemination of sperm incubated with antisperm antiserum led to reduction in the number of sperm in the vagina, uterus and oviduct; in sperm motility in vagina and oviduct; and in fertilization rate.	Immunization with seminal plasma had no effect.	Sperm motility and uterine leukocytes probably not significant factors.	Sperm immobilization.
Goldberg, 1973.	Female rabbits immunized with LDH-C4 in CFA.	Reduction in embryo/corpus luteum ratio from 88% (control) to 29%.	Untreated rabbits.	None.	Gel diffusion.

Reference	Treatment	Result	Control	Comment	Assay
Menge and Lieberman, 1974.	Female rabbits immunized with semen in CFA and/or poly A-U (i.m. and intravaginally).	Insemination of sperm incubated with uterine fluid (IgG or s IgA) led to reduced fertility.	Absorption of uterine fluid with sperm or antirabbit Ig abolished effect.	None.	Sperm immobilization, spermagglutination.
Russo and Metz, 1974.	Female rabbits immunized with rabbit sperm in CFA.	Ova from immunized rabbits with cumulus were less readily fertilized in vitro.	Rabbits immunized with CFA had no effect.	Antibody was probably carried over with the ova to the in vitro study.	None.
Kummerfield and Foote, 1976.	Female rabbits immunized with rabbit sperm in CFA.	Reduction of fertilization rate. Embryo and fetal mortality increased following embryo transplant into immunized rabbits.	CFA-immunized rabbits had no effect.	Infertility was temporary.	None.
Munoz and Metz, 1978.	Female rabbits immunized with sperm or MgCl$_2$-extracted molecules from sperm in CFA.	Reduction in fertilization rate of both groups. Part or all of cumulus of unfertilized eggs intact. Corona radiata intact.	CFA-immunized rabbits had no effect.	Sperm found near unfertilized eggs; thus sperm transport probably unaffected.	Sperm immobilization, spermagglutination, gel diffusion, antihyaluronidase antibody.
Menge et al., 1979.	Rabbit sperm treated with lithium diiodosalicylate; supernatant (SE-LIS), residue (SP-LIS). Female rabbits were immunized with SP-LIS or SE-LIS in CFA.	Reduction in fertility by day 13 after immunization with SE-LIS, but not with SP-LIS. Sperm treated with anti-SE-LIS but not with anti-SP-LIS failed to impregnate normal rabbits. Embryo loss after embryo transfer to SP-LIS but not SE-LIS immunized rabbits.	CFA immunized rabbits had no effect.	Antifertilization and antifertility effects may be due to immune responses to different sperm antigens.	Sperm immobilization, spermagglutination.
Moore, 1981.	GP anti-rabbit epididymal glycoproteins (R1, R2, or R3) in CFA.	Insemination of sperm with antiserum to R1, R2, or R3 or with Fab of anti-R1 significantly reduced fertilization rate (55-94% reduction).	Preimmune sera had no effect.	Sperm found near unfertilized cumulus, intact tubal eggs.	Spermagglutination, immunofluorescence.
O'Rand, 1981.	Female rabbits immunized with rabbit sperm antigen RSA1 in CFA.	Insemination of sperm incubated with antiserum Fab led to a 60% reduction in fertility rate.	CFA injected rabbits had no effect.	None.	None.

TABLE 18-6
Effect of Immunization of Female Mice with Spermatozoa or Teratocarcinoma on their Reproductive Performance

STUDY	MOUSE STRAIN	IMMUNIZATION	RESULTS	CONTROL	OTHER FINDINGS	ANTIBODY STUDY
McLaren, 1964	Q (5 mice per group).	Sperm, 3 injections per week, 7 weeks. (5×10^7 sperm per injection).	Reduction in cleaved eggs, in number of sperm in ampulla, in litter numbers in 25 days and six months, and in litter size.	Saline injected mice were normal.	Infertility correlated with antisperm antibody.	Spermagglutination.
Edward, 1964.	Outbred.	Sperm in CFA plus sperm alone.	No "consistent" reduction in fertility.	CFA alone.	None	Spermagglutination, sperm immobilization.
McLaren, 1966.	Q and C57Bl/6.	Sperm: sperm with pertussis; sperm with pertussis plus alum.	Reduction of litter number and size in 25 days in Q strain mice, regardless of adjuvant, but not in C57Bl/6.	Saline injection had no effect.	Pertussis increased and alum decreased antibody response; both reduced fertility.	Spermagglutination did not parallel infertility.
Bell, 1969b.	Q.	Sperm, 3 injections per week, 8 weeks. Repeat for 4 weeks. Compared i.p., vaginal, or both routes.	Reduction of litter size after first (26%) and second (52%) immunizations; and in six months. No reduction after vaginal immunization.	Saline injection had no effect.	Regression analysis indicated correlation of antibody titer with litter size reduction.	Spermagglutination, passive hemagglutination.
Bell and McLaren, 1970.	Q.	Sperm or supernatant of sperm extract (pressure cell) ± alum. Some supernatant was frozen-thawed.	Supernatant injections led to reduction of litter size in 38 days and 4 months. Also reduction in fertilization of tubal eggs.	Saline or sperm head injection had no effect.	Frozen-thawed supernatant sperm antigen induced infertility but not spermagglutinin.	Spermagglutination, passive hemagglutination.
Goldberg and Lerum, 1972.	Swiss Webster.	Rabbit antiserum to mouse testicular LDH-C4.	Transfer of sera to mated mice reduced pregnancy rate; maximum at 1-4 days, no effect after 10 days.	Normal rabbit serum or immune serum absorbed with LDH-C4 had no effect.	Effect was temporary.	None.

Reference	Strain	Treatment	Effect on fertility	Controls	Comments / antibody detection	Assay
		sperm conjugated with DNP (×3).	ters, and over 3 months.	no effect.		None.
Lerum and Goldberg, 1974.	Swiss Webster.	Mouse LDH-C$_4$ (×6) i.d. or i.p.	Pregnancy rate reduced to 40% of control. Litter size unchanged.	CFA immunization had no effect.		None.
Erickson et al., 1975.	Swiss Webster.	Mouse LDH-X in CFA or IFA (×3).	3/13 immunized mice had anti-LDH-X antibody and reduced litter size. Overall fertility unaltered.	None.		Gel diffusion.
Tsunoda and Chang, 1976.	Albino Webster	Sperm, 3 injections per week, 7 weeks. Supernatant of testis homogenate in CFA.	Teduction in litter size and fertilization rate. Eggs with cumulus from immunized mice were not readily fertilized in vitro.	PBS or PBS + CFA injection had no effect.		None.
Tung et al., 1979c.	Swiss Webster.	Sperm, 3 injections per week, 8 weeks. Repeated for 4 weeks.	Reduction in number of litters after 12 weeks of injection. No effect on fertilization or number of implantation sites.	PBS injection had no effect.	Sperm granuloma in peritoneal cavity.	Fluorescent anti-sperm antibody.
Hamilton et al., 1979.	C57Bl/6 or 129/- Ter Sv.	Multiple injections of irradiated F9 teratocarcinoma cells.	Reduction in pregnancy rate and litter size. Affected both strains.	Culture media injections had no effect.	Serum antibody (IgM and IgG) and uterine antibody (IgG and IgA) detected.	Radioimmunoassay.
Webb, 1980a,b.	C57Bl/6, 127/TER Sv. or 127/J.	Injection of teratocarcinoma, OTT 6050 prefixed in glutaraldehyde.	Reduction in pregnancy rate and in litter size.	Saline injection had no effect.	Rabbit anti-OTT 6050 antiserum reacted with sperm antigens by immunoprecipitation.	None.

409

TABLE 18-7
Effect of Immunization of Heifer, Guinea Pigs, Sheep, and Baboon with Testis or Spermatozoa on their Reproductive Performance

STUDY	IMMUNIZATION	RESULTS	CONTROL	OTHER FINDINGS	ANTIBODY STUDY
Menge, 1967	Heifer immunized with bull semen or testis + CFA (systemic or intrauterine injections).	Increased insemination/conception ratio in immunized heifer, with either systemic or local injection. Embryo loss frequent in pregnant heifer.	CFA injection had no effect.	Embryo loss was main cause of infertility. Correlation of infertility and antisperm antibody noted.	Spermagglutination.
Menge, 1969.	Heifer immunized with bull semen + CFA at different times after insemination.	Early embryo loss in heifer immunization up to 21 days after insemination. Increased incidence of delayed return to estrus (a finding indicative of embryo loss).	Seminal plasma + CFA injection had no effect.	Embryo loss was main cause of infertility.	None.
Menge, 1970b.	Heifer immunized with bull semen + CFA.	Sperm incubated in serum or cervical secretion of immunized cow did not incubate cow due to low fertilization rate.	CFA injection had no effect.	Antibodies to sperm detected in serum and cervical mucuc, but titres did not correlate with infertility.	Spermagglutination, sperm immmobilization.
Isojima et al., 1959.	Female GP immunized x3 with testis + CFA.	Reduction of pregnancy rate from 84% (control) to 24%.	Saline; saline, rat testis, or other tissues in CFA.	Species-specific. CMI, uterine anaphylaxis postulated as mechanisms.	Sperm immmobilization, gel diffusion, hemagglutination, skin test.
Katsh, 1959b.en, 1970.	Female GP immunized x 3 with testis + CFA.	Reduction of GP with litters from 83 to 100% (control)	Saline, rabbit sperm, bull sperm ± CFA.	Species-specific. Uterine anaphylaxis postulated as	None.

410

injections).		cent of fertilized ova developing into fetuses) reduced from 73% (control) to 46%.			
D'Almeida and Voisin, 1979.	Female GP repeatedly injected with S, P, T antigens, sperm, or sperm extract + CFA.	P antigen-immunized GP had more stillborns and prolonged mating period before pregnancy. Otherwise no reduction in fertility.	Spleen extract in CFA also had no effect.	Antibodies to sperm antigens in serum and cervical secretions.	Sperm immobilization, passive hemagglutination, delayed skin reaction.
Morton et al., 1979.	Female sheep immunized with Triton X-100 or $MgCl_2$ extracts of ram sperm + CFA.	Reduction of fertility in $MgCl_2$ extract immunized sheep by 39% as measured by fertilization rate or per cent sheep with litters.	CFA injection had no effect.	None.	Gel diffusion.
Morton and McAnulty, 1979.	Female sheep immunized with purified or partially purified ram acrosin or hyaluronidase + CFA.	No reduction in fertility rate.	CFA injection had no effect.	Nonpregnant ewes had significantly higher levels of anti-acrosin antibodies than pregnant ewes.	Radioimmunoassay, gel diffusion, spermagglutination.
Goldberg et al., 1981.	Female baboon immunized with murine and human $LDH-C_4$ in CFA or IFA.	Reduction of fertility from 72% (control) to 27% in 9 baboons (33 matings).	Injection of tetanus toxoid in CFA.	Infertility correlates with anti-$LDH-C_4$ and is reversible.	Enzyme inhibition assay.

411

Mechanisms of infertility induction

In what follows, seven parameters indicative of reproductive performance were analyzed (Table 18-8); in many of the studies reviewed here, more than one parameter was determined. Analysis of the data in Table 18-8 permits several conclusions to be drawn. First, it is clear that female infertility has been induced in rabbit, mouse, guinea pig, and heifer. Second, in rabbit and mouse, groups which have been adequately evaluated, the three parameters that were most consistently reduced were fertilization rate, embryo survival, and litter size. The lack of data on litter size of immunized rabbits probably reflects the extreme reduction in fertility in this species. Third, the significance of reduction of sperm transport, sperm motility, and cumulus dispersion has not been adequately evaluated. Similarly, the mechanism(s) of infertility in guinea pig and heifer also requires further studies. Elucidation of the mechanism(s) of infertility also comes from studies on *in vitro* fertilization and from studies comparing different antigens used for immunization.

Immunizing antigens

Three purified sperm-specific enzymes have been carefully evaluated for their antifertility properties. LDH-C_4 induced infertility in rabbit, mouse and baboon. As one might expect, immunization with heterologous LDH-C_4 resulted in a greater antibody response, and hence infertility, than immunization with homologous LDH-C_4 (Goldberg et al., 1981, and chapter 21). It was found that the antifertility effects influenced both preimplantation and postimplantation events (Goldberg and Lerum, 1972). Although hyaluronidase readily induced antibody responses in homologous females, fertility reduction was not observed (Morton and McAnulty, 1979). Similarly, immunization with purified acrosin in the sheep demonstrated at best a marginal effect on fertility rate (Morton and McAnulty, 1979).

TABLE 18-8

Mechanisms of Female Infertility in Rabbit, Mouse, Guinea Pig, and Heifer Immunized with Sperm, Testis, or Teratocarcinoma

PARAMETERS INVESTIGATED	RABBIT	MOUSE	GUINEA PIG	HEIFER
Overall fertility rate	5/5*	5/7	2/3	1/1
Fertilization rate	8/8	3/4	0/1	1/1
Embryo survival rate	3/3	1/1	1/1	1/1
Sperm transport in the female	2/2	1/1	–	–
Sperm motility in the female	1/1	–	–	–
Cumulus dispersion of eggs	1/1	–	–	–
Litter size	–	8/9	–	–

* No. of studies with reduction in the parameters investigated/total no. of studies.

– Not studied.

Based on immunization with whole sperm or testicular homogenates, the nature of the antifertility antigen(s) has been partially characterized:

1. It is species specific (guinea pig).
2. It first appears in secondary spermatocytes (rabbit).
3. Its activity resists treatment with pronase or chymotrypsin (rabbit).
4. Its activity resists freeze-thawing (mouse).
5. It is extracted from sperm by $MgCl^2$ but not by Triton X-100 (sheep and rabbit).

Recently, a cell surface sialoglycoprotein (RSA 1) with a molecular weight of around 15000 daltons was isolated from rabbit sperm by O'Rand and Porter (1979). Monovalent (Fab) antibodies to this antigen prevented fertilization *in vitro* and resulted in reduced fertility *in vivo* (O'Rand, 1981).

Since immunization with sperm led to reduction of both fertilization and embryo survival, studies were carried out to determine whether these reductions were due to immune responses to different sperm antigens. Menge et al. (1979) separated rabbit sperm into two crude fractions by extraction with lithium diiodosalicylate (LIS). They found that the immune response of rabbits to the LIS-extractable fraction (SP-LIS) led to a block in fertilization, whereas the immune response to the LIS-nonextractable (SE-LIS) fractions led to embryo loss.

An interesting finding was the antifertility effect of an immune response to two different teratocarcinoma cell lines (Hamilton et al., 1979; Webb, 1980a). Although the mechanism of infertility induction in this situation is unknown, IgG from antiserum to teratocarcinoma OTT 6050 was found to block fertilization *in vitro* (Webb, 1980a). Crossreaction between antigens of sperm, embryo, and teratocarcinoma has been reported (Goldberg and Tokuda, 1977; reviewed in Wiley, 1979). It will be of interest to determine whether these antigens are involved in fertilization events and whether an immune response to these antigens on sperm is responsible for embryo loss.

From the foregoing discussion, one may surmise that immune responses to several sperm antigens can lead to female infertility. These include the well-characterized sperm-specific enzyme LDH-C_4 and a membrane molecule on sperm surface. Some antigens may preferentially lead to an immune response that inhibits fertilization, whereas with others, embryonic loss is the result. Finally, some of these antigens may be shared among sperm, teratocarcinoma, and perhaps embryo.

Immunologic mechanisms that result in infertility

Some studies indicate that antibodies may be important in the induction of infertility. Both serum antibody and antibodies from secretions of the genital tract inhibit fertility. In the case of LDH-C_4, passive transfer of antiserum specifically prevents or terminates pregnancy. Less direct evidence has come from attempts to correlate antisperm antibodies with infertility.

In studies that used purified antigens, antibodies were measured by radioimmunoassay or enzyme inhibition assay. In two of these studies, a good correlation was found between antibody levels and the rate of antifertility (Morton et al., 1979; Goldberg et al., 1981). In studies where sperm or its fractions were used for immunization, antisperm antibodies were detected by several techniques (Tables 18-5, 18-6, and 18-7), although none of these assays used any defined antigens. While some earlier studies suggested a correlation between antibody levels and reduced fertility or fecundity (Menge, 1968; Bell, 1969a, 1969b), these findings were not confirmed in subsequent studies. Since spermagglutination has been used frequently for assessment of infertility in man, it was particularly disappointing that serum antibodies detectable by this assay rarely correlated with induction of infertility. Menge and Protzman (1967) observed that rabbits immunized with seminal plasma produced spermagglutinin but remained fertile. Bell and McLaren (1970) showed that freeze-thawing of soluble sperm antigens had no effect on fertility in the mouse, but that this procedure completely destroyed the antigen's capacity to induce an antisperm antibody response detectable by spermagglutination. In comparing immune responses of mice to sperm injected with different adjuvants, McLaren (1966) found that spermagglutinin was induced when sperm were injected with pertussis, but not with alum. However, infertility ensued regardless of the antibody titers of the immunized animals. Menge (1970a) found that treatment of a sperm homogenate with pronase had no effect on its capacity to induce infertility, but destroyed the antigen responsible for induction of spermagglutinin. Menge et al. (1979) studied the effect of immunization with two testicular antigenic preparations. Only one induced infertility, yet both raised comparable levels of spermagglutinin. Finally, the lack of correlation between spermagglutinin and infertility applied also to studies on antibodies in cervical secretions (Menge, 1970b; D'Almeida and Voisin, 1979). These data, collectively, would indicate that detection of serum spermagglutinin merely reflects in a general way the existence of immune responses to sperm antigens.

IN VITRO EFFECT OF AUTOANTIBODIES OR ISOANTIBODIES TO SPERM ANTIGENS ON PREFERTILIZATION AND FERTILIZATION EVENTS

Apart from the elucidation of mechanisms of infertility resulting from antisperm antibodies, these in vitro studies provide opportunities for studying a number of cellular changes that are of general biological interest, such as cellular recognition, cell adhesion, membrane fusion, and exocytosis. Antiserum to sperm has been shown to affect fertilization in the rabbit, guinea pig, hamster, and mouse and between human sperm and zona-free hamster eggs. With two exceptions, antisera to whole sperm were used. Dunbar et al. (1976) showed that isoantibodies to partially purified rabbit hyaluronidase inhibited cumulus dispersion of eggs in the presence of sperm. O'Rand (1981) provided evidence that Fab antiserum to RSA-1 prevented sperm from binding to zona pellucida and sperm fusion with the vitellus of the egg. Most studies used antiserum produced by female animals of the same species or by orchiectomized syngeneic males. A recent study demonstrated clearly that antisperm autoantibodies from vasectomized but not from sham-vasectomized guinea pigs could block many steps of fertilization *in vitro* (Huang et al., 1981). This finding is relevant to infertility following vasovasostomy with adequate sperm count.

SUMMARY AND CONCLUSIONS

Studies on EAO in different species have provided evidence that multiple pathogenetic mechanisms are operative. A T cell-mediated immune mechanism is involved since

1. typical lesions of EAO in the guinea pig are adoptively transferred with purified T lymphocytes from peritoneal exudates or lymph node cells from sensitized syngeneic donors,
2. athymic mice do not develop EAO,
3. in EAO induction, the sequence of events is a proliferative response of lymph node cells to aspermatogenic antigens, followed by lymphokine-producing peritoneal cells that respond to aspermatogenic antigens, and then testicular disease, and
4. rats have testis-reactive T lymphocytes capable of transferring EAO.

However, it is equally clear that mechanisms other than a T cell-mediated immune mechanism are also important since

1. antisperm antibodies transfer lesions of EAO to recipients primed with CFA,
2. lymph node cells depleted of T cells can transfer EAO in the guinea pig and mouse, and
3. EAO in rabbits is associated with sperm immune complexes in the testis.

EAO is no longer an experimental model in search of clinical relevance since there are several models of spontaneous orchitis with possible autoimmune etiology and since EAO is known to follow vasectomy (Bigazzi et al., 1976; Tung, 1978).

Spontaneous orchitis has been described in the "A" line beagle dogs, backcross mice with the T/t^{w18} genotype, and the black mink. Evidence of autoimmune response to sperm and testis antigens has not been documented in the diseased dogs or mice. However, dog orchitis was frequently associated with lymphocytic thyroiditis, and the diseased mice had some immunologic abnormalities in their germ cells. In the black mink, autoantibodies to sperm and sperm immune complexes within the testis are associated with orchitis and aspermatogenesis. However, it is apparent that sperm autoimmunity is not the only mechanism of mink infertility. There is evidence that sperm autoimmunity is but one of a complex series of pathogenic mechanisms leading to the infertile state.

Multiple mechanisms may exist which normally circumvent the development of testicular autoimmunity. In the testis, the well-defined blood-testis barrier at the site of autoantigen production (seminiferous tubules) undoubtedly sequesters these antigens from the immune system. Less certain is the completeness of the tissue barrier at the excurrent duct system. There is evidence that germ cells themselves may have the capacity to activate suppressor T cells. There is also circumstantial evidence for the existence of systemic suppressor mechanisms. Thus, neonatal thymectomy may not permit the generation of suppressor T cells or their precursors which are specific for antigens of the gonads. Aspermatogenesis and orchitis that occur in adult life may be due to unopposed effector T cells. Experimental evidence for antigen-mediated suppression of EAO can also be found in the literature. However, the physiologic relevance of these findings remains to be determined, and the key question of testicular antigen presence outside the testis has yet to be answered.

With regard to sperm autoimmunity and isoimmunity, studies of the past 20 years leave little doubt that female infertility will follow hyperimmunization with sperm, testis, or even teratocarcinoma antigens. Infertility can be due to interference with fertilization or to increased embryo loss. It is possible that these two pathologic phenomena result from immune responses to different antigens in the sperm. It is clear that immune responses to different and multiple sperm antigens can lead to infertility, and these antigens include the sperm-specific enzyme LDH-C$_4$, a membrane glycoprotein (RSA-1), antigens shared between sperm and teratocarcinoma, and perhaps other antigens. With the exception of LDH-C$_4$, the immunologic mechanisms responsible for the infertile state remain elusive. Spermagglutinin, an antisperm antibody directed to undefined sperm antigen(s), is unlikely to be an important agent for causing the infertile state.

Recent studies on *in vitro* fertilization have indicated that autoantigens specific to the sperm are involved in essentially all steps of prefertilization and fertilization. Thus, antisperm antiserum (or either Fab or whole IgG) can disperse guinea pig sperm rouleaux, prevent cumulus dispersion in the presence of sperm, block capacitation and/or the acrosome reaction, and interfere with the attachment of sperm (acrosome-reacted or otherwise) to the zona pellucida and the fusion between acrosome-reacted sperm and the vitelline membrane of the egg. It is possible that the antibodies interfere by competing for the specific sperm "receptors," by steric hindrance, or by crosslinking membrane molecules. Regardless of the precise mechanism involved, these findings support the concept that a sufficient concentration of antisperm antibodies can block fertilization in several well-defined steps by a mechanism akin to blocking by antireceptor antibodies.

Whether any of the experimental models of sperm and testis autoimmunity have any clinical relevance is unknown. In part, this is because the reality of the clinical entities themselves (Table 18-1) is not yet clarified. In part, this is because many of the experimental models are of questionable physiologic significance. A case in point is the hyperimmunization regime required to induce female infertility. One is left with the impression that meaningful results are more likely to come from analysis of the natural models of male and female infertility with an immunologic basis. It should be noted that the experimental models of spontaneous male infertility may also result in female infertility. Also important will be the elucidation of physiologic mechanisms that normally prevent autoimmunity in the male and isoimmunity in the female against sperm antigens. Spontaneous immunologic infertility may be due to subtle defects in these regulatory mechanisms rather than to hyperimmunization to sperm antigens. Finally, further clarification of the mechanisms of sperm and testis autoimmunity is likely to come from studies based on well-defined antigens involved in the induction of testicular autoimmunity, in the steps leading to fertilization, and in the rejection of implanted embryos.

ACKNOWLEDGMENTS

I wish to thank Ryuzo Yanagimachi and LeGrand Ellis and their colleagues for the many fine collaborative experiments. The many studies from my laboratory are the results of the thoughtful and diligent efforts of Cory Teuscher, Aniko Meng, B. L-P. Han, and Gaynor Wild. I thank Alan Menge and Michael O'Rand for useful discussions during the preparation of this chapter. I am grateful to Linda Lloyd for outstanding secretarial assistance. This study was supported by grants HD 14504, HD 14142, and HD 12247 from the Division of Child Health and Human Development of the National Institutes of Health.

REFERENCES

Alexander, N.J. 1972. Vasectomy: Long-term effects in the rhesus monkey. *J. Reprod. Fert.* 31:399.

Alexander, N.J. 1973. Autoimmune hypospermatogenesis in vasectomized guinea pigs. *Contraception* 8:147.

Alexander, N.J., B.J. Wilson, and G.D. Paterson. 1974. Vasectomy: Immunologic effects in rhesus monkeys in men. *Fertil. Steril.* 25:149.

Alexander, N.J., and K.S.K. Tung. 1977. Immunological and morphological effects of vasectomy in the rabbit. *Anat. Rec.* 188:339.

Ansbacher, R., K. Keung-Yeung, and J.C. Wurster. 1972. Sperm antibodies in vasectomized men. *Fertil. Steril.* 23:640.

Asherson, G.L. 1967. Antigen-mediated depression of delayed hypersensitivity. *Brit. Med. Bull.* 23:24.

Behrman, S.J., and M. Nakayama. 1965. Antitestis antibody: Its inhibition of pregnancy. *Fertil. Steril.* 16:37.

Bell, E.B. 1969a. Iso-antibody formation against rabbit spermatozoa and its effect on fertility. *J. Reprod. Fert.* 20:519.

Bell, E.B. 1969b. Immunologic control of fertility in the mouse: A comparison of systemic and intravaginal immunization. *J. Reprod. Fert.* 18:183.

Bell, E.B., and A. McLaren. 1970. Reduction of fertility in female mice iso-immunized with a sub-cellular sperm fraction. *J. Reprod. Fert.* 22:345.

Bernard, C.C.A., G.F. Mitchell, J. Leydon, and A. Bargerbos. 1978. Experimental autoimmune orchitis in T-cell-deficient mice. *Int. Arch. Allergy Appl. Immunol.* 56:156.

Bigazzi, P.E., L.L. Kosuda, K.C. Hsu, and G.A. Andres. 1976. Immune complex orchitis in vasectomized rabbits. *J. Exp. Med.* 143:382.

Bigazzi, P.E., L.L. Kosuda, and L.L. Harnick. 1977. Sperm autoantibodies in vasectomized rats of different inbred strains. *Science* 197:1282.

Born, W., and H. Wekerle. 1981. Do Leydig cells contribute to the immunologically privileged status of the testis? *J. Reprod. Immunol.* In press.

Brannen, G.E., A.M. Kwart, and D.S. Coffey. 1974. Immunologic implications of vasectomy: I. Cell-mediated immunity. *Fertil. Steril.* 25:508.

Brown, P.C., and L.E. Glynn. 1969. The early lesion of experimental allergic orchitis in guinea-pigs: An immunological correlation. *J. Pathol.* 98:277.

Brown, P.C., L.E. Glynn, and E.J. Holborwow. 1967. The dual necessity for delayed hypersensitivity and circulating antibody in the pathogenesis of experimental allergic orchitis in guinea pigs. *J. Path. Bact.* 86:505.

Cantor, H., and E.A. Boyse. 1977. Lymphocytes as models for the study of mammalian cellular differentiation. *Immunol. Rev.* 33:103.

Chutna. J. 1970. Study of mechanisms of specific inhibition of delayed sensitivity and IgM antibodies in guinea pigs immunized with organ-specific antigens. *Int. Arch. Allergy Appl. Immunol.* 37:278.

Chutna, J., and M. Rychlikova. 1964. Prevention and suppression of experimental autoimmune aspermatogenesis in adult guinea pigs. *Folia Biol.* (Praha) 10:177.

Clagett, J.A., C.B. Wilson, and W.O. Weigle. 1974. Interstitial immune complex thyroiditis in mice. The role of autoantibody to thyroglobulin. *J. Exp. Med.* 140:1439.

D'Almeida, M., and G.A. Voisin. 1979. Resistance of female guinea pig fertility to efficient iso-immunization iwth spermatozoa autoantigens. *J. Reprod. Immunol.* 1:237.

Dooher, G.B., K. Artz, D. Bennett, and U. Hurtenbach. 1981. Spontaneous allergic orchitis in sterile mice carrying a recessive, lethal mutation at the T/t complex: Fine structural observations on the pathogenesis and *in vitro* analysis of the immunological basis for the disease. *J. Reprod. Fertil.*

Dunbar, B.S., M.G. Munoz, C.T. Cordle, and C.B. Metz. 1976. Inhibition of fertilization *in vitro* by treatment of rabbit spermatozoa with univalent isoantibodies to rabbit sperm hyaluronidase. *J. Reprod. Fertil.* 47:381.

Edwards, R.G. 1964. Immunological control of fertility in female mice. *Nature* 201:50.

Ehrig, G., and L. Mettler. 1978. The influence of sperm-antibodies on the *in vitro* fertilization rate of inbred mice. In *Immunology of Reproduction* K. Bratanov, V.H. Vuchanov, V. Dikov, B. Somlev, and L. Kozhouharova, eds. Sofia: Bulgarian Academy of Sciences, pp. 455-458.

Erickson, R.P. 1973. Decreased fertility in female mice immunized with a carrier followed by small numbers of spermatozoa. *J. Reprod. Fertil.* 32:295.

Erickson, R.P., D.P. Stites, and H. Spielmann. 1975. Effects on fertility of immunization against lactate dehydrogenase-X in female mice. *Contraception* 12:333.

Friend, D.S., and D.W. Fawcett. 1974. Membrane differentiations in freeze-fractured mammalian sperm. *J. Cell. Biol.* 63:641.

Fritz, T.E., R.C. Zeman, and M.R. Zelle. 1970. Pathology and familial incidence of thyroiditis in a closed beagle colony. *Exp. Mol. Pathol.* 12:14.

Fritz, T.E., L.S. Lombard, S.A. Tyler, and W.P. Norris. 1976. Pathology and familial incidence of orchitis and its relation to thyroiditis in a closed beagle colony. *Exp. Mol. Pathol.* 24:142.

Freund, J., M.M. Lipton, and G.E. Thompson. 1953. Aspermatogenesis in the guinea pig induced by testicular tissue and adjuvant. *J. Exp. Med.* 97:711.

Freund, J., G.E. Thompson. and M.M. Lipton. 1959. Aspermatogenesis, anaphylaxis and cutaneous sensitization induced in the guinea pig by homologous testicular extract. *J. Exp. Med.* 101:591.

Galle, J., and D.S. Friend. 1977. The fate of sperm after vasectomy in the guinea pig, fine structure, freeze-fracture and cytochemistry. *Lab. Invest.* 37:79.

Gill, H.K., and F.Y. Liew. 1978. Regulation of delayed-type hypersensitivity III. Effect of cyclophosphamide on the suppressor cells for delayed-type hypersensitivity to sheep erythrocytes in mice. *Eur. J. Immunol.* 8:172.

Goldberg. E. 1973. Infertility in female rabbits immunized with lactate dehydrogenase X. *Science* 8:458.

Goldberg, E., and J. Lerum. 1972. Pregnancy suppression by an antiserum to the sperm specific lactate dehydrogenase. *Science* 176:686.

Goldberg, E., T.E. Wheat, J.E. Powell, and V.C. Stevens. 1981. Reduction of fertility in female baboons immunized with lactate dehydrogenase-C_4. *Fertil. Steril.* 35:214.

Goldberg, E.H., and S. Tokuda. 1977. Evidence for related antigens on sperm, tumor and fetal cells in the mouse. *Transp. Proc.* 9:1363.

Gwatkin, R.B.L. 1976. Fertilization. In *Cell Surface Reviews*. G. Poste and G.L. Nicolson, eds. Volume 1. New York: North-Holland, pp. 1-143.

Hagopian, A., J.J. Jackson, D.J. Carlo, G.A. Limjuco, and E.H. Eylar. 1975. Experimental allergic aspermatogenic orchitis. III. Isolation of spermatozoal glycoproteins and their roles in allergic aspermatogenic orchitis. *J. Immunol.* 115:1731.

Hamilton, M.S., A.E. Beer, R.D. May,and E.S. Vitetta. 1979. The influence of immunization of female mice with F9 teratocarcinoma cells on their reproductive performance. *Transplant. Proc.* 9:1069.

Han, L-P.B., and K.S.K. Tung. 1979. A quantitative assay for antibodies to surface antigens of guinea pig testicular cells and spermatozoa. *Biol. Reprod.* 21:99.

Han, L-P.B., J. Hu, and K.S.K. Tung. 1979. A simple method for isolation of Fab from IgG. *Fed. Proc.* 38:940.

Hattori, M., and M.R. Brandon. 1977. Infertility in rats induced by neonatal thymectomy. In *Immunologic Influence on Human Fertility*. B. Boettcher, ed. New York: Academic Press, pp. 311-322.

Hojo, K., C. Hiramine, and M. Ishitaki. 1980. Lymphocyte proliferative response *in vitro* and its cellular dependency in guinea pigs with experimental allergic orchitis. *J. Reprod. Fertil.* 59:113.

Huang, T.F. Jr., K.S.K. Tung, and R. Yanagimachi. 1981. Autoantibodies from vasectomized guinea pig inhibit fertilization *in vitro*. Submitted for publication.

Hurtenbach, U., F. Morgenstern, and D. Bennett. 1980. Induction of tolerance *in vitro* by autologous murine testicular cells. *J. Exp. Med.* 151:827.

Isojima, S., R.M. Graham. and J.B. Graham. 1959. Sterility in female guinea pigs induced by injection with testis. *Science* 129:44.

Isui, S., P.H. Lambert, C.J. Fournie, H. Türler, and P.A. Miescher. 1977. Features of systemic lupus erythematosus in mice injected with bacterial lipopolysaccharides. Identification of circulating DNA and renal localization of DNA-anti-DNA complexes. *J. Exp. Med.* 145:1115.

Jackson, J.J., A. Hagopian, D.J. Carlo, G.A.Limjuco, and E.H. Eylar. 1975. Experimental allergic aspermatogenic orchitis. I. Isolation of a spermatozoal protein (AP1) which induced allergic aspermatogenic orchitis. *J. Biol. Chem.* 250:6141.

Johnson, M.H. 1973. Physiological mechanisms for the immunological isolation of spermatozoa. *Adv. Reprod. Physiol.* 6:279.

Johnson, M.H., and B.P. Setchell. 1968. The protein and immunoglobulin content of rete testis fluid of rams. *J. Reprod. Fertil.* 17:403.

Julius, M.H., E. Simpson, and L.A. Herzenberg. 1973. A rapid method for the isolation of functional thymus-derived murine lymphocytes. *Eur. J. Immunol.* 3:645.

Kantor, G.L., and J.F. Dixon. 1972. Transfer of experimental allergic orchitis with peritoneal exudate cells. *J. Immunol.* 108:329.

Kaplan, M.E., and C. Clark. 1974. An improved rosetting assay for detection of human T lymphocytes. *J. Immunol. Methods* 5:131.

Katsh, S. 1959a. Immunology, fertility and infertility: A historical survey. *Am. J. Obstet. Gynecol.* 77:946.

Katsch, S. 1959b. Infertility in female guinea pigs induced by injection of homologous sperm. *Am. J. Obstet. Gynecol.* 78:276.

Katz, D.H. 1977. Lymphocyte Differentiation, Recognition and Regulation. Academic Press, New York, pp. 247-342.

Kiddy, C.A., and R.M. Rollins. 1973. Infertility in female guinea pigs injected with testis. *Biol. Reprod.* 8:545.

Klassen, J., R.T. McCluskey, and F. Milgrom. 1971. Nonglomerular renal disease produced in rabbits by immunization with homologous kidney. *Am. J. Pathol.* 63:333.

Kolk, A.H.J., T. Samuel, and P. Rumke. 1974. Auto-antigens of human spermatozoa. I. Solubilization of a new auto-antigen detected on swollen sperm-heads. *Clin. Exp. Immunol.* 16:63.

Koshimies, A.I., M. Kormano, and A. Lahti. 1971. A difference in the immunoglobulin content of seminiferous tubule fluid and rete testis fluid of the rat. *J. Reprod. Fertil.* 27:463.

Kouda, L.L., and P.E. Bigazzi. 1978. Autoantibodies to acrosomal antigens of spermatozoa in vasectomized mice. *Invest. Urol.* 16:140.

Kummerfeld, H.L., and R.H. Foote. 1976. Infertility and embryonic mortality in female rabbits immunized with different sperm preparations. *Biol. Reprod.* 14:300.

Lerum, J.E., and E. Goldberg. 1974. Immunological impairment of pregnancy in mice by lactate dehydrogenase-X. *Biol. Reprod.* 11:108.

Lipscomb, H.L., P.J. Gardner, and J.G. Sharp. 1979. The effect of neonatal thymectomy on the induction of autoimmune orchitis in rats. *J. Reprod. Immunol.* 1:209.

Lipscomb, H.L., J.J Breitkreutz, and J.G. Sharp. 1981. Autoimmune orchitis in thymectomized Lewis rats. *Fed. Proc.* 40:972 Abstract 4201.

Marcus, Z.H., J.H. Freisheim, J.L. Houk, J.H. Herman, and E.V. Hess. 198. *In vitro* studies in reproductive immunology. I. Suppression of CMI response by human spermatozoa and fractions isolated from human seminal plasma. *Clin. Immunol. Immunopathol.* 9:318.

McLaren, A. 1964. Immunological control of fertility in female mice. *Nature* 201:582.

McLaren, A. 1966. Studies on the isoimmunization of mice with spermatozoa. *Fertil. Steril.* 17:492.

Menge, A.C. 1967. Induced infertility in cattle by iso-immunization with semen and testis. *J. Reprod. Fertil.* 13:445.

Menge, A.C. 1968. Fertilization, embryo and fetal survival rates in rabbits isoimmunized with semen, testis and conceptus. *Proc. Soc. Exp. Biol. Med.* 127:1271.

Menge, A.C. 1969. Early embryo mortality in heifers isoimmunized with semen and conceptus. *J. Reprod. Fertil.* 18:67.

Menge, A.C. 1970a. Use of polynucleotides with seminal antigens to iduce isoantibodies and infertility in rabbits. *Proc. Soc. Exp. Biol. Med.* 135:108.

Menge, A.C. 1970b. Immune reactions and infertility. *J. Reprod. Fertil.* Suppl. 10:171.

Menge, A.C. 1971. Effects of isoimmunization and isoantisera against seminal antigens on fertility process in female rabbits. *Biol. Reprod.* 4:137.

Menge, A.C., and C.S. Black. 1979. Effects of antisera on human sperm penetration of zona-free hamster ova. *Fertil. Steril.* 32:214.

Menge, A.C., and M.E. Lieberman. 1974. Antifertility effects of immunoglobulins from uterine fluids of semen-immunized rabbits. *Biol. Reprod.* 10:422.

Menge, A.C., land W.P. Protzman. 1967. Origin of the antigens in rabbit semen which induce antifertility antibodies. *J. Reprod. Fertil.* 13:31.

Menge, A.C., V. Schweitzer, A. Rosenberg, and C. Westhoff. 1978. Local immunity of the female reproductive tract and infertility in rabbits. In *Immunology of Reproduction.* K. Bratanov, V.H. Vuchanov, V. Dikov, B. Somlev, and L. Kozhouharova, eds. Sofia: Bulgarian Academy of Sciences, pp. 492-496.

Menge, A.C., H. Peegel, and M.L. Riolo. 1979. Sperm fractions responsible for immunologic induction of pre- and postfertilization infertility in rabbits. *Biol. Reprod.* 20:931.

Metz, D.B., and J. Anika. 1970. Failure of conception in rabbits inseminated with nonagglutinating, univalent antibody-treated semen. *Biol. Reprod.* 2:284.

Moore, H.D.M. 1981. Glycoprotein secretions of the epididymis in the rabbit and hamster: Localization on epididymal spermatozoa and the effect of specific antibodies on fertilization *in vivo. J. Exp. Zool.* 215:77.

Moore, H.D.M., and J.M. Bedford. 1978. Fate of spermatozoa in the male. I. Quantitation of sperm accumulation after vasectomy in the rabbit. *Biol. Reprod.* 17:784.

Morton, D.B., and P.A. McAnulty. 1979. The effect on fertility of immunizing female sheep with ram sperm acrosin and hyaluronidase. *J. Reprod. Immunol.* 1:61.

Morton, D.B., V.A. Curry, and G.D. Harkiss. 1979. The effect on fertility of vaccinating female sheep with biochemicaly defined fractions of ram spermatozoa. *Fertil. Steril.* 31:683.

Muir, V.Y., J.L. Turk, and H.G. Haley. 1976. Comparison of allergic aspermatoenesis with that induced by vasectomy. I. *In vivo* studies in the guinea-pig. *Clin. Exp. Immunol.* 24:72.

Munoz, M.G., and C.B. Metz. 1978. Infertility in female rabbits isoimmunized with subcellular sperm fractions. *Biol. Reprod.* 18:669.

Nagano, T., and K. Okumura. 1973. Fine structural changes of allergic aspermatogenesis in the guinea pig: II. Induced by the homologous parotid gland as antigen. *Virch. Arch. Zell Pathol.* 14:237.

Nagarkatti, P.S., and S.S. Rao. 1976. Cell-mediated immunity to homologous spermatozoa following vasectomy in the human male. *Clin. Exp. Immunol.* 26:239.

O'Rand, M.G. 1981. Inhibition of fertility and sperm-zone binding by antiserum to the rabbit sperm membrane autoantigen, RSA-1. *Biol. Reprod.* 25:621.

O'Rand, M.G., and J.P. Porter. 1979. Isolation of a sperm membrane sialoglycoprotein autoantigen from rabbit testes. *J. Immunol.* 122:1248.

Prakash, C., and R.W. Lang. 1980. Studies on immune infertility: A hypothesis on the etiology of immune infertility based on the biological role of seminal plasma immune response inhibitor. *Mt. Sinai J. Med.* 47:491.

Rumke, P. 1980. Auto- and isoimmune reactions to antigens of the gonads and genital tract. In *Immunology 80.* M. Sougereau and J. Dausset, eds. New York: Academic Press, pp. 1065-1092.

Russo, I., and C.B. Metz. 1974. Inhibition of fertilization *in vitro* by treatment of rabbit spermatozoa with univalent isoantibody. *J. Reprod. Fertil.* 38:211.

Schlesinger, D.H., G. Goldstein, M.P. Scheid, and E.A. Boyse. 1975. Chemical synthesis of a peptide fragment of Thymopoietin II that induces selective T-cell differentiation. *Cell* 5:367.

Schumacher, G.F.B. 1980. Humoral immune factors in the female reproductive tract and their changes during the cycle. In *Immunological Aspects of Fertility and Fertility Regulation.* D.S. Dhindsa, and G.F.B. Schumacher, eds. New York: Elsevier/North-Holland, pp. 93-141.

Shulan, S., E. Zappi, U. Ahmed, and J.E. Davis. 1972. Immunologic consequences of vasectomy. *Contracep.* 5:269.

Sotsiou, F., G.F. Bottazzo, and D. Doniach. 1980. Immunofluorescence studies on autoantibodies to steroid-producing cells, and to germline cells in endocrine disease and infertility. *Clin. Exp. Immunol.* 39:97.

Stadecker, M.J., G. Bishop, and H.H. Wortis. 1973. Rosette formation by guinea pig thymocytes and thymus-derived lymphocytes with rabbit red blood cells. *J. Immunol.* 111:1834.

Stites, D.P., and R.P. Erickson. 1975. Suppressive effect of seminal plasma on lymphocyte activation. *Nature* 253:727.

Suzuki, F., and T. Nagano. 1978. Regional differentiation of cell junctions in the excurrent duct epithelium of the rat testis as revealed by freeze-fracture. *Anat. Rec.* 191:503.

Swierkosz, J.E., and R.H. Swanborg. 1975. Suppressor cell control of unresponsiveness to experimental allergic encalomyelitis. *J. Immunol.* 115:631.

Sy, M.-S., A.R. Brown, B. Benacerraf, and M.I. Greene. 1980. Antigen- and receptor-driven regulatory mechanisms. III. Induction of delayed-type hypersensitivity to azo-benzenarsonate with anti-crossreactive idiotypic antibodies. *J. Exp. Med.* 151:896.

Teuscher, C., G.C. Wild, and K.S.K. Tung. 1981. Isolation of guinea pig sperm/testis antigens which elicit two distinctive histopathologic patterns of experimental allergic orchitis (EAO). *J. Reprod. Immunol.* In press.

Thomas, W.R., G. Morahan, I.D. Waker, and J.F.A.P. Miller. 1981. Induction of delayed-type hypersensitivity to azobenzenearsonate by a monoclonal anti-idiotype antibody. *J. Exp. Med.* 153:743.

Toullet, F., and G.A. Voisin. 1976. Passive transfer of autoimmune aspermatogenic orchiepididymitis (A1A0) by antispermatozoa sera. Influence of the type of autoantigen and of the class of antibody. *Clin. Exp. Immunol.* 26:549.

Toullet, F., G.A. Voisin, and M. Naminovsky. 1973. Histochemical localization of three guinea pig spermatozoal autoantigens. *Immunology* 24:635.

Tsunoda, Y., and M.C. Chang. 1976. Reporduction in rats and mice isoimmunized with homogenates of ovary or testis with epididymis or sperm suspensions. *J. Reprod. Fertil.* 46:379.

Tumboh-Oeri, A.G., and T.K. Roberts. 1978. The induction of cell-mediated immunity to spermatozoa folowing vasectomy in inbred mice. *Int. J. Androl.* 1:81.

Tung, K.S.K. 1975. Human sperm antigens and antisperm antibodies. I. Studies on vasectomy patients. *Clin. Exp. Immunol.* 20:93.

Tung, K.S.K. 1977. The nature of antigens and pathogenetic mechanisms in autoimmunity to sperm. In *Immunobiology of Gametes.* M. Edidin and M.H. Johnson, eds. Cambridge: Cambridge University Press, pp. 157-180.

Tung. K.S.K. 1978. Allergic orchitis lesions are adoptively transferred from vasoligated guinea pigs to syngeneic reciipients. *Science* 201:833.

Tung, K.S.K. 1980. Autoimmunity of the testis. In *Immunological Aspects of Infertility and Fertility Regulation.* D.H. Dhindsa and G.F.B. Schumacher, eds. New York: Elsevier/North Holland Press, pp. 33-91.

Tung, K.S.K., and N.J. Alexander. 1977. Immunopathologic studies on vasectomized guinea pigs. *Biol. Reprod.* 17:241.

Tung, K.S.K., and N.J. Alexander. 1980. Monocytic orchitis and aspermatogenesis in normal and vasectomized rhesus macaques (*Macaca mulatta*). *Amer. J. Pathol.* 100:163.

Tung, K.S.K., and A.J. Woodroffe. 1978. Immunopathology of experimental allergic orchitis in the rabbit. *J. Immunol.* 120:320.

Tung, K.S.K., E.R. Unanue, and F.J. Dixon. 1970. The immunopathology of experimental allergic orchitis. *Am. J. Pathol.* 60:313.

Tung, K.S.K., E.R. Unanue, and F.J. Dixon. 1971a. Pathogenesis of experimental allergic orchitis. I. Transfer with immune lymph node cells. *J. Immunol.* 106:1453.

Tung, K.S.K., E.R. Unanue, and F.J. Dixon. 1971b. Pathogenesis of experimental allergic orchitis. II. The role of antibody. *J. Immunol.* 106:1463.

Tung, K.S.K., C. Leong, and T.A. McCarty. 1977. Pathogenesis of experimental allergic orchitis. III. T lymphocyte requirement in local adoptive transfer by peritoneal exudate cells. *J. Immunol.* 188:1774.

Tung, K.S.K., R. Bryson, E. Goldberg, and L-P.B. Han. 1979a. Antisperm antibody in vasectomy: Studies in human and guinea pig. In *Vasectomy: Immunopathologic and Pathophysiologic Effects in Animals and Man.* I.H. Lepow and R. Crozier, eds. New York: Academic Press, pp. 267-284.

Tung, K.S.K., L-P.B. Han, and A.P. Evan. 1979b. Differentiation antigen of testicular cells and spermatozoa in the guinea pig. *Develop. Biol.* 68:224.

Tung, K.S.K., E.H. Goldberg, and E. Goldberg. 1979c. Immunobiological consequence of immunization of female mice with homologous spermatozoa: Induction of infertility. *J. Reprod. Immunol.* 1:145.

Tung, K.S.K., A. Okada, and R. Yanagimachi. 1980. Sperm autoantigens and fertilization. I. Effects of antisperm autoantibodies on rouleaux formation viability and acrosome reaction of guinea pig spermatozoa. *Biol. Reprod.* 23:877.

Tung, K.S.K., C. Teuscher, and A.L. Meng. 1981a. Autoimmunity to spermatozoa and the testis. *Immunol. Rev.* 55:217.

Tung, K.S.K., L. Ellis, C. Teuscher, A. Meng, J.C. Blaustein, S. Kohno, and R. Howell. 1981b. The black mink (*Mustela vison*): A natural model of immunologic male infertility. *J. Exp. Med. 154:1016.*

Tung, K.S.K., C. Teuscher, E.H. Goldberg, and C. Wild. 1981c. Genetic control of antisperm autoantibody response in vasectomized guinea pigs. *J. Immunol.* 127:835.

Tung, K.S.K., A. Meng, C. Teuscher, and E. Goldberg. 1981d. Genetic control and pathogenetic significance of autoimmune responses sperm in vasectomized guinea pig (GP). *J. Reprod. Immunol.* In press.

Tyler, A. 1961. Approches to the control of fertility based on immunological phenomena. *J. Reprod. Fertil.* 2:473.

Tzartos, S.J. 1979. Inhibition of *in vitro* fertilization of intact and denuded hamster eggs by univalent anti-sperm antibodies. *J. Reprod. Fertil.* 55:447.

Unanue, E.R., F.J. Dixon, and J.D. Feldman. 1967. Experimental allergic glomerulonephritis induced in the rabbit with homologous renal antigens. *J. Exp. Med.* 125:163.

Voisin, G.A., A. Delunay, and M. Barber. 1951. Lesions testiculaires provoquées chez le cobaye par injection d'extrait de testicule homologue. *C. R. Acad. Sci.* (Paris) 232:1264.

Waksman, B.Y. 1959. A histologic study of the auto-allergic testis lesion in the guinea-pig. *J. Exp. Med.* 109-311.

Webb, C.G. 1980a. Decreased fertility in mice immunized with teratocarcinoma OTT6050. *Biol. Reprod.* 22:695.

Webb, C.G. 1980b. Characterization of antisera against mouse teratocarcinoma OTT6050: Molecular species recognized on embryoid bodies, preimplantation embryos, and sperm. *Develop. Biol.* 76:203.

Wekerle, H. 1978. Immunological T-cell memory in the *in vitro*-induced experimental autoimmune orchitis, specificity of the reaction and tissue distribution of the autoantigens. *J. Exp. Med.* 147:233.

Wekerle, H., and M. Begemann. 1978. *In vitro* induction of experimental autoimmune orchitis: Characterization of a primary T-lymphocyte response against testicular self antigens. *Eur. J. Immunol.* 8:294.

Wiley, L.M. 1979. Early embryonic surface antigens as developmental probes. In *Current Topics in Developmental Biology*, vol. 13. New York: Academic Press, pp. 167-197.

World Health Organization Workshop on Immunological Response of the Female Reproductive Tract. 1976. B. Cinader and A. de Weck, eds. Copenhagen: Scriptor.

Yanagimachi, R. 1978. Calcium requirement for sperm-egg fusion in mammals. *Biol. Reprod.* 19:949.

Yanagimachi, R. 1981. Mechanisms of fertilization in mammals. In *Fertilization and Embryonic Development* in vitro. L. Mastroniai, J.D. Biggers, and W. Sadler, eds. New York: Plenum Press. In press.

Yanagimachi, R., A. Okada, and K.S.K. Tung. 1981. Sperm autoantigens and fertilization. II. The effects of antisperm autoantibodies on the interaction between guinea pig spermatozoa with guinea pig and hamster ova. *Biol. Reprod.* 24:512.

Chapter 19

SIMILARITIES BETWEEN IMMUNOPATHOLOGIC CHANGES AFTER VASECTOMY AND EXPERIMENTAL IMMUNE COMPLEX DISEASE

NANCY J. ALEXANDER

Studies in my laboratory and in those of my colleagues suggest that vasectomy in animal models and experimentally induced serum sickness may cause similar changes in several body systems. In this chapter I shall present correlative data that demonstrate some of these similarities.

IMMUNE-COMPLEX DISEASE

The term "serum sickness" was first introduced by von Pirquet and Schick in 1905 when they observed that a variety of clinical symptoms including urticaria, edema, neutropenia, and enlarged lymph nodes developed in a large percentage of persons receiving injections of horse serum produced as a diphtheria antitoxin. They correctly identified the symptoms as a response to circulating foreign proteins. After large injections of any heterologous protein or serum, characteristic lesions develop in the glomeruli, arteries, and endocardium. However, even when the protein is given in large amounts, if the antigen exposure is a single event, the effects are transient. The symptoms defined as serum sickness have been shown to occur not only with an injection of foreign protein, but also as a result of viral antigens or autoantibodies. Based on these findings, Leber and McCluskey (1974) suggested that the term "serum sickness" be changed to "immune-complex disease."

Immune-complex disease (experimental serum sickness) has been studied in several laboratory species, but most extensively in rabbits. In 1947, Hawn and Janeway reported that arterial and glomerular lesions could be produced by one injection of plasma proteins (bovine serum albumin [BSA] or bovine gamma globulin [BGG]). This finding made possible an analysis of the relationship between the fate of injected antigen, the appearance of antibody, and the development of disease. Investigators have demonstrated that certain processes occur after the injection of a foreign protein. For a day or two, there is equilibrium between the intra- and extravascular spaces; catabolism of the foreign protein continues at the same rate as catabolism of the animal's own albumin. With the development of antibodies, approximately six days later, a much more rapid decrease in antigen occurs and small, soluble antigen-antibody complexes are formed. The disease state coincides with the appearance of these soluble immune complexes. At this juncture, serum complement levels drop as they attach to the Fc portion of the complexes and immune complexes are deposited (Germuth, 1953; Dixon et al., 1958; Dixon, 1963). With continued antibody production, the ratio of antibody to antigen

427

increases and circulating antigen is eliminated. As larger immune complexes are formed, they become more rapidly eliminated. The final stage is the disappearance of antigen. It is during immune elimination of antigen that complexes of antigen and antibody are circulating and can cause disease in arteries, glomeruli, joints, and pericardium.

Initially it was thought that specific antigens settled in an area and later were joined by immunoglobulins to form complexes. Although this process may occur when antigens leak into adjacent tissue and are exposed to free, circulating antibodies, it is more common for immune complexes to be formed within the circulation and later become tissue associated, usually in the renal glomerulus or arterial wall. In fact, injection of preformed soluble immune complexes results in damage as early as 36 hours later, an indication that an animal need not make an immune response for damage to occur (McCluskey et al., 1962). The reverse experiment can also be done. If large amounts of antigen and then antibody are injected into rabbits, disease symptoms again occur in 36 hours, a further indication that the combination of antigen and antibody causes the effects.

The formation of immune complexes is not limited to laboratory experiments. Everyone is exposed to disease organisms, and during the immune response, complexes may be formed. The formation of such complexes, which are then phagocytized, is an important component of the body's defenses against pathogens and foreign antigens. The presence of complexes in the circulation is usually short-lived. Injection of radiolabeled immune complexes into animals has revealed that these complexes almost entirely disappear within 24 hours. Many components of the immune system aid in their removal, including circulating leukocytes and phagocytic cells common in the lung, liver, and spleen. Kupffer cells of the liver, as well as other phagocytes, have Fc receptors with the capacity to bind the immunoglobulin portion of certain immune complexes (particularly those involving IgG_1 and IgG_3) and thereby activate various cell functions, e.g., phagocytosis (Figure 19-1) (Huber et al., 1971). Binding sites for activated C3 are also present on macrophages (Lay and Nussenzweig, 1968) and polymorphonuclear neutrophils (PMNs) (Henson and Cochrane, 1971). Specific C3b receptors appear to aid in the actual binding of the complexes and the Fc region activates phagocytosis (Griffin et al., 1975; Goldstein et al., 1976). One can demonstrate the importance of an intact Fc region by injecting alkylated or reduced complexes that tend to remain in circulation (Mannik et al., 1971).

The formation and biological activities of circulating immune complexes depend upon the type of antibody and antigen. Such factors as the size and tissue affinity of the antibody (Unanue and Dixon, 1967; Cochrane and Koffler, 1973), as well as of the antigen itself, affect the fate of immune complexes. Some antigens have an affinity for certain tissues—for example, DNA and the glomerular basement membrane (Nydegger, 1979).

The valence, size, and chemical composition of antigens determine whether they can have detrimental effects. Whether the antigen is small or large influences the size of the resultant complex. Monovalent antigens do not form lattices with their associated antibodies. For antibodies, their class and the subclass that determines valency as well as their ability to bind complement are important characteristics. Macrophage receptors most actively bind to IgG_1 and IgG_3, and thus clearance of these immunoglobulins is rapid. IgG_3 disappears rapidly because it has a short half-life and it is the most potent complement-activating subclass, followed by IgG_1 and IgG_2 (Nydegger, 1979).

Smaller complexes formed in great antigen excess stay in circulation much longer because they do not fix complement and do not initiate inflammatory processes. Immune complexes containing more than two IgG molecules are quickly removed from circulation, whereas complexes with one or two antibody molecules persist (Mannik et al., 1971). In fact, the solubility of the complexes increases progressively as the amount of antigen increases (Scherzer and Ward, 1978). Complexes formed in antibody excess have extensive lattices, are rapidly phagocytosed, and thus do little damage. The most pathogenic complexes are of an intermediate size, large enough to fix complement but small enough to remain soluble and therefore in circulation.

The duration of antigen exposure is an important consideration. If complexes remain in circulation for a limited period, the effects will be

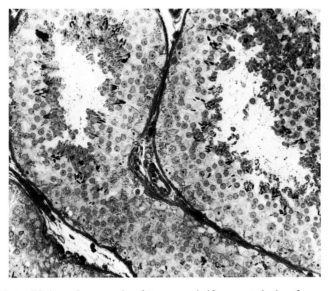

FIGURE 19-1. Light micrograph of two seminiferous tubules from a monkey that had received a vasectomy four years previously.
All stages of spermatogenesis are visible. Toluidine blue stain. Magnification: 250x.

transient. If there is a constant antigen supply (as in chronic infection or autoantigen production), chronic disease is a possibility.

TISSUE-ASSOCIATED DEPOSITION

Immune-complex disease, particularly in the case of autoimmune disease, may involve localizaton in specific tissues. If antigens are present in a tissue either as fixed cellular antigens or as a secretion, antibody diffuses from the vascular compartment and a localized complex develops.

Alternatively, soluble antigens may circulate, combine with antibody, and become deposited. In this case, there is no relationship between the antigen origin and the site of tissue injury. Deposition usually occurs in renal glomeruli, arterial walls, lung, choroid plexuses, and joints.

The hallmark of immune-complex disease is deposition of antigen in association with antibody. The simultaneous presence of antigen or, more commonly, antibody plus complement, strongly suggests immune complexes (Wilson and Dixon, 1974). Most frequently, fluorescein isothiocyanate conjugated to immunoglobulin or complement and then incubated with frozen tissue sections is used for such demonstrations; with an electron microscope, electron-dense deposits can be seen at higher magnifications (usually along aspects of the basement membrane). Further definition of specific complexes by electron microscopy can be accomplished with immunoferritin techniques (Andres et al., 1963). If enough diseased tissue is available, deposits of complexes can be eluted with low-pH buffers or chaotropic agents. After such treatments, antibodies and sometimes antigens can be recovered and identified.

VASECTOMY

Vasectomy results in blockage of the excurrent ducts. Although spermatozoa no longer have an exit passage, spermatogenesis continues relatively unabated, and a large mass of spermatozoa is stored in the epididymis. It seems likely that sperm antigens become exposed to the immune system as a result of the leakage of soluble products or the breakdown of duct integrity and subsequent granuloma formation. Sperm antigens readily elicit an autoimmune response. Spermatozoa are complex cells; as many as eight specific antigens have been revealed by immunofluorescence to be associated with the acrosome, head, midpiece, or tail (Figure 19-2). Sperm-specific antigens first appear at puberty and thereafter are sequestered from the immune system by

430

the blood-testis barrier, as well as by epithelial cell barriers in the rete testis, efferent ducts, and epididymis. Exposure to such antigens by immunization (either experimentally, by duct obstruction, or by infection) can result in the development of specific antibodies and/or orchitis.

Vasectomy has caused the production of antisperm antibodies in every species thus far studied. The response may vary in the length of initiation time as well as in magnitude. Studies have revealed that in monkeys the antibodies are first of the IgM class and subsequently of the IgG class; this primary-to-secondary immune response indicates exposure to hitherto sequestered antigens.

As revealed by sperm agglutination, sperm immobilization, and immunofluorescence tests, circulating antisperm antibodies develop in about 50 per cent of vasectomized men and animals. Why these antibodies are not present in all vasectomized mammals is not clear. One reason may be the amount of available antigen. The development of antibody in vasectomized monkeys is correlated with a high sperm count before vasectomy (Alexander, 1977). There are similar data on men (Linnet and Hjort, 1977). An alternative explanation is that antibodies to sperm develop in all vasectomized mammals, but that currently used laboratory methods detect only free antibodies, not those joined with antigens. Thus, the absence of free antibodies may reflect antigenemia. Support for this concept comes from the finding that after orchiectomy, levels

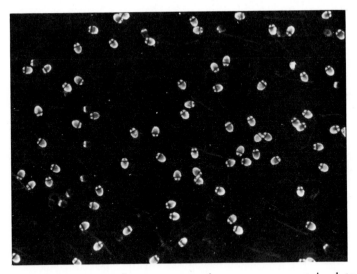

FIGURE 19-2. Spermatozoa from a cynomologus macaque, stained to reveal IgM antibodies.
Sperm-specific antibodies to the acrosome and postacrosomal region are seen. Magnification: 500x.

of circulating immune complexes in vasectomized rabbits, previously immunized with sperm antigens, fall precipitously and free antibody levels rise abruptly (Tung et al., 1979).

VASECTOMY AND TISSUE-ASSOCIATED DEPOSITION

If antigens leak from the reproductive system after vasectomy, then evidence of immune-complex deposition should occur at sites where antigens come in contact with antibodies against sperm products. There is evidence that immune deposition does occur after vasectomy. Deposition of immunoglobulins and complement is found around the seminiferous tubules of rabbits (Alexander and Tung, 1979a) and the excurrent duct system of monkeys (Alexander and Tung, 1979b). We found that 45 per cent of the control non-vasectomized monkeys exhibited immune deposition associated with the efferent ducts and caput epididymides, whereas 91 per cent of the vasectomized animals did. Furthermore, deposits were also found in the rete testis and cauda epididymides and focally in the seminiferous tubules of the vasectomized group, but not the controls (Tung and Alexander, 1980). Such findings suggest that vasectomy causes increased leakage of sperm antigens from the excurrent duct.

GLOMERULONEPHRITIS
EXPERIMENTAL IMMUNE COMPLEX DISEASE

In immune-complex disease, glomerulonephritis is characterized by endothelial swelling. This results in an enlargement of the entire glomerulus and narrowing or obliteration of the capillary lumens. With this type of disease, there is a little leukocytic infiltration. Proteinuria, but usually not hematuria, occurs. With electron microscopy, swelling of the endothelial cells and thickening of the capillary walls are evident, especially when the disease is chronic. Recent studies suggest that macrophages, attracted by chemotactic substances, may be important in the pathogenesis of lesions because they cause tissue injury (Hunsicker et al., 1979; Holdsworth et al., 1980).

When BSA is used to induce immune-complex disease experimentally, large amounts, as much as 20 μg of BSA per kidney, can be found (Wilson and Dixon, 1970). Suggestive information on the complex size is indicated by the fact that more molecules of antibody than of antigen are found (4.4 x 10^8 molecules of BSA per glomerulus and 1.3 x 10^9 molecules of antibody [Unanue and Dixon, 1967]). Whenever complexes of more than 10^6 daltons are present, glomerulonephritis occurs. Precisely why the glomerulus is especially vulnerable to immune complexes is not known, but a reasonable assumption is that the glomerular basement membrane is a filtering layer, uniquely subjected to a higher pressure gradient, in which macromolecules may be trapped.

In experimental immune-complex disease, the size of the antigen and the dosage of injections play important roles. Glomerulonephritis due to a one-time, high-dose injection is quickly reversed, but long-term injections may have more detrimental effects. Chronic glomerulonephritis will develop in any immunologically responsive rabbit if the quantity of antigen each day is in excess of the antibody production. With increasing antigen doses, rabbits with large amounts of circulating antibodies are affected first, since many small complexes may be formed. Because antibodies of different classes, subclasses, and avidities may develop in different rabbits, the amount of involvement may vary.

Immunofluorescence can be used to determine the locations of complexes. Characteristically, they are seen as a granular pattern within the glomerular basement membrane. Complex size, in part, determines the site of deposition—larger complexes are deposited near the vascular lumen of the glomerular capillary, whereas smaller aggregates are deposited closer to the urinary space (Germuth and Rodriguez, 1973).

POSTVASECTOMY GLOMERULAR CHANGES

If vasectomy results in circulating immune complexes, immune complexes should be visible within the renal glomeruli after vasectomy. Vasectomized rabbits exhibit glomerular changes that include an increased amount of mesangial matrix and cellular proliferation. Immune-complex deposition has been revealed by immunofluorescence techniques and electron microscopy (Bigazzi et al., 1976). In fact, Bigazzi and associates have shown that elution with an acidic buffer results in the removal of an antibody that can be shown to be sperm specific—further evidence that an immune complex is associated with the sperm-specific antigen. Deposition of both IgM and IgG in a granular pattern occurs in vasectomized monkeys and in some control animals. However, 30 per cent of vasectomized rhesus monkeys had glomerular C3 deposits, whereas none of the controls exhibited such deposition (Tung and Alexander, 1980). Although all animals do not respond to a similar degree, such findings suggest postvasectomy similarities to immune complex disease.

ARTERITIS

EXPERIMENTAL IMMUNE-COMPLEX DISEASE

Factors that lead to glomerular deposition differ from those that lead to complex deposition in blood vessels. Arterial lesions usually develop later than glomerular lesions (Germuth, 1953). My colleagues and I have confirmed this finding; we found arteritis when rabbits were kept in BSA excess for 34 days, but not when rabbits were injected once with BSA and then necropsied 9 to 14 days later, as soon as serum antibodies were first detected. Probably a key factor in immune-complex-mediated arteritis is the interaction of the complexes with PMNs (Wedmore and Williams, 1981). Although immune complexes can cause direct damage, they often involve other inflammatory mediators. Polymorphonuclear granulocytes release from their lysosomal system a variety of lytic agents that can degrade cartilage, elastin, and nucleic acids (Weissmann and Dukor, 1970) and thus result in inflammation and arterial wall damage. Like macrophages, PMNs have Fc receptors as well as receptors for C3b. Use of cobra venom factor to deplete complement prevents the involvement of PMNs, and although complex deposition may occur, inflammation does not result (Henson and Cochrane, 1971).

Since platelets serve as a reservoir of vasoactive amines, they play an important role in immune-complex-mediated arteritis. In the presence of immune complexes of all IgG classes, platelets aggregate. One mechanism for the release of amines depends upon the platelet response to an activating factor released from basophils (Wiggins and Cochrane, 1981). Immune complexes can react directly with platelets with or without complement. Platelets also have Fc (Israels et al., 1973) and C1q (Wautier et al., 1976) receptors. Upon activation, platelets release a variety of factors, including serotonin. In fact, antagonists of histamine or serotonin hinder the formation of lesions (Kniker and Cochrane, 1968). When platelets are removed from circulation, the development of arterial and glomerular lesions is suppressed even though immune complexes are deposited (Kniker and Cochrane, 1968).

Complement activation is a major factor in immune-complex disease. Complement receptors on the cell membrane nonspecifically amplify the inflammatory system. The Fc portion of the immunoglobulin molecule of the complex is involved. All circulating blood elements except erythrocytes have Fc receptors. Thus PMNs, basophils, and platelets become activated and release products, and macrophages may become activated and more phagocytic. B cells may become inhibited, and T cells may be stimulated to produce helper or suppressor factors (Theofilopoulos and Dixon, 1979). Immune complexes, then, can be important regulators of the immune system.

The size of the complexes is important to the development of arteritis. When we immunized rabbits, by means of a long-term regimen, with BSA (60 000 to 80 000 daltons), BGG (150 000 to 180 000 daltons), heat-aggregated BSA (250 000 daltons), and heat-aggregated BGG (500 000 daltons), we found that arteritis was more common with the smaller antigens. About 60 per cent of rabbits undergoing long-term immunization with BSA exhibited IgG deposition, C3 deposition, or both, adjacent to the internal lamina; only 30 per cent of those immunized with heat-aggregated BSA or BGG exhibited such deposition.

Arteritis occurs most commonly at coronary outflow areas, points where the aterial tree branches. Lesions will occur at the area of narrowing when an aorta is experimentally constricted. As a result of hemodynamic forces, lesions are more common at the entrance of the coronary arteries, at the branching of the aorta, and in the pulmonary arteries. Arteritis is first seen as a mild proliferation of intimal endothelial cells. Vascular permeability increases, and the complexes penetrate the media of the blood vessel wall. PMNs resulting from the presence of complexes enter and degrade the internal elastic lamina and thus gain entrance to the medium. As was mentioned, one can inhibit arteritis by impeding the neutrophils (Kniker and Cochrane, 1965). The interaction between PMNs and immune complexes is the most important event in immune-complex-induced injury of the arterial wall. Such reactants are quickly removed by these phagocytic cells (Cochrane and Koffler, 1973). Later mononuclear cells enter the inflammation site, and finally fibrinoid necrosis of focal areas of the media is apparent. Particularly at the onset of an injury, fluorescence microscopy of vessels reveals antigen, immunoglobulin, or C3 in the vicinity of the internal elastic lamina. Later, antigen and immunoglobulin can both be revealed throughout the media (Figures 19-3 to 19-6). Usually only immunoglobulins and complement can be demonstrated in the arterial wall, but antigens coupled with antibody have been found in hepatitis (Gocke et al., 1971) and streptococcal infection (Parish and Rhodes, 1967).

Problems associated with the study of experimentally induced arteritis include the focal nature of the lesions and the variable responses of individuals, probably due to genetic differences. None of the lesions is highly reproducible, and even with the most successful method lesions fail to develop in a fair percentage of individuals. Two general immunization schedules have been used to study the development of arteritis. With the short-term schedule, the host's exposure to antigen is of limited duration. Large amounts of antibodies develop in about 50 per cent of immunized rabbits—this immediately leads to removal of the antigen. Another 30 per cent fail to respond, and 20 per cent respond with poorly precipitating antibodies that form complexes which persist in cir-

FIGURE 19-3. Aortic arch stained to reveal IgG from a rabbit chronically immunized with BSA and fed an atherogenic diet.
Deposition is located in the upper one third of the medium of the vessel wall. Magnification: 500x.

FIGURE 19-4. Thoracic aorta stained to reveal IgM from a rabbit chronically immunized with BSA and fed an atherogenic diet.
These complexes, being larger, are more localized near the internal elastic lamina. Magnification: 500x.

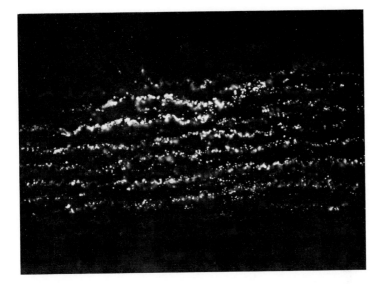

FIGURE 19-5. Pulmonary artery stained to reveal BSA deposition from a rabbit chronically immunized with BSA and fed an atherogenic diet.
Deposition throughout the medium of the vessel is seen. Magnification: 320x.

culation for some period of time. During the acute phase of the disease, complement levels may be depressed. Clinical manifestations are transient.

Under the long-term regimen, almost daily injections are given for one to several months. Usually blood samples are collected daily to ascertain whether the animal is in a state of antigen excess. If antigen is found, the daily dose is not changed. If antibody is found, the daily antigen dose is increased. Under these conditions, antigen remains in the circulation for extended periods of time, as would be the case with chronic infections or circulating autoantigens.

VASECTOMY

Is there any evidence of similarities between the effects of chronic foreign protein injections and vasectomy? Is there any evidence for arteritis in vasectomized animals? As a result of the difficulty in locating foci of arteritis and the variability of responses, such studies require a great many animals. In those with vasculitis, inflammation is often not apparent when the vessel is viewed grossly, and thus the localization of putative lesions depends upon evaluation of innumerable histological sections of arterial wall, most of which

437

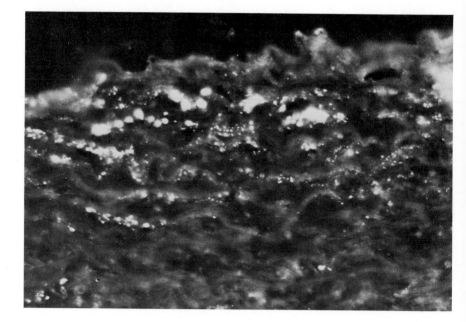

FIGURE 19-6. Aortic arch stained to reveal IgG from a rabbit repeatedly immunized with BGG and fed an atherogenic diet.
Magnification: 500x.

appear normal.

    With chronic arteritis, could there be changes in arteriolar function? The eye provides a unique chamber for evaluating vascular changes in a noninvasive manner. If vasectomy causes vascular changes, perhaps they can be seen in the retinal fundi. To examine this hypothesis we studied 159 men, 30 per cent of whom had been vasectomized. We confined the study to men over 30 and under 60 years of age. The researcher who evaluated the retinal changes was not aware of the vasectomy status of the patients. Arteriolar constriction, albeit mild, was more common in young vasectomized men than in control subjects over 40 years of age (Fahrenbach et al., 1980). This study indicating

438

arteriolar changes in some vasectomized young men was a preliminary one, and further studies involving a large number of individuals and the use of photographic evaluation will be necessary to validate this initial finding.

ATHEROSCLEROSIS

A useful technique to intensify and localize arteritis is the addition of another risk factor. It is possible to exacerbate the arterial injury with high levels of circulating serum cholesterol. Investigators have demonstrated a potentiation of atherogenesis in animals receiving a diet containing high levels of cholesterol plus injections of foreign proteins. In fact, Levy (1967) found that only one injection of BSA plus two weeks of a high-cholesterol diet caused a marked increase in atheromatous lesions. Furthermore, he found no decrease in the incidence of atheromatous lesions up to eight weeks after the termination of the high-cholesterol diet in rabbits receiving foreign protein plus atherogenic diet; when diet alone was the regimen, he found a markedly decreased incidence. These data suggest that a combination of inflammatory injury and lipid deposition in arteriolar walls changes vascular structure.

Use of particulate antigens rather than soluble proteins can yield similar results. We immunized rabbits with either sonicated human or sonicated rabbit sperm. We injected both the particulate and the soluble fractions (two subcutaneous injections one week apart and then daily intraperitoneal injections [each of 5×10^8 sperm]). The rabbits were maintained on a diet containing 0.50 mg of cholesterol per Calorie for two months prior to, as well as during, the experiment. The arteries of 11 of 13 showed plaques, and 12 of 13 rabbits had immune-complex deposition associated with plaques or at the internal elastic lamina, as revealed by staining with fluorescein-conjugated anti-IgG, IgM, and C3 (Figure 19-7). Staining with a fluorescein-conjugated antisperm IgG demonstrated that plaques from the arch and pulmonary artery of one animal had sperm antigens associated with the immune complexes. This finding strongly suggests that the sperm antigen:sperm antibody complexes are involved in the plaque development. Eleven of 13 kidneys exhibited glomerulonephritis. Examination of the testes of immunized rabbits revealed immune complexes around the seminiferous tubules, an indication that antigens do normally leak through the blood-testis barrier and that, in the presence of antibody, complexes can form. Furthermore, antibodies were formed to rhesus and human sperm that cross-reacted with rabbit sperm antigens (Clarkson and Alexander, 1979).

We postulate that as with immunization, vasectomy results in circulating antigen that may combine with antibody to cause circulating immune

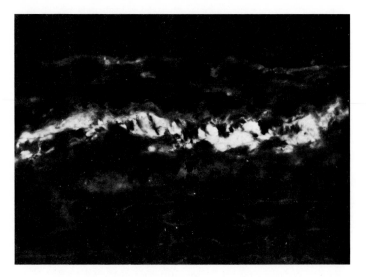

FIGURE 19-7. Aortic arch stained to reveal C3 from a rabbit repeatedly immunized with rabbit epididymal sperm and fed an atherogenic diet.
Deposition is adjacent to the internal elastic lamina at the base of the plaque. Magnification: 320x.

complexes. These complexes cause immunologic injury that, in association with elevated circulating cholesterol, can enhance the development of atherosclerosis. Figure 19-8, a thoracic aorta from a vasectomized rabbit, depicts immune deposition at the base of a plaque. The deposition is very similar to the type seen in immunized rabbits (Figure 19-7).

In one study to determine whether vasectomy exacerbates atherosclerosis, we used ten cynomolgus macaques (*Macaca fascicularis*) maintained on a diet in which 42 per cent of the calories were derived from butter and which contained 0.50 mg of cholesterol per Calorie. The animals were maintained on this atherogenic diet for six months, after which they were divided into two equivalent groups on the basis of their plasma cholesterol levels. Animals in one group were vasectomized, and those in the other group underwent sham vasectomies. Within one month, antibodies against spermatozoa had developed in all of the vasectomized animals but in none of the control animals. After ten months, circulating antisperm antibodies were detected in only three of five vasectomized animals. The sham and vasectomized groups were equivalent in plasma cholesterol concentration before the surgical procedures, as well as during the course of the study. Furthermore, plasma triglyceride levels were similar in both groups. The experiment was terminated, and the animals were killed 10 months after the surgical procedure or 16 months after initiation of the atherogenic diet. Tissues were taken for immunologic and

FIGURE 19-8. Thoracic aorta stained to reveal C3 from a vasectomized rabbit fed an atherogenic diet.
Immune deposits are seen adjacent to the internal elastic lamina. Magnification: 320x.

pathologic evaluation, as well as for chemical analyses of cholesterol and cholesterol ester concentrations of the arterial tree. The total cholesterol and cholesterol ester concentrations in the carotid, abdominal, iliac, and femoral arteries were significantly higher in the vasectomized monkeys than in the control monkeys (p < 0.001). Atherosclerosis was more extensive in the vasectomized than in the sham-vasectomized animals. Deposits of C3 were more commonly seen in the vasectomized group than in the control group. Vasectomy appeared to have striking effects, particularly on the abdominal aorta, carotid arteries, and iliac and femoral arteries.

Vasectomy also resulted in plaque deposition in the intercranial cerebral arteries—a phenomenon previously observed only in hypertensive monkeys (Alexander and Clarkson, 1978).

This initial study involved a cholesterol-rich diet that resulted in high circulating levels of plasma cholesterol. Another study was performed to determine whether animals fed Monkey Chow (devoid of cholesterol and low in fat) would also exhibit more atherosclerosis if vasectomized. In this study, animals that had been vasectomized 9 to 14 years earlier were matched with control (intact) animals that were given the same diet but not vasectomized. Even though the animals had a normal diet and low levels of plasma lipids, the vasectomized monkeys still had more extensive and severe atherosclerosis than the controls. Atherosclerotic plaques were not seen in the thoracic aortas of control animals, but large plaques were found in seven of ten vasectomized animals.

441

The extent of atherosclerosis in the abdominal aortas was also considerably greater in the vasectomized animals (Clarkson and Alexander, 1980).

If vasectomized rhesus macaques without demonstrable circulating antibodies were in a state of antigenemia, an evaluation of the correlation between atherosclerosis severity and the presence or absence of free antisperm antibodies would be appropriate. We found that atherosclerosis was more severe among vasectomized monkeys without circulating free antisperm antibodies than among monkeys with such antibodies. These differences in severity were particularly dramatic in the abdominal aorta (Clarkson and Alexander, 1980).

DETECTION OF IMMUNE COMPLEXES

Since endogenous and exogenous antigens can trigger immune-complex disease, the demonstration of complexes, both tissue-bound and soluble, is important. Tissue-bound complexes can be evaluated immunohistochemically and electron microscopically. If the antigen is known, staining with a conjugated antibody can be done. As mentioned previously, if enough tissue is available, elution with low-pH buffers and recovery of the antigen or more frequently, the antibody, can allow quantification and identification. Elution procedures, however, are selective only for certain complexed molecules, and such procedures may cause a change in functional characteristics (Nydegger, 1979). IgM, for example, is particularly susceptible. When materials are deposited in a granular, discrete pattern, it is probable that the individual has an immune-complex disorder.

Soluble complexes can be detected by a variety of methods. The provocative antigen of the pathogenic response is rarely known; therefore, assays that are antigen nonspecific are commonly used. There is no universal reagent that will detect all unknown immune complexes in sera. Since all current assays have distinct specificities, pathologic sera are usually analyzed by more than one assay. Assays are based on general characteristics of complexes. Complexes are not the same size as circulating immunoglobulin; thus analytical sucrose density ultracentrifugation can be used to determine the presence of complexes. Solubility can be used to evaluate complexes. Some are cryoglobulins that precipitate under cold conditions (Griswold et al., 1973). The solubility of complexes can be changed by the addition of polyethylene glycol (molecular weight, 6000); the precipitate can then be evaluated further. Other nonspecific methods, involving agglutination or electrical charge, can be used.

Immune complexes have distinct biological properties that are often used in their evaluation. A variety of tests that depend upon Fc receptors or complement activation have been developed.

In collaboration with Kenneth Tung, we have evaluated circulating immune complexes in vasectomized and control monkeys in order to determine whether complexes are more prevalent in vasectomized monkeys. We have found that 28 per cent (8 out of 29) were positive at 2 standard deviations or more above the mean for circulating immune complexes when a C1q solid-phase assay was used. Only 1 of 24 control monkeys had a similar reaction (Alexander, 1981).

There is a great deal that we do not know about circulating immune complexes after vasectomy. We have not yet determined whether there is any time-dependency after vasectomy. We do not know whether certain persons are prone to develop such complexes. On the basis of the varied responses in experimental immune-complex disease, we would expect such a susceptibility. We do not know whether vasectomy leads to circulating immune complexes in some men. Some studies (Tung et al., 1979) indicate that only a low percentage of vasectomized men have circulating immune complexes, whereas other studies (Witkin et al., 1980) reveal a much larger percentage of individuals with circulating immune complexes. We must know the amount and size of these complexes before we can determine whether some types formed after vasectomy are more detrimental than others.

Three studies funded by the National Institutes of Health are under way to assess whether atherosclerosis is more extensive in vasectomized men. One study will evaluate coronary occlusion in vasectomized and intact men who have had angiography. Another study will compare data on vasectomized men who appear to have cardiovascular disease based on electrocardiographic evidence with data on men who do not. Finally, the last study will be a multi-center effort to determine the frequency of vasectomy in men under 55 years of age who are admitted to hospitals with confirmed diagnoses of first myocardial infarction. In Canada, too, some studies have been initiated to evaluate the prevalence of circulating immune complexes in vasectomized and nonvasectomized men who have experienced a myocardial infarction. When the results of such studies have accumulated, we shall have a better idea of whether vasectomy in some men can result in immune-complex disease.

ACKNOWLEDGMENTS

I acknowledge and appreciate the collaborative efforts of T.B. Clarkson, Bowman Gray School of Medicine, and K.S.K. Tung, University of New

Mexico, and thank D.L. Fulgham, M. Joseph, P. Kimzey, L.D. Marsh, and B.A. Mixon for their excellent assistance.

The work described in this chapter, written as Publication No.1161 of the Oregon Regional Primate Research Center, was supported by the Program for Applied Research on Fertility Regulation (PARFR 107N under AID/csd-3608 and PARFR 222 under AID/DSPE-C-0035), National Institutes of Health Grant RR-00163, National Institute of Child Health and Human Development Contract N01-HD-8-2827, and National Institutes of Health Biomedical Research Support Grant RR-05694-11.

REFERENCES

Alexander, N.J. 1977. Vasectomy and vasovasostomy in rhesus monkeys. The effect of circulating antisperm antibodies on fertility. *Fertil. Steril.* 28:562.

Alexander, N.J. 1981. Primates: Their use in research on vasectomy. *Am. J. Primatol.* 1:167.

Alexander, N.J., and T.B. Clarkson. 1978. Vasectomy increases the severity of diet-induced atherosclerosis in *Macaca fascicularis*. *Science* 201:538.

Alexander, N.J., and K.S.K. Tung. 1979a. Vasectomy in the rabbit: Immunological and morphological effects. In *Vasectomy: Immunologic and Pathophysiologic Effects in Animals and Man*. I.H. Lepow and R. Crozier, eds. New York: Academic Press, pp. 355-377.

Alexander N.J., and K.S.K. Tung. 1979b. Effects of vasectomy in rhesus monkeys. In *Vasectomy: Immunologic and Pathophysiologic Effects in Animals and Man*. I.H. Lepow and R. Crozier, eds. New York: Academic Press, pp. 423-458.

Andres, G.A., B.C. Seegal, K.C. Hsu, M.S. Rothenberg, and M.L. Chapeau. 1963. Electron microscopic studies of experimental nephritis with ferritin-conjugated antibody. Localization of antigen-antibody complexes in rabbit glomeruli following repeated injections of bovine serum albumin. *J. Exp. Med.* 117:691.

Bigazzi, P.E., L.L. Kosuda, K.C. Hsu, and G.A. Andres. 1976. Immune complex orchitis in vasectomized rabbits. *J. Exp. Med.* 143:382.

Clarkson, T.B., and N.J. Alexander. 1979. Effect of vasectomy on diet-induced atherosclerosis. In *Vasectomy: Immunologic and Pathophysiologic Effects in Animals and Man*. I.H. Lepow and R. Crozier, eds. New York: Academic Press, pp. 121-162.

Clarkson, T.B., and N.J. Alexander. 1980. Long-term vasectomy: Effects on the occurrence and extent of atherosclerosis in rhesus monkeys. *J. Clin. Invest.* 65:15.

Cochrane, C.G., and D. Koffler. 1973. Immune complex disease in experimental animals and man. *Adv. Immunol.* 16:185.

Dixon, F.J. 1963. The role of antigen-antibody complexes in disease. *Harvey Lect.* 58:21.

Dixon, F.J., J.J. Vazquez, W.O. Weigle, and C.G. Cochrane. 1958. Pathogenesis of serum sickness. *Arch. Pathol.* 65:18.

Fahrenbach, H.B., N.J. Alexander, J.W. Senner, D.L. Fulgham, and L.J. Coon. 1980. Effect of vasectomy on the retinal vasculature of men. *J. Androl.* 1:299.

Germuth, F.G., Jr. 1953. A comparative histologic and immunologic study in rabbits of induced hypersensitivity of the serum sickness type. *J. Exp. Med.* 97:257.

Germuth, F.G. Jr., and E. Rodriguez. 1973. *Immunopathology of the Renal Glomerulus*. Boston: Little, Brown and Company.

Gocke, D.J., K. Hsu, C. Morgan, S. Bombardieri, M. Lockshin, and C.L. Christian. 1971. Vasculitis in association with Australia antigen. *J. Exp. Med.* 134:330s.

Goldstein, I., H.B. Kaplan, A. Radin, and M. Frosch. 1976. Independent effects of IgG and complement upon human polymorphonuclear leukocyte function. *J. Immunol.*

117:1282.

Griffin, F.M., C. Bianco, and S.C. Silverstein. 1975. Characterization of the macrophage receptor for complement and demonstration of its functional independence from the receptor for the Fc portion of IgG. *J. Exp. Med.* 141:1269.

Griswold, W.R., K.C. Hsu, and R.M. McIntosh. 1973. Cryoprecipitates and immune complexes: Decrease in antibody bound ^{131}I-labeled BSA antigen after cryoprecipitation in rabbit serum. *Proc. Soc. Exp. Biol. Med.* 142:1292.

Hawn, C.V., and C.A. Janeway. 1947. Histological and serological sequences in experimental hypersensitivity. *J. Exp. Med.* 85:571.

Henson, P.M., and C.G. Cochrane. 1971. Acute immune complex disease in rabbits. The role of complement and of a leukocyte dependent release of vasoactive amines from platelets. *J. Exp. Med.* 133:554.

Holdsworth, S.R., T.J. Neale, and C.B. Wilson. 1980. The participation of macrophages and monocytes in experimental immune complex glomerulonephritis. *Clin. Immunol. Immunopathol.* 15:510.

Huber, H., S.D. Douglas, J. Nusbacher, S. Kochwa, and R.E. Rosenfield. 1971. IgG subclass specificity of human monocyte receptor sites. *Nature* 229:419.

Hunsicker, L.G., T.P. Shearer, S.B. Plattner, and D. Weisenburger. 1979. The role of monocytes in serum sickness nephritis. *J. Exp. Med.* 150:413.

Israels, E.D., F. Paraskevas, and L.G. Israels. 1973. Immunological studies of coagulation factor XIII. *J. Clin. Invest.* 52:2398.

Kniker, W.T., and C.G. Cochrane. 1965. Pathogenic factors in vasular lesions of experimental serum sickness. *J. Exp. Med.* 122:83.

Kniker, W.T., and C.G. Cochrane. 1968. The localizaion of circulating immune complexes in experimental serum sickness. The role of vasoactive amines and hydrodynamic forces. *J. Exp. Med.* 127:119.

Lay, H.W., and V. Nussenzweig. 1968. Receptors for complement on leukocytes. *J. Exp. Med.* 128:991.

Leber, P.D., and R.T. McCluskey. 1974. Complement and the immunohistology of renal disease. *Transplant. Proc.* 6:67.

Levy, L. 1967. A form of immunological atherosclerosis. In *The Reticuloendothelial System and Atherosclerosis.* N.R. Di Luzio and R. Paoletti, eds. New York: Plenum Press, pp. 426-432.

Linnet, L., and T. Hjort. 1977. Sperm agglutinins in seminal plasma and serum after vasectomy. Correlation between immunological and clinical findings. *Clin. Exp. Immunol.* 30:413.

Mannik, M., W.P. Arend, A.P. Hall, and B.C. Gilliland. 1971. Studies on antigen antibody complexes. I. Elimination of soluble complexes from rabbit circulation. *J. Exp. Med.* 133:713.

McCluskey, R.T., B. Benacerraf, and F. Miller. 1962. Passive acute glomerulonephritis induced by antigen-antibody complexes solubilized in hapten excess. *Proc. Soc. Exp. Biol. Med.* 111:764.

Nydegger, U.E. 1979. Biologic properties and detection of immune complexes in animal and human pathology. *Rev. Physiol. Biochoem. Pharmacol.* 85:63.

Parish, W.E., and E.L. Rhodes. 1967. Bacterial antigens and aggregated gamma globulin in the lesions of nodular vasculitis. *Br. J. Dermatol.* 79:131.

Scherzer, H., and P.A. Ward. 1978. Lung injury produced by immune complexes of varying composition. *J. Immunol.* 121:947.

Theofilopoulos, A.N., and F.J. Dixon. 1979. The biology and detection of immune complexes. *Adv. Immunol.* 28:89.

Tung, K.S.K., and N.J. Alexander. 1980. Monocytic orchitis and aspermatogenesis in normal and vasectomized rhesus macaques *(Macaca mulatta). Am. J. Pathol.* 101:17.

Tung, K.S.K., R.K. Bryson, L.-P.B. Han, and L.C. Walker. 1979. Circulating immune complexes in vasectomy. In *Vasectomy: Immunologic and Pathophysiologic Effects in Animals and Man.* I.H. Lepow and R. Crozier, eds. New York: Academic Press, pp. 301-335.

Unanue, E.R., and F.J. Dixon. 1967. Experimental glomerulonephritis: Immunological events and pathogenic mechanisms. *Adv. Immunol.* 6:1.

445

von Pirquet, C.F., and B. Schick. 1905. *Die Serumkrankheiten.* Wien: F. Deuticke.

Wautier, J.L., G.M. Tobelem, A.P. Peltier, and J.P. Caen. 1976. C1 and human platelets. II. Detection by immunological methods and role. *Immunology* 30:459.

Wedmore, C.V., and T.J. Williams. 1981. Control of vascular permeability by polymorphonuclear leukocytes in inflammation. *Nature* 289:646.

Weissmann, G., and P. Dukor. 1970. The role of lysosomes in immune responses. *Adv. Immunol.* 12:283.

Wiggins, R.C., and C.G. Cochrane. 1981. Immune-complex-mediated biologic effects. *N. Engl. J. Med.* 304:518.

Wilson, C.B., and F.J. Dixon. 1970. Antigen quantitation in experimental immune complex glomerulonephritis. I. Acute serum sickness. *J. Immunol.* 105:279.

Wilson, C.B., and F.J. Dixon. 1974. Immunopathology and glomerulonephritis. *Annu. Rev. Med.* 25:83.

Witkin, S.S., S.K. Shahani, S. Gupta, R.A. Good, and N.K. Day. 1980. Demonstration of IgG Fc receptors on spermatozoa and their utilization for the detection of circulating immune complexes in human serum. *Clin. Exp. Immunol.* 41:441.

Chapter 20

IMMUNOINTERCEPTION OF FERTILITY

G.P. TALWAR
A. MULLICK
S. RAMAKRISHNAN
C. DAS
S.K. GUPTA
A.TANDON
R.K. NAZ
N. SHASTRI
S.K. MANHAR
O.M. SINGH

One area of research in reproductive immunology is the use of antibodies and immune cells to specifically counteract a naturally occurring hormone or protein of critical importance to reproduction. The overall aim is to develop alternative immunological methods for the control of fertility, especially for those parts of the world where population growth continues unabated. Although immunological procedures require periodic boosting, they can be delivered by the infrastructure available in most developing countries. The immunological approach should be cost-effective and also free from the risk of user failure.

POSSIBLE TARGETS FOR IMMUNOINTERCEPTION

Mammalian reproduction requires the contribution of gametes from two sexes. In both the female and the male, gametogenesis is subject to control by a number of hormones, including LHRH, FSH, LH, and the sex steroids. Antibodies that interfere with the action of any one of these hormones can theoretically cause infertility. It is uncertain if any of these are suitable for human use, but animal applications are a reasonable possibility. The reasons for this conclusion are:

1. unacceptable physiological hormone deficiencies, especially of sex steroids, which would be produced by blockage of the activity of pituitary gonadotropins; and
2. the danger of tissue damage from immune complexes formed between circulating antibodies and hormones continuously secreted.

Thus immunological interference with hormones normally present in males and nonpregnant females seems inadvisable as a means of contraception.

It is more rational to intercept an event unique to conception. For example, anti-zona pellucida antibodies prevent the attachment of the sperm and block fertilization of the egg.

During its early development, the embryo passes through stages in which it expresses oncofetal antigens crossreactive with teratocarcinomas and other tumors. Several of these antigens are stage specific, and do not persist in adult differentiated tissues. They may also constitute suitable targets for immunological attack. Oncofetal antigens are discussed elsewhere in this volume (see *Introduction* and chapters 6 and 7) and will not be elaborated on here.

Around the time of implantation, the blastocyst secretes several proteins. At least one of these is well characterized and has served as an index of pregnancy for the last 54 years, ever since Ascheim and Zondek (1928) found evidence for a gonadotropic activity in pregnancy. We shall summarize the current status of work in our laboratory on this antigen, human chorionic gonadotropin (hCG).

We shall also discuss another model in which cell-mediated immune reactions are mobilized to interrupt spermatogenesis. The targets here are non-hormonal proteins that appear at the onset of spermatogenesis. These proteins are recognized as foreign by the immune system and are therefore autoantigenic.

Luteinizing Hormone Releasing Hormone
LHRH is a decapeptide which is a "self" hormone in most species, but which can be rendered immunogenic by conjugation with a carrier protein, most commonly bovine serum albumin (BSA). Effective immunization was achieved only with Freund's complete adjuvant (Frazer et al., 1977). LHRH immunization has been shown to impair the fertility of bulls (Robertson et al., 1979), marmosets (Hodges and Hearn, 1977), and sheep (Frazer et al., 1977).

We attempted to develop an immunization procedure by which anti-LHRH antibodies can be induced with agents acceptable for use in humans. LHRH was conjugated to a number of carriers including BSA, which proved to be the poorest. Better antibody responses were obtained by chemically linking LHRH to tetanus toxoid or keyhole limpet hemocyanin (KLH) (Figure 20-1).

Anti-LHRH immunization is suitable for use in both males and females. Gametogenesis is blocked in both instances, accompanied by a deficiency in sex steroid hormone production. This approach may therefore be restricted to animal sterilization. It may also control aggressiveness in bulls and in other farm animals by lowering androgen levels or by interfering with the direct action of LHRH in the central nervous system.

Pituitary gonadotropins
Immunization of monkeys with B ovine leuteinizing hormone leads to the formation of antibodies that crossreact with both pituitary and chorionic gonadotropins. Animals can be rendered infertile by either active or passive immunization (Moudgal et al., 1978; Thau et al., 1979). It remains to be determined whether active immunization against this hormone is devoid of tissue reaction and thus safe. The production of progesterone in the female and testosterone in the male is diminished by high-titer anti-LH antibodies.

Recent studies have shown the involvement of follicle-stimulating hormone (FSH) in spermatogenesis in the male, as well as in the development of the follicle in the female. More than one laboratory is currently trying to develop anti-FSH immunization as a means of fertilty control (Murty et al., 1980; Sairam and Madhwa Raj, 1980). The impact of such immunization on hormonal status is not yet fully understood.

Sex steroids
Estrogens and progesterone in the female and testosterone in the male are important in the maturation of gametes, as well as in the conditioning of the reproductive tract to support pregnancy. Antibodies neutralizing these hormones can theoretically interfere with reproduction. There are, however, complications. Inhibition of steroid hormones would result in hypersecretion of pituitary gonadotropins by the feedback mechanism and, if persistent, might

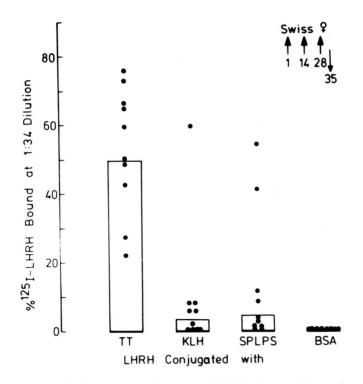

FIGURE 20-1. Antibody response of mice to LHRH decapeptide conjugated to either tetanus toxoid, Keyhole Limpet Hemocyanin, sodium phthalyl lipopolysaccharide of *S. enterides*, or bovine serum albumin.
The importance of the carrier for the response generated is evident. From Shastri et al., 1981; reprinted with permission.

451

lead to pituitary pathology. Also, antibodies may not neutralize steroid hormone activity despite their binding to the hormone. Hormone bound to the antibody might have a diminished metabolic clearance and a longer biological half-life (Nieschlag et al., 1975).

We have recently developed in our laboratory hybridomas making antiprogesterone antibodies. Work is in progress to see whether passive administration of these antibodies can terminate pregnancy. The vital role of progesterone in sustaining pregnancy is well established from several biological studies.

Human chorionic gonadotropin

Human chorionic gonadotropin is primarily a hormone of pregnancy. Its levels increase rapidly and attain peak values of approximately 4.6 mg per liter of serum between weeks 8 and 10 of pregnancy (Braunstein et al., 1976). There is controversy over whether hCG synthesis starts in the preimplantation period. Gonadotropin-like activity has been reported in the preimplantation period (Saxena et al., 1979), but others believe that secretion of this hormone begins at, or soon after, implantation. It is clearly detectable in circulation by eight to ten days after fertilization (Catt et al., 1975).

The biological role of this hormone is not yet fully established. It is thought to sustain the corpus luteum and maintain ovarian progesterone production. A strong argument for this role is provided by the observation that hCG is required to maintain blood progesterone levels in women who have undergone medical termination of pregnancy (Garner and Armstrong, 1977). There are, however, several gaps in our knowledge. Repeatedly administered to nonpregnant females, hCG delays menstruation by only a few days despite persistent high hCG levels. It is known that the interaction of large quantities of certain hormones with receptors reduces the tissue response, and hCG falls into this category (Hsueh et al., 1976, 1977; Sharpe, 1976). The level of hCG continues to rise sharply during early pregnancy, and this should thus result in a loss of target tissue sensitivity. This does not happen, however. The sensitivity of the corpus luteum to hCG is apparently maintained over the first seven weeks of pregnancy (Talwar, 1979).

Whatever the gaps in our knowledge of hCG, its important role in maintaining early pregnancy is undeniable, and this is best illustrated by the ability of anti-hCG antibodies to terminate pregnancy. The results of a representative experiment, in which antibodies generated by the vaccine Pr-β-hCG-TT were passively administered to a pregnant baboon, are presented in Figure 20-2. Chorionic gonadotropin activity in blood came down, accompanied by a fall in progesterone level. The animal aborted 72 hours later, and evacuation was complete. The baboon reverted to a normal hormonal pattern in the following cycle, without apparent side effects. Fertility was also regained. The termination was due to the presence of hCG-neutralizing antibodies, as an equivalent amount of gamma globulins from a control nonimmunized monkey failed to terminate pregnancy. The pregnancy-terminating role of anti-hCG antibodies has also been demonstrated in marmosets (Hearn et al, 1975), baboons (Stevens, 1976), and other animals.

Passive termination of pregnancy by anti-hCG antibodies provides strong evidence for antibody, as opposed to cell-mediated, immunity as the operative mechanism in active immunization, although cell-mediated immunity is by no means ruled out. The manner in which antibodies mediate this effect is, however, not fully clear, and more than one site may be involved. Antibodies could prevent hCG from signalling the corpus luteum for progesterone production. Atul Tandon in our laboratory has observed that anti-hCG antibodies given to mice daily from days 3 to 7 of pregnancy diminished the progesterone levels from 65.4 ± 15.3 ng/ml to 9.2 ± 3.1 ng/ml.

The physicochemical and antigenic similarities between mouse and human chorionic gonadotropins have been described by Wide and Wide (1979). Pregnant mice given anti-hCG treatment on day 8 did not show resorption of fetuses upon laparotomy. Anti-ovine luteinizing hormone (OLH) antibodies lowered the progesterone levels to nearly the same extent as in mice

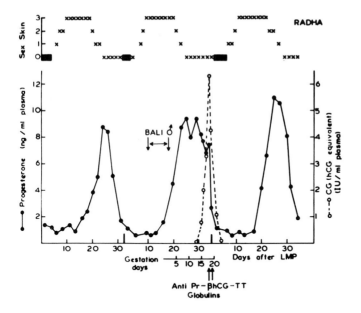

FIGURE 20-2. Ability of anti-β-hCG-TT antibody to terminate pregnancy in the baboon.

Progesterone values and sex skin swelling in three cycles are presented, one preceding and the other following the fertile cycle in which termination of pregnancy was carried out with the antibody.

given anti-hCG antibodies, and the anti-OLH-treated animals showed complete resorption of the embryos when laparotomized on day 8. Therefore the reduction in progesterone levels alone does not account for the antifertility effect of the anti-OLH antibodies. Animals treated with anti-hCG antibodies did not deliver their litters on the expected days, despite normal embryo implantation on day 8, as seen by laparotomy. Further investigation revealed that resorption took place sometime between days 10 and 14 of pregnancy. Incidentally, this time coincides with the period of maximum chorionic gonadotropin levels during pregnancy in the mouse and is also when the anti-hCG reaction is most pronounced in the feto-placental unit, when studied by immunofluorescence and immunoperoxidase techniques. All of these results suggest the involvement of more than one mechanism in the antifertility action of the antigonadotropin antibodies. Antibodies generated against whole native hCG have, by and large, maximal capacity for neutralizing the biological activity of the hormone. The antibodies produced are primarily of the conformational reading type. Procedures that alter hCG conformation drastically, such as breaking of the disulfide bonds by reduction and alkylation, yield molecules with a markedly reduced ability to bind the antibodies. It is also noteworthy that antibodies raised against native hCG rarely manifest binding with the carboxyl-terminal portion of the beta subunit of hCG.

Isolation of the hCG beta-subunit by immunological techniques preserves hCG crossreactive conformations, as long as intrachain disulfide bonds in β-hCG are not broken. This subunit has 12 half-cystines, but no free SH group can be titrated in the subunit in the native state. There are thus six intrachain disulfide bonds, of which only one has been mapped (Pierce et al., 1976). Nonetheless, they are important, from the biological as well as the immunological point of view; stepwise reduction and alkylation result in alterations of both biological and immunological activity.

Recognition by target tissue receptors of determinants in β-hCG
In the final analysis, immunointerception demands interference with the interaction of hormone with receptors on the target tissues. Thus it is necessary to delineate the regions of the molecule that are important for biological activity of the hormone. Although the hormone is 30 per cent carbohydrate by weight, these residues do not play an active role in its biological activity. Removal of the terminal sialic acid residues, however, alters the biological half-life of the hormone, as a result of the enhanced clearance. However, desialylation does not affect binding with receptors or *in vitro* biological activity (Canfield et al., 1971). Sequential removal of sugar residues up to mannose does not interfere with the steroidogenic potency of the hormone *in vitro* (Bahl, 1977). As the alpha subunit is exchangeable between hCG and other hormones of pituitary origin (TSH, LH, FSH), it is clearly the beta subunit which is responsible for the biological specificity of these hormones (Pierce

et al., 1971; Reichert et al., 1974; Aloj and Ingham, 1977). Human chorionic gonadotropin is most potent in its native associated form, and dissociation of the hormone into alpha and beta subunits results in a large decrease in biological activity (Canfield et al., 1971). The residual biological activity associated with the beta subunit was found to be intrinsic to that subunit (Ramakrishnan et al., 1978b). In the sensitive Leydig cell assay system, immunochemically purified β-hCG (Pr-β-hCG) stimulated steroidogenesis in a dose-dependent manner, similar to the native hormone. In comparative terms, the beta subunit had 400-fold lower biological activity than hCG, and this slight activity was abolished by reduction and alkylation. To delineate the structural sequences of the beta subunit involved in receptor recognition, synthetic peptides corresponding to different regions of the subunit were used. The three C-terminal synthetic peptides (115-145, 111-145, and 101-145) failed to stimulate testosterone production by mouse Leydig cells at concentrations of up to 1 nM. The lack of involvement of this portion of the hormone in receptor binding was further confirmed in a radioreceptor assay using goat corpus luteum receptors (Talwar et al., 1978). The synthetic peptide making up the sequence 39 to 71 was found to stimulate steroidogenesis in this system in a dose-dependent manner, but a shorter peptide, conforming to the sequence of 39 to 56, showed no biological activity (Ramakrishnan et al., 1978b). Thus the region between 57 and 71 alone, or in association with residues 39 to 56, may constitute one of the biologically active sites of the hormone recognized by tissue receptors. This does not preclude an equally important role for other parts of the molecule, which may serve to mold the biologically active site into an optimal conformation for interaction with the receptors.

Anti carboxyl-terminal peptide antibodies
Amino acid analyses of β-hCG and β-hLH have revealed the presence of a unique structural sequence of about 30 amino acid residues at the C-terminal end of β-hCG. It is therefore logical to use this unique region to raise specific antibodies against hCG, as these would not crossreact with hLH. Antibodies raised against the 23-amino acid C-terminal peptide of an asialo derivative of β-hCG were found to be highly specific for hCG, but could not neutralize the biological activity of the hormone (Louvet et al., 1974). Similar observations were made by Matsuura et al. (1978, 1979) on antibodies raised against the synthetic peptide making up the sequence 116 to 145 of β-hCG. By using a longer peptide (33 to 35 amino acid residues), Stevens (1976) obtained specific neutralizing antibodies in baboons. Antibodies were generated against a 45-amino acid carboxyl-terminal peptide of β-hCG in our laboratory after the peptide was conjugated to tetanus toxoid. Studies in monkeys and rabbits indicated that the peptide-TT conjugate had much poorer immunogenicity than β-hCG-TT. Quantitatively, the antisera showed a greater (about 120-fold higher) binding capacity for the peptide than for the native hCG

(Ramakrishnan et al., 1979). These antibodies were highly specific to hCG in a radioimmunoassay system. Unlike antisera raised against 23- and 30-amino acid C-terminal peptides, anti-45-CTP neutralized the biological activity of hCG in a Leydig cell bioassay system, although the neutralization capacity was much lower than that of an antibody of comparable titer raised against β-hCG-TT. This poor neutralization capacity can be attributed, at least in part, to the limited number of antigenic determinants present in the C-terminal peptides. Sucrose density gradient analysis of hCG-anti-45-CTP immune complexes indicated that the antigen-antibody complex has a ratio of 1:1, whereas in the case of anti-β-hCG antisera at least three antibody molecules bind on average with each molecule of hCG. Furthermore, the affinity of the anti-CTP antibodies was one order of magnitude lower than that of the anti-β-CTP antibodies (Chen et al., 1980). Recently Birken et al. (1980) investigated the binding characteristics of antibodies raised against the enzyme-cleaved 23-amino acid peptide, compared with those raised against native hCG and asialo-hCG. These studies indicate that the antipeptide antibodies recognized native hCG poorly relative to its asialo derivative. Thus the carbohydrates present at the C-terminal end of β-hCG do play a role, either directly or indirectly, in antibody recognition, and the absence of carbohydrate groups in the synthetic peptides could contribute to the poor recognition of the native hormone by anti-CTP antibodies. Thus the use of C-terminal peptides as immunogens has many limitations, although it does yield highly specific antibodies.

Properties of a vaccine based on β-hCG

Although native hCG has the most desirable conformation, the beta subunit of hCG conserves several desirable characteristics. It is a glycoprotein of 145 amino acids, and there is a large sequence homology between β-hCG and β-hLH. There are, however, differences in 51 amino acid residues, 30 in the C-terminal part and 21 elsewhere in the molecule. In spite of many similarities, hCG differs from hLH both biologically and immunologically. For example, hCG has FSH- and TSH-like activity (Siris et al., 1978; Nisula et al., 1974; Pekonen and Weintraub, 1980), as well as hLH-like activity. Immunologically, the beta subunit of hCG generates antibodies that preferentially react with hCG but not hLH (Das et al., 1978; Prem Mohini et al., 1978; Ramakrishnann et al., 1978a). Vaitukaitis et al. (1972) produced their well-known serum, SB_6, which enabled them to assay for hCG without interference from the hLH normally present in circulation. The monoclonal antibody P_3W_{80} demonstrates good reactivity with hCG and does not crossreact with hLH. The hybridoma was prepared by fusion of splenocytes of mice immunized with hCG (Gupta and Talwar, 1980). These antibodies do not recognize the C-terminal peptides. Thus the core of β-hCG has conformational antigenic determinants unique to hCG.

A vaccine, Pr-β-hCG-TT, was developed by chemical coupling of the immunochemically purified beta subunit of hCG with tetanus toxoid (Talwar et al., 1975). The rationale for conjugation with a bacterial carrier protein was to render an otherwise "self" hormone immunogenic within the species (Talwar et al., 1976). This vaccine stimulated the production of antibodies which bound with hCG and had the capacity to neutralize its biological activity (Das et al., 1976). Immunization did not have any demonstrable toxicity or other side effects (Gupta et al., 1976; Prasad et al., 1976).

Phase I clinical trials with this vaccine were carried out on 63 women at six centers located in five countries, under the direction of established clinical scientists. A carefully drawn protocol calling for frequent clinical and laboratory investigation was followed. Most of the results have been published

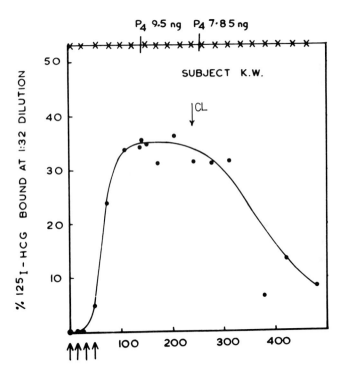

FIGURE 20-3.
Kinetics of antibody response to hCG and tetanus toxoid in one of the women investigated in phase I clinical trials with the vaccine Pr-β-hCG-TT. Immunization was carried out with four fortnightly injections of 80 μg of β-hCG linked to 10 mg of tetanus toxoid (see arrows). The X's on the upper abscissa mark the menstruation episodes, and P$_4$ represent the plasma progesterone values in the luteal phase, which are indicative of the ovulatory cycles.

(Hingorani and Kumar, 1979; Shahani et al., 1979; Nash et al., 1980). Almost all of the women responded by forming both anti-hCG and antitetanus antibodies. Representative data on the time kinetics of the antibody response in one of the subjects are shown in Figure 20-3. The antibodies were of the neutralizing type *in vivo* (Figure 20-4) and *in vitro* (Das et al., 1976). All women continued to ovulate and had regular menstrual cycles as well as unchanged libido. The hormonal profiles were normal, a finding confirmed and extended by Pala et al. (1976). The antibody levels eventually declined in each case, demonstrating the reversibility of the procedure.

Limitations and requirements for improvement of Pr-β-hCG-TT vaccine
Preliminary clinical pharmacology trials indicated wide variability in antibody titers in immunized individuals. This is to be expected from any vaccine and is therefore not a peculiar trait of this antigen. It was noted, however, that pregnancy was not prevented in women with low titers. There is thus a need to improve the present vaccine, and research to develop potent adjuvants devoid of toxicity and permissible for human use is being actively pursued. Improved immunization strategies may also be necessary to elicit a good response in all recipients. Since the carrier mobilizes helper-cell activity, and since carriers

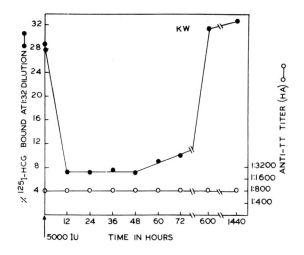

FIGURE 20-4. The results of immune clearance using 5000 IU of hCG in humans.
The hormone injected was bound by circulating antibodies, resulting in a temporary fall in anti-hCG titers. These returned to about the preclearance values in the following month, when the blood was again analyzed for antibodies. Exposure to hCG did not result in a secondary booster response, which was obtained only when β-hCG linked to tetanus toxoid was injected. These studies demonstrate the ability of circulating antibodies to handle hCG without hCG becoming an interval booster. The hormone-antihormone complex was devoid of bioactivity in the Leydig cell assay system. From Talwar et al., 1976; reprinted with permission.

differ in their properties (Shastri et al., 1981), the use of polyvalent antigens with mixed carriers may evoke a better overall response than would a vaccine with a single carrier (Talwar, 1980).

ALTERNATE STRATEGIES

Passive immunization using hybridoma antibodies
Since the antifertility effector in the present case is anti-hCG antibody, one way to ensure universal effectiveness would be to deliver preformed antibodies of known characteristics in adequate amounts. Passive immunization has not been deleterious in any primates tested so far. Hybridoma technology has permitted production of antibodies of extremely high titers in almost pure form. Clones making neutralizing antibody with no crossreaction with other hormones and body tissues can be selected by screening. In our laboratory, and perhaps in others, several clones fulfilling these characteristics have been developed (Gupta and Talwar, 1980), and an abundant supply of antibodies can be obtained at reasonable cost. These are mouse myelomas, however, and their products can be used only for studies of efficacy and safety in mice and subhuman primates. They can also be used in radioimmunoassays and immunodiagnostic kits. In the near future, it should be possible to develop similar clones using human cells, and reports on the successful development of two human hybridomas producing anti-DNP (Oslon and Kaplan, 1980) and anti-measles virus (Croce et al., 1980) antibodies are encouraging.

Autoantigens
We now describe a situation in which a virtually complete immune response can be obtained in all test animals by exploiting the potential of autoantigens. We experimentally analyzed the process of spermatogenesis in mammalian species. The basic structure of the testis is laid down during fetal life. Some cells, such as the Leydig cells, are functional even during a short period in midgestation, although active development of the germinal epithelium takes place at puberty. A number of proteins which first appear with spermatogenesis are intrinsically "foreign" to the immune system, and a response can be evoked against them if lymphoid cells are given access to the testicular compartment. Intratesticular injection of Bacillus Calmette Guerin (BCG) produces mononuclear infiltration primarily in the interstitium (Figure 20-6). If the bacilli are given in moderate amounts at a low density, granuloma formation does not take place. The relationship between the bacterial density and the type of reaction has been discussed elsewhere (Mustafa et al., 1981). The mononuclear cells present in the interstitial spaces do not damage the Leydig cells morphologically or functionally. Blood androgen levels are not reduced,

459

and the response to gonadotropin stimulation is retained (Talwar et al., 1979).

This method has been tested in a number of species, including rats, mice, rabbits, guinea pigs, dogs, rams, rhesus monkeys, baboons, and bulls. In those species where semen could be collected, the sperm count declined following the injection of BCG and, within four to six weeks reached a stage where no motile sperm were seen in the semen (Talwar et al., 1979; Naz and Talwar, 1981, and unpublished data). In all animals, libido was maintained. Testicular histology showed partially atropied tubules. The basement membrane and the peritubular cells appeared normal. Germinal cell development was blocked at different stages, ranging from the spermatogonial to the secondary spermatocyte stage. No formed elements were seen in the lumens, and some of the tubules contained eosinophilic debris (Figure 20-6).

BCG need not be used in live form: killed BCG or even extracts of the bacilli can produce a similar effect (Naz and Talwar, 1981). Killed BCG is eliminated by the system concomitant with the regeneration of tubular function and the return of sperm in the semen. Return of fertility has been observed in monkeys, baboons, and dogs. Live BCG does not colonize all species, and even in those in which it persists for long periods, clearance can be achieved by treatment with antimycobacterial drugs. The procedure is thus reversible.

The mechanism by which BCG induces azoospermia is not completely understood. BCG injected into juvenile dogs does not interfere with normal

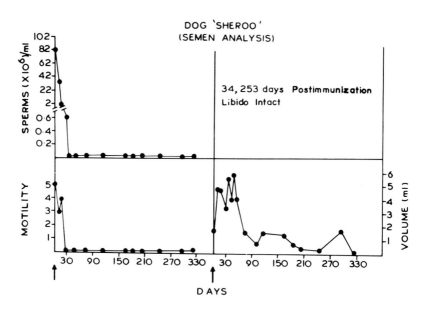

FIGURE 20-5. Sperm count and motility in semen of a dog injected intratesticularly with 110 units of BCG.

sperm production. Injection of BCG into one of the testes does not transfer azoospermia to the other, indicating that the phenomenon is a localized effect which is not transmitted through the general circulation. No antisperm antibodies were detectable in the blood. Cell-mediated immune reactions may be primarily involved in induction of the effect. This is of particular interest in light of the possible hazards of immune complexes in vasectomized monkeys (Alexander and Clarkson, 1978).

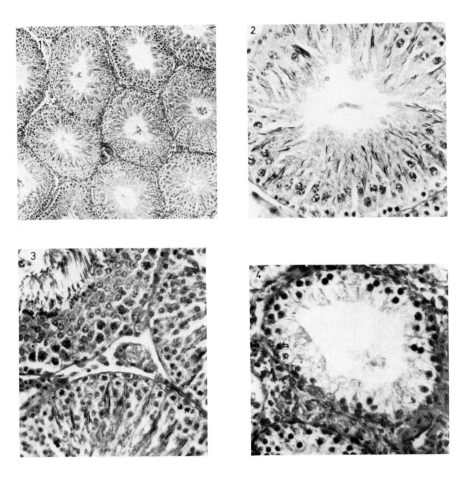

FIGURE 20-6. Histology of rat testis immunized intratesticularly with 7.5 units of BCG.

Photomicrographs at different magnifications show the mononuclear infiltration in the interstitium and the integrity of the basement membrane and cytoarchitecture of tubules. The development of germinal cells was blocked, however. From Talwar et al., 1979.

461

SUMMARY AND CONCLUSIONS

The immune system is competent to react against selected proteins and hormones which have a critical role in reproduction. Some of the vulnerable sites for immunointerception have been identified, and evidence is available as to the efficacy of immunological approaches against these antigens. They can be broadly classified into four types:

1. hormones of the hypothalamic-pituitary-gonad axis;
2. antigens specific to egg and sperm;
3. stage-related antigens appearing during early embryonic development, many of which are crossreactive with tumors (oncofetal antigens); and
4. hormones and proteins peculiar to peri-implantation events and essential for the establishment and maintenance of pregnancy.

Immunological approaches against the first type of antigen may have utility only in the control of fertility in animals and are not likely to be safe or acceptable for eventual human application. Antigens of types 2 and 3 would, in principle, constitute highly desirable targets for immunointerception. Many of these require extensive study and development before they can be utilized effectively.

Human chorionic gonadotropin is a type 4 antigen obtainable in large amounts in a highly purified state. The structure of the hormone and the primary sequence of its subunits were determined some years ago (Bellisario et al., 1973; Morgan et al., 1973, 1975). It is thus one of the better characterized molecules and has been the focus of extensive studies for the development of an antifertility vaccine. The efficacy of active and passive immunization against this hormone has been demonstrated in several families of subhuman primates.

Our group has developed an immunogenic vaccine based on the immunochemically purified beta subunit of hCG covalently linked to tetanus toxoid. This vaccine stimulated antibodies against both hCG and tetanus toxoid. These antibodies were primarily of a conformational type and had a much greater capacity for neutralization of the biological activity of hCG *in vitro* and *in vivo* than did immunogens based on synthetic carboxyl terminal peptides. The antibodies showed preferential reactivity with hCG and their low crossreactivity with hLH did not interfere with ovulation or normal hormonal profiles. In contrast, carboxyl terminal peptides of 31, 35, 45, and 53 amino acid residues were far poorer immunogens than β-hCG. The antibodies produced by the carboxyl terminal peptides recognized the primary amino acid sequence, were essentially monovalent in character, and had a much lower capacity for neutralization of hCG activity.

Phase 1 clinical trials with the Pr-β-hCG-TT vaccine have been carried out on 63 women in six centers. Investigations using a carefully planned protocol have shown the procedure to be safe and free of side effects. The vaccine induced the formation of anti-hCG antibodies with the potential to neutralize hCG activity *in vivo* and *in vitro*. The response was, however, widely variable. In some cases the antibody titers were high, and in others they were low. Since an antifertility vaccine for general use would be primarily used by women of proven fertility, further research is necessary to develop a potent and safe adjuvant to enhance the general antibody titers. There is also a need to devise novel strategies of immunization to minimize individual variability.

As an interim measure, preformed antibodies could be passively administered in amounts adequate to interrupt pregnancy. Anti-hCG antibodies can terminate pregnancy in subhuman primates, such as the baboon. Hybridoma technology provides new possibilities for obtaining antibodies of high titer and consistent quality. One of the clones developed has been found to be effective in terminating pregnancy in mice. If our present experiments in chimpanzees are successful, a model system will be available for the use of such antibodies in abortion. Given the consistent quality of monoclonal antibodies and the correct dosage, the problem of variability in effectiveness will disappear.

This chapter has also briefly reviewed another approach to infertility, a model system with which we have achieved 100 per cent sterility in male animals. This system is based on the immunogenic potential of autoantigens. Many proteins appearing at the onset of spermatogenesis could fall within this category. Intratesticular injection of moderate doses of BCG, distributed evenly, brings about mononuclear infiltration in the interstitium. The lymphoid cells do not damage the Leydig cells or reduce their output of testosterone. Spermatogenesis is blocked, however. The method works in almost all mammalian species and has been tested in the usual laboratory animals—rats, mice, rabbits—as well as in rams, dogs, monkeys, and baboons. Although the mechanism by which azoospermia is induced is not fully clear, the absence of antisperm antibodies in the circulation and the lack of effect on the contralateral uninjected testis indicate the localized nature of the phenomenon, the primary involvement of cell-mediated immune reactions, and a minimal hazard of immune complexes.

ACKNOWLEDGMENTS

The unit received research grants from the Indian Council of Medical Research, the International Development Research Centre (Canada), the Rockefeller Foundation, and the Family Planning Foundation of India. Collaborative studies for contraceptive development are ongoing with the International Committee for Contraception Research. K. K. Sarin rendered invaluable assistance. H. K. Sharma and Shri Kalra rendered valuable technical assistance.

REFERENCES

Alexander, N.J., and T.B. Clarkson. 1978. Vasectomy increases the severity of diet-induced atherosclerosis in *Macaca fascicularis*. *Science* 201:538.

Aloj, S.M., and K.C. Ingham. 1977. Kinetics of subunit interactions in glycoprotein hormones. In *Endocrinology Proceedings. 5th International Congress on Endocrinology.* V.H.T. James, ed. Amsterdam: Excerpta Medica, p. 108.

Ascheim, S., and B. Zondek. 1928. Die schwangerschafts diagnose aus dem harn durch nachewis des hypophyenvorderslappen-hormone. *Klin. Wchnschr* 7:1404.

Bahl, O.P. 1977. Human chorionic gonadotropin, its receptor and mechanism of action. *Fed. Proc.* 36:2119.

Bellidrio, R., R.B. Calsen, and O.P. Bahl. 1973. HGC: linear amino acid sequence of the β subunit. *J. Biol. Chem.* 248:6797.

Birken, S., R.E. Canfield, R. Lauer, G. Agosto, and M. Gabel. 1980. Immunochemical determinants unique to hCG: Importance of sialic acid for antisera generated to the hCG beta subunit COOH terminal peptide. *Endocrinology* 106:1659.

Braunstein, G.D., J. Rasor. D. Adler, H. Danzer, and M.E. Wade. 1976. Serum human chorionic gonadotropin levels throughout normal pregnancy. *Am. J. Obstet. Gynecol.* 126:677.

Canfield, R.E., F.J. Morgan, S. Kammerman, J.J. Bell, and G.M. Agasto. 1971. Studies of human chorionic gonadotropin. *Recent Prog. Horm. Res.* 27:121.

Catt, K.J., M.L. Dufau, and J.L. Vaitukaitis. 1975. Appearance of human chorionic gonadotropin in pregnancy plasma following the initiation of implantation of the blastocyst. *J. Clin. Endocr. Metab.* 40:537.

Chen, H.C., S. Matsuura, and M. Ohashi. 1980. Limitations and problems of hCG-specific antisera. In *Chorionic Gonadotropin.* S.J. Segal, ed. New York and London: Plenum Press, pp.231-252.

Croce, C.M., A. Linnenbach, W. Hall. Z. Steplewski, and H. Koprowski. 1980. Production of human hybridomas secreting antibodies to measles virus. *Nature* 288:488.

Das, C., M. Salahuddin, and G.P. Talwar. 1976. Investigations on the ability of antisera produced by Pr-B-hCG-TT to neutralise the biological activity of hCG. *Contraception* 13:171.

Das, C., G.P. Talwar, S. Ramakrishnan, M. Salahuddin, S. Kumar, V. Hingorani, E. Coutinho, H. Croxatto, E. Hemmingson, E. Johansson, T. Shahani, K. Sundaram, H. Nash, and S.J. Segal. 1978. Discriminatory effect of anti Pr-B-hCG-TT antibodies on the neutralisation of the biological activity of placental and pituitary gonadotropins. *Contraception* 18:35.

Frazer, H.M., K.J. Clarke, and A.S. McNeilly. 1977. Inhibition of ovulation and absence of positive feedback in ewes immunized against LHRH. *J. Endocrinol.* 75:45.

Garner, B.R., and D.T. Armstrong. 1977. The effect of hCG and estradiol 17β on the maintenance of the human corpus luteum of early pregnancy. *Am. J. Obstet. Gynecol.* 128:469.

Gupta, L., S.K. Dubey, and G.P. Talwar. 1976. Investigation on pharmacopoeial safety, microbial sterility and pyrogens of Pr-B-hCG-TT. *Contraception* 13:183.

Gupta, S.K., and G.P. Talwar. 1980. Development of hybridomas secreting antihuman chorionic gonadotropin antibodies. *Indian J. Exp. Biol.* 18:1361.

Hearn, J.P., R.V. Short, S.R. Lunn. 1975. The effects of immunising marmoset monkeys against the β-subunit of hCG. In *Physiological Effects of Immunity Against Hormones.* R.G. Edwards and M.H. Johnson, eds. Cambridge: Cambridge University Press, pp. 239-247.

Hingorani, V., and S. Kumar. 1979. Anti-hCG immunisation. In *Recent Advances in Reproduction and Regulation of Fertility.* G.P. Talwar, ed. Amsterdam: Elsevier/North-Holland Biomedical Press, pp. 467-472.

Hodges, J.K., and J.P. Hearn. 1977. Effects of immunisation against LHRH on reproduction in the marmoset monkey *callithrix jacchus. Nature* 265:746.

Hsueh, A.J., M.L. Dufau, K.J. Catt. 1976. Regulation of luteinizing hormone receptors in testicular interstitial cells by gonadotropin. *Biochem. Biophys. Res. Comm.* 72:1145.

Hsueh, A.J., M.L. Dufau, and K.J. Catt. 1977. Gonadotropin-induced regulation of luteinizing hormone receptors and desensitization of testicular 3',5'-cyclic AMP and testosterone responses. *Proc. Nat. Acad. Sci. U.S.A.* 74:592.

Louvet, J.P., G.T. Ross, S. Birken, and R.E. Canfield. 1974. Absence of neutralizing effect of antisera to the unique structural region of human chorionic gonadotropin. *J. Clin. Endocrinol. Metab.* 39:1155.

Matsuura, S., H.C. Chen, and G.D. Hodgen. 1978. Antibodies to the carboxy-terminal fragment of human chorionic gonadotropin beta subunit. Characterization of antibody recognition sites using synthetic peptide analogues. *Biochemistry* 17:575.

Matsuura, S., M. Ohashi, H.C. Chen, and G.D. Hodgen. 1979. A hCG specific antiserum against synthetic peptide analogues to the C-terminal peptide of its beta-subunit. *Endocrinology* 104:396.

Morgan, F.J., S. Birken, and R.E. Canfield. 1973. hCG: a proposal for the amino acid sequence. *Mol. Coll. Biochem.* 2:97.

Morgan, F.J., S. Birken, and R.E. Canfield. 1975. The amino acid sequence of hCG: the α-subunit and β-subunit. *J. Biol. Chem.* 250:5247.

Muniyappa, K., and P.R. Adiga. 1979. Occurrence and functional importance of a riboflavin-carrier protein in the pregnant rat. *FEBS Letters* 110:209.

Murty, G.S.R.C., C.S. Sheela Rani, and N.R. Moudgil. 1980. The role of FSH in regulating testicular function—a study involving the use of specific FSH antibodies. In *Endocrinology.* I.A. Cumming, J.W. Funder, and F.A.O. Mendelsohn, eds. Canberra: Australian Academy of Science. In press.

Mustafa, A.S., R.K. Naz, and G.P. Talwar. 1981. Relationship between bacterial index and type of tubular damage in BCG induced aspermatogenesis. *Ind. J. Exp. Biol.* 19:1011.

Moudgil, N.R., V.R. Mukku, S. Prahlada, G.S. Murty, and C.H. Li. 1978. Passive immunisation with an antibody to β subunit of ovine luteinising hormone as a method of early abortion—A feasibility study in monkeys *(Mocca radiata). Fertil. Steril.* 30:223.

Nash, H., G.P. Talwar, S. Segal, T. Luukkainen, E.D.B. Johansson, J. Vasquez, E. Coutinho, and K. Sundaram. 1980. Observation on the antigenicity and clinical effects of a candidate antipregnancy vaccine: β subunit of hCG linked to tetanus toxoid. *Fertil. Steril.* 34:328.

Naz, R.K., and G.P. Talwar. 1981. Ability of autoclaved BCG and extracts to induce aspermatogenesis. *Ind. J. Med. Res.* 74:251.

Nieschlag, E., K.H. Usadel, E.J. Wickings, H.K. Kley, and W. Wuttke. 1975. Effects of active immunisation with steroids on endocrine and reproductive functions in male animals. In *Immunization with Hormones in Reproduction Research.* E. Nieschlag, ed. Amsterdam: North Holland, pp.155-172.

Nisula, B.C., F.J. Morgan, and R.E. Canfield. 1974. Evidence that chorionic gonadotropin has

465

intrinsic thyrotropic activity. *Biochem. Biophys. Res. Commun.* 59:86.

Oslon, L., and H.S. Kaplan. 1980. Human-human hybridomas producing monoclonal antibodies of predefined antigenic specificity. *Proc. Nat. Acad. Sci.* 77:5429.

Pala, A., M. Ermini, L Caranza, and G. Benegiano. 1976. Immunization with hapten-coupled hCG-β subunit and its effect on the menstrual cycle. *Contraception* 14:579.

Pekonen, F., and B.D. Weintraub. 1980. Interaction of crude and pure chorionic gonadotropin with the thyrotropin receptor. *J. Clin. Endocr. Metab.* 50:280.

Pierce, J.G., O.P. Bahl, J.C. Cornell and N. Swaninathan. 1971. Biologically active hormones prepared by recombination of the α-chain of hCG and hormone specific chain of bovine thyrotropin or of bovine luteinizing hormone. *J. Biol. Chem.* 246:2321.

Pierce, J.G., M.R. Faith, L.C. Giudica, and J.R. Reeve. 1976. Structure and structure-function relationships in glycoprotein hormones. In *Polypeptide Hormones: Molecular and cellular aspects.* Ciba Foundation Symposium 41. Amsterdam: Elsevier Excerpta Medica/ North-Holland, pp. 225.

Prasad, C.R., R.C. Srimal, and B.N. Dhawan. 1976. Acute toxicity and pharmacology of β-hCG conjugated tetanus toxoid (Pr-β-hCG-TT). *Contraception* 13:189.

Prem Mohini, T.N. Chapekar, Joseph Raj Benjamin, N. Shastri, S.K. Dubey, and G.P. Talwar. 1978. Differences between the discriminatory activity of antisera raised against the total gonadotropin and the Pr-β-hCG-TT for neutralisation of hCG and LH action. *Contraception* 18:59.

Ramakrishnan, S., C. Das, and G.P. Talwar. 1978a. Progesterone levels in monkeys immunised with Pr-β-hCG-TT after injection of hLH and hCG during luteal phase. *Contraception* 18:51.

Ramarishnan, S., C. Das, and G.P. Talwar. 1978b. Recognition of the beta-subunit of human chorionic gonadotropin and sub-determinants by target tissue receptors. *Biochem. J.* 176:599.

Ramakrishnan, S., C. Das, S.K. Dubey, M. Salahuddin, and G.P. Talwar. 1979. Immunogenicity of three C-terminal synthetic peptides of the beta-subunit of hCG and properties of the antibodies raised against 45-amino acid C-terminal peptide. *J. Reprod. Immunol.* 1:249.

Reichert, L.E. Jr., C.G. Trowbridge, V.K. Bhalla, and G.M. Lawson. 1974. The kinetics of formation and biological activity of native and hybrid molecules of human follicle-stimulating hormone. *J. Biol. Chem.* 249:6472.

Robertson, K.S., J.C. Wilson, and H.M. Fraser. 1979. Immunological castration in male cattle. *Vet. Rec.* 105:556.

Sairam, M.P. 1980. Effect of immunization against follicle stimulating hormone on spermatogenesis in the rat and the monkey. In *Non Human Primate Models for Study of Human Reproduction.* T.C. Anand Kumar, ed. Basel: S. Karger, pp. 176-189.

Saxena, B.B. 1979. Current studies of a gonadotropin like substance in the preimplantation rabbit blastocyst. In *Recent Advances in Reproduction and Regulation of Fertility.* G.P. Talwar, ed. Amsterdam: Elsevier/North-Holland Biomedical Press, pp. 319-332.

Shaha, S.M., P.P. Kulkarni, and K.L. Patel. 1979. Evaluation of immunological and safety data in women treated with Pr-β-hCG-TT vaccine. In *Recent Advances in Reproduction and Regulation of Fertility.* G.P. Talwar, Amsterdam: Elsevier/North-Holland, Biomedical Press, pp. 473-476.

Sharpe, R.M. 1976. hCG-induced decrease in availability of rat testis receptors. *Nature* 264:664.

Shastri, N., S.K. Manhar, and G.P. Talwar. 1981. Important role of the carrier in the induction of antibody response without Freund's Complete Adjuvant against a "self" peptide hormone LHRH. *Am. J. Reprod. Immunol.* 1:262.

Siris, E.S., B.C. Nisula, K.J. Catt, K. Horner, S. Birken, R.E. Canfield, and G.T. Ross. 1978. New evidence for intrinsic FSH-like activity in hCG and LH. *Endocrinology* 102:1356.

Stevens, V.C. 1976. Perspectives of development of a fertility control vaccine from hormonal antigens of the trophoblast. In *Development of Vaccines for Fertility Regulation.* Copenhagen: Scriptor, p. 93.

Talwar, G.P. 1979. Human chorionic gonadotropin and ovarian and placental steroidogenesis. *J. Steroid Biochem.* 11:27.

Talwar, G.P. 1980. Vaccines based on the β-subunit of hCG. In *Immunological Aspects of Reproduction and Fertility Control.* J.P. Hearn, ed. Lancaster, U.K: M.T.P. Press. pp. 217-277.

Talwar, G.P., R.K. Naz, and C.Das. 1979. A practicable immunological approach to block spermatogenesis without loss of androgens. *Proc. Nat. Acad. Sci. U.S.A.* 76:5882.

Talwar, G.P. S. Ramakrishnan, C. Das, S.K. Gupta, M. Salahuddin, M.K. Viswanathan, P.D. Gupta, P. Pal, K. Buckshee, and J. Fric. 1978. In *Proceedings of the 6th Asia Oceania Congress on Endocrinology.* Vol. 2. Singapore, pp. 472-476.

Talwar, G.P., N.C. Sharma, S.K. Dubey, C. Das, M. Salahuddin, Nah., I. Ramakrishnan, S. Kumar, V. Hingorani, and B. Bloom. 1975. Immunological studies on human placental trophoblast components of potential utility for fertility regulation. In *8th World Congress on Fertility and Sterility.* Buenos Aires: Excerpta Medica Series No. 394, pp. 224-232.

Talwar, G.P., N.C. Sharma, S.K. Dubey, M. Salahuddin, C. Das, N. Ramakrishnan, K. Kumar, and V. Hingorani. 1976. Isoimmunization against human chorionic gonadotropin with conjugates of processed β-subunit of the hormone and tetanus toxoid. *Proc. Nat. Acad. Sci. U.S.A.* 73:218.

Thau, R.B., K. Sundaram, Y.S. Thermton, and L.S. Seidman. 1979. Effects of immunisation with the β-subunit of ovine luteinizing hormone on corpus luteum function in the rhesus monkey. *Fertil. Steril.* 31:200.

Vaitukaitis, J.L., G.D. Braunstein, and G.T. Ross. 1972. A radioimmunoassay which specifically measures human chorionic gonadotropin in the presence of human luteinizing hormone. *Am. J. Obstet. Gynecol.* 113:751.

Wide, L., and M. Wide. 1979. Chorionic gonadotropin in the mouse from implantation to term. *J. Reprod. Fertil.* 57:5.

Chapter 21

IMMUNE RESPONSES TO CARBOXYL-TERMINAL PEPTIDES OF THE hCG BETA SUBUNIT

VERNON C. STEVENS
JOHN E. POWELL
ARTHUR C. LEE

Biochemical studies identifying the subunits and amino acid sequence of the human chorionic gonadotropin (hCG) molecule have contributed significantly to our understanding of its hormonal function and thus to its potential manipulation for clinical application. Particularly noteworthy is the observation that the hCG beta subunit (β-hCG) contains a carboxy-terminal amino acid sequence completely different from those of other glycoprotein hormones, especially human luteinizing hormone (hLH). The hLH beta subunit (β-hLH) has extensive sequence homology in the N-terminal region of the molecule with the β-hCG subunit, but lacks the 30-amino-acid chain at the carboxyl end that is present on β-hCG (Carlsen et al., 1973; Morgan et al., 1973). Also, the last three to five amino acid residues on the β-hLH subunit are dissimilar to analogous residues in β-hCG (Closset et al., 1973; Shome and Parlow, 1973; Sairam and Li, 1975). Thus, the peptide chain representing the region of β-hCG beyond the 111 aspartic acid (residues 112 to 145) is different from any chain found in β-hLH.

Since fragments of varying lengths are obtained from this portion of β-hCG when different enzymes or chemical manipulations of the molecule are used some confusion and misunderstanding has resulted regarding the characteristics of the "unique" carboxyl-terminal peptide of β-hCG. Of course, it is a matter of semantics whether one considers the unique region to be the sequence beyond the length of β-hLH (residue 115) or those beyond residue 111, where no chemical homology to β-hLH is found. We have chosen the latter definition for describing the peptide unique to β-hCG.

Peptides containing 2 to 45 amino acids have been obtained from either enzymatic cleavage of β-hCG or synthesis of certain sequences. These peptides have been coupled to immunogenic carriers and a variety of animal species immunized with the conjugates. Assessment of immune responses to these antigens has involved, in the main, evaluations of the characteristics of antibodies produced from such immunizations. While work in this area is still quite limited, much has been learned about the immunogenicity of these peptides, the specificity of antibodies raised, the affinity and class of antipeptide globulins, the immunogenic regions of carboxyl-terminal peptides, and antihormone actions of antisera *in vitro* and *in vivo*. In this chapter, the data available on these subjects will be summarized.

IMMUNOGENICITY OF PEPTIDES

Immunogenicity of an antigen can be defined as the smallest quantity of that antigen which will elicit a measurable antibody response or by the maximum level of antibodies that can be raised to that antigen. We shall address the immunogenicity of COOH β-hCG peptides (CTP) in terms of the latter definition, since it is the most common and perhaps most useful description of relevant data.

Most of the CTP's studied to date are weakly immunogenic when they are injected as free monomer molecules. As with other small molecules, these antigens are more immunogenic when conjugated to larger carrier molecules. Reasons for this increase in immunogenicity are:

1. increased macrophage and/or lymphocyte recognition,
2. carrier-molecule T-lymphocyte "helper" effect, and
3. slower rate of antigen clearance.

Several studies have shown that this procedure is effective for producing significant levels of antibodies to CTP's. The method used for chemically coupling a CTP to a carrier has been shown to be a very important factor in determining the immunogenicity of a given CTP. Classical methods for coupling small molecules to carriers using carbodiimides, glutaraldehyde, or isocyanates as reagents for conjugation result in complex mixtures of cross-linked carrier, cross-linked peptide, and random conjugates of both components, as well as hydrolyzed coupling reagents on both components where the functional group of the reagent did not react with either carrier or antigen. Since one cannot predict from conjugate to conjugate, or accurately assess the composition of these conjugates, it is difficult to evaluate the relative immunogenicity of different CTP conjugates when these methods are used. Therefore, in our view it is not possible to judge the immunogenicity of an antigen when conditions for presenting its determinants to the reticuloendothelial system of an animal are not optimal. Considering this factor, there are few data published on the relative immunogenicity of CTP's.

Another factor that is vital to the establishment of the immunogenicity of any antigen is the method used for quantitating the immune response. In the case of CTP's, almost all experiments on antibody levels attained to conjugates containing these have used radioimmunoassay methods with iodinated intact hCG as the antigen for antibody detection. This procedure indicates the presence of antibodies that are reactive with the whole hormone molecule, not those raised to the antigen used for immunization, and therefore measures a combination of immunogenicity of the antigen and reactivity of antisera with intact hCG. Reactivity of antipeptide sera with intact hCG is not technically a crossreaction, but, for lack of a better term, will be used in this chapter to refer to reactivity to any antigen not used as the immunogen. As will be shown later in the chapter, the crossreactivity of antisera to CTP antigens of different chain lengths varies considerably. Therefore, it is inappropriate to report the immunogenicity of an antigen when the method used to quantitate the response is based upon an undefined crossreaction with an antigen not used for immunization.

Still more basic to the question of CTP immunogenicity determination is the design of assays used to quantitate antibody levels in sera. Many reported methods used only a single dose of iodinated hCG (sometimes only indicated by a certain number of cpm at a single dilution of the serum. Rarely was there any indication that binding of antigen to antibodies was at equilibrium. Certainly such experiments did not provide an antigen excess for saturation of antibody binding sites, despite the expression of "per cent bound" of tracer antigen as well below 100 per cent. While antigen binding, in the absence of definitive data regarding the number of epitopes on CTP antigens, is a valid method for quantitating antibody levels to CTP's and therefore an index of their immunogenicity, care must be exercised in interpreting data from studies where the methods used were not truly quantitative.

It is not our intent to introduce this section with a review of basic methods in immunology, but rather to point out the numerous errors, including some from our laboratory, that have been made in studies assessing the immune response to CTP's. A few reports, intended to relate only qualitative responses to these peptides, have been interpreted as quantitative in some reviews. We shall attempt to summarize the available data on antibody levels attained to conjugates of CTP's in light of the factors discussed above, and let the reader examine individual reports for interpretation of findings and conclusions.

Antisera reactive to ^{125}I-hCG have been raised to the natural fragment of β-hCG representing residues 123 to 145 obtained by tryptic cleavage of the subunit (Louvet and Ross, 1974; Chen et al., 1976). These antisera were obtained by coupling the peptides to bovine serum albumin (BSA) with carbodiimide and immunizing rabbits with a conjugate containing 14 peptides per mole of BSA. Serum titers were reported as those usable in a radioimmunoassay using ^{125}I-hCG at a dilution of 1:12000. Following tryptic digestion of the β-hCG subunit, Birken and Canfield (1980) reduced and carboxymethylated a peptide fragment representing the 115-145 region of the subunit before conjugation to thyroglobulin using carbodiimide as the coupling reagent. These authors noted that 10 per cent or less of the peptide was conjugated and that the major component was modified free peptide. Immunization of rabbits with this conjugate mixture resulted in antibody production in six of eight rabbits with four of the best antisera able to bind hCG suitably for radioimmunoassay at dilutions of 1:2000 to 1:10000. Higher levels of antibody were obtained by

Powell et al. (1980) from a fragment representing β-hCG residues 109 to 145 obtained by cleavage of the subunit with chymotrypsin (Keutmann and Williams, 1977). These sera, obtained from immunization of rabbits with a conjugate of the peptide to tetanus toxoid using a predictable and controllable coupling procedure (Lee et al., 1980), elicited antibody binding from 200 to 700 moles/liter x 10^{-10} to the 109-145 peptide with binding to intact hCG at approximately 90 per cent of this level.

There have been more studies in which the CTP antigens used for immunization were derived from synthetic peptides of the β-hCG COOH sequence than those using naturally cleaved fragments. Significant levels of antibodies were obtained from immunizations of rabbits with conjugates of peptide 111-145 (Stevens, 1975) that showed a high crossreactivity with a natural 109-145 peptide, and native β-hCG and hCG. However, no quantitative data were obtained in this study. Somewhat more meaningful data were obtained from immunizations with synthetic peptides 109-145, 111-145, and 126-145 (Stevens 1976). Antisera to the 109-145 peptide bound ^{125}I-hCG at a level of 200 to 600 ng/ml, whereas those to the 126-145 peptide bound only 30 to 70 ng/ml. Matsuura et al. (1979) immunized five rabbits with a conjugate containing synthetic β-hCG peptide 116-145 and compared the response of one rabbit's serum to the response of a rabbit serum obtained from immunization of rabbits with a natural desialylated peptide fragment (123-145). Similar levels of antibody were obtained for the natural and synthetic peptides, but only with respect to their use in radioimmunoassays. A synthetic peptide representing β-hCG peptide 101-145 was used by Ramakrishnan et al. (1979) to generate antisera following its coupling to tetanus toxoid with carbodiimide. Three species (rabbits, monkeys, and baboons) were immunized. One baboon raised antibody levels to peptide of 576 ng/ml of serum. Only three of ten rabbits and four of eight monkeys produced significant antibody levels reactive to hCG. The highest levels were in a monkey with hCG binding of 29.18 ng/ml. However, when the level of binding to ^{125}I-peptide was measured in this animal's serum, a peak level of 3000 ng/ml was observed, indicating a highly immunogenic response to peptide but a low crossreactivity of the antisera with intact hCG (about 1 per cent). The same discrepancy in the binding of antisera to peptide and hCG was observed when hCG and peptide antibody levels were compared after immunization with β-hCG peptide 138-145 (Stevens, 1981). The factors that regulate the crossreactivity of peptide antisera with hCG will be discussed later in this chapter. However, it is clear that binding of sera to hCG is not an accurate indicator of the immunogenicity of CTP molecules. Careful quantitation of the antibody levels to CTP conjugates was performed by Thanavala et al. (1979) and Powell et al. (1980). Using multiple antigen levels to establish antibody binding at equilibrium for both hCG and peptide, these authors have established estimates of the immunogenicity of representative CTP's. Mean binding levels of 30 to 700 moles/liter x 10^{-10} were found to labeled peptides and 10-600 moles/liter x 10^{-10} to iodinated hCG, depending upon whether the peptide used as immunogen was 125-145, 115-145, or 109-145. Compared with levels to other antigens raised in our laboratory, these antibody levels are lower than antibody levels raised to intact hCG and bovine gamma globulin; equivalent to antibody levels to such antigens as follicle-stimulating hormone, bovine serum albumin, and other serum proteins; and much higher than antibody levels raised to other protein-conjugated antigens, such as estradiol and progesterone. Whether one classifies CTP immunogenicity as high, low, or intermediate depends upon a relative comparison to other reference antigens.

SPECIFICITY OF ANTISERA

The specificity of antisera to CTP's has been tested mainly by reacting sera *in vitro* with ^{125}I-labeled antigens, particularly the pituitary glycoprotein hormones. No significant binding of CTP antisera to purified hormones other than hCG has been reported for any serum (Chen et al., 1976; Stevens, 1976; Matsuura et al., 1979; Ramakrishnan et al., 1979). Further, no inhibitory effect of CTP antisera on hLH activity *in vivo* could be found (Stevens, 1976; Ramakrishnan et al., 1979).

Despite this apparent hormone specificity of CTP antisera, when these sera are used in radioimmunoassays, crude pituitary and urinary extracts produce positive results. It is certainly possible that these responses are truly indicative of small quantities of hCG present in these extracts since these reacting materials have many of the physical, chemical, and immunological properties of hCG purified from pregnancy urine (Chen et al., 1976; Robertson et al., 1978; Matsuura et al., 1980). Other reports, which describe immunoassays with antisera to the entire β-hCG subunit, have suggested that hCG is present in a variety of "normal" tissues and body fluids (Braunstein et al., 1975; Yoshimoto et al., 1977; Borkowski and Muquardt, 1979; Yoshimoto et al., 1979). The interpretation of these data is difficult. The bulk of the evidence supports the notion that hCG is produced by normal cells. It is appealing to suggest that the CTP antisera are specific to the hCG molecule and will not react with any other human tissue or secretory components.

Other data have been obtained that raise some question of the specificity of these antisera. Since there is a four amino acid sequence in the β-hCG COOH region (residues 133 to 136, Arg-Leu-Pro-Gly) that is also present in the core of the subunit (residues 68 to 71) and this latter region has a common sequence with β-hLH, peptide fragments from the β-hLH subunit were reacted with CTP antisera as a test of specificity. No antiserum

475

reactivity was found with intact α-hLH, intact β-hLH, trypsin-digested α-hLH, and β-hLH fragment 60-74 (containing the Arg-Leu-Pro-Gly). However, reactivity was demonstrated with a chymotryptic digest of β-hLH and fragment 17-37 from β-hLH and β-hCG (Figure 21-1).

Since very little sequence homology exists between the 17-37 region of the hCG and hLH subunits and the hCG CTP's, crossreactivity of CTP antisera with both β-subunit fragments 17-37 is puzzling. One could argue that the hCG fragment reacted to CTP antisera as a result of contamination of the fragment with CTP peptides. However, this explanation cannot be used to explain the reactivity of the hLH peptide. Since no reactivity of this hLH fragment with CTP antiserum occurs when it is present in the intact subunit, one must presume that a disrupted β-hLH molecule produces a masked determinant, probably structural in nature, capable of reacting to antisera directed to an hCG peptide with little sequence homology. Such hormone fragments could be present in the serum or urine of normal persons following partial metabolism of hLH. Until more data are available to clarify these observations, nonspecific reactions of antisera to CTP peptides cannot be ruled out.

The effect of the carbohydrate moiety of β-hCG CTP's on their immune responses has not been studied extensively. Also, the limited number of studies conducted using peptides with and without the serine-linked sugar moieties intact have not used similar methods for preparing immunogens,

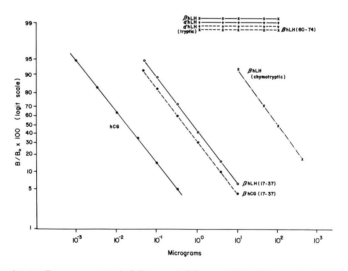

FIGURE 21-1. Response to hCG, an hCG peptide fragment, hLH α and β subunits, enzymatic digests of hLH subunits, and hLH peptide fragments in a radioimmunoassay using an antiserum to COOH β-hCG peptide 111-145 and ^{125}I-hCG.

immunizing animals, or assessing immune responses. Antisera raised by Chen et al. (1976) to CTP 123-145 used a desialylated fragment following tryptic digestion of β-hCG. Antibody levels and antibody specificity were studied in radioimmunoassays by binding antisera to ^{125}I-intact hCG. Birken and Canfield (1980) have demonstrated that these sera do not bind intact hCG as well as asialylated hCG and that dose-response lines of hCG have different slopes in radioimmunoassays using hCG or desialo-hCG as tracer antigens. Some antisera to desialo-β-hCG peptides recognize the remaining carbohydrate moiety (probably galactose), and their reactivity to hCG is quite different from that of antisera recognizing the peptide chain.

Birken and Canfield (1980) also found that antisera raised to β-hCG peptide 115-145, with the carbohydrate moiety intact, did not react with sugar residues on hCG. Also, their binding to desialo-hCG was at a lower level than their binding to intact hCG. These studies suggest that when the terminal sugar residue on the CTP carbohydrate moiety (sialic acid) is intact, few antibodies are raised to the carbohydrate, but when this residue is removed by neuraminidase, the remaining sugars are sometimes immunogenic.

Studies have been conducted in the authors' laboratory using β-hCG CTP 109-145 obtained by chymotryptic digestion of β-hCG. A portion of the peptide was treated with hydrofluoric acid to remove part of the carbohydrate from the molecule. These two peptides, as well as a synthetic peptide with the same amino acid sequence, were conjugated to a carrier in the same manner, and rabbits were immunized with the three different immunogens. No significant differences in the level of antibodies reactive to intact hCG were found from these immunizations. However, when the three types of antisera were tested for binding to ^{125}I-labeled synthetic peptide or dose-response lines compared in radioimmunoassays using different antisera bound to ^{125}I-hCG, differences in the reactivity of the antisera were detected. Antisera to the CHO-denuded peptide or the synthetic peptide had higher levels of antibodies reactive to carbohydrate-free antigens than antisera to the intact native peptide. Also, the native peptide was not able to compete with labeled hCG as well as the CHO-free antigens in radioimmunoassays where these antisera were used. On the other hand, antisera raised to the native peptide with the carbohydrate intact had no higher reactivity to native peptide or intact hCG than to synthetic or CHO-denuded peptide. These studies were interpreted as indicating that the carbohydrate moieties on β-hCG CTP's are not normally immunogenic and that few, if any, antibodies are raised to them following immunization. Further, since antisera to CHO-free antigens are more reactive to themselves than to CHO-intact peptides, it is likely that the carbohydrate moiety prevents some areas of the peptide chain, adjacent to their attachment, from expressing their immunological properties by folding or otherwise shielding them from recognition.

AFFINITY AND CLASS OF ANTIBODIES

Affinity measurements using antisera raised to CTP molecules ranging in length from 21 to 37 amino acid residues have indicated that affinity constants varied from 0.8 to 47.0 x 10^9 liters/mole for selected sera (Powell et al., 1980) in terms of binding the homologous peptide used for immunization. Powell et al. found that the affinity constants for antisera to a 21 amino acid residue peptide were slightly lower than those for antisera raised to longer peptides. Affinity constants for antisera to peptides containing 21 to 37 amino acid residues, in terms of intact hCG binding, varied from 5.5 x 10^8 to 1.0 x 10^{10} liters/mole with an average of around 3.5 x 10^9 liters/mole (Louvet and Ross, 1974; Stevens, 1976; Powell et al., 1980; Thanavala et al., 1980). No clear correlation has been found between the length of peptide used to raise antisera and affinity to hCG, although Powell et al. (1980) found affinities of sera from rabbits immunized with a 21-residue peptide lower than those of sera from animals injected with peptides 31 and 37 residues in length. These same authors also reported that a comparison of affinities to peptide and to hCG showed slightly higher affinities to peptide than to hCG, but this difference was not usually significant. The affinity of antibodies to CTP's is distinctly lower than the affinity of antibodies raised to intact hCG or the β-hCG subunit (Stevens, 1976; Thanavala et al., 1978). The reason(s) for a lower affinity response to CTP's than to larger molecules is not known, but probably is related to the fact that fewer determinants are available on the smaller molecules.

The class of antibodies raised to CTP's has been studied very little. Powell et al. (1980) demonstrated that rabbit sera collected 70 days after the primary immunization with several CTP's contained antibodies to peptide and hCG almost totally of the IgG class. No work has been reported regarding the changes in antibody class over a period of time following immunization or the subclass(es) of IgG antibodies raised. This latter work should be pursued in view of the confusing reports on the actions of CTP antibodies *in vivo*.

IMMUNOLOGICAL DETERMINANTS ON CARBOXYL TERMINAL PEPTIDES

The amino acid sequences of CTP's that bind antisera raised to β-hCG peptides 123-145 and 116-145 conjugated to BSA with carbodiimide have been carefully characterized by Matsuura et al. (1978, 1979). These workers studied the inhibition of ^{125}I-hCG binding to antisera by a series of synthetic peptides of increasing length, beginning with the two C-terminal residues (Pro-Gln) of β-hCG and extending the length by adding residues to the peptide N-terminus. A series of 32 peptides, analogous to the sequence of the native subunit, was used to ascertain which ones had the ability to compete with intact hCG for binding to antisera. The dipeptide Pro-Gln significantly inhibited hCG binding to antisera, and these workers suggested that these residues constituted the primary recognition site of the CTP of β-hCG. Increased binding competition was observed as the peptide chain length was increased to 15 amino acid residues. The greatest increase in binding capacity was attained with the addition of residues 133 to 136 to the peptide. Peptides longer than 15 amino acids did not compete to any greater extent with hCG for binding antibodies than did the 131-145 sequence. Thus, these authors concluded that the two C-terminal residues Pro-Gln constituted the primary recognition site of the β-hCG CTP and that the complete determinant(s) of the "unique" CTP of β-hCG was a linear sequence and resided between residues 131 and 145 of the subunit. These workers further concluded that there was no difference in the immunogenicity or immunological reactivity to hCG of peptides cleaved from the natural subunit and peptides produced synthetically.

This experiment was repeated in the authors' laboratory using an antiserum raised to synthetic peptides representing β-hCG sequences 125 to 145 and the same radioimmunoassay procedures used by Matsuura et al. (1978). Findings obtained in our studies were similar to those previously reported by Matsuura et al.

Since our antisera raised to peptide 125-145 behaved quite differently *in vivo* from antisera raised to peptide 111-145, an experiment was conducted in the same manner as the ones described above using an antiserum to the longer peptide. The same pattern of peptide reactivity was observed as in earlier experiments with CTP's up to 15 amino acids in length (Figure 21-2). In fact, peptide 131-145 competed with ^{125}I-hCG, on a mole/mole basis, better than unlabeled hCG for binding to antisera. This could be accounted for either by a better reactivity of the antisera with peptide than with hCG or the fact that the low molecular weight peptides reached equilibrium with the antibody faster than did hCG, because of the reaction conditions used. However, when peptides longer than 15 residues were used to compete with hCG, less reactivity with antibodies was observed. A slow but continuous decline in binding competition was found with peptides containing 18 to 35 amino acid residues. This experiment was repeated several times and similar findings were obtained each time. An explanation for this phenomenon was not apparent, but we speculated that CTP's up to 15 residues were reactive to the same determinant-specific antibodies described by Matsuura et al.(1978), but that when longer peptides were tested, folding of the peptide (perhaps by the Pro-Pro-Pro sequence at 124-126) resulted in inhibition by the N-terminal segment of some of the

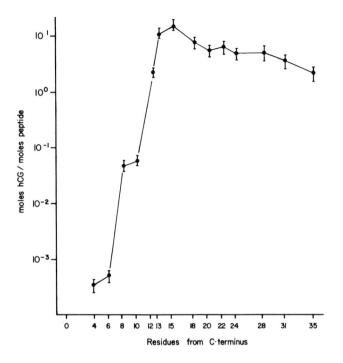

FIGURE 21-2. The relative reactivity of peptides of increasing length representing the C-terminal residues of β-hCG to an antiserum to peptide 111-145.

Peptides with 4 to 35 amino acids and native hCG were tested for their ability to compete with ^{125}I-hCG for binding to an antiserum. Results were expressed as moles of hCG required to compete equally with ^{125}I-hCG (B/F = 0.5) for antibody binding divided by the number of moles of peptide required to compete equally with ^{125}I-hCG binding.

binding capacity of the C-terminal 15 residues. Another, but unlikely, possibility is that the kinetics of the antigen-antibody reaction of the peptides longer than 15 residues were progressively slower, and after a fixed incubation time, short of equilibrium, the longer peptides competed less effectively than the shorter ones, although their antibody binding sites were the same. A definitive explanation must await further study.

The above evaluation of antisera to peptide 111-145 did not explain why they were able to neutralize hCG action *in vivo*, but antisera to peptide 125-145 were not. If this peptide possessed an additional determinant beyond the 15 residue C-terminus, competition of peptides 16-35 residues in length with hCG binding to anti-111-145 sera should have been greater than, and not less than, the 15 residue peptide. In an attempt to ascertain whether antisera to peptide 111-145 has a determinant other than the one on the C-terminal 15 residues, an antiserum was reacted with ^{125}I-peptides 111-130 and 133-145. Both labeled peptides bound the antiserum significantly. When peptides of increasing length from the C-terminus of β-hCG were tested for competition with the binding of the two labeled peptides, the existence of at least two immunogenic regions on peptide 111-145 was revealed (Figure 21-3).

Competition of binding to ^{125}I-labeled peptide 133-145 by CTP's up to 15 residues followed a pattern that was nearly identical to the pattern found when they were in competition with ^{125}I-hCG. Since the labeled peptide was 13, and not 15, residues long, maximum reactivity was attained with peptide 133-145. Longer peptides had neither increased nor decreased reactivity with the antiserum. On the other hand, no significant competition of ^{125}I-labeled peptide 111-130 binding to antibodies was found until a peptide with 28 residues (118-145) was incubated with labeled peptide and antiserum. Peptide reactivity further increased as the peptide length was extended to 31 and 35 residues. We interpreted these data as indicating that a population of antibodies was also raised to sequence 111-118 of the 111-145 peptide. Should this interpretation of the data be correct, the question logically follows, why was maximum competition with hCG for 111-145 antiserum binding found with the 15 residue CTP with a decreased reactivity as longer peptides were tested? Should there be two populations of antibodies in the sera, one would expect an increased reactivity when peptides with 28 to 35 amino acids were tested. Again, we can only speculate about this paradox. It may be possible that, although the 111-118 region of the longer peptides is competitive with hCG binding to the antisera, the inhibitory effect of peptide folding on the binding by the 131-145 region exceeded binding by the 111-118 region and the net reactivity of the longer peptides was less.

Further evidence for two determinants on peptide 111-145 has been provided by the observation that antisera raised to peptides 111-118 and 138-145 both react with ^{125}I-hCG and ^{125}I-labelled peptide 111-145. No competition of this binding was observed with any peptides except those used for immunization, indicating that a linear sequence in each peptide is exposed on the surface of intact hCG. In order to demonstrate this conclusion more clearly, antisera to both peptide 111-118 and peptide 138-145 were mixed in concentrations such that equal binding capacity to ^{125}I-hCG was present. The ability of peptides with increasing length from the C-terminus to compete with ^{125}I-hCG binding to the mixed antisera was tested. As shown in Figure 21-4, peptides representing the CTP 138-145 (eight residues) competed well with hCG for antisera binding, but additional reactivity was found when peptides of 10, 13, and 15 residues were tested. The reason for this increased reactivity to peptide region 131-138 is not clear since this sequence was not contained in the peptide used for immunization.

Peptides of 20 to 28 residues exhibited no significant increase in reactivity over the 15-residue peptide. However, 31- and 35-residue peptides

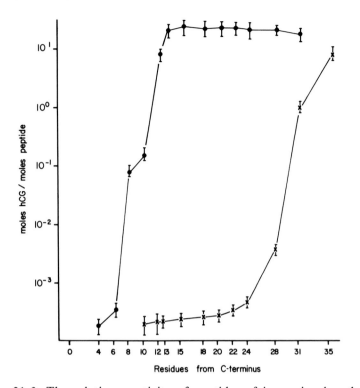

FIGURE 21-3. The relative reactivity of peptides of increasing length representing the C-terminal residues of β-hCG to an antiserum to peptide 111-145. In separate radioimmunoassays, peptides were tested for their ability to compete with either ^{125}I-labeled peptide 133-145 (●) or ^{125}I-labeled peptide 111-130 (-x-) for binding to antibodies. Results were expressed in terms of peptide binding competition relative to binding competition by hCG.

showed significantly greater ability than CTP 131-145 to compete with hCG. While it may be said that this experiment had little relevance to the determinants on the entire 111-145 peptide, it did demonstrate that increased reactivity of peptides longer than 15 residues can be observed when two separate sequence-dependent antibody populations are available to react either with peptides or with hCG. Also, this experiment strengthens the argument that peptide 111-145 has structural properties that do not permit all of its surface sequence regions to behave as immunologically independent segments.

Since some evidence from these immunological studies suggests a secondary structure for CTP's and since there are numerous proline and serine residues in the β-hCG CTP, studies of peptide structure were conducted by obtaining circular dichroic spectra of 35- and 37-residue peptides. In aqueous solutions or in trifluoroethanol, spectra were consistent with the occurrence of

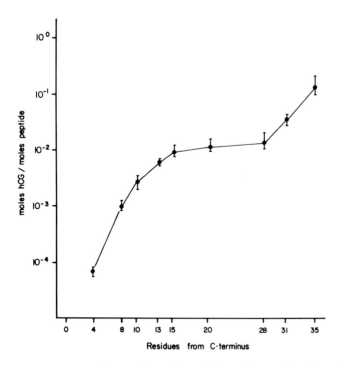

FIGURE 21-4. The relative reactivity of peptides of increasing length, representing the C-terminal residues, of β-hCG to a mixture of antisera raised to peptides 111-118 and 138-145.
Peptides and hCG were tested for their ability to compete with ^{125}I-hCG in radioimmunoassay. Results were expressed as described in Figures 21-2 and 21-3.

type II β-turns but no evidence was found for the existence of α-helicity or β-structure (David Puett, personal communication). The single phenylalanine residue at position 115 contributed to the circular dichroic spectra and may be involved in an immunological determinant in the 111-118 region. Structural predictive rules and a computer program show that predicted β-turns have the greatest probability of occurring between residues 110-115, 117-122, 125-128, 130-133, and 134-140 (Robert J. Ryan, personal communication). Should these type II β-turns actually exist on the 111-145 peptide, our data would suggest that the N- and C-terminal turns contain the most immunogenic sites on the peptide molecule.

The observations described above provide strong evidence that CTP's with 35 or more amino acids have a structural configuration that affects their immunological properties. However, with the data now available it cannot be demonstrated conclusively that these peptides contain structurally related immunological determinants.

483

NEUTRALIZATION OF hCG BY ANTISERA TO β-hCG COOH PEPTIDES

Antisera to intact hCG have been shown to be capable of blocking the activity of hCG or hLH exogenously administered to laboratory animals (Rao and Shanhani, 196l; Snook and Cole, 1965; Robyn and Diczfalusy, 1968). Other work has demonstrated the ability of antisera raised to the β-hCG subunit to also neutralize both hormones but to be less effective in the inhibition of hLH than hCG (Stevens, 1975; Ramakrishnan et al., 1979; Pandian et al., 1980). A report by Louvet and Ross (1974) indicated that antisera to a β-hCG COOH peptide did not have the ability to inhibit *in vivo* biological activity of hCG. Since that time, other studies have been conducted with antisera to a variety of CTP preparations with some conflicting results.

Major reasons for different findings regarding the hCG neutralizing capacity of anti-CTP sera are:

1. a variety of peptides have been used to generate antisera,
2. different methods have been used to couple these peptides to carriers prior to immunization, and
3. different methods have been used to assay antigonadotropin actions.

As discussed above, CTP peptides representing fewer than 25 amino acids probably contain only one immunological determinant, and peptides of at least 35 residues certainly represent at least two determinants. Further, when peptides are conjugated to carriers via their amino groups (the N-terminus or the side-chain NH_2 group site, depending upon which peptide is used), determinant regions may be blocked from an immune response by the linkage to carrier. In some cases, NH_2 to NH_2 coupling may result in conjugation of peptide polymers to carrier. These undefined antigens may not present the peptide to the animal's reticuloendothelial system in a manner capable of eliciting a response to determinants in that region of the molecule adjacent to the point of conjugation.

The method of assessing antigonadotropic activity *in vivo* is of critical importance, particularly for quantitation of antiserum potencies. Most workers have, for convenience, incubated hCG with antisera *in vitro* prior to testing by an *in vitro* or *in vivo* bioassay system. Some have precipitated immune complexes formed before testing (Louvet and Ross, 1974). While these procedures often will yield meaningful qualitative data, misleading findings can be obtained from such experiments. It is very tempting to use *in vitro* bioassays for testing antisera because of the relative ease and speed of these methods compared with *in vivo* studies. Caution should be used in interpreting data when antisera are used to test blocking of hormone binding to crude target receptor organ extracts or when they are used to inhibit steroid production by hormones in incubated organs or tissues. In these tests, unless the target cells are adequately dispersed, antibody-hormone complexes may block steroidogenesis by preventing the hormone from reaching its cell receptor and thereby provide a false positive result. Even the injection of antibody-hormone complexes formed *in vitro* into assay animals is not a true indication of what action antibodies will have if the complexes are formed *in vivo*. The conditions in which complexes are formed and the resultant effects on *in vivo* hormone action are best studied if the antiserum is injected separately from the hormone into the animal and the complexes allowed to form in the circulation. All of these factors should be considered when examining the reports of antigonadotropic activity of antisera to CTP's of β-hCG.

Louvet and Ross (1974) showed that antisera to β-hCG 123-145 would not reduce the hCG-stimulated increase in weight of the ventral prostate of the hypophysectomized immature male rat. In fact, although not claimed by these authors, antisera actually enhanced hCG activity in this assay. Stevens (1976) showed an inhibition of hCG-stimulated increases in immature mouse uterine weight by antisera to CTP 111-145 or 109-145. However, antisera to CTP 126-145 or 111-130 failed to inhibit the biological activity of hCG. Thanavala et al. (1980) demonstrated that antisera to CTP 109-145 were capable of blocking hCG-induced multiple ovulation in rats, and Chang et al. (1978) reported that an antiserum to CTP 101-145 inhibited hCG stimulation of rat uterine weight in immature females. Using this same assay method, Matsuura et al. (1979) failed to demonstrate any neutralizing activity in antisera raised to CTP 116-145.

Most of the discrepancies in the data reported can be explained by the different peptides used for raising antibodies. No worker has suggested that a peptide with fewer than 30 C-terminal residues elicited antibodies with significant neutralizing capacity, although Ohashi et al. (1980) reported weak inhibition of hCG activity *in vitro* by an antiserum raised to CTP 123-145. Also, no report has suggested that antisera to peptides representing 35 or more of the C-terminal β-hCG residues will not neutralize hCG biological activity. There is some discrepancy about actions of antisera to intermediate-length peptides. Stevens (1976) reported that antisera to CTP 115-145 had weak capacity to inhibit hCG, but Matsuura et al. (1979) found no such capacity of an antiserum raised to CTP 116-145. Since only one serum was tested at one concentration in the latter report, animal variation in antibody production could account for the failure to detect neutralizing activity.

The importance of the methods and scope of studies on the antigonadotropin effects of antisera can be illustrated by the results of an experiment in which several doses of antisera raised to CTP 109-145 were incubated with a fixed quantity of hCG before the resulting complexes were injected into assay animals. Very small doses of antisera showed no effect on the biological response to hCG, but slightly higher doses caused an enhancement. Still larger doses reduced the response in a dose-dependent fashion. In this study, three conclusions could have been reached if only one dose of antiserum had been used:

1.   no effect on hCG activity,
2.   potentiation of hCG activity, or
3.   inhibition of hCG activity.

The progonadotropic effect of antisera to hCG when the intact molecule was used for immunization has been reported by Ahern et al. (1972), Hopkins and Clayton-Hopkins (1972), and Cole et al. (1975). These workers speculated that those antisera with a single population of antibodies, not reacting with the hormone's target organ receptor site, prolonged the clearance of injected hCG and permitted the hormone to be retained in the circulation for a longer time. Current data clearly indicate that the antisera to β-hCG CTP's from immunizations with peptides of 35 or more residues will efficiently neutralize the biological action of hCG (Figure 21-5). The mechanism of the neutralizing action has not been demonstrated, although it most likely involves two or more immunoglobulins binding one hCG molecule, which results in a rapid clearance of the complex from the circulation. Ohashi et al. (1980) suggested that the relatively low affinity (10^9 M^{-1}) of non-neutralizing antisera permitted the hCG-antibody complex to be dissociated in the circulation, allowing the freed hCG to have biological effects, whereas antisera of high affinity (10^{12} M^{-1}) maintained the hCG-antibody complex and the hCG was not biologically active. Data from our laboratory dispute this hypothesis since some antisera to CTP 109-145, with even lower affinity than the one tested by Ohashi et al., are very potent inhibitors of hCG biological activity. We suggest, since it is generally agreed that CTP antisera do not bind the hormone receptor binding site (Bahl, 1977), that binding of hCG by antisera containing at least two different populations of antibodies causes formation of immune complexes that are rapidly cleared from the circulation, or that binding of hCG by the population of antibodies to the N-terminal region of longer CTP's (perhaps 109-118 residues) blocks hCG binding to its target organ receptor by steric hindrance.

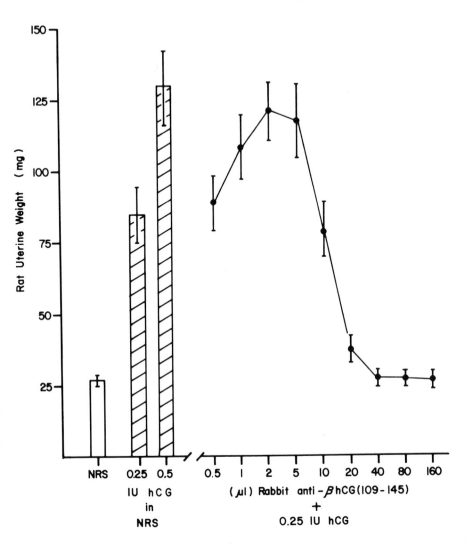

FIGURE 21-5. Effects of different doses of an antiserum raised to β-hCG peptide 109-145 on the biological activity of hCG.

Groups of immature female rats (N = 5) were injected with either normal rabbit serum (NRS), two doses of hCG in NRS, or 0.25 IU hCG incubated for two hours with various amounts of a rabbit antiserum before injection. Five injections were given over three days and 68 hours after the first injection, the animals were sacrificed, and their uteri removed and weighed.

ACKNOWLEDGMENTS

Our cited research received financial support from the Special Programme for Research in Human Reproduction, World Health Organization. Special thanks are extended to G. Tregear, H. Keutmann, H. Niall, and J. Stewart for preparing hCG peptides used in these studies.

REFERENCES

Ahern, C., H.H. Cole, I.I. Geschwind, and L. Itze. 1972. Physiological studies on the identity of the gonadotropic and progonadotropic substance(s) in sera of ewes immunized against hCG. *Endocrinology* 90:1619.

Bahl, O.P. 1977. Human chorionic gonadotropin, its receptor and mechanism of action. *Fed. Proc.* 36:2119.

Birken, S., and R.E. Canfield. 1980. Chemistry and immunochemistry of human chorionic gonadotropin. In *Chorionic Gonadotropin*. S.J. Segal, ed. New York: Plenum Publishing, pp 65-88.

Borkowski, A., and C. Muquardt. 1979. Human chorionic gonadotropin in the plasma of normal nonpregnant subjects. *N. Engl. J. Med.* 301:298.

Braunstein, G.D., J. Rasor, and M.E. Wade. 1975. Presence in normal human testes of a chorionic-gonadotropin-like substance distinct from human luteinizing hormone. *N. Engl. J. Med.* 293:1339.

Carlsen, R.B., O.P. Bahl, and N. Swaninathan. 1973. Human chorionic gonadotropin: linear amino acid sequence of the β-subunit. *J. Biol. Chem.* 248:6810.

Chang, C.C., Y.Y. Tsong, J.D. Rone, S.J. Segal, D. Chang, and K. Folkers. 1978. A highly specific antiserum to synthetic C00H terminal peptide of human chorionic gonadotropin β-subunit. *Fertil. Steril.* 29:250 Abstract.

Chen, H.-C., C.D. Hodgen, S. Matsuura, L.J. Lin, E. Gross, L.E. Reichert Jr., S. Birken, R.E. Canfield, and G.T. Ross. 1976. Evidence for a gonadotropin from nonpregnant subjects that has physical, immunological, and biological similarities to human chorionic gonadotropin. *Proc. Nat. Acad. Sci. USA* 73:2885.

Closset, J., G. Hennen, and R.M. Lequin. 1973. Human luteinizing hormone: the amino acid sequence of the β-subunit. *FEBS Lett.* 29:97.

Cole, H.H., R. Dewey, I.I. Geschwind, and M. Chapman. 1975. Separation of progonadotropic activities in ovine and equine hCG antisera. *Biol. Reprod.* 12:516.

Hopkins, T.F., and J.A. Clayton-Hopkins. 1972. Biphasic effects of antiserum to hCG-plus gonadotropin-enhancing factor(s) on hCG response of the immature mouse (36456). *Proc. Soc. Exp. Biol. Med.* 140:346.

Keutmann, H.T., and R.M. Williams. 1977. Human chorionic gonadotropin: amino acid sequence of the hormone-specific C00H-terminal region. *J. Biol. Chem.* 252:5393.

Lee, A.C.J., J.E. Powell, G.W. Tregear, H.D. Niall, and V.C. Stevens. 1980. A method for preparing β-hCG C00H peptide-carrier conjugates of predictable composition. *Mol. Immunol.* 17:749.

Louvet, J.P., and G.T. Ross. 19974. Absence of neutralizing effect of antisera to the unique structural region of human chorionic gonadotropin. *J. Clin. Endocrinol. Metab.* 39:ll55.

Matsuura, S., H.-C. Chen, and G.D. Hodgen. 1978. Antibodies to the carboxyl-terminal fragment of human chorionic gonadotropin β-subunit; characterization of antibody recognition sites using synthetic peptide analogs. *Biochem.* 17:575.

Matsuura, S., M. Ohashi, H.-C. Chen, and G.D. Hodgen. 1979. A human chorionic gonadotropin-specific antiserum against synthetic peptide analogs to the carboxyl-terminal peptide of its β-subunit. *Endocrinology* 104:396.

Matsuura, S., M. Ohashi, H.-C. Chen, R.C. Shownkeen, A. Stockell Hartree, L.E. Reichert Jr., V.C. Stevens, and J.E. Powell. 1980. Physico-chemical and immunological characterization of an hCG-like substance from human pituitary glands. *Nature* 286:740.

Morgan, F.J., S. Birken, and R.E. Canfield. 1973. Human chorionic gonadotropin: a proposal for the amino acid sequence. *Mol. Cell. Biochem.* 2:97.

Ohashi, M., S. Matsuura, H.-C. Chen, and G.D. Hodgen. 1980. Comparison of *in vivo* and *in vitro* neutralization of human chorionic gonadotropin (hCG) activities by antisera to hCG and a carboxy-terminal fragment of the β-subunit. *Endocrinol.* 107:2034.

Pandian, M.R., R. Mitra, and O.P. Bahl. 1980. Immunological properties of the β-subunit of human chorionic gonadotropin (hCG). II. Properties of a hCG-specific antibody prepared against a chemical analog of the β-subunit. *Endocrinol.* 107:1564.

Powell, J.E., A.C. Lee, G.W. Gregear, H.D. Niall, and V.C. Stevens. 1980. Characteristics of antibodies raised to carboxy-terminal peptides of hCG beta subunit. *J. Reprod. Immunol.* 2:1.

Ramakrishnan, S., C. Das, S.K. Dubey, M. Salahuddin, and G.P. Talwar. 1979. Immunogenicity of three C-terminal synthetic peptides of the beta subunit of human chorionic gonadotropin and properties of the antibodies raised against 45-amino acid C-terminal peptide. *J. Reprod. Immunol.* 1:249.

Robertson, D.M., H. Suginami, H. Hernandez-Montes, C.P. Puri, S.K. Choi, and E. Diczfalusy. 1979. Studies on a chorionic gonadotropin-like material present in nonpregnant subjects. *Acta Endocr.* 59:261.

Sairam, M.R. and C.H. Li. 1975. Human pituitary luteotropin. Isolation, properties, and the complete amino acid sequence of the β-subunit. *Biochim. Biophys.* Acta, 412:70.

Shome, B., and A.F. Parlow. 1973. The primary structure of the hormone specific, beta subunit of human pituitary luteinizing hormone (hLH). *J. Clin. Endocrinol. Metab.* 36:618.

Stevens, V.C. 1975. Antifertility effects from immunization with intact subunits, and fragments of hCG. In *Physiological Effects of Immunity Against Reproductive Hormones.* R.G. Edwards and M.H. Johnston, eds. Cambridge: Cambridge University Press, p. 249.

Stevens, V.C. 1976. Actions of antisera to hCG-β: *in vitro* and *in vivo* assessment. In *Proceedings of the V International Congress of Endocrinology.* V.H.T. James, ed. Amsterdam: Excerpta Medica, pp 379-385.

Stevens, V.C. 1981. Current status of research for development of a hCG antifertility vaccine. In *Immunological Factors in Human Reproduction.* Serono Symposium, March. In press.

Thanavala, Y.M., F.C. Hay, and V.C. Stevens. 1977. Immunological control of fertility: measurement of affinity of antibodies to human chorionic gonadotropin. *Clin. Exp. Immunol.* 33:403.

Thanavala, Y.M., F.C. Hay, and V.C. Stevens. 1980. Affinity, crossreactivity and biological effectiveness of rabbit antibodies against a synthetic 37 amino acid C-terminal peptide of human chorionic gonadotropin. *Clin. Exp. Immunol.* 39:112.

Yoshimoto, Y., A.R. Wolfsen, and W.D. Odell. 1977. Human chorionic gonadotropin-like substance in nonendocrine tissues of normal subjects. *Science.* 197:575.

Yoshimoto, Y., A.R. Wolfsen, F. Hirose, and W.D. Odell. 1979. Human chorionic gonadotropin-like material: presence in normal human tissues. *Am. J. Obstet. Gynecol.* 134:729.

Chapter 22

DEVELOPMENT OF A CONTRACEPTIVE VACCINE BASED ON SYNTHETIC ANTIGENIC DETERMINANTS OF LACTATE DEHYDROGENASE-C$_4$

ERWIN GOLDBERG
THOMAS E. WHEAT
VICTORIA GONZALES-PREVATT

While the antigenicity of spermatozoa and testes has been known since the early 1900's, precise definition of the immune response to male reproductive tissue requires the isolation and characterization of specific antigens. A well-defined sperm antigen could also be useful in the development of an immunological contraceptive technology that would be applicable in both sexes. Indeed, the fertility in females of several species is reduced following immunization with sperm or testis extracts. In addition, circulating antisperm antibodies have been reported in association with human idiopathic infertility. These reports are consistent with the development of a contraceptive vaccine. In order to maintain control of the immune response and to minimize side effects, a well-defined antigen is required. These considerations have stimu-lated a series of investigations on the testis-specific isozyme lactate dehydrogenase-C_4 (LDH-C_4). Considerable progress has been made in under-standing the structure, localization, and antigenic properties of this important sperm enzyme.

All vertebrate tissues contain lactate dehydrogenase (LDH). This tetrameric enzyme occurs in multiple molecular forms, or isozymes, randomly assembled from two types of subunits, A and B. The tissue-specific pattern of five isozymes results from the differential regulation of the Ldh-a and Ldh-b genes (Markert, 1963; Markert et al., 1975). A sixth isozyme, LDH-X or LDH-C_4, was discovered by electrophoretic analysis of extracts of human spermatozoa and testes (Figure 22-1)(Goldberg, 1963; Blanco and Zinkham, 1963). It represents the most abundant LDH isozyme in spermatozoa and is readily detectable in mature testes displaying active spermatogenesis. The synthesis of this isozyme ceases when spermatogenesis is interrupted experi-mentally, pathologically, or seasonally.

The molecular basis of this sixth LDH isozyme has now been well established. It has been purified to crystalline homogeneity and biochemically characterized (Goldberg, 1972). It differs from the other LDH isozymes in amino acid composition and in its catalytic properties (Goldberg, 1972; Wheat and Goldberg, 1975). The three-dimensional structure of this isozyme has been determined by x-ray crystallography (Musick and Rossmann, 1979a, 1979b), and some amino acid sequence data have been presented (Wheat et al., 1977a; Pan et al., 1980). These data demonstrate that LDH-X is a homotetramer of a unique subunit type, C, and that this LDH-C_4 iso-zyme is homologous to the other isozymes of LDH. Genetic analyses of allelic variants at the Ldh-c locus are consistent with this hypothesis (Blanco et al., 1964; Wheat and Goldberg, 1977a).

There is a remarkable spatial and temporal regulation of the Ldh-c locus. Cytochemical and immunofluorescent techniques enabled us (Hintz

493

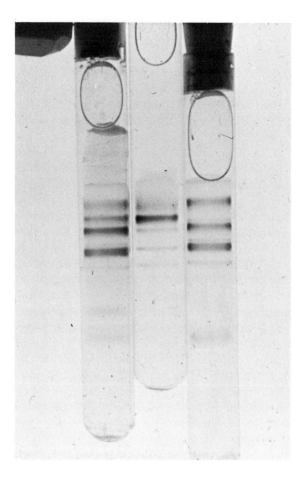

FIGURE 22-1. LDH isozyme pattern of human testes and spermatozoa.
Each extract was resolved by electrophoresis on polyacrylamide disc gels which were then incubated in reaction mixture to localize enzyme activity (Goldberg, 1963). The gel in the center shows a prominent zone of LDH-C_4 activity in sperm, which is also present in mature testes (left gel) but not in prepubertal testes (right gel).

and Goldberg, 1977) to demonstrate the presence of LDH-C_4 in spermatocytes, spermatids, and spermatozoa. The distribution of specific immunofluorescence reveals that the enzyme is absent from the nongerminal elements of the testis, such as interstitial cells and Sertoli cells, as well as from spermatogonia. In addition, the initial appearance of LDH-C_4 has been localized precisely to the midpachytene primary spermatocyte during prophase of the first meiotic division (Wheat et al., 1977a). This is readily apparent in a cross-section of an

494

immature mouse testis, in which only tubules containing midpachytene spermatocytes show the specific immunofluorescence due to the presence of LDH-C₄. Gonocytes, spermatogonia, preleptotene, and leptotene spermatocytes do not contain detectable levels of LDH-C₄. From these data, it may be concluded that the Ldh-c gene is first activated during midpachytene. Also, there is increased fluorescence in the apical germinal elements, suggesting that synthesis of the C subunit continues throughout spermatogenesis. This has been confirmed by direct measurement of LDH-C₄ biosynthesis in spermatids (Meistrich et al., 1977). Furthermore, A and B subunit synthesis in spermatocytes must be reduced or repressed at the time C subunit synthesis commences, as evidenced by the lack of heterotetramer formation. LDH-C₄ is not found in prepubertal testes or in any other tissue of the male and is also completely absent from the female (Lerum and Goldberg, 1974; Hintz, 1975). Thus, LDH-C₄ is antigenic in both the male and the female of both homologous and heterologous species. Rabbit antiserum to mouse LDH-C₄ reacts with this isozyme from any mammalian species, but does not crossreact with any isozymes composed of LDH-A and -B subunits (Goldberg, 1971). The availability of gram quantities of mouse LDH-C₄ (Lee et al., 1977; Wheat and Goldberg, 1977b) facilitates rigorous analysis of the effect of immunization with LDH-C₄ on fertility.

Female mice, rabbits, and baboons were immunized with LDH-C₄ and then mated, as summarized in Table 22-1 (Goldberg, 1973, 1974; Lerum and Goldberg, 1974; Goldberg et al., 1981). This treatment significantly reduces fertility in each case. The extent of this fertility reduction is approximately the same as in mice immunized with sperm or sperm fractions (McLaren, 1964; Bell and McLaren, 1970; Tung et al., 1979) and in rabbits injected with semen or testis extracts (Menge, 1968).

The results obtained with female baboons (Goldberg et al., 1981) underscore the potential utility of LDH-C₄ in a contraceptive vaccine. The fertility of these animals was reduced, and this effect was directly related to the magnitude and the duration of the immune response. Conceptions which did occur during the experiment yielded normal fetuses and the immunocontra-

TABLE 22-1
Fertility Reduction in Females After
Immunization with LDH-C₄

		% PREGNANT	
ANIMAL	NUMBER OF MATINGS	CONTROL	IMMUNIZED
Mouse	50	96	60
Rabbit	10	90	32
Baboon	30	80	27

ceptive effect was reversible. Thus, this treatment apparently does not impair embryonic development or damage the female reproductive system.

It seems likely that the immunocontraceptive effect is mediated by the blocking of fertilization in the fallopian tube, as illustrated in Figure 22-2. Thus, serum immunoglobulins enter the oviduct and prevent sperm from reaching the egg by agglutination or by complement-dependent cell lysis. Oviductal fluids of rabbits immunized with LDH-C$_4$ showed antibody levels directly related to serum titers (Kille and Goldberg, 1979). In addition, sperm transport through the oviduct was impaired in LDH-C$_4$ immunized rabbits compared with nonimmunized controls (Kille and Goldberg, 1980). This is consistent with observations that anti-LDH-C$_4$ agglutinates mouse sperm and enhances complement-mediated cytotoxicity *in vitro*. Thus, manipulation of antigen delivery to produce sustained high titers of circulating antibody should more completely suppress fertility at a level practical for widespread use. Further, the effects of local or secretory immunoglobulin production and cell-mediated immunity have not been fully exploited in the immunosuppression of fertility by LDH-C$_4$.

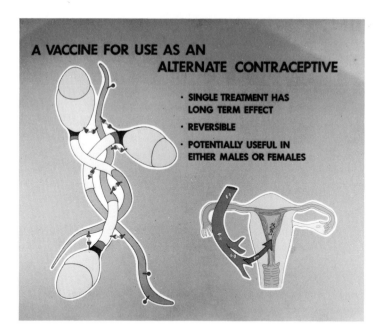

FIGURE 22-2. A model depicting the potential mechanism of action of a contraceptive vaccine.
Spermatozoa are agglutinated by interaction of specific antibodies with surface-associated antigens (viz., LDH-C$_4$) and prevented from reaching and fertilizing the ovum. See text for details.

Exploitation of this phenomenon in a widely used contraceptive vaccine is inhibited by the use of a natural product antigen. Problems of both supply and homogeneity would be considerable. Therefore, an alternative strategy has been developed in which one or more small peptides amenable to chemical synthesis are used to provoke antibodies directed against LDH-C$_4$. Essentially, this involves chemical or enzymatic digestion of LDH-C$_4$ into peptide fragments; isolation and chemical characterization of those peptides which bind anti-LDH-C$_4$; synthesis of appropriate peptides, which are then conjugated to carrier molecules for immunization; tests of antipeptide antibodies for binding to LDH-C$_4$; and finally, assessment of immunocontraceptive properties of the synthetic determinants.

Antibody binding by peptides in various digests of LDH-C$_4$ is shown in Table 22-2. The unresolved mixture of tryptic peptides clearly retains a large fraction of the antibody-binding activity of the native enzyme. When the tryptic digestion is limited to arginyl residues by reacting lysine with citraconic anhydride (Wheat et al., 1977b), the smaller number of peptides is still more active, suggesting that cleavage at lysine disrupts some antigenic determinants. When the same digest is tested after removal of the citraconic acid blocking groups, there is an increase in antibody-binding activity. In sharp contrast, the unfractionated mixture of chymotryptic peptides binds little or no antibody. Thus, the ability of peptide fragments to bind anti-LDH-C$_4$ is established, as is the capacity of this assay to distinguish active and inactive peptides. The tryptic peptides were selected for further analysis since these active peptides are likely to be of a size ultimately appropriate for chemical synthesis.

Peptides were isolated from a tryptic digest of reduced, carboxymethylated LDH-C$_4$ by chromatography on a column of PA-35 resin, a strong cation

TABLE 22-2
Solid Matrix Radioimmunoassay of Antibodies to
LDH-C$_4$ and its Peptide Fragments

	INCUBATION MEDIUM	
IMMOBILIZED ANTIGEN	AGG*	CGG†
Native enzyme (10^{-6} M)	8200‡	466
Tryptic peptides (10^{-4} M)	4530	306
Chymotrypic peptides (10^{-4} M)	534	240
Limited tryptic peptides (10^{-4} M)	6458	364
Limited tryptic peptides (Blocking groups removed, 10^{-4}M)	7130	276

* Rabbit antibodies to LDH-C$_4$ purified by ammonium sulfate fractionation; Equivalence Point (enzyme inhibition) 1.4 U/ml; 0.39 mg protein/ml.

† Rabbit control antibodies purified by ammonium sulfate fractionation; no enzyme inhibition; 0.48 mg protein/ml.

‡ Counts per minute: ^{125}I-labeled second antibody (Goat-anti-rabbit gamma globulin).

exchanger. One fraction contained a single peptide with substantial antibody-binding activity and in sufficient yield for further analysis. Improved resolution of peptides was accomplished by reverse-phase, high-pressure liquid chromatography on C-18 silica. Ten fractions contained antibody-binding activity, and nine of these each contained a single peptide fragment (Table 22-3). In addition, six pure peptides with little or no antibody-binding capacity could be clearly distinguished from the active fractions (Table 22-3). Amino acid composition and sequence data allowed assignment of the active peptides to their respective positions in the C subunit (Table 22-3). The amino acid sequences are presented in Table 22-4 and the position of each peptide in the three-dimensional structure of the C subunit is shown diagrammatically in Figure 22-3.

The antibody-binding peptide that was isolated on the PA-35 column is designated $MC_{152-159}$. It forms a loop at the surface of the molecule connecting helix $\alpha 1F$ and sheet βF (Musick and Rossmann, 1979a), as illustrated by the model in Figure 22-4. The properties of $MC_{152-159}$ fit well with those expected of an antigenic determinant peptide, as derived from studies of other proteins (Atassi, 1979a, 1979b). It is at least large enough and does lie on the surface of the molecule. It carries amino acid substitutions, including changes involving charged side chains (Table 22-4). It forms a loop connecting two ordered secondary structures. Most important, it does bind antibody directed against the native LDH-C4 molecule, and it can provoke antibody that will bind to native LDH-C4, as discussed below.

Peptide MC_{5-16} contains a 12-residue sequence of amino acids that

TABLE 22-3
Antibody Binding: Tryptic Peptides of
Mouse LDH-C4

HPLC FRACTION NUMBER	ANTIBODY BINDING*	POSITION
10	1110	Mixture
11	1706	211 → 220
12	1354	Unknown
16	1140	Unknown
17	1478	Unknown
18	1360	5 → 16
20	960	44 → 58
24	918	61 → 77
25	1032	180 → 210
26	1811	282 → 317
Control peptides†	284	

* Antibody binding = (mean counts per 30 sec with anti-LDH-C4 gamma globulins) – (mean counts per 30 sec with control gamma globulins); 5 nmoles of peptide per well in solid matrix radioimmunoassay.

† Control peptides = average of 5 LDH-C4 tryptic peptides which are larger than 5 residues and were obtained in yields comparable to the antigenic peptides.

includes most of the intersubunit arm. This peptide is entirely on the surface of the tetramer and contains many more amino acid substitutions than the other LDH isozymes (Table 22-4). Extreme chemical variability and exposed position are quite consistent with expected properties of antigenic determinants.

Most of MC_{44-58} is internal with residues 45 to 54 in a β-sheet and 55 to 58 beginning the α-helix C (Figure 22-3). The sequence (Table 22-4) shows that this peptide is largely invariant except at the C-terminal end, where residues 54 to 56 form a bend or loop at the surface of the subunit near the edge of the coenzyme-binding site (Figure 22-3). Significant substitutions may occur at positions 55 (Asp or Asn for Met or Leu) and 56 (Thr for Glu) (Pan et al., 1980).

The peptide comprising residues MC_{61-77} has been partially sequenced (Table 22-4) and makes up most of α-helix C (Figure 22-3). Although the helix and loop are buried in a subunit contact region, this peptide, when isolated as a linear sequence, binds substantial anti-LDH-C₄ antibody (Table 22-3). The ten residue peptide $MC_{211-220}$ is relatively conservative (Table 22-4). However, a reduction in polarity and net charge would be expected from the substitutions at positions 215 and 216, which form a loop on the surface of the tetramer (Figure 22-3).

SCHEMATIC VIEW OF ONE LDH SUBUNIT

LOOKING ALONG THE MOLECULAR Q AXIS
TOWARDS THE CENTER OF THE MOLECULE
(from Adams etal (1972))

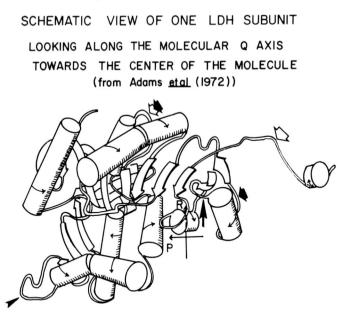

FIGURE 22-3. Schematic drawing of an LDH subunit, looking along the molecular Q axis toward the center of the molecule.
Peptides which bind anti-LDH-C₄ correspond to the sequences in Table 22-4. From Adams et al., 1972; reprinted with permission.

While each of the peptides is likely to contain an antigenic determinant of LDH-C$_4$, the most stringent requirement of this strategy is that a putative determinant provoke antibody which will react with the native protein. Peptide MC$_{152-159}$ was chemically synthesized and was conjugated to bovine serum albumin in a stoichiometric ratio ranging from 15 to 22 moles of peptide per mole BSA (Gonzales-Prevatt et al., in preparation). Rabbits immunized with this conjugate produced antibody which bound to the peptide. This binding was inhibited by free peptide and by the BSA-peptide conjugate. These antibodies to MC$_{152-159}$ were purified and tested for binding to ^{125}I-LDH-C$_4$. The results, in Table 22-5, show that purified anti-MC$_{152-159}$does indeed bind to LDH-C$_4$. Furthermore, the specificity of this binding is confirmed by competition of unlabeled LDH-C$_4$ for both antibodies (Table 22-6). As expected, only

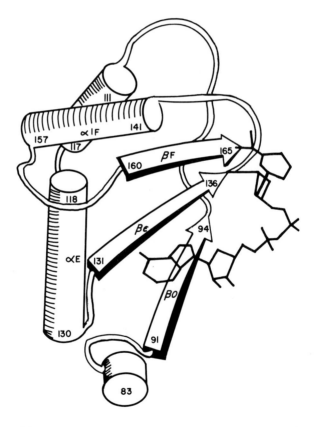

Figure 22-4. Diagrammatic representation of a portion of the LDH subunit containing MC$_{152-159}$.
From Holbrook et al., 1975; reprinted with permission.

TABLE 22-4
Antibody Binding Peptides of Mouse LDH-C$_4$

SPECIES	RESIDUES 5→16 (HPLC #18)
MC	GLU-GLN-LEU-ILE-GLN-ASN-LEU-VAL-PRO-GLU-ASP-LYS
DA	ASP-LYS-LEU-ILE-GLY-HIS-LEU-ALA-THR-SER-GLN-GLU
PA	ASP-GLN-LEU-ILE-HIS-ASN-LEU-LEU-LYS-GLU-GLU-HIS
CA	ASP-HIS-LEU-ILE-HIS-ASN-VAL-HIS-LYS-GLU-GLU-HIS
PB	GLU-LYS-LEU-ILE-ALA-PRO-VAL-ALA-GLN-GLN-GLU-THR
CB	GLU-LYS-LEU-ILE-THR-PRO-VAL-ALA-ALA-GLY-SER-THR

	RESIDUES 44→58 (HPLC #20)
MC	GLY-LEU-ALA-ASP-GLU-LEU-ALA-LEU-VAL-ASP-ALA
DA	ASP-LEU-ALA-ASP-GLU-VAL-ALA-LEU-VAL-ASP-VAL-MET-GLU-ASP-LYS
PA	GLU-LEU-ALA-ASP-GLU-ILE-ALA-LEU-VAL-ASP-VAL-MET-GLU-ASP-LYS
CA	ASX-LEU-ALA-ASX-GLX-LEU-THR-LEU-VAL-ASX-VAL-VAL-GLX-ASX-LYS
PB	SER-LEU-THR-ASP-GLU-LEU-ALA-LEU-VAL-ASP-VAL-LEU-GLU-ASP-LYS
CB	GLY-LEU-CYS-ASP-GLU-LEU-ALA-LEU-VAL-ASP-VAL-LEU-GLU-ASP-LYS

	RESIDUES 61→77 (HPLC #24)
MC	GLY-GLU-ALA-LEU-ASP-LEU
DA	GLY-GLU-MET-MET-ASP-LEU-GLN-HIS-GLY-SER-LEU-PHE-LEU-HIS-THR-ALA-LYS
PA	GLY-GLU-MET-MET-ASP-LEU-GLN-HIS-GLY-SER-LEU-PHE-LEU-ARG-THR-PRO-LYS
CA	GLY-GLU-MET-MET-ASP-LEU-GLN-HIS-GLY-SER-LEU-PHE-LEU-LYS-THR-PRO-LYS
PB	GLY-GLU-MET-MET-ASP-LEU-GLN-HIS-GLY-SER-LEU-PHE-LEU-GLN-THR-PRO-LYS
CB	GLY-GLU-MET-MET-ASP-LEU-GLN-HIS-GLY-SER-LEU-PHE-LEU-GLN-THR-PRO-LYS

	RESIDUES 152→159 (PA 35)
MC	ILE-SER-GLY-PHE-PRO-VAL-GLY-ARG
DA	LEU-SER-GLY-LEU-PRO-MET-HIS-ARG
PA	ILE-SER-GLY-PHE-PRO-LYS-ASN-ARG
CA	ILE-SER-GLY-PHE-PRO-GLY-HIS-ARG
PB	LEU-SER-GLY-LEU-PRO-LYS-HIS-ARG
CB	LEU-SER-GLY-LEU-PRO-LYS-HIS-ARG

	RESIDUES 211→220 (HPLC #11)
MC	SER-LEU-ASN-PRO-ALA-ILE-GLY-THR-ASP-LYS
DA	GLU-LEU-HIS-PRO-GLU-LEU-GLY-THR-ASN-LYS
PA	ASN-LEU-HIS-PRO-GLU-LEU-GLY-THR-ASP-ALA
CA	ASN-LEU-HIS-PRO-ASP-MET-GLY-THR-ASX-ALA
PB	GLN-LEU-ASN-PRO-GLU-MET-GLY-THR-ASP-ASN
CB	GLN-LEU-ASN-PRO-ALA-MET-GLY-THR-ASP-LYS

TABLE 22-5
Binding of ^{125}I-LDH-C_4 to Antibodies Elicited by Native
LDH-C_4 and by BSA-MC$_{152-159}$

AMOUNT ANTIBODY (µg/ml)		% ^{125}I-LDH-C_4 BOUND	
ANTI-LDH-C_4	ANTI-MC$_{152-159}$	ANTI-LDH-C_4	ANTI-MC$_{152-159}$
0.01	0.01	6.7	4.6
0.02	0.03	14.8	14.9
0.05	0.06	26.8	25.2
0.10	0.16	48.2	48.8
0.19	0.32	71.0	63.4
0.39	0.63	84.7	69.7
0.78	1.26	88.5	68.2
1.56	1.89	88.9	68.4

a small portion of the antibodies elicited by the peptide-BSA conjugate bind to LDH-C_4 (Gonzales-Prevatt et al., in preparation). From these data, we can conclude that peptide MC$_{152-159}$ contains an antigenic determinant of LDH-C_4. Thus, our strategy for development of synthetic determinants which provoke an immune response to native LDH-C_4 has proven to be sound. In order to identify those determinants which elict the most useful immune response, similar studies of the other peptides in Table 22-4 are in progress. Additional peptides are being isolated from alternative digestion procedures. Ultimately, one or more of these synthetic determinants could form the basis of a contraceptive vaccine where the antigen may be synthesized in large quantities. This, in turn, should allow the studies necessary to establish with certainty whether immunocontraception is a viable and effective alternative technology for fertility control.

TABLE 22-6
Competition of Unlabeled LDH-C_4 with ^{125}I-LDH-C_4
for Anti-LDH-C_4 and Anti-MC$_{152-159}$

AMOUNT LDH-C_4 ADDED (pmol/ml)	% ^{125}I-LDH-C_4 BOUND	
	ANTI-LDH-C_4	ANTI-MC$_{152-159}$
0	62.3	48.0
0.05	44.2	28.5
0.10	34.7	22.1
0.20	26.3	12.2
0.40	17.2	5.5
0.80	11.4	2.2
1.60	7.6	0.4
3.20	4.7	

ACKNOWLEDGEMENTS

Gratitude is expressed to the many students and technologists who spent time in this laboratory during the past decade and who provided both research expertise and intellectual stimulation. Various aspects of the research done in our laboratory were supported by grants from the National Institutes of Health (HD05863; ST32 HD07068; 1F32 HD05973) and by contracts with the World Health Organization and the Northwestern University Program for Applied Research in Fertility Regulation.

REFERENCES

Adams, M.J., M. Buehner, K. Chandrasekhar, G.C. Ford, M.L. Hackert, A. Liljas, P. Lentz, S.T. Rao, M.G. Rossman, I.E. Smiley, and J.L. White. 1972. Subunit interaction in lactate dehydrogenase. In *Protein-Protein Interactions*. R. Jaenicke and E. Helmreich, eds. Berlin and New York: Springer-Verlag, pp. 139

Atassi, M.Z. 1979a. The antigenic structure of myoglobin and initial consequences of its precise determination. *CRC Critical Reviews in Biochemistry* 6:337.

Atassi, M.Z. 1979b. Determination of the entire antigenic structure of native lysozyme by surface-simulation synthesis, a novel concept in molecular recognition. *CRC Critical Reviews in Biochemistry* 6:371.

Bell, E.B., and A. McLaren. 1970. Reduction of fertility in female mice isoimmunized with a subcellular sperm fraction. *J. Reprod. Fertil.* 22:356.

Blanco, A., and W.H. Zinkham. 1963. Lactate dehydrogenase in human testes. *Science* 139:601.

Blanco, A., W.H. Zinkham, and L. Kupchyk. 1964. Genetic control and ontogeny of lactate dehydrogenases in pigeon testes. *J. Exp. Zool.* 156:137.

Eventoff, W., M.G. Rossman, S.S. Taylor, H.J. Torff, H. Meyer, W. Keil, and H.H. Kiltz. 1977. Structural adaptations of lactate dehydrogenase isozymes. *Proc. Nat. Acad. Sci. USA* 74:2677.

Goldberg, E. 1963. Lactic and malic dehydrogenases in human spermatozoa. *Science* 139:602.

Goldberg, E. 1971. Immunochemical specificity of lactate dehydrogenase-X. *Proc. Nat. Acad. Sci. USA* 68:349.

Goldberg, E. 1972. Amino acid composition and properties of crystalline lactate dehydrogenase-X from mouse testes. *J. Biol. Chem.* 247:2044.

Goldberg, E. 1973. Infertility in female rabbits immunized with lactate dehydrogenase-X. *Science* 181:458.

Goldberg, E. 1974. Effects of immunization with LDH-X on fertility. *Acta Endocr.* 78:202.

Goldberg, E., T.E. Wheat, J.E. Powell, and V.C. Stevens. 1981. Reduction of fertility in female baboons immunized with lactate dehydrogenase-C₄. *Fertil. Steril.* 35:214.

Hintz. M. 1975. Immunohistochemical and immunocytochemical localization of LDH isozymes in the mouse testis. Ph.D Dissertation. Evanston, Illinois: Northwestern University.

Holbrook, J.J., A. Liljas, S.J. Steindel, and M.G. Rossman, 1975. Lactate dehydrogenase. In *The Enzymes. XI.* P. Boyer, ed. New York: Academic Press, pp. 191- .

Hintz, M., and E. Goldberg. 1977. Immunohistochemical localization of LDH-X during spermatogenesis in mouse testes. *Dev. Biol.* 57:375.

Kille, J.W., and E. Goldberg. 1979. Female reproductive tract immunoglobulin responses to a purified sperm-specific antigen (LDH-C₄). *Biol. Reprod.* 20:863.

Kille, J.W., and E. Goldberg. 1980. Inhibition of oviductal sperm transport in rabbits immunized against sperm-specific lactate dehydrogenase (LDH-C₄). *J. Reprod. Immunol.* 2:15.

Lee, C.Y., B. Pegoraro, J. Topping, and J.H. Yuan. 1977. Purification and partial characterization of lactate dehydrogenase-X from mouse. *Mol. Cell. Biochem.* 18:49.

Lerum, J.E., and E. Goldberg. 1974. Immunological impairment of pregnancy in mice by lactate dehydrogenase-X. *Biol. Reprod.* 11:108.

Markert, C.L. 1963. Lactate dehydrogenase isozymes: Dissociation and recombination of subunits. *Science* 140:1329.

Markert, C.L., J.B. Shaklee, and G.S. Whitt. 1975. Evolution of a gene. *Science* 189:102.

McLaren, A. 1964. Immunological control of fertility in female mice. *Nature* 201:582.

Meistrich, M.L., P.K. Trostle, M. Frapart, and R.P. Erickson. 1977. Biosynthesis and localization of lactate dehydrogenase-X in pachytene spermatocytes and spermatids of mouse testes. *Dev. Biol.* 60:428.

Menge, A.C. 1968. Fertilization, embryo and fetal survival rates in rabbits isoimmunized with semen, testis and conceptus. *Proc. Soc. Exp. Biol. Med.* 127:1271.

Musick, W.D.L., and M.G. Rossmann. 1979a. The structure of mouse testicular lactate dehydrogenase isoenzyme C₄ at 2.9 A resolution. *J. Biol. Chem.* 254:7611.

Musick, W.D.L., and M.G. Rossmann. 1979b. The tentative amino acid sequencing of lactate dehydrogenase C₄ by x-ray diffraction analysis. *J. Biol. Chem.* 254:7621.

Pan, Y.C., S. Huang, J.P. Marciniszyn, C.Y. Lee, and S.S.L. Li. 1980. The preliminary amino acid sequence of mouse testicular lactate dehydrogenase. *Hoppe-Seyler's Z. Physiol. Chem.* 361:795.

Tung, K.S.K., E.H. Goldberg, and E. Goldberg. 1979. Immunobiological consequence of immunization of female mice with homologous spermatozoa: induction of infertility. *J. Reprod. Immunol.* 1:145.

Wheat, T.E., and Goldberg, E. 1975. LDH-X: The sperm-specific C₄ isozyme of lactate dehydrogenase. In *Isozymes, Vol. III: Developmental Biology.* C.L. Markert, ed. New York: Academic Press, pp. 325-346.

Wheat, T.E., and E. Goldberg. 1977a. An allelic variant of the sperm-specific lactate dehydrogenase c₄ (LDH-X) isozyme in humans, *J. Exp. Zool.* 202:425.

Wheat, T.E., and E. Goldberg. 1977b. Isolation of the sperm-specific lactate dehydrogenase from mouse, rabbit and human testis and human spermatozoa. In *Immunological Influence on Human Fertility.* B. Boettcher, ed. Sydney, Australia: Harcourt, Brace, Jovanovich, pp. 221-227.

Wheat, T.E., M. Hintz, E. Goldberg, and E. Margoliash. 1977a. Analyses of stage-specific multiple forms of lactate dehydrogenase and of cytochrome *c* during spermatogenesis in the mouse. *Differentation* 9:37-41.

Wheat, T.E., E. Goldberg, and E. Margoliash. 1977b. Primary structure of the essential thiol peptide from the lactate dehydrogenase C subunit. *Biochem. Biophys. Res. Commun.* 74:1066:1070.

Chapter 23

ANTIBODIES TO ZONA PELLUCIDA ANTIGENS AND THEIR ROLE IN FERTILITY

BONNIE S. DUNBAR

Interest in the biochemistry and immunology of the zona pellucida stems from the potential contraceptive use of antibodies to zona antigens. Immunization with zona pellucida antigens has distinct advantages over other proposed immunological contraceptive methods for the following reasons:

1. It is not abortive but instead inhibits fertilization.
2. Only low titers of antibodies are needed to block fertilization.
3. The zona antigens studied to date are tissue specific.
4. The zona of a variety of animal species are immunologically crossreactive.

Also, antigenic material is readily available for purification and characterization (Shivers, 1974, 1975, 1977; Dunbar and Shivers, 1976; Sacco and Shivers, 1978).

Some infertile women have been shown to have antibodies to zona antigens, and these antibodies have therefore been implicated as a cause of infertility (Shivers and Dunbar, 1977; Mori et al., 1978, 1979).

Before we can critically evaluate the use of antibodies to zona pellucida antigens for immunological contraception or evaluate the clinical implications of antibodies to zona antigens in infertility, it is imperative that the biochemical and immunochemical nature of the zona be better understood.

BIOLOGICAL ROLES OF THE ZONA PELLUCIDA

The zona pellucida is the extracellular structure surrounding the mammalian oocyte. Spermatozoa must first bind to and penetrate it before coming into contact with the oocyte itself (Hartmann and Gwatkin, 1971; Hartmann et al., 1972; Hartmann and Hutchison, 1974; Saling et al., 1978). The sperm presumably binds to the zona via specific sperm "receptor" sites (Gwatkin and Williams, 1976a, 1978), which are thought to play an important role in the species specificity of sperm-egg interaction (Austin, 1961; Hanada and Chang, 1972; Bedford, 1977; Yanagimachi, 1977; Petersen et al., 1980; Schmell and Gulyas, 1980; Swenson and Dunbar, 1981). The sperm receptor sites of hamster zona pellucida are presumably protein since they are sensitive to trypsin, chymotrypsin, and trypsin-like acrosin preparations from hamster, ram, and boar spermatozoa (Gwatkin, 1977; Gwatkin et al., 1977), whereas the sperm receptors of rabbit (Overstreet and Bedford, 1975) or mouse zona (Aitken and Richardson, 1979) are not destroyed by such treatment. Oikawa et al. (1975) have shown, however, that treatment with these enzymes causes profound modifications in terminal oligosaccharide residues of zona glycopeptides. It still remains to be determined whether species differences in the specificity of sperm-zona interaction are due to variation in sperm membrane composition, to zona composition, or to both.

After binding, the sperm must penetrate the zona by limited proteolytic digestion by sperm enzymes (Srivastava et al., 1965; reviews by McRorie and Williams, 1974, and Stambaugh, 1978). The zona remains intact after fertilization, to ensure proper embryonic development and perhaps to prevent embryo fusion in the oviduct (Mintz, 1962). The zona is also thought to be responsible for movement of the embryo in the oviduct (Modlinski, 1970).

Finally, the zona pellucida plays a role in the block to polyspermy. In some mammalian species, fertilization alters sperm binding to the zona and its resistance to proteolytic digestion (Austin and Braden, 1956; Barros and Yanagimachi, 1971, 1972; Gwatkin et al., 1973), changes referred to as the zona reaction (Braden et al., 1954). These are thought to be induced by the release of components from cortical granules after sperm-oocyte plasma membrane fusion (Austin and Braden, 1956; Barros and Yanagimachi, 1972; Gwatkin et al., 1973; review by Yanagimachi, 1977). Again, species variations have been observed in the zona reaction. Rabbit zona, unlike mouse and rat zona (Yanagimachi, 1977), remains penetrable by spermatozoa long after fertilization (Austin, 1961), even though the zona is more resistant to proteolytic enzyme digestion (Overstreet and Bedford, 1974). Chemical changes in rabbit zona proteins after fertilization have also been reported (Gould et al., 1971).

PHYSICAL PROPERTIES OF THE ZONA PELLUCIDA

Although the functions of egg envelopes of different species are analogous, their size relative to the oocyte varies dramatically (Austin, 1961; Monroy, 1965). Figure 23-1 illustrates the dramatic differences in size of the zona pellucida from a mature mouse oocyte and those of rabbit and pig follicular oocytes in the same microscopic field.

Zonae of a variety of mammalian oocytes have similar external surfaces when examined by scanning electron microscopy (Zamboni, 1971; Dudkiewicz et al., 1976; Gwatkin et al., 1976; Jackowski and Dumont, 1979; Phillips and Shalgi, 1980). The outer surface of the zona appears to be fenestrated, and the inner surface has a smoother, less distinct appearance (Phillips and Shalgi, 1980; Phillips and Dunbar, unpublished observations).

The zona is permeable to a variety of molecules, including IgG and IgM (molecular weight, 150 000 to 900 000) and ferritin (molecular weight,

400 000), although it is impermeable to heparin (molecular weight, 16 000), as observed by Sellens and Jenkinson (1975). These investigators suggested that the ability of molecules to pass through the zona does not depend on the molecular weight so much as on the configuration and/or charge properties of the molecule.

BIOCHEMICAL PROPERTIES OF THE ZONA PELLUCIDA

Early studies of the physicochemical properties and the immunogenicity of zonae pellucida required methods that could be adapted to the analysis of small amounts of material, since few mammalian oocytes could readily be obtained.

Soupart and Noyes (1964) used cytochemical methods to detect sialic acid as a component of the rabbit zona pellucida. Similar studies showed that the zona pellucida was a complex of sulfated acid and neutral mucopolysaccharide and protein in the form of glycopeptide units stabilized by disulfide bonds, hydrophobic interactions, or salt bridges (Piko, 1969). Additional evidence for carbohydrate composition of zonae was given by autoradiographic studies, which showed that radiolabeled carbohydrates, such as (N-acetyl-^{3}H)D-glucosamine, are incorporated into newly synthesized mouse zonae (Oakberg and Tyrell, 1975).

The binding distribution of lectins to rodent eggs has been used to show an asymmetric distribution of sugars in rodent zonae (Oikawa et al., 1973; Nicholson et al., 1975). Similar studies have shown that there is also asymmetric lectin binding to porcine zonae, but that binding of these lectins to rabbit zonae is different (Dunbar, 1980; Dunbar et al., 1980). Although it

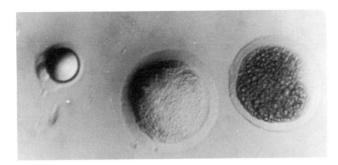

FIGURE 23-1. Size comparison of three mammalian eggs and zonae pellucidae.
Ovulated mouse ova recovered from the oviduct (left); follicular rabbit oocyte (center); follicular porcine oocyte (right).

is possible that some species differences may in part be due to the methods used to determine the binding (e.g., immunofluorescence versus electron microscopy), these studies collectively imply that there are species differences in carbohydrate distribution in the zonae.

Recently, methods allowing large-scale isolation of porcine zonae for detailed physicochemical and immunochemical characterization have been developed (Dunbar et al., 1978a, 1979b; Oikawa, 1978; Dunbar et al., 1980). These methods have also been used to isolate large numbers of rabbit (Dunbar, 1980; Wood et al., 1981) and bovine (Gwatkin and Williams, 1980) zonae, although they are unsuitable for the isolation of rodent zonae. Our inability to isolate rodent zonae is most likely due to variability in size and the fragility of rodent follicular oocytes (Dunbar, 1980, and unpublished observations; Figure 23-1). The original methods of large-scale oocyte isolation have now been improved upon with an apparatus that allows isolation of as many as 400000 porcine oocytes and zonae in six hours (Wood and Dunbar, 1981). In studies by Dunbar et al. (1980), zona preparations were estimated to be 93 per cent pure by electron microscopic, enzymatic, and chemical criteria. The porcine zona has been shown to be predominantly protein (71 per cent) and carbohydrate (19 per cent) (Dunbar et al., 1980). It contains 33 ng of protein, a figure similar to those reported by Gwatkin (1980). We found that the porcine zona exhibited extreme microheterogeneity on one-dimensional gel electrophoresis (Dunbar et al., 1980), as have other investigators (Sacco and Palm, 1977; Menino and Wright, 1979; Gwatkin et al., 1980).

High-resolution, two-dimensional gel electrophoresis has improved the separation of zona proteins for detailed characterization. Hedrick and Wardrip (1980) have shown that the porcine zona is made up of three major glycoprotein species, characterized by size and charge heterogeneity. Dunbar et al. (1981) used both internal isoelectric focusing and molecular weight standards in two-dimensional electrophoresis to show that the porcine zona comprises three predominant glycoprotein species (molecular weights of 49000 to 119000, 70000 to 101000, and 95000 to 118000), each characterized by charge and size heterogeneity (Dunbar and Sammons, 1980; Dunbar et al., 1981; also, Figure 23-2). Such charge and size heterogeneity is also characteristic of other glycoproteins, including plasma proteins (Anderson and Anderson, 1977). The rabbit zona also contains three species of proteins (molecular weights of 68000 to 125000, 81000 to 100500, and 100000 to 132000), although the charge and size distribution differs from that of porcine zonae (Dunbar et al., 1981). We further compared the two-dimensional electrophoresis patterns of porcine and rabbit zonae with those of serum, follicular fluid, and follicular cells to demonstrate the uniqueness of zona proteins.

Studies by Repin and Akimova (1976) suggested that the mouse zona was composed of five macromolecules. A series of studies of the mouse zona pellucida have shown it to contain 4.8 μg of protein (approximately one seventh as much as the porcine zona), representing 17 per cent of the total oocyte protein (Bleil and Wassarman, 1980a, 1980b, 1980c). Mouse zona was further characterized by one- and two-dimensional gel electrophoresis and was found to be composed of three proteins having molecular weights of 200 000, 120 000, and 83 000.

Variations between zona proteins from different species may be responsible for differences observed in the solubility of zonae. These differences may in part be due to post-translational modification of zona proteins. Wright et al. (1977) have shown that dissolution of bovine and ovine zonae by pronase takes longer (263 minutes, cow; 110 to 143 minutes, ewe) than dissolution of

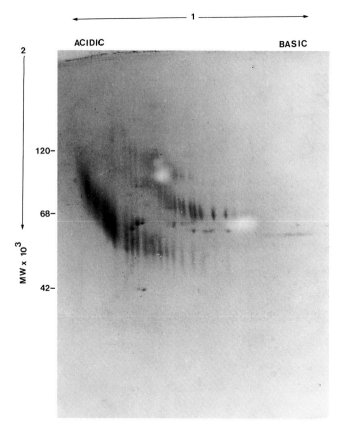

FIGURE 23-2. Two-dimensional gel electrophoresis pattern (Coomassie blue stain) of porcine zonae pellucidae (approximately 6000 zonae). Isoelectric focusing in the first dimension followed by sodium dodecylsulfate gel electrophoresis in the second dimension.

the mouse zona (34 minutes). Furthermore, the mouse zona is soluble at pH 4 (Gwatkin, 1964), whereas porcine zonae are not readily solubilized at low pH (Dunbar et al., 1980).

METHODS OF DETECTION OF ANTIBODIES TO ZONA PELLUCIDA ANTIGENS

Most investigators to date have used microscopic methods to study antibodies to zona pellucida antigens since only limited numbers of isolated zonae have been available. More recently, the methods for isolating large numbers of oocytes from the pig, cow, and rabbit have allowed detailed studies of antibodies to zona antigens (Dunbar, 1980; Dunbar et al., 1980; Gwatkin et al., 1980). Microscopic methods continue to be the most acceptable for detection of antibodies to zona antigens in rodent species (Dunbar, 1980).

Zona Precipitation Reaction
The zona precipitation reaction (ZPR) is a change in light-scattering properties of the zona detected by light- or dark-field microscopy and is illustrated in Figure 23-3. Figure 23-2A is a light micrograph of a porcine oocyte treated with antiserum to heat-solubilized porcine zona (demonstrating the ZPR), compared with an oocyte (Figure 23-2B) treated with control antiserum to Freund's complete adjuvant (FCA). The ZPR has been attributed

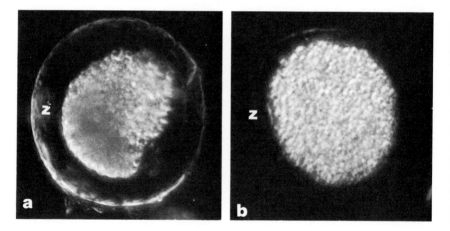

FIGURE 23-3. Zona precipitation reaction.
(a) Porcine oocyte treated with rabbit antisera to heat-solubilized porcine zonae. (b) Porcine oocyte treated with control sera from rabbit immunized with Freund's complete adjuvant. Z, the zona pellucida.

to binding of antibodies to the outer region of the zona pellucida (Ownby and Shivers, 1972; Garavagno et al., 1974; Dudkiewicz et al., 1976; Dunbar and Shivers, 1976; Sacco, 1976; Tsunoda et al., 1979). Dudkiewicz et al. (1976) used scanning electron microscopy to demonstrate changes in the surfaces of pig zonae after treatment with antibodies to pig ovary. Although Garavagno et al. (1974) demonstrated with light microscopy that the precipitate was present only on the outside of the hamster zona, Flechon and Gwatkin (1980) showed, using transmission electron microscopy, that antibody to cow zonae localizes at the internal, as well as the external, surface of zona. The immunochemical phenomena responsible for the surface precipitation reaction are not clearly understood, although a recent report by Ahuja and Tzartos (1981) shows that the ZPR is absent if Fab fragments from rabbit antisera to hamster ovary homogenate are used instead of intact immunoglobulins.

Indirect Immunofluorescence

Indirect immunofluorescence (II) methods have been used routinely to detect antibodies to zona antigens in the sera of animals immunized with ovarian antigens, as well as in the sera of infertile women. These methods generally involve incubation of isolated oocytes in the primary antiserum followed by treatment with a fluorescein-conjugated second antibody.

Although immunofluorscence has been the principal method for detecting antibodies to zona antigens, there are reports of positive binding reactions in sera from men as well as from fertile and infertile women (Mori et al., 1978, 1979; Sacco and Moghiss, 1979; Dakhno et al., 1980). Extensive absorption with tissue homogenates or with red blood cells is therefore required before specific antibodies to zona antigens can be detected.

More recently, it has been shown that a variety of hyperimmune sera to protein antigens frequently give positive immunofluorescence with zonae, despite the lack of any immunological crossreactivity with zona antigens, either by electrophoresis or by the more sensitive radioimmunoassay methods (Gerrity et al., 1981). The immunofluorescence observed with sera of hyperimmunized animals is evenly distributed throughout the zona, in contrast to the rim fluorescence of antisera against intact or heat-solubilized porcine zonae. Positive fluorescence may result from trapping of immunoglobulins (possibly immune complexes) in the lattice-like structure of the zona. Immunization with serum proteins in FCA can induce serum sickness, accompanied by increases in serum immune complexes (Haalenstad and Mannik, 1979). This may be particularly true for antisera prepared against ovarian homogenates, since this tissue contains many serum proteins also present in follicular fluid (Dunbar et al., 1981). Other studies have shown that rabbits with induced serum sickness have immune complexes localized specifically in the zona pellucida (Albini et al., 1979). Mori et al. (1979, 1980) have shown that absorption of sera with red blood cells will remove binding activity as observed by immunofluorescence. Since blood group substances are present in sera (Kabat, 1955) and can be absorbed from circulating fluids onto cell surfaces (Race and Sanger, 1968), these factors may also affect immunofluorescence results.

Positive immunofluorescence may also be the result of nonspecific charge binding, a property that can be reduced by purification of fluorescein-conjugated immunoglobulins by DEAE-affinity chromatography (Kawamura, 1977). Further immunofluorescence studies using Fab fragments and purified fluorescein conjugated immunoglobulins, along with parallel studies with other specific antibody detection methods (e.g., radioimmunoassay), may resolve some of the previous inconsistencies in reports of the presence of antibodies to zona antigens.

Resistance of Zona Dissolution to Proteolytic Enzymes

Light microscopic studies by Shivers et al. (1972) and Garavagno et al. (1977) showed that pretreatment of hamster zonae with antibodies to hamster ovarian tissue blocks zona digestion by trypsin. More recently, this antibody-induced resistance to proteolytic digestion by pronase has been used to demonstrate the presence of antibodies to zona antigens (Tsunoda and Chang, 1978a, 1978b). Since the rodent zonae (mouse, rat, hamster) are easily removed with proteolytic enzymes, this method has been routinely used to detect antibodies to zona antigens in these species. Although these methods are used to detect antibodies to zona antigens, there are species differences in susceptibility to dissolution by proteolytic enzymes (Wright et al., 1977) and there has been no standardization of the type or purity of proteolytic enzyme (e.g., trypsin, pronase). It is difficult, therefore, to directly compare the results of studies in different laboratories using these methods of antibody detection.

Passive Hemagglutination Assay

Tsunoda and Chang (1976c) have used a passive hemagglutination assay as described by Herbert (1977) to detect ovarian antigens. Antibodies to ovary homogenate were reported to be detectable in serum dilutions up to 2^{12}, although exact details of the experimental methods were not given.

Immunodiffusion and Immunoelectrophoresis

Many early studies used antibodies to whole ovarian tissue in immunoprecipitation studies (Ownby and Shivers, 1972; Sacco and Shivers, 1973; Tsunoda and Chang, 1976a, 1976c). Because of the immunogenic complexity of ovarian tissue and the limited sensitivity and resolution of immunodiffusion methods (Ouchterlony and Nilsson, 1978), it is difficult to determine which ovarian antigens were identified in these studies. It is now possible to prepare antisera containing precipitating antibodies to isolated porcine and rabbit zonae (Dunbar et al., 1980; Dunbar and Raynor; 1980) and bovine zonae (Gwatkin et al., 1980). Tsunoda et al. (1980) have used immunoelectrophoretic methods to demonstrate a specific antigen in ovarian homogenate recognized by goat antiserum to isolated bovine zonae. Recently, antibodies have been raised to purified zona proteins isolated from both one-dimensional (Dunbar and Raynor, 1980) and two-dimensional (Wood and Dunbar, 1981) electrophoretic gels. Conventional techniques such as rocket immunoelectrophoresis and crossed immunoelectrophoresis (Weeke, 1973a, 1973b) have been used to better characterize zona antigens (Dunbar and Raynor, 1980; Wood and Dunbar, 1981). These studies clearly demonstrate that there are multiple zona-specific antigens associated with porcine and rabbit zonae pellucida. These studies have also allowed the characterization of specific zona proteins.

Radioimmunoassay Methods
Radioimmunoassay methods (RIA) to detect antibodies to zona antigens have been described by Palm et al. (1979) and Gerrity et al. (1981). These assays are far more sensitive than microscopic methods since antibody can be detected in serum dilutions of up to 5×10^5. Both have been used to detect antibodies specific to zona antigens.

IMMUNOCHEMICAL CHARACTERIZATION OF OVARIAN AND ZONA PELLUCIDA ANTIGENS

The majority of studies to date have used ovarian tissue as a source of immunogenic material (Porter, 1965; Shahini et al., 1970; Ownby and Shivers, 1972; Sacco and Shivers, 1972; Tsunoda and Chang, 1976a, 1976b, 1976c, 1976d; Sacco, 1977). When porcine ovarian proteins are analyzed by high-resolution, two-dimensional gel electrophoresis, there is a predominance of serum proteins (Dunbar et al., 1981). When recently developed silver stain methods are used (Adams and Sammons, 1981), hundreds of other proteins can be identified (Dunbar, unpublished results). Since serum albumin alone contains at least six nonrepeating antigenic determinants (Benjamin and Teale, 1979), it would be expected that vast numbers of antibodies of different specificities will be generated when an animal is exposed to ovarian tissue. Much of the complexity of this system can be simplified by absorption of immune sera with a variety of tissues (Ownby and Shivers, 1972; Sacco and Shivers, 1973; Tsunoda and Chang, 1976a, 1976b, 1976c; Sacco, 1977). Since the operational immunogenicity of one antigen might be quite different when injected with hundreds of others, it is difficult to directly compare the response obtained when whole ovarian tissue is used as immunogen instead of isolated zonae.

The characteristic zona proteins have never been observed when the ovarian tissue is analyzed by two-dimensional gel electrophoresis. This is not surprising since it can be estimated that there would be approximately 16 μg of zona protein per 5 g of ovarian tissue (Dunbar et al., 1980; Gwatkin et al., 1980). Even with sensitive methods of protein detection, the relative amount of zona protein would be 1×10^6-fold lower than total ovarian protein. Given this, significant titers of antibodies to zona antigens would not be expected when ovarian tissue is used as immunogen. Although it has yet to be determined whether the follicular cell or the oocyte, or both, are responsible for secretion of zona proteins (Odor, 1960; Kang, 1974; Oakberg and Tyrell, 1975; Bleil and Wassarman, 1980; Hedrick and Fry, 1980), it is possible that unsecreted zona proteins may be present in cells of ovarian homogenates. As such, they would not have undergone the full post-translational modification which gives rise to the typical microheterogeneity of zona proteins. Antibodies have been raised to isolated mouse and hamster eggs (Tsunoda, 1977; Tsunoda and Chang, 1978) and to human oocytes (Takai et al., 1981). The characterization of the antibodies generated to human oocytes has, to date, been restricted to microscopic methods. If the protein content of the human zona is comparable to that of the porcine zona, the amount of total zona protein used for immunization of the rabbit in the studies by Tahai et al. (1981) would be 3 to 4 μg. This is a small amount of antigen compared with the quantities necessary to obtain antibody titers sufficient for detailed immunochemical characterization of antigens (Dunbar and Raynor, 1980; Gerrity et al., 1981).

Detailed studies have been carried out on the immunogenicity of isolated porcine zonae in rabbits, and of rabbit zonae in rabbits and guinea pigs. Studies by Dunbar and Raynor (1980) have shown that the immunogenicity of the porcine zona is complex, as the immunological response depends on the amount of zona protein (300 μg versus 100 μg) as well as on the solubilization conditions used to prepare the zonae prior to immunization. These studies demonstrated that as many as four porcine zona-specific antigens were detectable by crossed-immunoelectrophoresis, using rabbit antisera to intact or heat-solubilized zonae (0.01 mM Na_2CO_3, 60°C, 60 minutes; Dunbar et al., 1980). These studies also suggest that antibodies which recognize a conformational antigenic determinant are induced when intact zonae are used as the immunogen. This type of antigenic determinant is similar to that described for the recognition of structural antigens of the H3-H4 complex of calf thymus histones, as well as for other protein conformations (see Habeeb, 1977 and Cinader, 1977, for reviews). An antibody which appears to recognize a conformational determinant has also been detected by radioimmunoassay (Gerrity et al., 1981).

In more recent studies, antibodies to rabbit zona antigens have given crossed immunoelectrophoresis patterns similar to those obtained with antisera to porcine zonae (Wood and Dunbar, 1981). Purified zona proteins have been obtained from two-dimensional electrophoresis gels, and antibodies prepared against these proteins give precipitation reactions in immunoelectrophoretic studies (Wood and Dunbar, 1980). Antisera to two purified rabbit zona proteins have now been used in radioimmunoassay as well as in immunoelectrophoresis studies to show directly two immunologically crossreactive antigens between pig and rabbit zonae pellucida. Crossed-line immunoelectrophoresis studies have shown further that the two antigens are immunochemically identical. Gwatkin and Williams (1980) have obtained one immunoprecipitin arc with rocket immunoelectrophoresis methods when a rabbit antiserum to bovine zonae was used. Tsunoda et al. (1980) have made goat antisera to bovine zonae. One immunoprecipitation line was obtained when immunoelectrophoresis was carried out against ovarian homogenate; no precipitin lines were observed against other tissues.

THE EFFECTS OF ANTISERA TO OVARIAN ANTIGENS ON SPERM BINDING TO ZONAE OR ON *IN VITRO* FERTILIZATION

Since the early observations of Porter et al. (1965, 1970) and Shahini et al. (1972) that antisera prepared against whole ovarian tissue can reduce fertility, there have been numerous studies to determine the causes of this effect. Most of these studies have concentrated on antigens associated with the zona pellucida since antibodies to these antigens could be readily detected by immunofluorescence and/or the zona precipitation reaction.

Many of these earlier studies have been reviewed by Shivers (1977, 1978), Tsunoda and Sugie (1979), and Aitken and Richardson (1979). In the present overview, the data of these earlier studies, and of more recent ones, are collected and tabulated.

THE EFFECTS OF ANTIBODIES TO OVARIAN AND ZONA PELLUCIDA ANTIGENS ON FERTILITY

Many studies have used ovarian tissue as immunogen for the preparation of antisera because it is difficult to isolate sufficient numbers of zonae pellucida (see Tables 23-1 and 23-3). It was observed that hetero- or alloantisera to ovarian tissues dramatically reduce or inhibit fertility *in vitro* by preventing sperm binding to the zona (Table 23-1). In these studies, antibodies to zona antigens were detected with the microscopic methods described above. These antibodies were presumed to be responsible for the reduction in fertilization

517

TABLE 23-1
The Effect of Antisera to Ovarian Homogenates on Sperm Binding and/or Penetration of Zonae or on Fertilization *In Vitro*

ANTISERA SOURCE	ANTIGEN (OVARY) SOURCE	METHOD OF ANTIBODY DETECTION[a]	SOURCE OF EGGS	INHIBITION OF SPERM BINDING BY ANTIBODY[b]	FERTILIZATION (%) CONTROL SERA	FERTILIZATION (%) IMMUNE SERA	REFERENCE
Rabbit	Hamster	ZPR RPD	Hamster	+	48 (58)	(0) (64)	Shivers, 1972
Rabbit	Hamster	ZPR RPD	Hamster	+	100 (20)	0 (20)[c]	Garavango et al., 1974
Rabbit	Hamster	II ZPR RPD	Hamster	+	100 (103)	0 (82)	Oikawa and Yanagimachi, 1975
Rabbit	Mouse	II	Mouse	+	71-94 (219)	0-7.2 (280)[f]	Jilek and Pavlok, 1975
Mouse	Mouse	ID ZPR[g]	Mouse	+	38 (115)	12 (229)	Tsunoda and Chang, 1976a
Rabbit	Rat	ID PH	Rat Mouse Hamster	+	39 (83) 88 (58) 90 (48)	2 (106) 0 (127) 0 (60)	Tsunoda and Chang, 1976b
Rabbit	Hamster	PH ZPR ID	Hamster	+	59-89 (146)	0 (93)	Tsunoda and Chang, 1976c
Rabbit	Mouse	RPD ZPR ID	Mouse Hamster	+	63-94 (118) 90 (48)	0 (81) 88 (91)	Tsunoda and Chang, 1977
Rabbit	Dog	ZPR	Dog	+			Mahi and Yanagimachi, 1979
Rabbit	Hamster	ZPR II RPD	Hamster	+ — +	100 (45) 100 (28) 95-100 (41)	0-6 (92) 96 (30) 0 (60)	Ahuja and Tzartos, 1981

a ZPR = zona precipitation reaction; RPD = resistance to proteolytic enzyme digestion; II = indirect immunofluorescence; ID = immunodiffusion; PH = passive hemagglutination.
b + = inhibition of sperm binding; − = no inhibition.
c Control sera or immune absorbed sera.
d Value in parentheses indicates number of intact ova used in study.
e Penetration of zonae evaluated by the presence of sperm in perivitelline space.
f Dependent on concentration of immune sera and time of incubation.

rates since absorption of antisera with other tissues did not affect the inhibition of sperm binding or fertilization. These observations are of particular interest because this method, if applied to contraception, would inhibit fertilization and would not be abortive.

More recently, these *in vitro* studies have been extended to examine antibodies prepared against isolated eggs with or without cumulus cells, or to isolated zonae pellucida (Table 23-2). Again, these studies show that heteroantisera prepared against eggs or zonae inhibit sperm binding to zonae or reduce *in vitro* fertilization (Tsunoda, 1977). This reduction of fertilization *in vitro*, however, was found to depend on the species of oocytes used. Other studies in mice and hamsters showed isologous antizona antisera to have little effect on the *in vitro* fertilization rate (Gwatkin et al., 1977).

PASSIVE IMMUNIZATION WITH ANTIBODIES TO OVARIAN ANTIGENS

Passive immunization with heteroantisera to ovarian tissue can reduce or inhibit fertility. This effect was found to depend on the time of immunization before mating as well as on the quantities of sera or immunoglobulins administered. Details of these studies are summarized in Table 23-3. Reductions in fertility were generally thought to be due to antibodies to zonae pellucida since microscopic methods could detect such antibodies and since absorption of antisera with nonovarian tissues had no effect on the fertility reduction. As with fertilization studies *in vitro*, the reduction in fertility was found to be less pronounced when animals (mice or rats) were passively immunized with rabbit antisera to heterologous ovarian tissue (hamster). Heteroantisera to isolated eggs or zonae pellucida also inhibit fertilization *in vivo* in animals of the same species as the egg or zona donors (Table 23-4).

ACTIVE IMMUNIZATION WITH OVARIAN TISSUE OR WITH ISOLATED ZONAE PELLUCIDA

There have been fewer studies using active immunization (summarized in Table 23-5) with zonae than with passive immunization. Studies in which animals were immunized with ovarian homogenates have demonstrated a reduction in fertility, although it is difficult to conclude from such studies which immunogens are responsible. The observations of Dunbar et al. (1981) show not only that the effect on fertility is more pronounced when zonae

TABLE 23-2
The Effect of Antisera to Isolated Oocytes or Zonae Pellucida on Sperm Binding or Fertilization

ANTISERA SOURCE	IMMUNOGEN SOURCE	METHOD OF ANTIBODY DETECTION[a]	SOURCE OF EGGS	INHIBITION OF SPERM BINDING BY ANTIBODY[b]	FERTILIZATION (%)		REFERENCE
					CONTROL SERA	IMMUNE SERA	
Mouse	Mouse zonae	II[d]	Mouse		No effect[c]		Gwatkin et al., 1977
Hamster	Hamster zonae	II[d]	Hamster		No effect[c]		Gwatkin et al., 1977
Rabbit	Mouse eggs	RPD ZPR	Mouse	+	51 (61)[f]	0 (57)	Tsunoda, 1977
Rabbit	Mouse eggs	ZPR RPD	Rat[g]	+	45 (87)	44 (127)	Tsunoda and Chang, 1978a
Rabbit	Mouse zonae	ZPR RPD	Rat[g]	+	62 (47)	47 (95)	Tsunoda and Chang, 1978a
Rabbit	Mouse zonae	ZPR RPD	Rat[h]	+	51 (164)	6 (174)	Tsunoda and Chang, 1978a
Rabbit	Mouse eggs	ZPR RPD	Hamster[g]	+	96 (45)	84 (45)	Tsunoda and Chang, 1978a
Rabbit	Mouse zonae	ZPR RPD	Hamster[g]	+	100 (87)	60 (93)	Tsunoda and Chang, 1978a
Rabbit	Mouse zonae	ZPR RPD	Hamster[h]	+	74 (263)	57 (223)	Tsunoda and Chang, 1978a
Rabbit	Mouse zonae	ZPR RPD	Mouse[g] / Mouse[h]	+ / +	63 (94) / 49 (51)	0 (31) / 0 (102)	Tsunoda and Sugie, 1979
Rabbit	Denuded mouse	ZPR	Mouse[g]		51 (61)	0 (57)	Tsunoda and Sugie, 1979

RPD	Mouse[h]	85 (78)	86 (66)

a Abbreviations as in Table 23-1.
b + = inhibition of sperm binding.
c Control sera or absorbed immune sera.
d Not detectable.
e Details not given.
f Value in parentheses indicates number of ova used in study.
g Eggs with cumulus.
h Eggs without cumulus.

TABLE 23-3
The Effect of Passive Immunization with Antibodies to Ovarian Tissue on Fertility

ANTISERA SOURCE	ANTIGEN (OVARY) SOURCE	ANIMAL PASSIVELY IMMUNIZED	METHOD OF ANTIBODY DETECTION[a]	FERTILIZATION (%)		REFERENCE
				CONTROL SERA[b]	IMMUNE SERA	
Rabbit	Mouse	Mouse	None	88-94 (16)[c]	0-25 (8)	Shahini et al., 1972
Rabbit	Hamster	Hamster	II ZPR	98-100 (22)	0-96 (30)[d]	Oikawa and Yanagimachi, 1975
Rabbit	Hamster	Hamster	^{131}I-IgG	—	—	Yanagimachi et al., 1976
Rabbit	Rat	Rat	ZPR PH ID	98-100 (12)	0-50 (27)[d]	Tsunoda and Chang, 1976b
Rabbit	Rat	Rat	ZPR	75-79 (9)	4-93 (11)[d]	Tsunoda and Chang, 1976b
Rabbit	Hamster	Hamster	ZPR[e]	92-100 (9)	0-10 (11)[d]	Tsunoda and Chang, 1976b
Mouse	Mouse	Mouse	ZPR[e]	97 (3)	94 (7)	Tsunoda and Chang, 1976b
Rat	Rat	Rat	ZPR[e]	100 (4)	97 (8)	Tsunoda and Chang, 1976b
Rabbit	Hamster	Mouse	ZPR[e]	100 (4)	57 (5)	Tsunoda and Chang, 1976c
Rabbit	Hamster	Rat	ZPR[e]	73 (3)	85 (3)	Tsunoda and Chang, 1976c
Rabbit	Rat	Rat	ID PH	75 (5)	4 (6)	Tsunoda and Chang, 1976a
Rabbit	Rat	Mouse	ID PH	100 (4)	12 (7)	Tsunoda and Chang, 1976c
Rabbit	Rat	Hamster	ID PH	91 (4)	82 (4)	Tsunoda and Chang, 1976a
Rabbit	Mouse	Mouse	ZPR			

| Rabbit | Mouse | Rat | ZPR ID | ?? | (3) | 100 | (4) | Tsunoda and Chang, 1977 |
| Rabbit | Mouse | Hamster | ZPR ID | 98 | (4) | 100 | (4) | Tsunoda and Chang, 1977 |

[a] Abbreviations as in Table 23-1.

[b] Control sera and/or absorbed immune sera.

[c] Value in parentheses indicates number of animals used in study.

[d] Dependent on the time of passive immunization before mating and/or on the amount of Ig given.

[e] Not detected.

TABLE 23-4
The Effect of Passive Immunization with Antibodies to Isolated Oocytes or Isolated Zonae Pellucidae on Fertility

ANTISERA SOURCE	ANIMAL IMMUNIZED	EGG OR ZONA SOURCE	METHOD OF ANTIBODY DETECTION[a]	FERTILIZATION (%)		REFERENCE
				CONTROL SERA	IMMUNE SERA	
Rabbit	Mouse	Mouse eggs	ZPR	79 (3)	0 (5)[d]	Tsunoda and Sugie, 1977
Rabbit	Mouse	Mouse zonae	ZPR	94-100 (11)	0-5 (6)[d]	Tsunoda and Change, 1978a
Rabbit	Mouse	Mouse eggs	ZPR	94-100 (11)	0-92 (26)[d]	Tsunoda and Sugie, 1979
Rabbit	Mouse	Mouse eggs	ZPR	95 (5)	0 (4)[d]	Tsunoda, 1977
Rabbit	Mouse	Mouse zonae	ZPR	86 (8)	0-28 (27)[d]	Sacco, 1979
			IP	75-100 (35)[e]	0 (28)	
Rabbit	Rat	Mouse zonae	ZPR	69 (3)	16 (4)	Tsunoda and Chang, 1978b
Rabbit	Rat	Mouse eggs	ZPR	70-85 (7)	0-56 (9)[d]	
Rabbit	Hamster	Mouse zonae	ZPR RPD	100 (2)	100 (2)	Tsunoda and Chang, 1978b
Rabbit	Hamster	Mouse eggs	ZPR	100 (2)	100 (2)	
Goat	Cow	Cow zonae	CIE	83 (5)	0-50 (5)	Tsunoda et al., 1981

[a] Abbreviations as in Table 23-1; also CIE = crossed immunoelectrophoresis.

[b] Control sera or immune absorbed sera.

[c] Value in parentheses indicates number of animals used in study.

[d] Dependent on time of passive immunization before mating or on the amount of immunoglobulin given.

[e] Term fertility study.

TABLE 23-5
The Effect of Active Immunization with Ovarian Tissue or with Isolated Zonae on Fertility

ANIMAL ACTIVELY IMMUNIZED	IMMUNOGEN SOURCE	METHOD OF ANTIBODY DETECTION[a]	FERTILIZATION (%) CONTROL	FERTILIZATION (%) IMMUNE	REFERENCE
Mouse	Mouse ovary	ZPR	62-77 (10)[b] 49-100 (10)[c]	0 (3) 4-74 (7)[c]	Tsunoda and Chang, 1976
Rat	Rat ovary	Not given	91 (11)	60 (12)	Tsunoda and Chang, 1976
Rat	Mouse ovary	II ISB	88 (7)	4 (6)	Aitken and Richardson, 1979
Mouse	Hamster zonae	RPD II	100 (14)[d]	7 (4)	Gwatkin et al., 1977
Rabbit	Bovine zonae	II	100 (4)	25 (4)	Gwatkin and Williams, 1978
Rabbit	Porcine zonae	II ZPR	100 (7)	12 (8)[e]	Gwatkin and Williams, 1980
Rabbit	Porcine zonae Heat-solubilized Intact	IE, RIA IE, RIA	92-94 (6)[f] 92-94 (6)[f]	0 (4)[f,g] 0 (4)[f,g]	Wood et al., 1981
	Detergent-solubilized	IE, RIA	92-94 (6)[f]	31 (4)[f]	
Rabbit	Rabbit zonae Heat-solubilized Intact	IE, RIA IE, RIA	92-94 (6)[f] 92-94 (6)[f]	30 (4)[f] 30 (4)[f]	Wood et al., 1981
	Detergent-solubilized	IE, RIA	92-94 (6)[f]	48 (4)[f]	

[a] Abbreviations as in Table 23-1; also, ISB = inhibition of sperm binding; RIA = radioimmunoassay; IE = immunoelectrophoresis

[b] Value in parentheses indicates number of animals used in study.

[c] *In vitro* fertilization of ova from immunized females.

[d] Dependent on time after immunization as well as on number of immunizations.

[e] Alum used as adjuvant instead of CFA.

[f] % fertilization determined by number of cleaving embryos/number of ovulation sites.

[g] Ovulation inhibited.

from a heterologous species are used (e.g., porcine versus rabbit zonae) but also that the methods of zona solubilization markedly affect the inhibition of fertility. Even more dramatic was the observation that the rabbits heteroimmunized with porcine zonae did not ovulate normally in response to injections of human chorionic gonadotropin (hCG) or cervical stimulation (during artifical insemination). This appears to depend on antibodies to a "conformational antigenic determinant" (discussed above), since zonae solubilized with detergents did not exhibit this effect.

SPECIES CROSSREACTIVITY OF ZONA PELLUCIDA ANTIGENS

Immunodiffusion methods have shown that hamster ovarian antigens are immunologically crossreactive with rat and mouse ovarian antigens (Tsunoda and Chang, 1976a). In these studies, however, relatively few antigens were identified, presumably because of the limited resolution of immunodiffusion methods, and specific zona antigens could not be identified. Sacco and Shivers (1973) and Garavagno et al. (1974) found no immunological crossreactivity with mouse, rat, guinea pig, or rabbit ovary when heteroantisera to hamster or rabbit ovary were used in immunodiffusion studies. Numerous studies using indirect immunofluorescence methods have shown that a variety of species have zona pellucida antigens which are immunologically crossreactive (mouse and hamster, Gwatkin et al., 1977; pig and human, Shivers and Dunbar, 1977; Sacco, 1978; human and marmoset, Shivers et al., 1978; cow, pig, and rhesus monkey, Gwatkin, 1980). The studies demonstrating the inhibition of fertility by heteroantibodies to zona antigens have been used to confirm the immunofluorescence findings that zona antigens are immunologically crossreactive between species. In view of recent concerns about the specificity of immunofluorescence techniques (Sacco and Moghiss, 1979; Gerrity et al., 1981) and the need for extensive tissue absorption to obtain specificity (Mori et al., 1980; Dakhno, 1981), alternative methods have been used to demonstrate the immunological crossreactivity of the zona antigens of different species. Recent studies by Wood and Dunbar (1981) have shown that the porcine and rabbit zonae have two separate, but immunologically identical antigens. These studies have used antisera to purifed zona proteins in rocket, crossed, and crossed-line immunolectrophoresis, as well as in radioimmunoassay.

ANTIBODIES TO ZONA PELLUCIDA ANTIGENS IN THE SERA OF INFERTILE WOMEN AND IN AGING ANIMALS

Several studies have used immunofluorescence methods to demonstrate that antibodies which react with zonae pellucida are found in the sera of some infertile women (Shivers and Dunbar, 1977; Mori et al., 1978). These studies are possible because porcine zona has been shown to be immunochemically similar to human zona (Shivers and Dunbar, 1977; Sacco, 1977). Using immunofluorescence methods and sperm binding assays, Trounson et al. (1979) and Tsunoda and Chang (1979) have claimed that autoantibodies to zona antigens are present in sera of aging women and aging animals. These investigators have suggested that autoantibodies to zona antigens might occur naturally in aging animals. Such antibodies, however, have yet to be characterized. Nishimoto et al. (1980) have recently reported lower levels of zona binding activity in sera of aging women after absorption with red blood cells.

When immunofluorescence methods are used, zona-binding activity is also observed in the sera of fertile males and fertile females (Sacco and Moghissi, 1979; Dakhno et al., 1980). Studies by Dakhno et al. (1980) found that not all the antizona activity of the human sera tested could be absorbed with red blood cells. Mori et al. (1980) reported, however, that after absorption with red blood cells to remove hemagglutinins, some human sera still gave positive immunofluorescence reactions. Neither Palm et al. (1979) nor Gerrity et al. (1981) have observed any crossreactivity with porcine red blood cells or other porcine tissues, and it is not clear why this crossreactivity is seen in human sera. The significance of these observations is unknown at this time. Since microscopic methods of antibody detection have been the only studies suggesting the presence of antibodies to zona antigens in human sera, it will be essential to confirm these observations using alternative methods now available, such as radioimmunoassay.

SUMMARY AND CONCLUSIONS

The studies on antibodies to ovarian and/or zona antigens described in this chapter collectively suggest a dramatic reduction of fertility in animal models. If antibodies to zona antigens are to be accepted as a contraceptive method, the considerations discussed by Mitchison (1974) on the long-term hazards of immunological fertility control will have to be considered. In the future the following problems will have to be resolved:

1. More detailed information on zona synthesis (both the cell type and the time of synthesis) is needed to establish the effect of antibody to zona antigens on ovarian function.
2. The antigenic determinant(s) giving rise to specific antibodies that block fertility will have to be characterized in detail.
3. Sufficient amounts of purified zona proteins which elicit the proper immune response will have to be made available.

Recently, new techniques have been developed which should prove invaluable in resolving these problems.

1. Advances in hybridoma technology for producing monoclonal antibodies (Kohler and Milstein; 1975; Kennett et al., 1980) allow the production of antibody to specific antigenic determinants. These specific antibodies can be used to better characterize zona antigens, as well for use in precise fertility studies (passive immunization).
2. High-resolution, two-dimensional gel electrophoresis methods can now be used to characterize antibodies to zona antigens, both by immunoprecipitation of specific antigens followed by two-dimensional gel electrophoretic separation (Anderson and Anderson, 1978a, 1978b) and alternatively, by immunotransfer methods for proteins isolated by two-dimensional gel electrophoresis for the characterization of specific antisera (Renart et al., 1979).

It is anticipated that these methods, as well as the zona isolation and radioimmunoassay methods for detecting specific antibodies to zona antigens, will help to determine the feasibility of such a contraceptive method.

ACKNOWLEDGEMENTS

The author wishes to thank David Wood, Dan Drell, and Bernard Erlanger for their helpful suggestions in preparing this chapter. The current work discussed in this chapter was supported by a grant to B.S. Dunbar from the National Institute of Child Health and Development HD-12357.

REFERENCES

Adams, L.D., and D.W. Sammons. 1981. A unique silver staining procedure for color characterization of polypeptides. In *Electrophoresis 81*. R.C. Allen and P. Arnaud, eds. New York: Walter deGruyter.

Ahuja, K.K., and S.J. Tzartos. 1981. Investigation of sperm receptors in hamster zona pellucida by using univalent (Fab) antibodies to hamster ovary. *J. Reprod. Fertil.* 61:257.

Aitken, R.J., and D.W. Richardson. 1980. Immunization against zona pellucida antigens. In *Immunological Aspects of Reproduction and Fertility Control*. J.P. Heam, ed. Lancaster, U.K.: MTP Press Ltd.

Albini, B., E. Ossi, C. Newland, B. Noble, and G. Andres. 1979. Deposition of immune complexes in the ovarian follicle of rabbits with experimental chronic serum sickness. I. *Immunopathology. Lab. Investigation* 41:446.

Anderson, L., and N.G. Anderson. 1977. High resolution two-dimensional electrophoresis of human plasma proteins. *Proc. Nat. Acad. Sci. USA* 74:5421.

Anderson, N.G., and L. Anderson. 1978a. Analytical techniques for cell fractions. XXI. Two-dimensional analysis of serum and tissue proteins: multiple isoelectric focusing. *Anal. Biochem.* 85:331.

Anderson, N.G., and N.L. Anderson. 1978b. Analytical techniques for cell fractions. XXII. Two-dimensional analysis of serum of tissue proteins: multiple gradient slab electrophoresis. *Anal. Biochem.* 85:341.

Anderson, N.G., N.L. Anderson, and S.L. Tollaksen. 1979. Operation of the ISO-DALT system. ANL-BIM-79-2, Argonne, Ill.

Austin, C.R. 1961. *The Mammalian Egg.* Oxford: Blackwell.

Austin, C.R., and A.W.H. Braden. 1956. Early reactions of the rodent egg to spermatozoan penetration. *J. Exp. Biol.* 33:358.

Barros, C., and R. Yanagimachi. 1972. Polyspermy-preventing mechanisms in the golden hamster egg. *J. Exp. Zool.* 180:251.

Barros, C., and R.Y. Yanagimachi. 1971. Induction of zona reaction in golden hamster eggs by cortical granule material. *Nature* 233:268.

Bedford, J.M. 1977. Sperm/egg interaction: The specificity of human spermatozoa. *Anat. Rec.* 188:477.

Benjamin, D.C., and J.M. Teale. 1978. The antigenic structure of bovine serum albumin. *J. Biol. Chem.* 253:8087.

Bleil, J.D., and P.M. Wassarman. 1980a. Mammalian sperm-egg interaction: Identification of a glycoprotein in mouse egg zonae pellucida possessing receptor activity for sperm. *Cell* 20:873.

Bleil, J.D., and P.M. Wassarman. 1980b. Structure and function of the zona pellucida; identification and characterization of the proteins of the mouse zona pellucida. *Dev. Biol.* 76:185.

Bleil, J.D., and P.M. Wassarman. 1980c. Synthesis of zona pellucida proteins by denuded and follicle-enclosed mouse oocytes during culture *in vitro. Proc. Nat. Acad. Sci. USA* 77:1029.

Braden, A.W.H., C.R. Austin, and H.A. David. 1954. The reaction of the zona pellucida to sperm penetration. *Austral. J. Biol. Sci.* 7:391.

Cinader, B. 1977. Enzyme-antibody interactions. In *Methods in Immunology and Immunochemistry.* Vol. 44. C.A. Williams and M.W. Chase, eds. New York: Academic Press, pp. 313-335.

Dakhno, F.V., T. Hjort, and V.I. Grischenko. 1980. Evaluation of immunofluorescence on pig zona pellucida for detection of anti-zona antibodies in human sera. *J. Reprod. Immunol.* 20:281.

Dudkiewicz, A.B., C.A Shivers, and W.L. Williams. 1976. Ultrastructure of hamster zona pellucida treated with zona-precipitating antibody. *Biol. Repr.* 14:175.

Dunbar, B.S. 1980. Model systems to study the relationship between antibodies to zonae and infertility. Comparison of rabbit and porcine zona pellucida. In *9th International Con-*

gress on Animal Reproduction and Artificial Insemination. Vol. 2. Madrid, Spain, June 16-20. Madrid: Editorial Garsi, pp. 191-199.

Dunbar, B.S., C. Liu, and D.W. Sammons. 1981. Identification of the three major proteins of porcine and rabbit zonae pellucidae by high resolution two dimensional gel electrophoresis: Comparison with serum, follicular fluid, and ovarian cell proteins, *Biol. Reprod.* In press.

Dunbar, B.S., and D.W. Sammons. 1980. Comparison of the macromolecular structure of the zona pellucida of porcine and rabbit oocytes. *J. Cell Biol.* 87:143a.

Dunbar, B.S., and C.A. Shivers. 1976. Immunological aspects of sperm receptors on the zona pellucida of mammalian eggs. In *Immunology of Receptors.* B. Cinader, ed. New York: Marcel Dekker, pp. 509-519.

Dunbar, B.S., N. Wardrip, and J.L. Hedrick. 1978a. Large scale isolation and characterization of the porcine zona pellucida. *Biol. Reprod.* (Suppl. D) 18: Abstract 23.

Dunbar, B.S., N.J. Wardrip, and J.L. Hedrick. 1978b. Isolation and physicochemical properties of porcine zona pellucida. *J. Cell. Biol.* 79:F916.

Dunbar, B.S., N.J. Wardrip, and J.L. Hedrick. 1980. Isolation, physicochemical properties and the macromolecular composition of the zona pellucida from porcine oocytes. *Biochemistry* 19:356.

Flechon, J.E., and R.B.L. Gwatkin. 1980. Immunochemical studies on the zona pellucida of cow blastocysts. *Gamete Res.* 3:141.

Garavagno, A., J. Posada, C. Barros, and C.A. Shivers. 1974. Some characteristics of the zona pellucida antigen in the hamster. *J. Exp. Zool.* 189:37.

Gerrity, M., E. Niu, and B.S. Dunbar. 1981. A specific radioimmunoassay for evaluation of serum antibodies to zona pellucida antigens. *J. Reprod. Immunol.* In press.

Gould, K., L.J.D. Zaneveld, P.N. Srivastava, and W.L. Williams. 1971. Biochemical changes in the zona pellucida of rabbit ova induced by fertilization and sperm enzymes. *Proc. Soc. Exp. Biol. Med.* 136:6.

Gurkin, M., and E.A. Kallet. 1971. An instrument for quantitative microspectrofluorometry. *Amer. Lab.* (Oct Issue):1.

Gwatkin, R.B.L. 1964. Effect of enzymes and acidity on the zona pellucida of the mouse egg before and after fertilization. *J. Reprod. Fertil.* 7:99.

Gwatkin, R.B.L. 1977. *Fertilization Mechanisms in Man and Mammals.* New York: Plenum Press.

Gwatkin, R.B.L., O.F. Anderson, and D.T. Williams. 1980. Large scale isolation of bovine and pig zonae pellucida: chemical, immunological and receptor properties. *Gamete Res.* 3:217.

Gwatkin, R.B.L., H.W. Carter, and H. Patterson. 1976. Association of mammalian sperm with the cumulus cells and the zona pellucida studied by scanning electron microscopy. Part VI. Proceedings of the Workshop on SEM in Reproductive Biol. O. Johari and R.P. Becker, eds. Chicago: IITRES. Inst. Vd. pp. 379-384.

Gwatkin, R.B.L., and D.T. Williams. 1976. Receptor activity of the solubilized hamster and mouse zona pellucida before and after the zona reaction. *J. Reprod. Fertil.* 19:55.

Gwatkin, R.B.L., and D.T. Williams. 1978a. Bovine and hamster zona solutions exhibit receptor activity for capacitated but not for noncapacitated sperm. *Gamete Res.* 1:259.

Gwatkin, R.B.L., and D.T. Williams. 1978b. Immunization of female rabbits with heat-solubilized bovine zonae: Production of anti-zona antibody and inhibition of fertility. *Gamete Res.* 1:19.

Gwatkin, R.B.L., D.T. Williams, and O.F. Anderson. 1973. Zona reaction of mammalian eggs: properties of the cortical granule protease (Cortin) and its receptor substrate in hamster eggs. *J. Cell. Biol.* 59:128a.

Gwatkin, R.B.L., D.T. Williams, and D.J. Carlo. 1977. Immunization of mice with heat-solubilized hamster zonae; production of anti-zona antibody and inhibition of fertility. *Fertil. Steril.* 28:871.

Haalenstad, A.O., and Manik, M. 1977. The biology of the immune complexes. In *Autoimmunity.* N. Talal, ed. New York: Academic Press, pp. xxx-xxx.

Habeeb, A.F.S.A. 1977. Influence of conformation on immunochemical properties of proteins. In

Immunochemistry of Proteins. Vol. 2. M.Z. Atassi, ed. New York: Plenum, pp. 163-230.

Hanada, A., and M.C. Chang. 1972. Penetration of zona-free eggs by spermatozoa of different species. *Biol. Reprod.* 6:300.

Hartmann, J.F., and R.B.L. Gwatkin. 1971. Alteration of sites on the mammalian sperm surface following capacitiation. *Nature* 234:479.

Hartmann, J.F., R.B.L. Gwatkin, and C.F. Hutchison. 1972. Early contact interactions between mammalian gametes *in vitro*: Evidence that the vitellus influences adherence between sperm and zona pellucida. *Proc. Nat. Acad. Sci.* 69:2767.

Hartmann, J.F., and C.F Hutchison. 1974. Nature of the pre-penetration contact interaction between hamster gametes *in vitro*. *J. Reprod. Fertil.* 36:49.

Hedrick, J.L., and G.N. Fry. 1980. Immunocytochemical studies on the porcine zona pellucida. *J. Cell. Biol.* 87:FE1025.

Hedrick, J.L., and N. Wardrip. 1980. The macromolecular composition of the porcine zona pellucida. *Fed. Proc.* 39:Abstract 2516.

Herbert, W.J. 1977. Passive haemagglutination. In *Handbook of Experimental Immunology.* D.M. Weir, ed. Oxford and Edinburgh: Blackwell Scientific Publications, pp. 720-747.

Jackowski, S., and J.N. Dumont. 1979. Surface alterations of the mouse zona pellucida and ovum following *in vitro* fertilization: correlation with cell cycle. *Biol. Reprod.* 20:150.

Jilek, F., and A. Pavlok. 1975. Antibodies against mouse ovaries and their effect on fertilization *in vitro* and *in vivo* in the mouse. *J. Reprod. Fertil.* 42:377.

Kabat, E.A. 1955. *Blood Group Substances: Their Chemistry and Immunochemistry.* New York: Academic Press, pp. 100-105.

Kang, Y. 1974. Development of the zona pellucida in the rat oocyte. *J. Anat.* 139:535.

Kawamura, A. 1977. *Fluorescent antibody techniques and their applications.* 2nd ed. Baltimore: University Park Press.

Kohler, G., and C. Milstein. 1975. Continuous cultures of fused cells secreting antibody of predefined specificity. *Nature* 256:495.

Mahi, C.A., and R. Yanagimachi. 1979. Prevention of *in vitro* fertilization of canine oocytes by anti-ovary antisera: A potential approach to fertility control in the bitch. *J. Exp. Zool.* 210:129.

McDevitt, H.O., J.H. Peters, L.W. Pollard, J.G. Harter, and A.H Coons. 1963. Purification and analysis of fluorescein labeled antisera by column chromatography. *J. Immunol.* 90:634.

McRorie, R.A., and W.L. Williams. 1974. Biochemistry of mammalian fertilization. *Ann. Rev. Biochem.* 43:777.

Menino, A.R., and R.W. Wright, Jr. 1979. Characterization of porcine oocyte zonae pellucida by polyacrylamide gel electrophoresis. *Proc. Soc. Exp. Biol. Med.* 160:499.

Mintz, B. 1962. Experimental study of the developing mammalian egg: Removal of the zona pellucida. *Science* 138:594.

Mitchison, N.A. 1974. Long-term hazards in immunological methods of fertility control. In *Immunological approaches to fertility control.* E. Diczfalusy, ed. Karolinska Symposium No. 7. Stockholm: Karolinska Inst.

Modlinski, J.A. 1970. The role of the zona pellucida in the development of mouse eggs *in vivo*. *J. Embryol. Exp. Morph.* 23:539.

Monroy. A. 1965. *Chemistry and physiology of fertilization.* New York: Holt, Rinehart and Winston.

Mori, T., T. Nishimoto, M. Katagawa, Y. Noda, T. Nishimura, and T. Oikawa. 1978. Possible presence of autoantibodies to zona pellucida in infertile women. *Experentia* 34:797.

Nicholson, G.L., R. Yanagimachi, and H. Yanagimachi. 1975. Ultrastructural localization of lectin-binding sites on the zona pellucida and plasma membrane of mammalian eggs. *J. Cell. Biol.* 66:263.

Nishimoto, T., T. More, I. Yamada, and T. Nishimura. 1980. Autoantibodies to zona pellucida in infertile and aged women. *Fertil. Steril.* 34:522.

Oakberg, E.F., and P.D. Tyrell. 1975. Labeling the zona pellucida of the mouse oocyte. *Biol. Reprod.* 12:477.

Odor, D.L. 1960. Electron microscopic studies on ovarian oocytes and unfertilized tubal ova in the rat. *J. Biophys. Biochem. Cytol.* 7:567.

Oikawa, T. 1978. A simple method for the isolation of a large number of ova from pig ovaries. *Gamete Res.* 1:265.

Oikawa, T., G.L. Nicholson, and R. Yanagimachi. 1975. Trypsin-mediated modification of the zona pellucida glycopeptide structure of hamster eggs. *J. Reprod. Fertil.* 43:133.

Oikawa, T., and R. Yanagimachi. 1975. Block of hamster fertilization by anti-ovary antibody. *J. Reprod. Fertil.* 45:487.

Oikawa, T., R. Yanagimachi, and G.L. Nicholson. 1973. Wheat germ agglutinin blocks mammalian fertilization. *Nature* 241:256.

Ouchterlony, O., and L.A. Nilsson. 1977. Immunodiffusion andimmunoelectrophoresis. In *Handbook of Experimental Immunology.* D.M. Weir, ed. Oxford and London: Blackwell Scientific Publications.

Overstreet, J.W., and J.M. Bedford. 1974. Comparison of the permeability of the egg vestments in follicular oocytes, unfertilized and fertilized ova of the rabbit. *Devel. Biol.* 41:185.

Overstreet, J.W., and J.M. Bedford. 1975. The penetrability of rabbit ova treated with enzymes or anti-progesterone antibody: a probe into the nature of a mammalian fertilization. *J. Reprod. Fertil.* 44:273.

Ownby, C.L., and C.A. Shivers. 1972. Antigens of the hamster ovary and effects of anti-ovary serum on eggs. *Biol. Reprod.* 6:310.

Palm, U.S., A.G. Sacco, F.N. Syner, and M.G. Subramanian. 1979. Tissue specificity of porcine zona pellucida antigen(s) tested by radioimmunoassay. *Biol. Reprod.* 21:709.

Petersen, R.N., L. Russell, D. Bundman, and M. Freund. 1980. Sperm-egg interaction: evidence for boar sperm plasma membrane receptors for porcine zona pellucida. *Science* 107:73.

Phillips, D.M., and R.M. Shalgi. 1980. Surface properties of the zona pellucida. *J. Exp. Zool.* 213:1.

Piko, L. 1969. Gamete structure and sperm entry in mammals. In *Fertilization.* C.B. Metz and A. Monroy, eds. New York: Academic Press: pp. 325-403.

Porter, C.W. 1965. Ovarian antibodies in female guinea pigs. *Int. J. Fertil.* 10:257.

Porter, C.W., D. Highfill, and R. Winovich. 1970. Guinea pig ovary and testis: demonstration of common gonad specific antigens in the ovary and testis. *Int. J. Fertil.* 15:171.

Race, R.R., and R. Sanger. 1968. *Blood Groups in Man.* 5th edition. Philadelphia: F.A. Davis, pp. 39-43.

Renart, J., J. Reiser, and F.R. Stark. 1979. Transfer of proteins from gels to diazobenzyloxymethyl-paper and detection with antisera: A method for studying antibody specificity and antigen structure. *Proc. Nat. Acad. Sci. USA* 76:3116.

Repin, V.S., and I.M. Akimova. 1976. Microelectrophoretic analysis of protein composition of zonae pellucidae of mammalian oocytes and zygotes. *Biokhimiya (USSR)* 41:50.

Sacco, A.G. 1977. Antigenic cross-reactivity between human and pig zona pellucida. *Biol. Reprod.* 16:164.

Sacco, A.G., and K.S. Moghiss. 1979. Anti-zona pellucida activity in human sera. *Fertil. Steril.* 31:503.

Sacco, A.G., and V.S. Palm. 1977. Heteroimmunization with isolated pig zonae pellucidae. *J. Reprod. Fertil.* 51:165.

Sacco, A.G., and C.A. Shivers. 1973. Comparison of antigens in the ovary, oviduct and uterus of the rabbit and other mammalian species. *J. Reprod. Fertil.* 32:421.

Sacco, A.G., and C.A. Shivers. 1978. Immunologic inhibition of development. In *Methods in Mammalian Reproduction.* J.C. Daniel, Jr., ed. New York: Academic Press, pp. 203-228.

Saling, P.M., B.T. Storey, and D.P. Wolf. 1978. Calcium-dependent binding of mouse epididymal spermatozoa to the zona pellucida. *Devel. Biol.* 65:515.

Schmell, E.D., B.J. Gulyas. 1980. Mammalian sperm-egg recognition binding *in vitro.* Specificity of sperm interactions with live and fixed eggs in homologous and heterologous inseminations of hamster mouse and guinea pig oocytes. *Biol. Reprod.* 23:1075.

Sellens, M.H., and E.J. Jenkinson. 1975. Permeability of the mouse zona pellucida to immunoglobulin. *J. Reprod. Fertil.* 42:153.

Shahini, S.K., J.R. Padbidri, and S.S. Rao. 1972. Immunological studies with the reproductive organs, adrenals, and spleen of the female mouse. *Int. J. Fertil.* 17:161.

Shivers, C.A. 1974. Immunological interference with fertilization. In *Immunological Approaches to Fertility Control*. E. Diczfalusy, ed. Stockholm: Karolinska Institute, pp. 223-243.

Shivers, C.A. 1975. Antigens of ovum as a potential basis for the development of contraceptive vaccine. *3rd International Symposium on Immunology of Reproduction*. Copenhagen: Scriptor, pp. 881-891.

Shivers, C.A. 1977. The zona pellucida as a possible target in immunocontraception. In *Immunological Influence on Human Fertility*. Proc. Workship on Fertility in Human Reproduction. B. Boettcher, ed. New York: Academic Press. pp. xxx-xxx.

Shivers, C.A., A.B. Dudkiewicz, L.E. Franklin, and E.N. Fussell. 1972. Inhibition of sperm-egg interaction by specific antibody. *Science* 178:1211.

Shivers, C.A., and B.S. Dunbar. 1977. Autoantibodies to zona pellucida: a possible cause for infertility in women. *Science* 197:1187.

Shivers, C.A., N. Gengozian, S. Franklin, and C.L. McLaughlin. 1978. Antigenic cross-reactivity between human and marmoset zonae pellucidae, a potential target for immunocontraception. *J. Med. Primatol.* 7:242.

Soupart, P., and R.W. Noyes. 1964. Sialic acid as a component of the zona pellucida of the mammalian ovum. *J. Reprod. Fertil.* 8:251.

Srivastava, P.N., C.E. Adams, and E.F. Hartree. 1965. Enzymatic action of acrosomal preparation of the rabbit ovum *in vitro*. *J. Reprod. Fertil.* 10:61.

Stambaugh, R. 1978. Enzymatic and morphological events in mammalian fertilization. *Gamete Res.* 1:65.

Swenson, C.E., and B.S. Dunbar. 1981. Specificity of sperm-zona interaction. *J. Exp. Zool.* In press.

Takai, I., T. Mori, Y. Noda, and T. Nishimura. 1981. Heteroimmunization with isolated human ova. *J. Reprod. Fertil.* 61:19.

Trounson, A.D., C.A. Shivers, R. McMaster, and A. Lopata. 1980. Inhibition of sperm binding and fertilization of human ova by antibody to porcine zona pellucida and human sera. *Arch. Androl.* 4:29.

Tsunoda, Y. 1977. Inhibitory effect of anti-mouse egg serum on fertilization *in vitro* and *in vivo* in the mouse. *J. Reprod. Fertil.* 50:353.

Tsunoda, Y., and M.C. Chang. 1976a. Effect of anti-rat ovary antiserum on the fertilization of the mouse and hamster eggs *in vivo* and *in vitro*. *Biol. Reprod.* 14:354.

Tsunoda, Y., and M.C. Chang. 1976b. The effect of passive immunization with hetero- and isoimmune anti-ovary antiserum on the fertilization of mouse, rat, and hamster eggs. *Biol. Reprod.* 15:361.

Tsunoda, Y., and M.C. Chang. 1976c. *In vivo* and *in vitro* fertilization of hamster, rat and mouse eggs after treatment with anti-hamster ovary antiserum. *J. Exp. Zool.* 195:409.

Tsunoda, Y., and M.C. Chang. 1976d. Reproduction in rats and mice isoimmunized with homogenate of ovary or testis with epididymis, or sperm suspensions. *J. Reprod. Fertil.* 46:379.

Tsunoda, T., and M.C. Chang. 1977. Further studies of antisera on the fertilization of mouse, rat, and hamster eggs *in vivo* and *in vitro*. *Int. J. Fertil.* 22:129.

Tsunoda, Y., and M.C. Chang. 1978. Effect of antisera against eggs and zonae pellucida on fertilization and development of mouse eggs *in vivo* and in culture. *J. Reprod. Fertil.* 54:233.

Tsunoda, Y., and M.C. Chang. 1979. The suppressive effect of sera from old female mice on *in vitro* fertilization and blastocyst development. *Biol. Reprod.* 20:355.

Tsunoda, Y., T. Soma, and T. Sugie. 1981. Inhibition of fertilization in cattle by passive immunization with anti-zona pellucida serum. *Gamete Res.* 4:133.

Tsunoda, Y., and T. Sugie. 1977. Inhibition of fertilization in mice by anti-zona pellucida antiserum. *Jap. J. Zootech. Sci.* 48:784.

Tsunoda, Y., and T. Sugie. 1979. Inhibitory effect on fertilization *in vitro* and *in vivo* by zona pellucida antibody and the titration of this antibody. In *Recent advances in reproduction and regulation of fertility*. G.P. Dalway, ed. Amsterdam: Elsevier/North-Holland Biomedical Press, p. 123.

Tsunoda, Y., T. Sugie, and J. Mori. 1979. Quantitative determination of titers of anti-zona serum. *J. Exp. Zool.* 207:315.

Weeke, B. 1973a. General remarks on principles, equipment, reagents and procedures. In *A*

533

Manual of Quantitative Immunoelectrophoresis. Methods and Applications. (*Scand, J. Immunol.* Vol. 2, Suppl. 1). N.H. Axelson, J. Kroll, and B. Weeke, eds. Oslo: Universitetsforlaget, pp. 15-35.

Weeke, B. 1973b. Rocket immunoelectrophoresis. In *A Manual of Quantitative Immunoelectrophoresis. Methods and Applications.* (*Scand. J. Immunol.*), Vol. 2, Suppl. 1). N.H. Axelsen, J. Kroll, and B. Weeke, eds. Oslo: Universitetsforlaget, pp. 37-46.

Weeke, B. 1973c. Crossed immunoelectrophoresis. In *A Manual of Quantitative Immunoelectrophoresis. Methods and Applications. Scand. J. Immunol.*, Vol. 2, Suppl. 1. N.H. Axelsen, J. Kroll, and B. Weeke, eds. Oslo: Universitetsforlaget, pp. 47-56.

Wood, D.M., and B.S. Dunbar. 1981. Direct detection of two cross-reactive antigens between porcine and rabbit zonae pellucida by radioimmunoassay and immunoelectrophoresis. *J. Exp. Zool.* In press.

Wright, R.W. Jr., P.T. Cupps, C.T. Goskins, and J.K. Hillers. 1977. Comparative solubility properties of the zona pellucida of unfertilized murine, ovine and bovine ova. *J. Animal Sci.* 44:850.

Yanagimachi, R. 1977. Specificity of sperm-egg interaction. In *Immunobiology of Gametes*. M. Edidin and M.H. Johnson, eds. Cambridge: Cambridge Univ. Press, pp. 187-207.

Yanagimachi, R., J.L. Winkelhake, and G.L. Nicholson. 1976. Immunological block to mammalian fertilization: Survival and organ distribution of immunoglobulin which inhibits fertilization *in vivo. Proc. Nat. Acad. Sci. USA* 73:2405.